Lecture Notes in Mathematics 2153

Editors-in-Chief:
J.-M. Morel, Cachan
B. Teissier, Paris

Advisory Board:
Camillo De Lellis, Zürich
Mario di Bernardo, Bristol
Alessio Figalli, Austin
Davar Khoshnevisan, Salt Lake City
Ioannis Kontoyiannis, Athens
Gábor Lugosi, Barcelona
Mark Podolskij, Aarhus
Sylvia Serfaty, Paris and New York
Catharina Stroppel, Bonn
Anna Wienhard, Heidelberg

More information about this series at http://www.springer.com/series/304

Claude Mitschi • David Sauzin

Divergent Series, Summability and Resurgence I

Monodromy and Resurgence

Springer

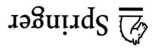

Claude Mitschi
Inst. de Recherche Mathématique Avancée
Université de Strasbourg et CNRS
Strasbourg, France

David Sauzin
CNRS UMR 8028 -- IMCCE
Observatoire de Paris
Paris, France

ISSN 0075-8434 ISSN 1617-9692 (electronic)
Lecture Notes in Mathematics
ISBN 978-3-319-28735-5 ISBN 978-3-319-28736-2 (eBook)
DOI 10.1007/978-3-319-28736-2

Library of Congress Control Number: 2016940058

Mathematics Subject Classification (2010): 34M30, 30E15, 30B40, 34M03, 34M40, 37F10, 34M35

Printed on acid-free paper

This Springer imprint is published by Springer Nature
The registered company is Springer International Publishing AG Switzerland

à la mémoire d'Andrey Bolibrukh, C.M.

à Lili, D.S.

Avant-Propos

Le sujet principal traité dans la série de volumes *Divergent Series, Summability and Resurgence* est la théorie des développements asymptotiques et des séries divergentes appliquée aux équations différentielles ordinaires (EDO) et à certaines équations aux différences dans le champ complexe.

Les équations différentielles dans le champ complexe, et dans le cadre holomorphe, sont un sujet très ancien. La théorie a été très active dans la deuxième moitié du XIX-ème siècle. En ce qui concerne les *équations linéaires*, les mathématiciens de cette époque les ont subdivisées en deux classes. Pour la première, celle des équations *à points singuliers réguliers* (ou *de Fuchs*), généralisant les équations hypergéométriques d'Euler et de Gauss, ils ont enregistré *"des succès aussi décisifs que faciles"* comme l'écrivait René Garnier en 1919. En revanche, pour la seconde, celle des équations dites *à points singuliers irréguliers*, comme l'écrivait aussi Garnier, *"leurs efforts restent impuissants à édifier aucune théorie générale"*. La raison centrale de ce vif contraste est que toute série entière apparaissant dans l'écriture d'une solution d'une équation différentielle de Fuchs est *automatiquement convergente* tandis que pour les équations irrégulières ces séries sont génériquement *divergentes* et que l'on ne savait qu'en faire. La situation a commencé à changer grâce à un travail magistral de Henri Poincaré entrepris juste après sa thèse, dans lequel il "donne un sens" aux solutions divergentes des EDO linéaires irrégulières en introduisant un outil nouveau, et qui était appelé à un grand avenir, la théorie des développements asymptotiques. Il a ensuite utilisé cet outil pour donner un sens aux séries divergentes de la mécanique céleste, et remporté de tels succès que presque tout le monde a oublié l'origine de l'histoire, c'est-à-dire les EDO ! Les travaux de Poincaré ont (un peu...) remis à l'honneur l'étude des séries divergentes, abandonnée par les mathématiciens après Cauchy. L'Académie des Sciences a soumis ce sujet au concours en 1899, ce qui fut à l'origine d'un travail important d'Émile Borel. Celui-ci est la source de nombre des techniques utilisées dans *Divergent Series, Summability and Resurgence*. Pour revenir aux EDO irrégulières, le sujet a fait l'objet de nombreux et importants travaux de G.D. Birkhoff et R. Garnier durant le premier quart du XX-ème siècle. On retrouvera ici de nombreux prolongements des méthodes de Birkhoff. Après 1940, le sujet a étrangement presque disparu, la théorie étant, je ne sais trop pourquoi, considérée comme achevée, tout comme celle des équations de Fuchs. Ces dernières ont réémergé au début des années 1970, avec les travaux de Raymond Gérard, puis un livre de Pierre Deligne. Les équations irrégulières ont suivi avec des travaux de l'école allemande et surtout de l'école française. De nombreuses techniques complètement nouvelles ont été introduites (développements asymptotiques Gevrey, k-sommabilité, multisommabilité, fonctions résurgentes...) permettant en particulier une vaste généralisation du *phénomène de Stokes* et sa mise en relation avec la théorie de Galois différentielle et le problème de Riemann-Hilbert généralisé. Tout ceci a depuis reçu de très nombreuses applications dans des domaines très variés, allant de l'intégrabilité des systèmes hamiltoniens aux problèmes de points tournants pour les EDO singulièrement perturbées ou à divers problèmes de modules. On en trouvera certaines dans *Divergent Series, Summability and Resurgence*, comme l'étude résurgente des germes de difféomorphismes analytiques du plan complexe tangents à l'identité ou celle de l'EDO non-linéaire Painlevé I.

Le sujet restait aujourd'hui difficile d'accès, le lecteur ne disposant pas, mis à part les articles originaux, de présentation accessible couvrant *tous les aspects*. Ainsi *Divergent Series, Summability and Resurgence* comble une lacune. Ces volumes présentent un large panorama des recherches les plus récentes sur un vaste domaine classique et passionnant, en pleine renaissance, on peut même dire en pleine explosion. Ils sont néanmoins accessibles à tout lecteur possédant une bonne familiarité avec les fonctions analytiques d'une variable complexe. Les divers outils sont soigneusement mis en place, progressivement et avec beaucoup d'exemples. C'est une belle réussite.

À Toulouse, le 16 mai 2014,

Jean-Pierre Ramis

Preface to the Three Volumes

This three-volume series arose out of lecture notes for the courses we gave together at a CIMPA[1] school in Lima, Peru, in July 2008. Since then, these notes have been used and developed in graduate courses held at our respective institutions, that is, the universities of Angers, Nantes, Strasbourg (France) and the Scuola Normale Superiore di Pisa (Italy). The original notes have now grown into self-contained introductions to problems raised by analytic continuation and the divergence of power series in one complex variable, especially when related to differential equations.

A classical way of solving an analytic differential equation is the power series method, which substitutes a power series for the unknown function in the equation, then identifies the coefficients. Such a series, if convergent, provides an analytic solution to the equation. This is what happens at an ordinary point, that is, when we have an initial value problem to which the Cauchy-Lipschitz theorem applies. Otherwise, at a singular point, even when the method can be applied the resulting series most often diverges; its connection with "actual" local analytic solutions is not obvious despite its deep link to the equation.

The hidden meaning of divergent formal solutions was already pondered in the nineteenth century, after Cauchy had clarified the notions of convergence and divergence of series. For ordinary *linear* differential equations, it has been known since the beginning of the twentieth century how to determine a full set of linearly independent formal solutions[2] at a singular point in terms of a finite number of complex powers, logarithms, exponentials and power series, either convergent or divergent. These formal solutions completely determine the linear differential equation; hence, they contain all information about the equation itself, especially about its analytic solutions. Extracting this information from the divergent solutions was the underly-

[1] Centre International de Mathématiques Pures et Appliquées, or ICPAM, is a non-profit international organization founded in 1978 in Nice, France. It promotes international cooperation in higher education and research in mathematics and related subjects for the benefit of developing countries. It is supported by UNESCO and IMU, and many national mathematical societies over the world.

[2] One says *a formal fundamental solution.*

ing motivation for the theories of summability and, to some extent, of resurgence. Both theories are concerned with the precise structure of the singularities.

Divergent series may appear in connection with any local analytic object. They either satisfy an equation, or are attached to given objects such as formal first integrals in dynamical systems or formal conjugacy maps in classification problems. Besides linear and non-linear ordinary differential equations, they also arise in partial differential equations, difference equations, q-difference equations, etc. Such series, issued from specific problems, call for suitable theories to extract valuable information from them.

A theory of *summability* is a theory that focuses on a certain class of power series, to which it associates analytic functions. The correspondence should be injective and functorial: one expects for instance a series solution of a given functional equation to be mapped to an analytic solution of the same equation. In general, the relation between the series and the function –the latter is called its sum– is *asymptotic*, and depends on the direction of summation; indeed, with non-convergent series one cannot expect the sums to be analytic in a full neighborhood, but rather in a "sectorial neighborhood" of the point at which the series is considered.

One summation process, commonly known as the Borel-Laplace summation, was already given by Émile Borel in the nineteenth century; it applies to the classical Euler series and, more generally, to solutions of linear differential equations with a single "level", equal to one, although the notion of level was by then not explicitly formulated. It soon appeared that this method does not apply to all formal solutions of differential equations, even linear ones. A first generalization to series solutions of linear differential equations with a single, arbitrary level $k > 0$ was given by Le Roy in 1900 and is called *k-summation*. In the 1980's, new theories were developed, mainly by J.-P. Ramis and Y. Sibuya, to characterize k-summable series, a notion a priori unrelated to equations, but which applies to all solutions of linear differential equations with the single level k. The question of whether any divergent series solution of a linear differential equation is k-summable, known as the *Turrittin problem*, was an open problem until J.-P. Ramis and Y. Sibuya in the early 1980's gave a counterexample. In the late 1980's and in the 1990's *multisummability theories* were developed, in particular by J.-P. Ramis, J. Martinet, Y. Sibuya, B. Malgrange, W. Balser, M. Loday-Richaud and G. Pourcin, which apply to all series solution of linear differential equations with an arbitrary number of levels. They provide a unique sum of a formal fundamental solution on appropriate sectors at a singular point.

It was proved that these theories apply to solutions of non-linear differential equations as well: given a series solution of a non-linear differential equation, the choice of the right theory is determined by the linearized equation along this series. On the other hand, in the case of difference equations, not all solutions are multisummable; new types of summation processes are needed, for instance those introduced by J. Écalle in his theory of resurgence and considered also by G. Immink and B. Braaksma. Solutions of q-difference equations are not all multisummable either; specific processes in this case have been introduced by F. Marotte and C. Zhang in the late 1990's.

Summation sheds new light on the *Stokes phenomenon*. This phenomenon occurs when a divergent series has several sums, with overlapping domains, which correspond to different summability directions and differ from one another by exponentially small quantities. The question then is to describe these quantities. A precise analysis of the Stokes phenomenon is crucial for classification problems. For systems of linear differential equations, the meromorphic classification easily follows from the characterization of the Stokes phenomenon by means of the *Stokes cocycle*. The Stokes cocycle is a 1-cocycle in non-abelian Čech cohomology. It is expressed in terms of finitely many automorphisms of the normal form, the *Stokes automorphisms*, which select and organize the "exponentially small quantities". In practice, the Stokes automorphisms are represented by constant unipotent matrices called the *Stokes matrices*. It turned out that these matrices are precisely the correction factors needed to patch together two contiguous sums, that is, sums taken on the two sides of a singular direction, of a formal fundamental solution.[3]

The theory of *resurgence* was independently developed in the 1980's by J. Écalle, with the goal of providing a theory with a large range of applications, including the summation of divergent solutions of a variety of functional equations, differential, difference, differential-difference, etc. Basically, resurgence theory starts with the Borel-Laplace summation in the case of a single level equal to one, and this is the only situation we consider in these volumes. Let us mention however that there are extensions of the theory based on more general kernels.

The theory focuses on what happens in the Borel plane, that is, after one applies a Borel transform. The results are then pulled back via a Laplace transform to the plane of the initial variable also called the Laplace plane. In the Borel plane one typically gets functions, called *resurgent functions*, which are analytic in a neighborhood of the origin and can be analytically continued along various paths in the Borel plane, yet they are not entire functions: one needs to avoid a certain set Ω of possible singular points and analytic continuation usually gives rise to multiple-valuedness, so that these Borel-transformed functions are best seen as holomorphic functions on a Riemann surface akin to the universal covering of $\mathbb{C} \setminus \{0\}$. Of crucial importance are the singularities[4] which may appear at the points of Ω, and *Écalle's alien operators* are specific tools designed to analyze them.

The development of resurgence theory was aimed at non-linear situations where it reveals its full power, though it can be applied to the formal solutions of linear differential equations (in which case the singular support Ω is finite and the Stokes matrices, hence the local meromorphic classification, determined by the action of finitely many alien operators). The non-linearity is taken into account via the convolution product in the Borel plane. More precisely, we mean here the complex convolution which is turned into pointwise multiplication when returning to the original variable by means of a Laplace transform. Given two resurgent functions, analytic

[3] A less restrictive notion of Stokes matrices exists in the literature, which patch together any two sectorial solutions with same asymptotic expansion, but they are not local meromorphic invariants in general.

[4] The terms *singularity* in Écalle's resurgence theory and *microfunction* in Sato's microlocal analysis have the same meaning.

continuation of their convolution product is possible, but new singularities may appear at the sum of any two singular points of the factors; hence, Ω needs to be stable by addition (in particular, it must be infinite; in practice, one often deals with a lattice in $\mathbb{C}$). All operations in the Laplace plane have an explicit counterpart in the Borel plane: addition and multiplication of two functions of the initial variable, as well as non-linear operations such as multiplicative inversion, substitution into a convergent series, functional composition, functional inversion, which all leave the space of resurgent functions invariant.

To have these tools well defined requires significant work. The reward of setting the foundations of the theory in the Borel plane is greater flexibility, due to the fact that one can work with an *algebra* of resurgent functions, in which the analysis of singularities is performed through *alien derivations*[5].

Écalle's important achievement was to obtain the so-called *bridge equation*[6] in many situations. For a given problem, the bridge equation provides an all-in-one description of the action on the solutions of the alien derivations. It can be viewed as an *infinitesimal version of the Stokes phenomenon*: for instance, for a linear differential system with level one it is possible to prove that the set of Stokes automorphisms in a given formal class naturally has the structure of a unipotent Lie group and the bridge equation gives infinitesimal generators of its Lie algebra.

Summability and resurgence theories have useful interactions with the algebraic and geometrical approaches of linear differential equations such as *differential Galois theory* and the *Riemann-Hilbert problem*. The local differential Galois group of a meromorphic linear differential equation at a singular point is a linear algebraic group, the structure of which reflects many properties of the solutions. At a "regular singular" point [7] for instance, it contains a Zariski-dense subgroup finitely generated by the monodromy. However, at an "irregular singular" point, one needs to introduce further automorphisms, among them the Stokes automorphisms, to generate a Zariski-dense subgroup. For linear differential equations with rational coefficients, when all the singular points are regular, the classical Riemann-Hilbert correspondence associates with each equation a monodromy representation of the fundamental group of the Riemann sphere punctured at the singular points; conversely, from any representation of this fundamental group, one recovers an equation with prescribed regular singular points.[8] In the case of possibly irregular singular points, the monodromy representation alone is insufficient to recover the equation; here too one has to introduce the Stokes automorphisms and to connect them via "analytic continuation" of the divergent solutions, that is, via summation processes.

---

[5] Alien derivations are suitably weighted combinations of alien operators which satisfy the Leibniz rule.

[6] Its original name in French is *équation du pont*.

[7] This means that the formal solutions at that point may contain powers and logarithms but no exponential.

[8] The Riemann-Hilbert problem more specifically requires that the singular points in this restitution be *Fuchsian*, that is, simple poles only, which is not always possible.

These volumes also include an application of resurgence theory to the first Painlevé equation. Painlevé equations are nonlinear second-order differential equations introduced at the turn of the twentieth century to provide new transcendents, that is, functions that can neither be written in terms of the classical functions nor in terms of the special functions of physics. A reasonable request was to ask that all the movable singularities[9] be poles and this constraint led to a classification into six families of equations, now called Painlevé I to VI. Later, these equations appeared as conditions for isomonodromic deformations of Fuchsian equations on the Riemann sphere. They occur in many domains of physics, in chemistry with reaction-diffusion systems and even in biology with the study of competitive species. Painlevé equations are a perfect non-linear example to be explored with the resurgent tools.

We develop here the particular example of Painlevé I and we focus on its now classical truncated solutions. These are characterized by their asymptotics as well as by the fact that they are free of poles within suitable sectors at infinity. We determine them from their asymptotic expansions by means of a Borel-Laplace procedure after some normalization. The non-linearity generates a situation which is more intricate than in the case of linear differential equations. Playing the role of the formal fundamental solution is the so-called *formal integral* given as a series in powers of logarithm-exponentials with power series coefficients. More generally, such expansions are called *transseries* by J. Écalle or *multi-instanton expansions* by physicists. In general, the series are divergent and lead to a Stokes phenomenon. In the case of Painlevé I we prove that they are resurgent. Although the Stokes phenomenon can no longer be described by Stokes matrices, it is still characterized by the alien derivatives at the singular points in the Borel plane (see O. Costin *et al.*). The local meromorphic class of Painlevé I at infinity is the class of all second-order equations locally meromorphically equivalent at infinity to this equation. The characterization of this class requires *all* alien derivatives in all higher sheets of the resurgence surface. These extra invariants are also known as *higher order Stokes coefficients* and they can be given a numerical approximation using the *hyperasymptotic theory* of M. Berry and C. Howls. The complete resurgent structure of Painlevé I is given by its *bridge equation* which we state here, seemingly for the first time.

Recently, in quantum field and string theories, the resurgent structure has been used to describe the instanton effects, in particular for quartic matrix models which yield Painlevé I in specific limits. In the late 1990's, following ideas of A. Voros and J. Écalle, applications of the resurgence theory to problems stemming from quantum mechanics were developed by F. Pham and E. Delabaere. Influenced by M. Sato, this was also the starting point by T. Kawai and Y. Takei of the so-called *exact semi-classical analysis* with applications to Painlevé equations with a large parameter and their hierarchies, based on isomonodromic methods.

---

[9] The fixed singular points are those appearing on the equation itself; they are singular for the solutions generically. The movable singular points are singular points for solutions only; they "move" from one solution to another. They are a consequence of the non-linearity.

Summability and resurgence theories have been successfully applied to problems in analysis, asymptotics of special functions, classification of local analytic dynamical systems, mechanics, and physics. They also generate interesting numerical methods in situations where the classical methods fail.

In these volumes, we carefully introduce the notions of analytic continuation and monodromy, then the theories of resurgence, k-summability and multisummability, which we illustrate with examples. In particular, we study tangent-to-identity germs of diffeomorphisms in the complex plane both via resurgence and summation, and we present a newly developed resurgent analysis of the first Painlevé equation. We give a short introduction to differential Galois theory and a survey of problems related to differential Galois theory and the Riemann-Hilbert problem. We have included exercises with solutions. Whereas many proofs presented here are adapted from existing ones, some are completely new. Although the volumes are closely related, they have been organized to be read independently. All deal with power series and functions of a complex variable; the words *analytic* and *holomorphic* are used interchangeably, with the same meaning.

This book is aimed at graduate students, mathematicians and theoretical physicists who are interested in the theories of monodromy, summability or resurgence and related problems.

Below is a more detailed description of the contents.

- Volume 1: *Monodromy and Resurgence* by C. Mitschi and D. Sauzin.
  An essential notion for the book and especially for this volume is the notion of analytic continuation "à la Cauchy-Weierstrass". It is used both to define the monodromy of solutions of linear ordinary differential equations in the complex domain and to derive a definition of resurgence.
  Once monodromy is defined, we introduce the Riemann-Hilbert problem and the differential Galois group. We show how the latter is related to analytic continuation by defining a set of automorphisms, including the Stokes automorphisms, which together generate a Zariski-dense subgroup of the differential Galois group. We state the inverse problem in differential Galois theory and give its particular solution over $\mathbb{C}(z)$ due to Tretkoff, based on a solution of the Riemann-Hilbert problem. We introduce the language of vector bundles and connections in which the Riemann-Hilbert problem has been extensively studied and give the proof of Plemelj-Bolibrukh's solution when one of the prescribed monodromy matrices is diagonalizable.
  The second part of the volume begins with an introduction to the 1-summability of series by means of Borel and Laplace transforms (also called Borel or Borel-Laplace summability) and provides non-trivial examples to illustrate this notion. The core of the subject follows, with definitions of resurgent series and resurgent functions, their singularities and their algebraic structure. We show how one can analyse the singularities via the so-called *alien calculus* in resurgent algebras; this includes the *bridge equation* which usefully connects alien and ordinary derivations. The case of tangent-to-identity germs of diffeomorphisms in the complex plane is given a thorough treatment.

- Volume 2: *Simple and Multiple Summability* by M. Loday-Richaud.
  The scope of this volume is to thoroughly introduce the various definitions of k-summability and multisummability developed since the 1980's and to illustrate them with examples, mostly but not only, solutions of linear differential equations. For the first time, these theories are brought together in one volume.
  We begin with the study of basic tools in Gevrey asymptotics, and we introduce examples which are reconsidered throughout the following sections. We provide the necessary background and framework for some theories of summability, namely the general properties of sheaves and of abelian or non-abelian Čech cohomology. With a view to applying the theories of summability to solutions of differential equations we review fundamental properties of linear ordinary differential equations, including the main asymptotic expansion theorem, the formal and the meromorphic classifications (formal fundamental solution and linear Stokes phenomenon) and a chapter on index theorems and the irregularity of linear differential operators. Four equivalent theories of k-summability and six equivalent theories of multisummability are presented, with a proof of their equivalence and applications. Tangent-to-identity germs of diffeomorphisms are revisited from a new point of view.
- Volume 3: *Resurgent Methods and the First Painlevé equation* by E. Delabaere.
  This volume deals with ordinary non-linear differential equations and begins with definitions and phenomena related to the non-linearity. Special attention is paid to the first Painlevé equation, or Painlevé I, and to its tritruncated and truncated solutions. We introduce these solutions by proving the Borel-Laplace summability of transseries solutions of Painlevé I. In this context resonances occur, a case which is scarcely studied. We analyse the effect of these resonances on the formal integral and we provide a normal form. Additional material in resurgence theory is needed to achieve a resurgent analysis of Painlevé I up to its bridge equation.

Acknowledgements. We would like to thank the CIMPA institution for giving us the opportunity of holding a winter school in Lima in July 2008. We warmly thank Michel Waldschmidt and Michel Jambu for their support and advice in preparing the application and solving organizational problems. The school was hosted by IMCA (Instituto de Matemática y Ciencias Afines) in its new building of La Molina, which offered us a perfect physical and human environment, thanks to the colleagues who greeted and supported us there. We thank all institutions that contributed to our financial support: UNI and PUCP (Peru), LAREMA (Angers), IRMA (Strasbourg), IMT (Toulouse), ANR Galois (IMJ Paris), IMPA (Brasil), Universidad de Valladolid (Spain), Ambassade de France au Pérou, the International Mathematical Union, CCCI (France) and CIMPA. Our special thanks go to the students in Lima and in our universities, who attended our classes and helped improve these notes via relevant questions, and to Jorge Mozo Fernandez for his pedagogical assistance.

Angers, Strasbourg, Pisa, November 2015

Éric Delabaere, Michèle Loday-Richaud, Claude Mitschi, David Sauzin

Introduction to this volume

This volume is the first of the three-volume book *Divergent Series, Summability and Resurgence*. It is composed of two parts, "Monodromy in Linear Differential Equations" by C. Mitschi, and "Introduction to 1-Summability and Resurgence" by D. Sauzin.

In the field of linear analytic ordinary differential equations, problems range from pure analysis to algebra, geometry and topology. The aim of these lecture notes is to show how one translates questions such as the existence and behavior of solutions of differential equations in the neighborhood of singular points, into questions of geometric topology, differential algebra and algebraic geometry. A central issue in this three-volume work is to give divergent power series an analytic meaning via summability and resurgence theory. Divergent solutions of a differential equation account for the presence of singular points which in general prevent local analytic solutions from extending as single-valued functions on the punctured complex plane. For linear differential equations, the monodromy and Stokes matrices 'measure' the multivaluedness of the solutions, depending on the regularity or irregularity of the singular points. In the regular singular case, the monodromy representation provides a geometric, topological description of the differential equation. In the irregular case, the Stokes matrices which arise from the formal divergent solutions are, together with the monodromy matrices, elements of the differential Galois group. This is a linear algebraic group, the algebraic structure of which reflects many properties, even analytic, of the differential equation: their solvability for instance, or the existence of transcendental solutions.

In the second part of the volume, power series are considered independently of any equation, differential or not, that they may satisfy; still, rather surprisingly, interesting structures can be identified. The central tool is the formal Borel transform, in terms of which we give definitions of 1-summability and resurgence (alternative definitions of summability will be given in the second volume [Lod16]). All convergent power series are both 1-summable and resurgent, but many divergent power series also satisfy one or both properties. Emphasis is placed on the differential algebra structure: 1-summable series form a space which is stable by multiplication and differentiation, and so do resurgent series. This is proved by studying the counterpart of differentiation and multiplication via the formal Borel transform; the former is elementary, whereas the latter requires a careful analysis of the analytic continuation of convolution products.

In contrast to convergent power series, for which the sum is a uniquely defined function analytic in a full neighborhood of the origin, a divergent 1-summable power series gives rise to several functions, called Borel-Laplace sums; these are analytic in appropriate sectorial neighborhoods of the origin and asymptotic to the original series. The relation between these functions depends on the singularities of the Borel transform; if moreover the series is resurgent, then this relation can be analyzed by means of Écalle's "alien calculus". In this volume, we develop alien calculus for the subclass of simple resurgent series: this is an algebra on which we define a family of derivations, the so-called alien derivations, which a priori have nothing to do with ordinary differentiation and allow us to describe the passage from one Borel-Laplace sum to the other. Numerous examples are given, in relation in particular with differential and difference equations, as for instance the Fatou coordinates of a tangent-to-identity germ of diffeomorphism.

These notes grew out of lectures given at our CIMPA school in Lima. Aiming at students with a diverse variety of backgrounds, we presented the elementary and introductory parts of the subject in more detail than we normally would have in a graduate course. We decided to reproduce these tutorial parts here, hoping they will give the beginners an easier access to the more specialized parts of the three volumes.

Acknowledgements The sections about the Riemann-Hilbert problem were inspired by several articles and books of Andrey Bolibrukh as well as by a beautiful graduate course he gave at the University of Strasbourg in 1998. C.M. thanks Viktoria Heu for sharing her notes of a graduate course Frank Loray gave in Rennes in 2006. She also thanks the anonymous referees for helpful and encouraging comments. D.S. thanks Fibonacci Laboratory (CNRS UMI 3483), the Centro Di Ricerca Matematica Ennio De Giorgi and the Scuola Normale Superiore di Pisa for their hospitality. C.M. and D.S. owe special thanks to Michèle Loday, who initiated the CIMPA project in Peru, for numerous useful exchanges. D.S.'s work has received funding from the European Community's Seventh Framework Program (FP7/2007–2013) under Grant Agreement n. 236346 and from the French National Research Agency under the reference ANR-12-BS01-0017.

C. Mitschi and D.Sauzin

Strasbourg and Pisa, November 2015

Contents

Part I
Monodromy in Linear Differential Equations

Claude Mitschi

To give divergent power series an analytic meaning via summability and resurgence theory is a central issue in this three volume work. Divergent solutions of a differential equation account for the presence of singular points which in general prevent local analytic solutions to extend as single-valued functions on the punctured complex plane. For linear differential equations, the monodromy and Stokes matrices 'measure' the multivaluedness of the solutions, depending on the regularity or irregularity of the singular points. In the regular singular case, the monodromy representation gives a geometric description of the differential equation. In the irregular case, the Stokes matrices which arise from formal divergent solutions, are together with the monodromy matrices elements of the differential Galois group. This is a linear algebraic group which provides an algebraic interpretation of the differential equation : of its solvability for instance, or the existence of transcendental solutions.

The first chapter of this volume is an elementary introduction to analytic continuation, monodromy and singular points, with detailed definitions and proofs.

The second chapter is devoted to differential Galois theory, with basic facts over general differential fields followed by the analytic theory over the field of complex rational functions. With a view to the direct problem of calculating differential Galois groups, we present two important density theorems: Schlesinger's theorem, which relates the differential Galois group of differential systems with regular singularities to their monodromy, and Ramis's density theorem for irregular singularities, which describes the differential Galois group in terms of more specific invariants than the monodromy.

The third chapter gives a short overview of inverse problems, from the Riemann-Hilbert problem to differential inverse Galois problems via the Tretkoff theorem. Recent developments of these problems are mentioned, with references.

The fourth chapter is an introduction to the Riemann-Hilbert problem from Bolibrukh's point of view. To produce his famous counterexample, Bolibrukh introduced specific methods to attack the still open problem of characterizing those monodromy representations which can be realized by Fuchsian systems. We explain these methods and give some accessible proofs, once the necessary and elementary material about fiber bundles is presented.

The following notes grew out of lectures given to students of the CIMPA school with a diverse variety of backgrounds. We therefore presented the introductory parts of the subject in more detail than we normally would have in a graduate course. We decided to reproduce these notes here, hoping they will give the beginners an easier access to the more specialized parts of the three volumes.

Chapter 1
Analytic continuation and monodromy

1.1 Basic tools in complex analysis

The study of linear ordinary differential equations in the complex domain involves specific tools of complex analysis such as asymptotic expansion and analytic continuation, which are briefly presented in this chapter. For general results in complex analysis and differential equations, we refer to any of the books [Ahl79], [Cha90], [CoLe55], [Ha64], [In56], [JS87], [La61], [NN01], [Ru86].

1.1.1 The Riemann sphere

Let $\overline{\mathbb{C}} = \mathbb{C} \cup \{\infty\}$ denote a compactification of $\mathbb{C}$, the field of complex numbers, by adding a point ∞ called *infinity*. The open subsets of $\overline{\mathbb{C}}$ consist of the open subsets of $\mathbb{C}$ (for the usual topology of $\mathbb{R}^2$) and the subsets $V \cup \{\infty\}$ where $V = \mathbb{C} \setminus K$ for some compact subset K of $\mathbb{C}$. There are different ways of thinking of $\overline{\mathbb{C}}$.

1. The *complex projective line* $\mathbb{P}^1_{\mathbb{C}}$ is the set of all lines ℓ of the $\mathbb{C}$-vector space $\mathbb{C}^2$. Let ℓ_v, for a non-zero $v \in \mathbb{C}^2$, denote the line generated by v in $\mathbb{C}^2$. There is a one-to-one correspondence
$$\iota_1 : \overline{\mathbb{C}} \to \mathbb{P}^1_{\mathbb{C}}$$
such that $\iota_1(z) = \ell_{(z,1)}$ if $z \in \mathbb{C}$, $\iota_1(\infty) = \ell_{(1,0)}$.
2. *The unit sphere* $\mathbb{S} = \{(x,y,t) \in \mathbb{R}^3 \mid x^2 + y^2 + t^2 = 1\}$ is another model for $\overline{\mathbb{C}}$. Here we identify $\mathbb{C}$ with the (x,y)-plane in $\mathbb{R}^3$, by representing a complex number $z = x + \mathrm{i}y$ as $(x,y,0) \in \mathbb{R}^3$. Let $N = (0,0,1)$ denote the 'North pole' of $\mathbb{S}$ and δ_p, for any $p \in \mathbb{R}^3$, $p \neq N$, the affine line defined by N and p in $\mathbb{R}^3$. The one-to-one correspondence
$$\iota_2 : \overline{\mathbb{C}} \to \mathbb{S}$$
is defined by $\iota_2(z) = \mathbb{S} \cap \delta_z$ if $z \in \mathbb{C}$, $\iota_2(\infty) = N$. Its inverse map

C. Mitschi, D. Sauzin, *Divergent Series, Summability and Resurgence I*,
Lecture Notes in Mathematics 2153, DOI 10.1007/978-3-319-28736-2_1

$$\mathbb{S} \to \overline{\mathbb{C}}$$

is given by the stereographic projection

$$\pi : \mathbb{S} \setminus \{N\} \to \mathbb{C}$$

from N defined by $\pi(p) = \delta_p \cap \mathbb{C}$, and the additional convention $\pi(N) = \infty$.

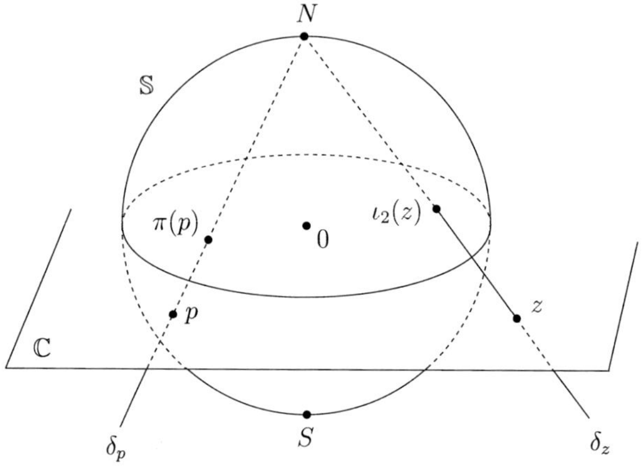

Fig. 1.1: The Riemann sphere

The space $\overline{\mathbb{C}}$ has the structure of a Riemann surface (cf. [ACMS53], [Fo81]); it is called the *Riemann sphere*.

Definition 1.1. A *Riemann surface* is a one-dimensional connected complex analytic variety, that is, a Hausdorff topological space X together with an *atlas* consisting of *charts* $(U_i, \varphi_i)_{i \in I}$ where

(a) $(U_i)_{i \in I}$ is an open covering of X,

(b) $\varphi_i : U_i \to \mathbb{C}$ maps U_i homeomorphically to $\mathbb{C}$ for all $i \in I$,

(c) if $U_i \cap U_j \neq \emptyset$ then $\varphi_{ij} = \varphi_i \circ \varphi_j^{-1} : \varphi_j(U_i \cap U_j) \to \mathbb{C}$ is holomorphic.

The map φ_i is called a *local coordinate* on U_i, for each $i \in I$.

The Riemann surface structure of $\overline{\mathbb{C}}$ is given by the atlas $\{(U_0, \varphi_0), (U_\infty, \varphi_\infty)\}$ where $U_0 = \mathbb{C}$, $U_\infty = \mathbb{C}^* \cup \{\infty\}$, $\varphi_0 = \mathrm{id}_{\mathbb{C}}$, $\varphi_\infty(z) = 1/z$ if $z \in \mathbb{C}^*$ and $\varphi_\infty(\infty) = 0$. The geometric description of this atlas is the following.

1. For the projective line we have $\mathbb{P}^1_{\mathbb{C}} = U_0 \cup U_\infty$ with

$$U_0 = \mathbb{P}^1_{\mathbb{C}} \setminus \{\ell^*_{(1,0)}\}, \quad U_\infty = \mathbb{P}^1_{\mathbb{C}} \setminus \{\ell^*_{(0,1)}\},$$

$$\varphi_0(\ell^*_{(z_1,z_2)}) = \frac{z_1}{z_2}, \quad \varphi_\infty(\ell^*_{(z_1,z_2)}) = \frac{z_2}{z_1}.$$

2. For the unit sphere we have $\mathbb{S} = U_0 \cup U_\infty$, with
$$U_0 = \mathbb{S} \setminus \{N\}, \;\; U_\infty = \mathbb{S} \setminus \{S\},$$
where $S = (0,0,-1)$ denotes the 'South pole'. The maps φ_0 and φ_∞ are the stereographic projections from N and S respectively.

Via this atlas, any complex function on $\overline{\mathbb{C}}$ can be viewed locally, in particular at ∞, as a function on $\mathbb{C}$, with the subsequent definitions of holomorphy and meromorphy. We will admit the following well-known fact (cf. [Cha90, Theorem 4 p. 135], [JS87, Theorem 1.4.1 p. 9]).

Theorem 1.2. *A function on $\overline{\mathbb{C}}$ is meromorphic at all $z \in \overline{\mathbb{C}}$ if and only if its restriction to $\mathbb{C}$ is a rational function.*

For a fixed $a \in \overline{\mathbb{C}}$, let $\mathscr{O}_a$ denote the set of functions defined each in a neighbourhood of a and holomorphic at a. The relation $f \overset{a}{\sim} g$ if $f = g$ in some neighbourhood of a, is an equivalence relation. An equivalence class is called a *germ* of holomorphic[1] function at a. The germ of $f \in \mathscr{O}_a$ will be denoted by $\overline{f}$ if there is no ambiguity about a.

Recall that a *domain* in $\overline{\mathbb{C}}$ is a nonempty open (pathwise) connected subset of $\overline{\mathbb{C}}$. The following theorem will be referred to as the Fundamental Uniqueness Theorem (FUT) for holomorphic functions.

Theorem 1.3. *If f and g are holomorphic functions on a domain $\mathscr{D} \subset \overline{\mathbb{C}}$ and $f \overset{a}{\sim} g$ for some $a \in \mathscr{D}$, then $f = g$ on $\mathscr{D}$.*

1.1.2 Analytic continuation. Monodromy

We recall that a *path* in $\overline{\mathbb{C}}$ is by definition a continuous map
$$\gamma : [0,1] \to \overline{\mathbb{C}}.$$

Note that any proper subset of $\overline{\mathbb{C}}$ can be considered as a subset of $\mathbb{C}$, using suitable local coordinates. For a given path γ we may therefore at convenience assume that $\mathrm{Im}(\gamma) \subset \mathbb{C}$ (we admit the fact that $\mathrm{Im}(\gamma)$ is a proper subset of $\overline{\mathbb{C}}$, which is for instance related to the Jordan-Schönfliess theorem, cf. [Cai51]). Originally in their discovery of multivalued functions, Weierstrass and Riemann used the innovative notion of a mathematical object consisting of a function together with a domain of holomorphy. We will follow this approach in the following definitions.

[1] We warn the reader that we may use the words 'holomorphic' and 'analytic' indistinctively in these lecture notes.

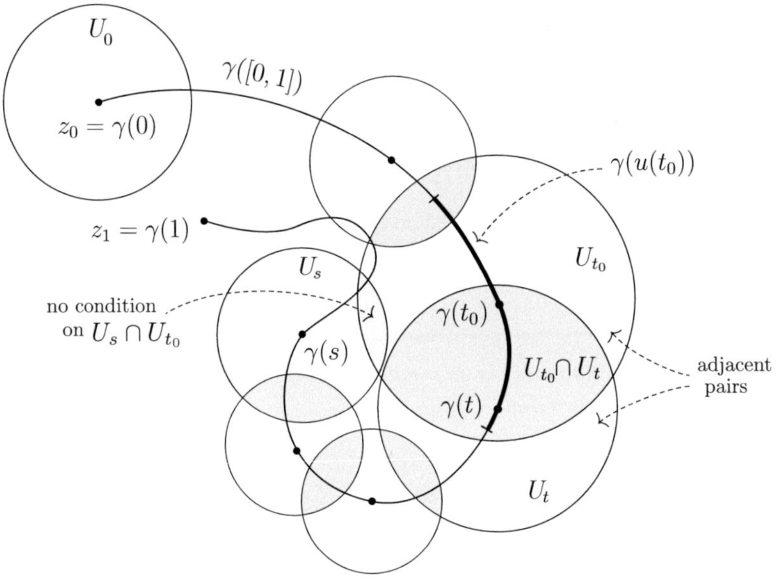

Fig. 1.2: A continuable pair (U_0, f_0) along γ

Definition 1.4. A *pair* (U, f) is a nonempty open disk $U \subset \mathbb{C}$ (of finite or infinite radius) together with a function f, holomorphic in U, such that the radius of U is the maximal radius of convergence of the series expansion of f at the center of U. The *center* of (U, f) is the center of U, or any point of $\mathbb{C}$ if $U = \mathbb{C}$.

Definition 1.5. Two pairs (U, f) and (V, g) are *adjacent* if $U \cap V \neq \emptyset$ and $f = g$ on $U \cap V$.

Definition 1.6. A pair (V, g) is an *analytic continuation* of the pair (U, f) if there is a *finite* sequence of pairs $((U_i, f_i))_{0 \le i \le n}$ with $(U_0, f_0) = (U, f)$, $(U_n, f_n) = (V, g)$, such that (U_i, f_i) and (U_{i+1}, f_{i+1}) are adjacent for $i = 0, \ldots, n-1$.

Notation 1.7 *For any open interval* $]a, b[\subset \mathbb{R}$ *let*

$$]a, b[_{[0,1]} =]a, b[\cap [0, 1]$$

denote the corresponding open subset of $[0, 1]$ *for the induced topology.*

Definition 1.8. Let γ be a path in $\overline{\mathbb{C}}$ and (U_0, f_0) a pair centered at $z_0 = \gamma(0)$. We say that the pair (U_0, f_0) *can be continued along* γ if there is a family $((U_t, f_t))_{t \in [0,1]}$ of pairs such that for each $t \in [0, 1]$

(a) $z_t = \gamma(t)$ is the center of U_t

(b) For any open subset $u(t) =]t - h, t + k[_{[0,1]}$ of $[0, 1]$ such that $\gamma(u(t)) \subset U_t$, the pairs (U_s, f_s) and (U_t, f_t) are adjacent for all $s \in u(t)$ (see Fig. 1.2).

In what follows, we aim to prove that the resulting germ $\overline{f_1}$ (of f_1 at z_1) in Definition 1.8 only depends on the homotopy class of the path.

Proposition 1.9. *With notation of Definition 1.8, if the pair (U_0, f_0) can be continued along γ, then the germ $\overline{f_1}$ of f_1 at z_1 does not depend on the family $((U_t, f_t))_{t\in[0,1]}$.*

Proof. Assume that (U_0, f_0) can be continued along γ by means of two families $((U_t, f_t))_{t\in[0,1]}$ and $((V_t, g_t))_{t\in[0,1]}$, with $V_0 = U_0,\ g_0 = f_0$, and let $W_t = U_t \cap V_t$ for each $t \in [0,1]$. Consider the subset

$$E = \{t \in [0,1] \mid f_t \overset{\gamma(t)}{\sim} g_t\}$$

of $[0,1]$, which is nonempty since $0 \in E$. To show that E is open in $[0,1]$, fix $t_0 \in E$. The functions f_{t_0} and g_{t_0} are analytic in W_{t_0} and define the same germ at $\gamma(t_0)$. The FUT implies that $f_{t_0} = g_{t_0}$ on W_{t_0} since W_{t_0} is a domain. Consider an open neighbourhood

$$u(t_0) =]t_0 - h, t_0 + h[_{[0,1]}$$

of t_0 in $[0,1]$ such that $\gamma(u(t_0)) \subset W_{t_0}$, which exists since γ is continuous. By Definition 1.8 we have $f_t = f_{t_0}$ on $U_t \cap U_{t_0}$ and $g_t = g_{t_0}$ on $V_t \cap V_{t_0}$, and in particular $f_t = f_{t_0}$ and $g_t = g_{t_0}$ on $W_t \cap W_{t_0}$. Since $f_{t_0} = g_{t_0}$ in W_{t_0} we have $f_t = g_t$ in $W_t \cap W_{t_0}$. The functions f_t and g_t are holomorphic on the connected open subset W_t and coincide on a nonempty open subset of W_t, hence $f_t = g_t$ on W_t and

$$f_t \overset{\gamma(t)}{\sim} g_t.$$

To show that E is closed, fix t_0 in the closure $\overline{E}$ of E, and an open connected neighbourhood $u(t_0) =]t_0 - h, t_0 + h[_{[0,1]}$ of t_0 in $\subset [0,1]$ such that $\gamma(u(t_0)) \subset W_{t_0}$. Let $t \in u(t_0) \cap E$. Then, by definition of E and Definition 1.8, we get

$$f_t \overset{\gamma(t)}{\sim} g_t, \quad f_t \overset{\gamma(t)}{\sim} f_{t_0}, \quad g_t \overset{\gamma(t)}{\sim} g_{t_0}.$$

The functions f_{t_0} and g_{t_0} are holomorphic on W_{t_0} and coincide on a neighbourhood of $\gamma(t)$ in W_{t_0}, hence $f_{t_0} = g_{t_0}$ on W_{t_0} and

$$f_{t_0} \overset{\gamma(t_0)}{\sim} g_{t_0},$$

that is, $t_0 \in E$. The connectedness of $[0,1]$ implies that $E = [0,1]$ and in particular that

$$f_1 \overset{\gamma(1)}{\sim} g_1$$

which ends the proof. □

Exercise 1.10. With the same notation, show that if a pair (U_0, f_0) can be continued along γ by means of a family $((U_t, f_t))_{t\in[0,1]}$ then the radius of U_t is either infinite for all $t \in [0,1]$ or a continuous function of $t \in [0,1]$.

Proposition 1.11. *If the pair (U_0, f_0) can be continued along γ by means of a family $((U_t, f_t))_{t\in[0,1]}$ then (U_1, f_1) is an analytic continuation of (U_0, f_0) in the sense of Definition 1.6.*

Proof. If the radius of all disks U_t is infinite (see Exercise 1.10) then obviously $U_0 = U_1$, $f_0 = f_1$. If not, the radius $R(t)$ of U_t is a continuous, non-vanishing function on the compact interval $[0,1]$, hence there is an $\varepsilon > 0$ with $R(t) \geq \varepsilon$ for all $t \in [0,1]$. By the uniform continuity of γ on $[0,1]$ there is a finite sequence $t_0 = 0 < t_1 < \ldots < t_n = 1$ such that $|\gamma(t') - \gamma(t'')| < \varepsilon/2$ for all $t', t'' \in [t_{k-1}, t_k]$, $k = 1, \ldots, n$, which implies that for some $\alpha > 0$,

$$\gamma(]t_{k-1} - \alpha, t_k + \alpha[_{[0,1]} \subset U_{t_k}$$

for $k = 1, \ldots, n$. It then follows from the asumption that the pairs $(U_{t_{k-1}}, f_{t_{k-1}})$ and (U_{t_k}, f_{t_k}) are adjacent for all $k = 1, \ldots, n$, which by Definition 1.6 means that (U_1, f_1) is an analytic continuation of (U_0, f_0), obtained via *finitely* many adjacent pairs. □

Notation 1.12 *If (V, g) is an analytic continuation of the pair (U, f) obtained by continuing (U, f) along a path γ as in Proposition 1.11, will write $g = f^{\gamma}$.*

Let us recall the definition of homotopy.

Definition 1.13. Let $\mathscr{D} \in \mathbb{C}$ be a domain, and $a, b \in \mathscr{D}$. Two paths γ_0, γ_1 from a to b such that $\mathrm{Im}(\gamma_0) \subset \mathscr{D}$, $\mathrm{Im}(\gamma_1) \subset \mathscr{D}$ are *homotopic in $\mathscr{D}$* if there is a continuous function $\gamma \colon [0,1]^2 \to \mathscr{D}$ such that

$$\gamma(s, 0) = a, \;\; \gamma(s, 1) = b$$

$$\gamma(0, t) = \gamma_0(t), \;\; \gamma(1, t) = \gamma_1(t)$$

for all $s, t \in [0,1]$. The map γ is called a homotopy map.

An important property of analytic continuation is its homotopy invariance.

Theorem 1.14 (Monodromy theorem). *Let γ_0, γ_1 be homotopic paths from a to b in $\mathscr{D}$, and γ_s for $s \in [0,1]$ the path*

$$\gamma_s(t) = \gamma(s, t) \quad \text{for} \quad t \in [0,1],$$

where γ is the homotopy map. If a pair (U, f) centered at a can be continued along each γ_s, $s \in [0,1]$, then

$$f^{\gamma_0} \overset{b}{\sim} f^{\gamma_1}.$$

Proof. Let us prove that the map $s \mapsto f^{\gamma_s}$ is locally constant on $[0,1]$. For a given $s \in [0,1]$ the continuation of (U, f) along γ_s is obtained via a family of pairs $((U_t^s, f_t^s))_{t\in[0,1]}$ which we write $(f_t^s)_{t\in[0,1]}$ for short (keeping in mind the corresponding maximal disk U_t^s of radius $R^s(t)$ for each t). By the uniform continuity of γ on $[0,1]$ there is an $\varepsilon > 0$ such that

(a) $R^s(t) \geq \varepsilon$ for all $t, s \in [0,1]$

(b) for each $s_0 \in [0,1]$ there is an open connected neighbourhood $u(s_0)$ of s_0 in $[0,1]$ such that

$$|\gamma_s(t) - \gamma_{s_0}(t)| < \frac{\varepsilon}{6} \text{ for all } s \in u(s_0),\ t \in [0,1].$$

Fix such an $\varepsilon > 0$, and $s_0 \in [0,1]$. As in the proof of Proposition 1.11, note that there is a finite sequence $t_0 = 0 < t_1 < \ldots < t_n = 1$ such that $|\gamma_{s_0}(t') - \gamma_{s_0}(t'')| < \varepsilon/6$ for all $t', t'' \in [t_{k-1}, t_k]$, $k = 1, \ldots, n$. It follows from b) that $|\gamma_s(t') - \gamma_s(t'')| < \varepsilon/2$ for all $t', t'' \in [t_{k-1}, t_k]$, $k = 1, \ldots, n$ and all $s \in u(s_0)$. For a given $s \in u(s_0)$ let us prove by induction on k that $f^s_{t_k}$ and $f^{s_0}_{t_k}$ are adjacent for all $k = 0, \ldots, n$. Note that if this holds, then $f^{\gamma_s} = f^{\gamma_{s_0}}$ for all $s \in u(s_0)$, that is, the map $s \mapsto f^{\gamma_s}$ is locally constant.

Let $z_k = \gamma_s(t_k)$, $z^0_k = \gamma_{s_0}(t_k)$ for all k (see Fig. 1.3).

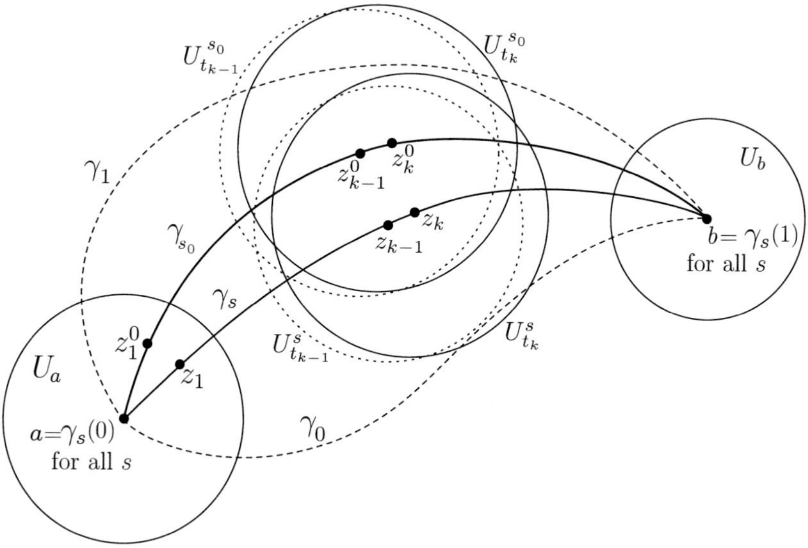

Fig. 1.3: Homotopy invariance of analytic continuation

If $k = 1$, (the pairs) $f^{s_0}_{t_1}$ and $f^s_{t_1}$ are adjacent. This follows from

$$|z^0_1 - a)| < \frac{\varepsilon}{2}, \quad |z_1 - a| < \frac{\varepsilon}{2}$$

which implies that $|z^0_1 - z_1| < \varepsilon$, hence that $W = U^{s_0}_{t_1} \cap U^s_{t_1} \cap U_a$ is nonempty (since all radii are $\geq \varepsilon$) and $f^s_{t_1}$ and $f^{s_0}_{t_1}$ coincide on W. The FUT implies that $f^s_{t_1}$ and $f^{s_0}_{t_1}$ coincide on $U^{s_0}_{t_1} \cap U^s_{t_1}$, and $f^s_{t_1}$ and $f^{s_0}_{t_1}$ are adjacent as pairs.

Now assume that $f^s_{t_{k-1}}$ and $f^{s_0}_{t_{k-1}}$ are adjacent for a given $k \geq 1$. Note that

$$|z_k - z_k^0)| < \frac{\varepsilon}{2},\ |z_k - z_{k-1})| < \frac{\varepsilon}{2},\ |z_k^0 - z_{k-1}^0)| < \frac{\varepsilon}{2}.$$

Since $|z_{k-1} - z_{k-1}^0)| < \frac{\varepsilon}{2}$, the center of each disk $U_{t_{k-1}}^s$ and $U_{t_{k-1}}^{s_0}$ belongs to the other, and hence $f_{t_{k-1}}^s$ and $f_{t_{k-1}}^{s_0}$ define the same germ at z_{k-1}^0. Note that $f_{t_k}^{s_0}$ and $f_{t_{k-1}}^{s_0}$ on one hand, $f_{t_k}^s$ and $f_{t_{k-1}}^s$ on the other hand, are adjacent. From $|z_k^0 - z_{k-1}^0| < \frac{\varepsilon}{2}$ and

$$|z_k - z_{k-1}^0| \leq |z_k - z_k^0| + |z_k^0 - z_{k-1}^0| < \varepsilon$$

it follows that $f_{t_k}^s$, $f_{t_{k-1}}^s$ and $f_{t_{k-1}}^{s_0}$ define the same germ at z_{k-1}^0, hence $f_{t_k}^s$ and $f_{t_k}^{s_0}$ are adjacent by the FUT. This holds for any $k = 0, \ldots, n$, in particular for $k = n$. Since $f_{t_n}^s = f_1^s$ and $f_{t_n}^{s_0} = f_1^{s_0}$ are adjacent and U_1^s, $U_1^{s_0}$ have the same center b these pairs are equal by the FUT. This proves that the map $s \mapsto \overline{f^{\gamma_s}}$ is locally constant, hence constant on $[0,1]$. In particular $f^{\gamma_0} \overset{b}{\sim} f^{\gamma_1}$. □

We end this section by proving that continuation via a finite number of steps induces continuation by a continuous family of pairs in the sense of Definition 1.8, which yields a converse to Proposition 1.11.

Proposition 1.15. *Let γ be a path in $\mathbb{C}$ and $t_0 = 0 < t_1 < \ldots < t_n = 1$, a subdivision of $[0,1]$. Assume that for each $i = 0, \ldots, n$ there is an open disk V_i with center $\gamma(t_i)$ such that $V_i \cap V_{i+1} \neq \emptyset$ and $\gamma([t_i, t_{i+1}]) \subset V_i$ for $i \leq n-1$. Let $(f_i)_{0 \leq i \leq n}$ be a family of functions f_i holomorphic in V_i and such that $f_i = f_{i+1}$ on $V_i \cap V_{i+1}$. Then the pair (U_0, f_0), centered at $\gamma(0)$ can be continued along γ in the sense of Definition 1.8, and $f_0^\gamma = f_n$.*

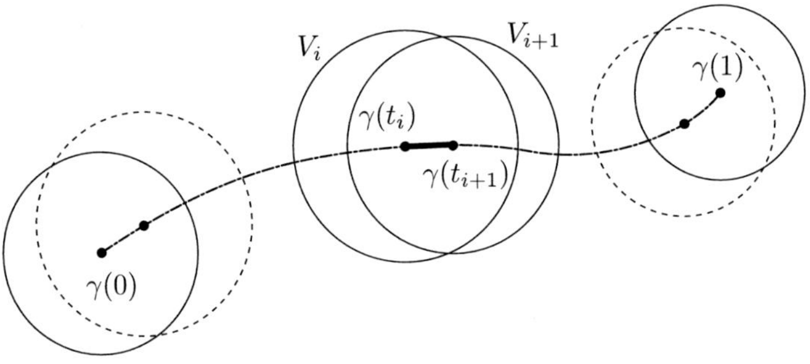

Fig. 1.4

Proof. (See Fig. 1.4) For each $t \in [0,1]$ we have $t \in [t_i, t_{i+1}]$, hence $\gamma(t) \in V_i$ for some i. Let U_t be the open disk with center $\gamma(t)$ and maximal radius of convergence for the series expansion of f_i at $\gamma(t)$, and g_t the sum in U_t of this series. In particular, $g_{t_i} = f_i$ in V_i for each i, and the family of pairs $((U_t, g_t))_{t \in [0,1]}$ clearly defines a continuation of the pair (U_0, f_0) resulting in the pair $(U_1, g_1) = (U_1, f_n)$. □

1.2 Linear differential systems

Notation 1.16 *Throughout the next sections* $\mathbb{C}(x)$ *denotes the field of complex rational functions, and* $\mathscr{M}(\mathscr{D})$ *(resp.* $\mathscr{O}(\mathscr{D})$*) the field (resp. ring) of meromorphic (resp.holomorphic) functions on a domain* $\mathscr{D} \subset \mathbb{C}$. *A vector function is said to be rational, holomorphic, meromorphic respectively, whenever its components are.*

A complex linear ordinary differential system, *linear system* for short, of *order p* is a system of p ordinary first order scalar differential equations

$$\begin{cases} \frac{dy_1}{dx} &= a_{11}(x)y_1 + \ldots + a_{1p}(x)y_p \\ \ldots \ldots & \ldots \quad \ldots \quad \ldots \\ \frac{dy_p}{dx} &= a_{p1}(x)y_1 + \ldots + a_{pp}(x)y_p \end{cases}$$

where, depending on the section, we will assume that $a_{ij} \in \mathbb{C}(x)$ or $a_{ij} \in \mathscr{M}(\mathscr{D})$ for all i, j, where $\mathscr{D}$ is a domain in $\overline{\mathbb{C}}$. It is more convenient to write this system in the matrix form (S)

$$\frac{dy}{dx} = A(x)y \tag{S}$$

where the $p \times p$ matrix $A = (a_{ij})_{1 \le i,j \le p}$ is called the *coefficient matrix of* $(S)$and y is an unknown vector function of x. Note that one hundred years ago the language of matrices for differential equations was not as common as it is today. In a letter to Gösta Mittag-Leffler (Nov. 21, 1915) George Birkhoff writes:

> For several years I have been much interested in this part of the theory of functions in which the matrix [2] of analytic functions instead of the single analytic function appears the natural element. It seems to me that this is the case in the theory of linear difference and differential equations.

1.2.1 Analytic continuation of solutions

In the matrix form of linear differential systems, a basis of solutions is given by a single invertible matrix.

Definition 1.17. A *fundamental solution* of (S) on a domain $\mathscr{D}$ is a $p \times p$ matrix with entries in some ordinary differential field extension K (see Definition 2.1 below) of $\mathscr{M}(\mathscr{D})$, whose columns are $\mathbb{C}$-linearly independent vector functions satisfying (S) on $\mathscr{D}$.

The only obstruction to analytic continuation of its solutions is the set of singular points of (S).

Definition 1.18. An element $a \in \mathscr{D} \subset \overline{\mathbb{C}}$ is a *singular point* of (S) if

[2] underlined by Birkhoff

(a) either $a \in \mathbb{C}$ and a is a pole of some entry of A

(b) or $a = \infty$ and $z = 1/x$ changes (S) into a system

$$\frac{dy}{dz} = B(z)y$$

with a singular point at 0.

Let $\Sigma = \{a_1, \ldots, a_n\} \subset \overline{\mathbb{C}}$ be the set of singular points of (S) and $U_\Sigma = \overline{\mathbb{C}} \setminus \Sigma$. If not otherwise specified, we will assume that $U_\Sigma \subset \mathbb{C}$. We first recall a 'miraculous' property of the Wronskian determinant.

Proposition 1.19. *Let $\mathscr{D}$ be a domain in U_Σ and W some $p \times p$ matrix solution of (S) with entries in $\mathscr{O}(\mathscr{D})$, solution of (S). The following are equivalent :*

(a) W is a fundamental solution of (S) in $\mathscr{D}$,

(b) $\det(W(x)) \neq 0$ for some $x \in \mathscr{D}$,

(c) $\det(W(x)) \neq 0$ for all $x \in \mathscr{D}$.

In other words, the Wronskian determinant is either $\neq 0$ for all $x \in \mathscr{D}$, or identically zero.

Proof. It is easy to see that (b) implies (a): since a $\mathbb{C}$-linear dependence relation on the columns of W holds for all values of x, it is trivial by (b). To show that (a) implies (c), fix $x_0 \in \mathscr{D}$ and an open disk D_0 in $\mathscr{D}$ with center x_0. Consider a $\mathbb{C}$-linear dependence relation

$$\sum_{j=1}^{p} \lambda_j C_j(x_0) = 0$$

on the columns $C_j(x_0)$ of $W(x_0)$. Restricted to D_0, the vector function

$$F(x) = \sum_{j=1}^{p} \lambda_j C_j(x)$$

is by linearity a solution of (S) with the same initial value 0 at x_0 as the trivial solution $0 \in \mathscr{O}(D_0)$ of (S). By Cauchy's theorem, for any initial value $y_0 \in \mathbb{C}^p$ there is a unique solution $y \in \mathscr{O}(D_0)$ of (S) such that $y(x_0) = y_0$, hence $F = 0$ on D_0. By the FUT 1.3 we have $F = 0$ on $\mathscr{D}$, hence $\lambda_j = 0$ for all j, by (a). Since (c) implies (b), the equivalence is proved. □

Let us state and prove the fundamental existence theorem, FET for short, of analytic fundamental solutions.

Theorem 1.20 (FET). *With notation and assumptions as before, let Ω be a simply connected domain in U_Σ. For any $x_0 \in \Omega$ and $Y_0 \in \mathrm{GL}(p, \mathbb{C})$ there is a unique fundamental solution Y of (S) with entries in $\mathscr{O}(\Omega)$ and initial condition $Y(x_0) = Y_0$. The subset $\mathrm{Sol}_\Omega((S)) \subset \mathscr{O}(\Omega)^p$ of holomorphic solutions of (S) on Ω is a $\mathbb{C}$-vector space of dimension p.*

Proof. Fix $x \in \Omega$ and an arbitraty path γ from x_0 to x in Ω. Let

$$\rho = \min_{1 \le i \le n} d(a_i, \Gamma)$$

be the minimum distance of the points a_i to the image $\Gamma = \mathrm{Im}\gamma$ of the path γ. By Cauchy's theorem, for any fixed $t \in [0,1]$ there is a unique solution of (S) which is holomorphic in the open disk $D(\gamma(t), \rho)$, with any given initial condition at $\gamma(t)$. By the uniform continuity of γ there is a sequence $t_0 = 0, < \ldots < t_n = 1$ such that

$$\gamma([t_k, t_{k+1}]) \subset D(t_k, \rho)$$

for all $k = 0, \ldots, n-1$. Let $x_k = \gamma(t_k)$ and let us construct a family of fundamental solutions (F_k) of (S) with each F_k holomorphic in $D(x_k, \rho)$, inductively as follows:

We define F_0 as the holomorphic matrix solution in $D(x_0, \rho)$ with initial condition $F_0(x_0) = Y_0$ and since Y_0 is invertible, so is F_0 by Proposition 1.19, that is, F_0 is a fundamental solution. Similarly, F_1 is defined as the holomorphic fundamental solution in $D(x_1, \rho)$ with initial condition $F_1(x_1) = Y_1$, where $Y_1 = F_0(x_1)$ (since Y_1 is invertible by Proposition 1.19, so is F_1). If $F_0, \ldots, F_k$ are fundamental solutions such that $F_j(x_j) = Y_j$ for all $0 \le j \le k$, with $Y_j = F_{j-1}(x_j)$ and each F_j is holomorphic in $D(x_j, \rho)$ respectively, then we define F_{k+1} as the holomorphic fundamental solution in $D(x_{k+1}, \rho)$ with initial condition $F_{k+1}(x_{k+1}) = Y_{k+1}$, where $Y_{k+1} = F_k(x_{k+1})$. The family

$$\Big(D(x_k, \rho), F_k\Big)_{0 \le k \le n}$$

or, more precisely, the corresponding family of pairs for each (i,j) entry of the matrices, satisfies the conditions of Proposition 1.15 and hence

$$F_n \overset{x_n}{\sim} F_0^{\gamma}$$

(meaning that this holds for the germs of each entry). In other words, the pairs defined by the entries of F_n are the analytic continuation each of the pairs defined by the corresponding entries of F_0 along γ. Since Ω is simply connected, all paths from x_0 to x are homotopic and since F_0 can be continued along any path γ from x_0 to x in Ω, the germ F_0^{γ} is independent of γ. If we define the function $Y : \Omega \to \mathrm{GL}(p, \mathbb{C})$ by

$$Y(x_0) = F_0(x_0) = Y_0, \quad Y(x) = F_0^{\gamma}(x)$$

for $x \neq x_0$ and any path γ from x_0 to x in Ω, then Y is the unique fundamental solution of (S) in Ω with initial condition $Y(x_0) = Y_0$. The uniqueness of Y easily follows from the FUT, and $\dim_{\mathbb{C}} \mathrm{Sol}_{\Omega}((S)) = p$ from Proposition 1.19. □

Exercise 1.21. Show that the analytic continuation of a germ of holomorphic function is $\mathbb{C}$-linear and commutes with the usual operations: derivation, sum, product, inverse.

From the proof of the FET and Exercise 1.21 we deduce the following result.

Corollary 1.22. *A germ F of holomorphic fundamental solution of (S) at $x_0 \in U_\Sigma$ can be analytically continued along any path γ in U_Σ starting at $\gamma(0) = x_0$. The germ F^γ at the endpoint of γ is a germ of fundamental solution of (S) which depends on the homotopy class only of γ in U_Σ.*

1.2.2 Singular points

In the study of differential equations, singular points play a key role. We will see how one recovers most information about the equation from the local and asymptotic analysis of its solutions at the singular points. We may, via a change of local coordinate, assume that the singular point lies in $\mathbb{C}$.

Definition 1.23. By a general *sectorial neighbourhood* of $a \in \mathbb{C}$ we mean a neighbourhood

$$S(a, \alpha, \theta, r) = \{x \in \mathbb{C}, |\arg(x-a) - \alpha| < \theta, 0 < |x-a| < r\}$$

with $\alpha \in \mathbb{R}$, $0 \leq \theta < \pi$ and $r > 0$, for a given determination of the argument.

Consider a linear system

$$\frac{dy}{dx} = A(x)y \tag{S}$$

with coefficients in $\mathbb{C}(x)$ or in $\mathscr{M}(\mathscr{D})$ for some domain $\mathscr{D}$.

Definition 1.24. The singular point $a \in \mathbb{C}$ is *regular singular* if in any sectorial neighbourhood V of a containing no other singular points of (S) there is an analytic fundamental solution Y of (S) of *moderate growth*, meaning that there are real constants λ_V, C_V such that each entry y_{ij} of Y satisfies

$$|y_{ij}| < C_V\,|x-a|^{\lambda_V}$$

for all $x \in V$. The singular point is *irregular singular* otherwise.

We may for simplicity assume that $a = 0$. Singular points are also distinguished by their Poincaré rank, defined as follows.

Definition 1.25. Assume 0 is singular. The *Poincaré rank* of the singular point 0 is the integer $r \in \mathbb{Z}$ such that $r+1$ is the order of the pole 0 of the matrix $A(x)$ of (S), that is,

$$A(x) = \frac{B(x)}{x^{r+1}}$$

where B is holomorphic at 0 with $B(0) \neq 0$. The singular point is said to be *Fuchsian* if it is a simple pole of A, equivalently if $r = 0$.

Note that the notions of regularity and Poincaré rank are unrelated, except when $r = 0$.

Proposition 1.26. *Fuchsian singular points are regular singular.*

Proof. Assume the Fuchsian singular point is 0. We can write the system (S)

$$\frac{dy}{dx} = \frac{B(x)}{x} y$$

with $B(0) \neq 0$. Let $D(0,\rho)$, $\rho > 0$, be an open disk containing no other singular points of (S). For any open disk D with center 0 let $\breve{D} = D \setminus \mathbb{R}_+$, where $\mathbb{R}_+ = [0,+\infty[$. Let y be a holomorphic vector solution of (S) in $\breve{D}(0,\rho)$. We may for convenience assume that $\rho > 1$ (via any change of variable $x \leftarrow x/\lambda$ with $\lambda < \rho$). Let $x_0 = \rho_0 \mathrm{e}^{\mathrm{i}\theta_0} \in \breve{D}(0,1)$, $x_1 = \mathrm{e}^{\mathrm{i}\theta_0}$ and consider a parametrization of the segment $[x_0, x_1]$ by $x(t) = t\mathrm{e}^{\mathrm{i}\theta_0}$, $\rho_0 \leq t \leq 1$. Consider the function

$$Y(t) = y(x(t))$$

on $[\rho_0, 1]$. For $h \in \mathbb{R}$ such that $x(t+h) \in]x_0, x_1[$, we have

$$\| Y(t+h) - Y(t) \| \geq -(\| Y(t+h) \| - \| Y(t) \|)$$

and as h tends to 0

$$\| \frac{dY}{dt} \| \geq -\frac{d \| Y \|}{dt}.$$

Since y is a solution of (S) we have

$$\| \frac{dY}{dt} \| = \| \frac{dy}{dx} \| \, |\frac{dx}{dt}| = \| \frac{dy}{dx} \| = \| \frac{B(x)y}{x} \| \leq \| B(x) \| \frac{\| y \|}{|x|} \leq K \frac{\| y \|}{|x|}$$

where $K = \sup_{x \in D(0,1)} \| B(x) \|$. Since $\| y \| = \| Y \|$ and $|x(t)| = |t| = t$, the function $Z(t) = \| Y(t) \|$ satisfies the inequality

$$\frac{dZ}{dt} \geq -K \frac{Z}{t}.$$

Multiplying both sides by t^K gives

$$\frac{d}{dt}(t^K Z) \geq 0,$$

hence (since $t^K Z$ is continuous)

$$Z(1) - Z(\rho_0)\, \rho_0^K \geq 0,$$

that is,

$$\| y(x_0) \| \leq \| y(x_1) \| \, \| x_0 \|^{-K}.$$

The solution y is a determination in $\breve{D}(0,\rho)$ of a multivalued function $\tilde{y}$ defined on the universal covering $\widetilde{D}$ of $D^* = D(0,\rho) \setminus \{0\}$, with base-point 1. Recall that this universal covering is simply connected and can either be defined as the set of homotopy classes of all paths starting from 1 in D^* (see Section 4.3.2 p. 105 for more details about universal coverings) or, more concretely, as the restriction to D^* of the covering $\exp : \mathbb{C} \to \mathbb{C}^*$ given by the exponential map, which is also called the *Riemann surface of the logarithm*, see Section 6.7 p. 195, and denoted $\widetilde{\mathbb{C}}$.

Any loop starting from 1 in D^* lifts uniquely to a path in $\widetilde{D}$. Let γ be the path lifting the circular loop $\theta \mapsto e^{i\theta}$ and let $C = \sup_{\mathrm{Im}(\gamma)} \| \tilde{y} \|$ (which exists since $\tilde{y}$ is holomorphic and $\mathrm{Im}(\gamma)$ is compact). Since $\tilde{y}(\tilde{x}) = y(\exp(\tilde{x}))$ for all $\tilde{x} \in \mathrm{Im}(\gamma)$, we have in particular proved that

$$\| y(x_0) \| \leq C \| x_0 \|^{-K}$$

where the constants K and C are independent of x_0. In other words, the solution y is of moderate growth as x_0 tends to 0 in the sectorial neighbourhood $\check{D}(0, \rho)$ of 0 (with maximal opening). □

Example 1.27. The differential system

$$\frac{dy}{dx} = \begin{pmatrix} 1/x & 1 \\ 0 & 0 \end{pmatrix} y$$

has a Fuchsian singular point at 0. In $\mathbb{C} \setminus \mathbb{R}_+$ (resp. $\mathbb{C} \setminus \mathbb{R}_-$) a fundamental solution is given by

$$y_+ = \begin{pmatrix} x & x\log x \\ 0 & 1 \end{pmatrix}, \quad \text{resp. } y_- = \begin{pmatrix} -x & -x\log(-x) \\ 0 & -1 \end{pmatrix},$$

which is of moderate growth as x tends to 0 in each of the sectors $\mathbb{C} \setminus \mathbb{R}_+$, $\mathbb{C} \setminus \mathbb{R}_-$, that is, 0 is as expected regular singular. These solutions are of moderate growth also as x tends to infinity. The system

$$\frac{dy}{dz} = \begin{pmatrix} -1/z & -1/z^2 \\ 0 & 0 \end{pmatrix} y$$

which we obtain by the change $x = 1/z$ has a pole of order 2 at ∞, which is regular singular but not Fuchsian.

Example 1.28. The differential equation

$$\frac{dy}{dx} = -\frac{y}{x^2}$$

has an irregular singular point at 0 since the general solution $y = Ce^{1/x}$ has an exponential growth as x tends to 0.

Example 1.29. A fundamental solution of

$$\frac{dy}{dx} = \begin{pmatrix} 0 & 1 \\ 1/x^2 & -1/x \end{pmatrix} y$$

is

$$y = \begin{pmatrix} x & 1/x \\ 1 & -1/x^2 \end{pmatrix},$$

which is of moderate growth as x tends to 0. The singular point 0 here is regular singular and non-Fuchsian since its Poincaré rank is 1, and the same holds for ∞ as is easily seen by the change $x = 1/z$.

1.2.3 The monodromy representation

Let Ω be a domain in U_Σ and $x_0 \in \Omega$. Let $\mathscr{L}(\Omega;x_0)$ denote the set of loops with base-point x_0 (that is, paths starting and ending at x_0) and let

$$\pi_1(\Omega;x_0) = \mathscr{L}(\Omega;x_0)/\sim$$

denote the quotient of $\mathscr{L}(\Omega;x_0)$ by the homotopy equivalence relation. This is a group with the following law.

Definition 1.30. The *product* of $\gamma_1, \gamma_2 \in \mathscr{L}(\Omega;x_0)$ is defined by

$$\gamma_2\gamma_1 = \begin{cases} \gamma_1(2t) & \text{if } 0 \le t \le 1/2, \\ \gamma_2(2t-1) & \text{if } 1/2 \le t \le 1 \end{cases}$$

and the *unit* loop $e_{x_0} \in \mathscr{L}(\Omega;x_0)$ by $e_{x_0}(t) = x_0$ for all $t \in [0,1]$. For any $\gamma \in \mathscr{L}(\Omega;x_0)$ let $\gamma^- \in \mathscr{L}(\Omega;x_0)$ denote the loop $\gamma^-(t) = \gamma(1-t)$ for all $t \in [0,1]$.

One easily proves the following properties.

Proposition 1.31. *For any $\gamma_1, \gamma_1', \gamma_2, \gamma_2', \gamma \in \mathscr{L}(\Omega;x_0)$ the following holds:*

(a) $\gamma_2\gamma_1 \sim \gamma_2'\gamma_1'$ if $\gamma_1 \sim \gamma_1'$ and $\gamma_2 \sim \gamma_2'$,
(b) $e_{x_0}\gamma \sim \gamma e_{x_0} \sim \gamma$,
(c) $\gamma\gamma^- \sim \gamma^-\gamma \sim e_{x_0}$,
(d) $(\gamma_3\gamma_2)\gamma_1 \sim \gamma_3(\gamma_2\gamma_1)$.

Let $[\gamma] \in \pi_1(\Omega;x_0)$ denote the homotopy class of a loop $\gamma \in \mathscr{L}(\Omega;x_0)$.

Proposition 1.32. *The quotient $\pi_1(\Omega;x_0)$ is a group with identity element $e = [e_{x_0}]$ and group law $[\gamma_2][\gamma_1] = [\gamma_2\gamma_1]$ for $\gamma_1, \gamma_2 \in \mathscr{L}(\Omega;x_0)$. The inverse of $[\gamma]$ is $[\gamma^-]$.*

The group $\pi_1(\Omega;x_0)$ is called the *fundamental group* of Ω with base-point x_0. Analytic continuation yields a specific representation of this group.

Consider a differential system

$$\frac{dy}{dx} = A(x)y \qquad (S)$$

of order p with coefficients in $\mathbb{C}(x)$. As before $U_\Sigma = \overline{\mathbb{C}} \setminus \Sigma$, where Σ is the set of singular points of (S). Let F be a germ of fundamental solution of (S) at $x_0 \in U_\Sigma$ and F^γ the analytic continuation of F along $\gamma \in \mathscr{L}(U_\Sigma; x_0)$. In an open disk Δ with center x_0 in which F and F^γ are represented by analytic matrix functions, the columns of F form a basis of the p-dimensional vector space of solutions of (S) in Δ, hence there is an invertible matrix $G_\gamma \in \mathrm{GL}(p,\mathbb{C})$, which does not depend on Δ, such that

$$F^\gamma = FG_\gamma.$$

Proposition 1.33. *With notation as above, $[\gamma] \mapsto G_\gamma$ defines a representation*

$$\mathrm{Mon} : \pi_1(U_\Sigma; x_0) \longrightarrow \mathrm{GL}(p,\mathbb{C}),$$

of the group $\pi_1(U_\Sigma; x_0)$ on $\mathbb{C}^p$.

The group homomorphism Mon is called the *monodromy representation*, the matrices G_γ the *monodromy matrices* and Im(Mon) the *monodromy group*

$$\mathrm{Mon}(\pi_1(U_\Sigma; x_0)) \subset \mathrm{GL}(p,\mathbb{C})$$

of (S) with respect to F and x_0.

Proof. Corollary 1.22 tells us that F^γ, hence G_γ, depend on the homotopy class only of γ, thus define the map Mon. Moreover, since analytic continuation is $\mathbb{C}$-linear, we have

$$F^{\gamma_2 \cdot \gamma_1} = (F^{\gamma_1})^{\gamma_2} = (FG_{\gamma_1})^{\gamma_2} = F^{\gamma_2} G_{\gamma_1} = FG_{\gamma_2} G_{\gamma_1} = FG_{\gamma_2 \cdot \gamma_1},$$

that is, $G_{\gamma_2 \cdot \gamma_1} = G_{\gamma_2} G_{\gamma_1}$. □

Let us consider again Example 1.27.

Example 1.34. The differential system

$$\frac{dy}{dx} = \begin{pmatrix} 1/x & 1 \\ 0 & 0 \end{pmatrix} y$$

has two singular points: 0 which is Fuchsian and ∞ which is regular singular and non-Fuchsian (this is easily seen via the change of variable $x \leftarrow 1/x$). Here $U_\Sigma = \mathbb{C}^*$ and $\pi_1(U_\Sigma; x_0)$ is isomorphic to $(\mathbb{Z}, +)$. It is generated by $[\gamma]$, where γ is any fixed loop with base-point x_0, enclosing 0 once, counterclockwise. The analytic continuation along γ of the fundamental solution

$$y = \begin{pmatrix} x & x\log x \\ 0 & 1 \end{pmatrix}$$

is the fundamental solution

$$y^\gamma = \begin{pmatrix} x & x\log x + 2\pi \mathrm{i} x \\ 0 & 1 \end{pmatrix} = \begin{pmatrix} x & x\log x \\ 0 & 1 \end{pmatrix} \begin{pmatrix} 1 & 2\pi \mathrm{i} \\ 0 & 1 \end{pmatrix},$$

that is,

$$G_\gamma = \mathrm{Mon}([\gamma]) = \begin{pmatrix} 1 & 2\pi\mathrm{i} \\ 0 & 1 \end{pmatrix}.$$

The monodromy group is the subgroup of $\mathrm{GL}(p,\mathbb{C})$ generated by G_γ

$$\mathrm{Mon}\big(\pi_1(U_\Sigma;x_0)\big) = \left\{ \begin{pmatrix} 1 & 2\pi\mathrm{i}k \\ 0 & 1 \end{pmatrix},\ k \in \mathbb{Z}. \right\}$$

1.2.4 Local solutions

Locally in the neighbourhood of singular points, one can choose fundamental solutions of a specific, useful form. Consider a differential system (S) with a singular point, say at 0, and coefficients in $\mathbb{C}(x)$ or $\mathbb{C}(\{x\})$, where $\mathbb{C}(\{x\})$ is the field of germs of meromorphic functions at 0 (that is, the fraction field of the ring of germs of holomorphic functions; prove as an exercise that this ring is an integral domain). Fix x_0, non-singular and close to 0, and an elementary loop γ with base-point x_0 around 0, that is, a loop enclosing 0 once, counterclockwise, and enclosing no other singular point than 0. We also fix a germ of analytic fundamental solution Y at x_0, which we will call (as well as its analytic continuations in open sectors of opening $< 2\pi$) a *local fundamental solution at* 0. Let G be the monodromy matrix with respect to γ, Y and x_0. We can conjugate G in $\mathrm{GL}(p,\mathbb{C})$ to its Jordan form

$$\mathrm{diag}(J_1,\ldots,J_r)$$

with Jordan blocks $J_1,\ldots,J_r$. Note that any Jordan block can be written as $J = \lambda I + N$ where N is nilpotent, and that $\log J$ (hence $\log G$) is well-defined, of the form

$$\log J = \log(\lambda I) + P(N)$$

where P is a polynomial of degree $\leq p-1$. Let

$$E = \frac{1}{2\pi\mathrm{i}} \log G \quad \text{and} \quad x^E = \exp\big((\log x)E\big).$$

Analytic continuation along γ yields

$$(x^E)^\gamma = \big(\exp((\log x)E)\big)^\gamma = \exp\big((\log x + 2\pi\mathrm{i})E\big) = x^E \exp(2\pi\mathrm{i}E) = x^E G. \quad (1.1)$$

Lemma 1.35. *Any local fundamental solution of* (S) *at* 0 *is of the form*

$$Y = M(x)\,x^E$$

where the entries of the matrix M *are single-valued and analytic in a punctured neighbourhood of* 0 *and*

$$E = \frac{1}{2\pi i} \log G$$

where G is the monodromy matrix with respect to Y and an elementary loop γ around 0.

Proof. Let Y be any germ of analytic fundamental solution at a fixed, non-singular point x_0 close to 0. By definition of G and by (1.1) we have

$$Y^\gamma = Y G, \qquad (x^{-E})^\gamma = G^{-1} x^{-E},$$

hence by Exercise 1.21

$$(Y x^{-E})^\gamma = Y x^{-E}.$$

In other words $M(x) = Y x^{-E}$ is *single-valued* at 0, defined in a full neighbourhood of 0. By analytic continuation, the entries of Y and x^E can be extended as analytic functions in any (punctured) sectorial neighbourhood of 0 of opening $< 2\pi$ and further as analytic functions on the universal covering $\widetilde{D}$ of an open, punctured disk $\mathring{D} = D \setminus \{0\}$ with center 0 (cf Section 4.3.2) containing no other singular point of (S). It follows from the single-valuedness of $Y x^{-E}$ that (the analytic continuation of) $M(x) = Y x^{-E}$ is analytic in the full punctured disk $\mathring{D}$. □

Note that the eigenvalues of E, which are of the form $(\log \lambda)/2\pi i$ where λ is an eigenvalue of G, are defined modulo $\mathbb{Z}$ only. We shall therefore normalize them by assuming that all eigenvalues ρ of E have a real part $\Re(\rho) \in [0, 1[$.

Exercise 1.36. Let $x^E = (a_{ij})$. Show that each a_{ij} is a finite sum

$$a_{ij} = \sum_{\ell \geq 1} x^{\rho_\ell} P_{ij\ell}(\log x)$$

where ρ_ℓ are the eigenvalues of E and $P_{ij\ell}$ are polynomials of degree $\leq p - 1$.

Corollary 1.37. *With notation from Lemma 1.35, the singular point* 0 *is regular singular if and only if $M(x)$ is meromorphic at* 0.

Proof. The sufficiency of the condition easily follows from Lemma 1.35 and Exercise 1.36. Indeed, if $M(x)$ is meromorphic at 0, it is easy to see that each entry of Y is of the form

$$y_{ij}(x) = \sum_{k,\ell} x^{\rho_k} f_{k\ell}(x) (\log x)^{b_\ell} \tag{1.2}$$

where the coefficients $f_{k\ell}(x)$ are meromorphic germs at 0 (depending on i, j), the expondents ρ_k are the (normalized) eigenvalues of E and b_ℓ are non-negative integers all $< p - 1$. We leave it as an exercise to prove that the sum, product or inverse of functions of moderate growth is again of moderate growth. This implies that each y_{ij} has moderate growth in any sectorial neighbourhood of 0 of opening $< 2\pi$, since each term in the sum (1.2) clearly has moderate growth in such a sector. In other words, 0 is a regular singular point following Definition 1.24 p. 14.

To prove that the condition is necessary, assume that 0 is a regular singular point, and let Y be a germ of analytic fundamental solution at some fixed, non-singular x_0

close to 0. We know that $Y = M(x)x^E$, where $M(x)$ by Lemma 1.35 is analytic in a full open punctured disk $\mathring{D}$ with center 0. It is a well-known fact that if a function is holomorphic in an annulus, in particular in a punctured disk with center 0, it has a Laurent series expansion $\sum_{n\in\mathbb{Z}} \lambda_n x^n$ (cf. [Cha90, Theorem 1, p.121]). Since by Definition 1.24 p. 14 the entries of Y, hence of Yx^{-E} are of moderate growth in any sectorial neighbourhood of opening $< 2\pi$, all entries of $M(x) = Yx^{-E}$ must have an expansion $\sum_{n\geq n_0} \lambda_n x^n$ by Exercise 1.38 below, that is, $M(x)$ is meromorphic at 0.□

For other characterizations of regular singular points, see Theorem 2.33 p. 38 and Remark 2.34 p. 40.

Exercise 1.38. Assume that a complex function $f(x)$, analytic in a punctured open disk $\mathring{D}$ with center 0, has a (convergent) Laurent series expansion

$$f(x) = \sum_{n\in\mathbb{Z}} a_n x^n$$

in $\mathring{D}$. Show that if f is of moderate growth in any open sector in $\mathring{D}$ of opening $< 2\pi$ and center 0, then the Laurent series expansion of f has a finite principal part, that is, $f(x) = \sum_{n\geq N} a_n x^n$ for some $N \in \mathbb{Z}$.

1.3 Solutions to exercises of Chapter 1

Exercise 1.10 p. 7 If $R(t_0) = \infty$ for some $t_0 \in I$ then obviously $R(t) = \infty$ for all $t \in I$. Assume $R(t_0)$ is finite. Since (U_0, f_0) can be analytically continued along γ we know that there is a neighbourhood $u(t_0) \subset I$ of t_0 such that (U_{t_0}, f_{t_0}) and (U_t, f_t) are adjacent for all $t \in u(t_0)$. The bounding circles of U_t and U_{t_0} either intersect in two points, or one disk is contained in the other and the circles are tangent. In both cases we have

$$|R(t) - R(t_0)| \leq |\gamma(t) - \gamma(t_0)|$$

(which is an equality in the second case) and the continuity of R easily follows from the continuity of γ.

Exercise 1.21 p. 13 We have proved that analytic continuation along a path γ can be achieved in finitely many steps. If we start with some germ f given by its power series expansion $f(z) = \sum_{n\geq 0} a_n (z - \gamma(0))^n$ at $\gamma(0)$, each step consists of a rescaling of this series via a change of variable $z \leftarrow z - \lambda$ for some $\lambda \in \mathbb{C}$. This procedure commutes with both algebraic operations and derivation.

Exercise 1.36 p. 20 One easily reduces the exercise to the case where the Jordan form of G consists of one single Jordan block. In this case, we can write $G = \lambda(I + N)$, where $\lambda \in \mathbb{C}^*$ and N is a nilpotent matrix. Note that since N is nilpotent, $Q(N)^p = 0$ for any polynomial Q, where p is the order of the system, and the size of the matrix G. We get

$$x^E = \exp(E\log x) = \exp\left(\frac{\log G\ \log x}{2\pi\mathrm{i}}\right) = \exp\left(\frac{\log\lambda\ \log x}{2\pi\mathrm{i}}I + \frac{\log(I+N)\ \log x}{2\pi\mathrm{i}}\right)$$

hence

$$x^E = x^\rho\ \left(\frac{Q(N)\log x}{2\pi\mathrm{i}}\right)$$

where $\rho = (\log\lambda\ /2\pi\mathrm{i})$ is an eigenvalue of G and Q is a polynomial of degree $\leq p-1$. Since $Q(N)^p = 0$, this shows that each entry of x^E is of the form

$$a_{ij} = x^\rho P_{ij}(\log x)$$

where P_{ij} is a polynomial of degree $\leq p-1$. This implies, in the general case, the formula

$$a_{ij} = \sum_{\ell\geq 1} x^{\rho_\ell} P_{ij\ell}(\log x)$$

for each entry a_{ij} of x^E, where ρ_ℓ are the eigenvalues of G and $P_{ij\ell}$ are polynomials of degree $\leq p-1$.

Exercise 1.38 p. 21 Consider a covering of the punctured disk $\mathring{D}$ by two open sectors V_1 and V_2 of opening $< 2\pi$. We may moreover assume that the radius of $\mathring{D}$ is < 1. Since the sum $f = \sum_{n\in\mathbb{Z}} a_n x^n$ of the Laurent series is single-valued, defined on the full $\mathring{D}$, and is of moderate growth on both sectors, it satisfies on each V_i an inequality

$$|f(x)| < C_{V_i}\,|x|^{\lambda_{V_i}}$$

by definition of the moderate growth (cf. Definition 1.24 p. 14). Let

$$C = \max_{i=1,2} C_{V_i}, \quad \lambda = \min_{i=1,2} \lambda_{V_i}$$

and let $N \in \mathbb{Z}$ be an integer such that $N < \lambda$. Then f satisfies an inequality

$$|f(x)| < C\,|x|^N$$

in $\mathring{D}$ and dividing both sides by $|x|^N$ we get

$$\left|\sum_{n\in\mathbb{Z}} a_n x^{n-N}\right| < C \tag{1.3}$$

on $\mathring{D}$. If we rewrite the Laurent series as

$$\sum_{n\in\mathbb{Z}} a_n x^{n-N} = \sum_{n\in\mathbb{Z}} b_n x^n$$

with $b_n = a_{n+N}$, then by definition of a Laurent series, its regular part $\sum_{n\geq 0} b_n x^n$ is convergent in the full disk D and its principal part $\sum_{n<0} b_n x^n$ is convergent in $\mathbb{C}^*$. This implies in particular that the sum of $\sum_{n\geq 0} b_n x^n$ is bounded as x tends to 0 and $\sum_{n<0} b_n x^n$ is bounded too by the inequality (1.3). Changing x into $z = 1/x$ we get that the series $\sum_{n>0} b_{-n} z^n$ is convergent and its sum bounded in $\mathbb{C}$. By Liouville's theorem it is constant, that is, $b_n = 0$ for all $n \in \mathbb{N}$, which implies that f has the expansion

$$f(x) = \sum_{n \geq N} a_n x^n,$$

that is, f is meromorpic at 0.

References

ACMS53. Ahlfors, L., Calabi, E., Morse, M., Sarlo, L., Spencer, D. (eds), Contributions to the Theory of Riemann Surfaces. Annals of Math. Studies, Princeton (1953)

Ahl79. Ahlfors, L.: Complex Analysis. Third Edition. MacGraw Hill, New York (1979)

Cai51. Cairns, S.: An elementary proof of the Jordan-Schoenflies theorem. Proceedings of the American Mathematical Society **2-6**, 860–867 (1951)

Cha90. Chabat, B.: Introduction à l'analyse complexe. Éditions MIR. Moscou (1990)

CoLe55. Coddington, E. A., Levinson, N.: Theory of Ordinary Differential Equations. McGraw-Hill (1955)

Fo81. Forster, O.: Lectures on Riemann surfaces. Graduate Texts in Mathematics vol. 81, Springer (1981).

Ha64. Hartman, P.: Ordinary differential equations. Joh. Wiley (1964)

In56. Ince, E. L.: Ordinary differential equations. Second edition. Dover Publications, Inc., New York (1956)

JS87. Jones, G. P., Singermann, D.: Complex Functions, An algebraic and geometric point of view. Cambridge University Press (1987)

La61. Lang, S.: Complex Analysis. Fourth Edition. Springer (1999)

NN01. Narashiman, R., Nievergelt, Y.: Complex Analysis in One Variable. Second Edition. Birkhäuser (2001)

Ru86. Rudin, W.: Real and Complex Analysis. Third edition. McGraw-Hill (1986)

Chapter 2
Differential Galois Theory

2.1 Differential Galois theory

In this chapter we show how differential Galois groups are related to monodromy. To learn about differential Galois theory we refer to the following authors: Crespo and Hajto [CH11], Kaplansky [Kap76], Magid [Mag94], Kolchin[Kol76] , van der Put and Singer [PSi01], Singer ([Sin99], [Sin09]).

2.1.1 *Differential fields, Picard-Vessiot extensions*

Definition 2.1. An ordinary *differential field* (k,∂) is a field k with a derivation, that is, a map $\partial : k \to k$ such that

1. $\partial(a+b) = \partial(a) + \partial(b)$,
2. $\partial(ab) = a\partial(b) + \partial(a)b$.

The *field of constants* of (k,∂) is $C_k = \{a \in k, \partial(a) = 0\}$. A *differential homomorphism* $\phi : (k_1,\partial_1) \to (k_2,\partial_2)$ is a field homomorphism from k_1 to k_2 such that $\phi \circ \partial_1 = \partial_2 \circ \phi$. The triple (k_1,ϕ,k_2) is called a *differential extension*, and k_2 itself is (abusively) called a differential extension of k_1 if there is no ambiguity about ϕ and the derivations.

Example 2.2. The following inclusions

$$\mathbb{C}(x) \subset \mathbb{C}(\{x\}) \subset \mathbb{C}((x))$$

between the field $\mathbb{C}(x)$ of rational functions, the field $\mathbb{C}(\{x\})$ of germs of meromorphic functions at 0 and the field $\mathbb{C}((x))$ of formal Laurent series are differential field extensions with respect to any of their common derivations $x^n d/dx$, $n \in \mathbb{Z}$. Note that $\mathbb{C}((x))$ here denotes the field of formal series $\sum_{n \geq n_0} a_n x^n$ with a *finite* principal

C. Mitschi, D. Sauzin, *Divergent Series, Summability and Resurgence I*,
Lecture Notes in Mathematics 2153, DOI 10.1007/978-3-319-28736-2_2

part, whereas the usual terminology 'Laurent series' in the literature applies to (not necessarily meromorphic) series $\sum_{n\in\mathbb{Z}} a_n x^n$.

Exercise 2.3. For any $n \in \mathbb{N}^*$ show that the field extensions

(a) $\mathbb{C}(x) \subset \mathbb{C}(t)$,
(b) $\mathbb{C}((x)) \subset \mathbb{C}((x))(t)$,
(c) $\mathbb{C}((x)) \subset \mathbb{C}((t))$,
(d) $\mathbb{C}(x) \subset \mathbb{C}((t))$,

with $t^n = x$ are in a unique way differential extensions with respect to the usual derivation d/dx of the base-fields.

Let (k, ∂) be a differential field of characteristic zero with an algebraically closed field of constants C_k. Consider a differential system

$$\partial y = A y \tag{S}$$

where A is a $p \times p$ matrix with entries in k and ∂y denotes the column-matrix obtained by applying ∂ to the p entries of an unknown vector y. We will call this indistinctively *a system over k*, or a system *with coefficients in k*.

Definition 2.4. A differential extension (L, ∂) of (k, ∂) is a *Picard-Vessiot extension* for (S) if there is a fundamental solution Y of (S) in L, that is

$$\partial Y = AY \quad \text{with} \quad Y \in \mathrm{GL}(p, L)$$

such that:

(a) L is generated, as a field extension of k, by the entries of Y and denoted $k(Y)$,
(b) $C_L = C_k$.

Theorem 2.5. *For any differential field (k, ∂) of characteristic zero with an algebraically closed field of constants, and any system (S): $\partial y = Ay$ over k, there exist Picard-Vessiot extensions of k for (S) and they are k-isomorphic.*

We refer to ([PSi01, Section 1.3]) or ([CH11, Section 5.6]) for the full proof of this theorem. One way of constructing a Picard-Vessiot extension is the following. Let $Y = (Y_{ij})_{ij}$ denote a $p \times p$ matrix with indeterminate entries Y_{ij}. Consider the differential ring

$$R = k\left[Y_{ij}, \frac{1}{\det Y}, \ 1 \le i, j \le p\right]$$

of polynomials in the indeterminates Y_{ij} and $1/\det Y$, where the derivation ∂ of k extends to R via $\partial(Y) = AY$. Let M be a maximal differential ideal of R, that is, an ideal M of R such that $\partial M \subset M$ and maximal with this property. Note that M is not, in general, a maximal ideal, but it is prime. The differential ring R/M is an integral domain and its fraction-field $L = \mathrm{Frac}(R/M)$, with the derivation ∂, provides a Picard-Vessiot extension of k generated by the fundamental solution $y = (y_{ij})$, where y_{ij} denotes the class in R/M of Y_{ij} . Note that R is the function ring, or co-ordinate ring, of $\mathrm{GL}(p,C)$ as an algebraic variety. Its elements can be viewed as the restrictions to $\mathrm{GL}(p,C)$ of the polynomial functions from $\mathfrak{gl}(p,C)$ (the affine space of $p \times p$ matrices, isomorphic to C^{p^2}) to C .

Remark 2.6. The existence of a Picard-Vessiot extension L of k for (S), which relies on the existence of a fundamental solution in L, implies in particular that the set of solutions of (S) in L is a C-vector space of full dimension p.

Exercise 2.7. Let $\alpha_1, \ldots, \alpha_m \in \mathbb{C}^*$. Consider the differential field

$$L = \mathbb{C}(x, \mathrm{e}^{\alpha_1 x}, \ldots, \mathrm{e}^{\alpha_m x})$$

endowed with the usual derivation ∂ such that

$$\partial x = 1,\ \partial \mathrm{e}^{\alpha_1 x} = \alpha_1 \mathrm{e}^{\alpha_1 x}, \ldots, \partial \mathrm{e}^{\alpha_m x} = \alpha_m \mathrm{e}^{\alpha_m x}.$$

Show that the differential field L is a Picard-Vessiot extension

(a) of $\mathbb{C}(x)$,
(b) of $\mathbb{C}$.

In each case write an explicit differential system $\partial y = Ay$ for which L is a Picard-Vessiot extension of the base-field.

Exercise 2.8. Let (k, ∂) be a differential field as above, C its field of constants and (L, ∂) a Picard-Vessiot extension of k with respect to a linear differential system(S). Show that any ℓ solutions $y_i \in L^p$ of (S), $1 \leq i \leq \ell \leq p$, are linearly independent over C if and only if they are linearly independent over L.

2.1.2 *The differential Galois group*

In analogy to algebraic field extensions in classical Galois theory, Picard-Vessiot extensions are best described by their group of automorphisms. Following Galois's point of view, this group often tells more about the differential equation than any numerical computation of its solutions.

Definition 2.9. Consider a differential field (k, ∂) and a linear differential system (S) over k. The *differential Galois group* of (S) over k is the group of differential k-automorphisms of L (meaning that their restriction to k is the identity) where L is any Picard-Vessiot extension of k for (S).

The differential Galois group is defined up to isomorphism and has the following property. As before, C denotes the field of constants of k.

Theorem 2.10. *Let (S) be a linear differential system of order p over k. The differential Galois group of (S) over k is a linear algebraic group over C.*

Proof. A linear algebraic group over C is by definition a subgroup of some $\mathrm{GL}(n,C)$ defined as the zero-set in $\mathrm{GL}(n,C)$ of a family of polynomials in n^2 variables. In other words, it is a closed subgroup of $\mathrm{GL}(n,C)$ for the Zariski topology. We refer to [Hu75], [Bor91], [PSi01],[CH11] for general definitions and properties about linear algebraic groups. Let us first prove that the differential Galois group G can be identified with a subgroup[1] of $\mathrm{GL}(p,C)$. Let L be a Picard-Vessiot extension of k for (S). By the uniqueness result of Theorem 2.5 p. 26, we can in particular choose $L = \mathrm{Frac}(R/M)$ where

$$R = k\left[Y_{ij}, \frac{1}{\det Y}, \ 1 \leq i,j \leq p\right]$$

is endowed with the derivation $\partial Y = AY$ extending the derivation of k with $Y = (Y_{ij})$ and M any maximal differential ideal of R.

We will for simplicity use the notation

$$K[T_{ij}, -] = K\left[T_{ij}, \frac{1}{\det(T_{ij})}, 1 \leq i,j \leq p\right]$$

for any field K and variables T_{ij}.

Note that k is a subfield of $L = \mathrm{Frac}(R/M)$. We can therefore consider $L[Y_{ij}, -]$ as the tensor product

$$L \otimes_k k[Y_{ij}, -]$$

of L by R over k (as k-algebras), and the ideal M of R as a subset of $L[Y_{ij}, -]$ via the natural inclusion

$$R = k[Y_{ij}, -] \subset L[Y_{ij} -]$$

induced by $k \subset L$. Note that the indeterminates Y_{ij} in $L[Y_{ij}, -]$ are now, via the tensor product, independent of the homonymous ones originally used to define L. The 'old' indeterminates Y_{ij} only survive via their images y_{ij} in $R/M \subset L$. Let us show that the group of k-differential automorphisms of $R = k[Y_{ij}, -]$ is $\mathrm{GL}(p,C)$. Since any such automorphim σ commutes with the derivation of R, it maps a fundamental solution of (S) in R to another, which implies that

$$\sigma(Y) = (\sigma(Y_{ij})) = Y\gamma_\sigma$$

for some invertible constant matrix $\gamma_\sigma \in \mathrm{GL}(p,C)$. Conversely, any $\gamma \in \mathrm{GL}(p,C)$ defines a k-automorphism σ_γ of R by $Y \mapsto Y\gamma$ and σ_γ is differential since

$$\sigma_\gamma(\partial Y) = \sigma_\gamma(AY) = A\sigma_\gamma(Y) = AY\gamma = (\partial Y)\gamma = \partial(Y\gamma) = \partial(\sigma_\gamma(Y))$$

---

[1] the author apologizes for using the same notation G for groups and monodromy matrices; it is in both cases a common notation in the literature.

by the k-invariance of σ_γ and the C-linearity of ∂. The differential Galois group G is the group of differential k-automorphisms of $L = \mathrm{Frac}(R/M)$. It is isomorphic to the group of differential k-automorphisms of R/M since indeed a differential k-automorphism of L is uniquely determined by its restriction to R/M, and any differential k-automorphism of R/M extends uniquely to L. This identifies G with the subgroup of $\mathrm{GL}(p,C)$ of those $\gamma \in \mathrm{GL}(p,C)$ for which $\sigma_\gamma(M) \subset M$, and this proves the first part of the theorem.

Let us show that G is Zariski closed in $\mathrm{GL}(p,C)$. We will sketch this part of the proof and refer to [PSi01, p. 20-21] or [CH11, p. 145-151] for more details. Let $C[Z_{ij},-]$ be the ring of polynomial functions of $\mathrm{GL}(p,C)$ over C. Let us prove that there is an ideal I of $C[Z_{ij},-]$ such that G is the zero-set of I in $\mathrm{GL}(p,C)$. This means, for such an I, that a matrix $\gamma \in \mathrm{GL}(p,C)$ belongs to G if and only if $f(\gamma)=0$ for all $f \in I$, in other words if any $f \in I$ is in the kernel of the C-morphism

$$C[Z_{ij},-] \longrightarrow C$$

defined by $Z \mapsto \gamma$. It is indeed equivalent to show that there is such an ideal I of $C[Z_{ij},-]$ or to prove that there is an ideal I_0 of $C[Z_{ij}]$, namely $I_0 = I \cap C[Z_{ij}]$, such that G is the intersection

$$V(I_0) \cap \mathrm{GL}(p,C),$$

in the affine space $\mathfrak{gl}(p,C) \simeq C^{p^2}$ of $p \times p$ matrices, of the zero-set $V(I_0)$ of I_0 and $\mathrm{GL}(p,C)$. The latter condition means that G is the intersection with $\mathrm{GL}(p,C)$ of a Zariski closed subset of $\mathfrak{gl}(p,C)$, that is, that G is closed in $\mathrm{GL}\,(p,C)$. Let $\gamma \in \mathrm{GL}(p,C)$ and let σ_γ be the corresponding differential k-automorphism of $R = k[Y_{ij},-]$. We leave it as an exercise to show the equivalence of the following conditions:

(a) $\gamma \in G$,

(b) $\sigma_\gamma(M) \subset M$,

(c) M is in the kernel of the k-morphism

$$R \longrightarrow L$$

defined by $Y \mapsto y\gamma$, where y as before denotes the image of Y in R/M,

(d) The ideal $J = ML[Y_{ij},-]$ of $L[Y_{ij},-]$ is in the kernel of the L-morphism

$$L[Y_{ij},-] \longrightarrow L$$

defined by $Y \mapsto y\gamma$,

(e) The ideal $J = ML[Y_{ij},-]$, considered as an ideal of the ring $L[Z_{ij},-] = L[Y_{ij},-]$ in the new indeterminates $Z = (Z_{ij})$ given by $Y = yZ$, is in the kernel of the L-morphism

$$L[Z_{ij},-] \longrightarrow L$$

defined by $Z \mapsto \gamma$.

Note that $\partial(Z) = 0$ since both $\partial Y = AY$ and $\partial y = Ay$. We have thus replaced the original differential indeterminates with non-differential ones, which allows us to define G as the zero-set of a (non-differential) ideal of $C[Z_{ij}, -]$. Indeed, by Exercise 2.11 below, J as an ideal of $L[Z_{ij}, -]$ is generated by

$$I = J \cap C[Z_{ij},)]$$

that is,

$$J = I\,L[Z_{ij}, -].$$

It follows that J is in the kernel of the L-morphism

$$L[Z_{ij}, -] \longrightarrow L$$

defined by $Z \mapsto \gamma$ if and only if the ideal I of $C[Z_{ij}, -]$ is in the kernel of the C-morphism

$$C[Z_{ij}, -] \longrightarrow C$$

defined by $Z \mapsto \gamma$. This ends the proof, since it shows that G is the zero-set of I in $\mathrm{GL}(p, C)$, by the equivalence of conditions (a) and (e) above. □

Exercise 2.11. With notation from the proof of Theorem 2.10, show that the ideal $J = M\,L[Y_{ij}, -]$ of $L[Y_{ij}, -]$, when considered as an ideal of $L[Z_{ij}, -] = L[Y_{ij}, -]$, is generated by the ideal

$$I = J \cap C[Z_{ij}, -]$$

of $C[Z_{ij}, -]$, where $Z = (Z_{ij})$ is defined by $Y = yZ$.

Remark 2.12. Note that in condition (d) of the proof of Theorem (2.10) above, the representation of G as a subgroup of $\mathrm{GL}(p, \mathbb{C})$ depends on the choice of a fundamental solution Y. The subgroup G is therefore defined up to conjugation only in $\mathrm{GL}(p, \mathbb{C})$.

Exercise 2.13. As in Exercise 2.7 above, consider the differential field

$$K = \mathbb{C}(x, \mathrm{e}^{\alpha_1 x}, \ldots, \mathrm{e}^{\alpha_m x})$$

with $\alpha_1, \ldots, \alpha_m \in \mathbb{C}^*$. Show that the differential Galois group of the Picard-Vessiot extension $\mathbb{C} \subset K$ is isomorphic to $\mathbb{C} \times (\mathbb{C}^*)^r$ for some $r \leq m$.

This exercise in particular relies on the following classical fact.

Exercise 2.14. If for some $r \in \mathbb{N}^*$ the complex numbers $\beta_1, \ldots, \beta_r$ are linearly independent over $\mathbb{Z}$, then the functions $\mathrm{e}^{\beta_1 x}, \ldots, \mathrm{e}^{\beta_r x}$ are algebraically independent over $\mathbb{C}$.

In analogy to classical Galois theory, there is a correspondence, for a given Picard-Vessiot extension $k \subset K$, between differential extensions L of k such that $k \subset L \subset K$ and Zariski closed subgroups of G. We refer to [PSi01, p. 25-26] and [CH11, p. 151-158] for a proof of the following result.

Theorem 2.15 (Differential Galois correspondence). *Let $k \subset K$ be a Picard-Vessiot extension with differential Galois group G. Let $\mathscr{I}$ denote the set of differential extensions L of k such that $k \subset L \subset K$, and $\mathscr{C}$ the set of Zariski closed subgroups of G. Then the following holds:*

(a) For any $L \in \mathscr{I}$, $L \subset K$ is a Picard-Vessiot extension (for the same differential system as $k \subset K$) and its differential Galois group, denoted $\phi(L)$, is a closed subgroup of G.

(b) For any $H \in \mathscr{C}$, the subfield $K^H = \{x \in K \mid \sigma(x) = x, \ \forall \sigma \in H\}$ is a differential extension of k such that $\phi(K^H) = H$.

(c) ϕ is a one-to-one correspondence between $\mathscr{I}$ and $\mathscr{C}$. In particular $K^G = k$.

(d) Given $L \in \mathscr{I}$, L is a Picard-Vessiot extension of k if and only if $\phi(L)$ is a normal subgroup of G, in which case the differential Galois group of $k \subset L$ is $G/\phi(L)$.

Example 2.16. Let G be a differential Galois group over k and let G^0 denote its *identity component*, that is, the connected component of G (for the Zariski topology) that contains the identity element. It is well known that G^0 is a normal, closed subgroup of G of finite index. The Galois correspondence tells us that G/G^0 is the differential Galois group, in this case the classical Galois group as well, of the finite algebraic extension K^{G^0} of k.

An important property of the differential Galois group relates its dimension to the transcendence of the solutions (cf. [PSi01, Corollary 1.30 p. 23]).

Proposition 2.17. *Let G be the differential Galois group of a Picard-Vessiot extension $k \subset K$. Then the dimension of G as an algebraic variety over C_k is equal to the transcendence degree of K over k. Moreover, if H is a subgroup of G, its Zariski closure $\overline{H}$ is equal to G if and only if $K^H = k$.*

2.1.3 Invariance modulo k-equivalence

Another important property of differential Galois groups is their invariance modulo the k-equivalence of linear differential systems, where k is a differential base-field containing their coefficients.

Definition 2.18. Two linear differential systems $\partial y = Ay$ and $\partial y = By$, where A and B are $p \times p$ matrices with entries in a differential field k, are said to be *k-equivalent* if

$$B = P^{-1}AP - P^{-1}\partial P$$

for some invertible $p \times p$ matrix P with entries in k, in other words if the system $\partial y = By$ is obtained by replacing y with Py in $\partial y = Ay$, which is sometimes referred to as $\partial y = By$ being the *gauge transform* of $\partial y = Ay$ by P.

Exercise 2.19. Let k be a differential field of characteristic zero with an algebraically closed field of constants. Show that two k-equivalent linear differential systems with coefficients in k have the same Picard-Vessiot extension and differential Galois group.

2.1.4 Differential Galois theory for scalar equations

In a similar way one associates a Picard-Vessiot extension and a differential Galois group with any scalar linear differential equation. Let us from now on write $(\ )'$ for $\partial(\ )$ if there is no ambiguity about the derivation ∂, and $(\ )^{(n)}$ for $\partial^n(\ )$.

Definition 2.20. Let (k,∂) be a differential field of characteristic zero, with an algebraically closed field of constants. The *Picard-Vessiot extension* of k and the *differential Galois group* of the scalar linear differential equation

$$x^{(p)} - \sum_{i=0}^{p-1} a_i x^{(i)} = 0, \quad a_i \in k \tag{E}$$

are the Picard-Vessiot extension of k and the differential Galois group respectively of the associated *companion system* $\partial y = Ay$ of (E), where $A = (\alpha_{ij})$ is the $p \times p$ matrix

$$\alpha_{ij} = \begin{cases} 0 & \text{if } i \neq p \text{ and } j \neq i+1 \\ 1 & \text{if } i \neq p \text{ and } j = i+1 \\ a_{j-1} & \text{if } i = p. \end{cases}$$

The differential Galois theory for systems and scalar equations is essentially the same. This is a consequence of the following result, known as the Cyclic vector lemma.

Theorem 2.21 (Cyclic vector lemma). *Let k be a differential field as in Definition 2.20 and such that k has at least one non-constant element. Then any linear differential system of order p with coefficients in k is k-equivalent to the companion system of some scalar linear differential equation of order p with coefficients in k.*

This result goes back to Cope[2] ([Cop34],[Cop36]) and Jacobson [Ja37]. For more recent proofs, we refer for instance to [BCL03] and the bibliography therein, or [Sin09, Theorem 1.2.8 p. 6], [CH11, Exercise 29 p. 139].

Exercise 2.22. Show that the differential field $\mathbb{C}(x, e^{x^2})$ endowed with the usual derivation d/dx is a Picard-Vessiot extension of $\mathbb{C}(x)$ but not of $\mathbb{C}$.

---

[2] Elizabeth Frances Cope, née Thorndike, was a former student of George D. Birkhoff.

2.1.5 Examples

The following examples show how the differential Galois group depends on the differential equation as much as on the base-field k.

Example 2.23. For any differential field k with constant field C as before, the differential Galois group G over k of the scalar differential equation

$$y' = ay, \quad a \in k^*$$

is C^* or a finite cyclic subgroup of C^*, or is trivial. To see this, we use the formal construction of the Picard-Vessiot extension as explained on p. 26. Consider the differential ring $R = k[T, T^{-1}]$ with the derivation $T' = aT$, and the fraction-field $K = \mathrm{Frac}(\overline{R})$ of the quotient $\overline{R}$ of R by a maximal differential ideal. Let t denote the class of T in $\overline{R}$. Then t is a solution of the equation in K. For any $\sigma \in G$ we have

$$\sigma(t)' = \sigma(t') = \sigma(at) = a\sigma(t),$$

that is, $\sigma(t)$ is also a solution in K of the equation. The solution space in K has dimension 1 over C and σ is injective, hence $\sigma(t) = \lambda t$ for some $\lambda \in C^*$, and G is isomorphic to a closed subgroup of C^*.

Assume that the differential equation $y' = nay$ has no solution in k^*, for any $n \in \mathbb{Z}, n \neq 0$. Then (cf. [PSi01, Example 1.19 p. 14]) R is a simple differential ring, with no differential ideals other than (0) and R (since R is a principal ring, it is sufficient to show that $(P(T))' \notin (P)$ for any $P \in k[T] \setminus k$). In this case, K is the fraction-field of R, that is, $K = k(T)$ endowed with the derivation $T' = aT$. For any $\lambda \in C^*$, $T \mapsto \lambda T$ induces a k-automorphism of K which commutes with the derivation and the differential Galois group is therefore C^* in this case. This also follows from the fact that $K = k(T)$ has transcendence degree 1 over k and C^* is the only closed subgroup of dimension 1 of C^*.

Assume that the above does not hold. Let $n > 0$ be minimal such that $y' = nay$ has a solution $y_0 \in k^*$. The ideal I generated by $F = T^n - y_0$ is then a maximal differential ideal of R. It is clearly differential since $F' = naF \in I$. Let us show that n is the minimal degree of the polynomials $\{P, P \in I \cap k[T], P(0) \neq 0\}$. Assume $P = \sum_{i=0}^{d} a_i T^i \in I$ is a monic polynomial of minimal degree d, $0 < d < n$, with (necessarily) $a_0 \neq 0$. Since the polynomial $(P(T))' - d\,aP(T)$ belongs to I, it must be 0 by the minimality of d, hence $a_0' = d\,a\,a_0$. This contradicts the assumption since $d < n$ and $a_0 \neq 0$. Let $\sigma \in G$. Then $\sigma(t) = \lambda t$ implies that $\lambda^n = 1$ since $t^n = y_0$ and $y_0 \in k^*$. The differential Galois group in this case is a finite cyclic subgroup of roots of unity of C^*.

Example 2.24. The differential Galois group over $\mathbb{C}(x)$ of the differential equation

$$y' = \frac{\alpha}{x} y, \quad \alpha \in \mathbb{C}$$

is $\mathbb{C}^*$ if $\alpha \notin \mathbb{Q}$, a finite subgroup (of roots of unity) of $\mathbb{C}^*$ else.

A fundamental solution is $y = x^\alpha$ and $K = \mathbb{C}(x, x^\alpha)$ is a Picard-Vessiot extension of $(\mathbb{C}(x), d/dx)$ for this equation. The differential Galois group G over $\mathbb{C}(x)$ is either $\mathbb{C}^*$ or a finite subgroup of $\mathbb{C}^*$. By Proposition 2.17, G is finite if and only if x^α is algebraic over $\mathbb{C}(x)$, that is, if $\alpha = m/n \in \mathbb{Q}$ in which case G is a group of n-th roots of unity.

Example 2.25. The differential Galois group over k of

$$y'' = c^2 y, \quad c \in C^*,$$

where k is a differential field and C its field of constants, is a closed subgroup of C^*. For example, the differential Galois group of

$$y'' = \frac{1}{4} y$$

over the particular fields $k = \mathbb{C}, \mathbb{C}(\mathrm{e}^x), \mathbb{C}(\mathrm{e}^{x/2})$ is $\mathbb{C}^*, \{1, -1\}, \{1\}$ respectively.

To prove this, let y_0 be a non-trivial solution of $y_0' = c y_0$ in some differential extension of k. Then y_0 and $1/y_0$ are two C-independent solutions of the given equation. They generate the Picard-Vessiot extension $K = k(y_0)$. The representation of the differential Galois group G as a subgroup of $\mathrm{GL}(2, C)$ with respect to the corresponding fundamental solution (of the companion system) is of the form

$$\sigma \mapsto \begin{pmatrix} \lambda_\sigma & 0 \\ 0 & \lambda_\alpha^{-1} \end{pmatrix},$$

that is, G is isomorphic to a closed subgroup of

$$C^* \simeq \left\{ \begin{pmatrix} \lambda & 0 \\ 0 & \lambda^{-1} \end{pmatrix}, \lambda \in C^* \right\} \subset \mathrm{SL}(2, C)$$

and by the discussion of Example 2.23 we conclude that it is isomorphic to C^* or a finite, cyclic subgroup (possibly trivial) of roots of unity of C^*.

Assume $c = 1/2$. The Picard-Vessiot extension of $y'' = y/4$ is $k \subset k(\mathrm{e}^{x/2})$. Over $k = \mathbb{C}$, the differential $\mathbb{C}$-automorphisms of $\mathbb{C}(\mathrm{e}^{x/2})$ are given by $\mathrm{e}^{x/2} \mapsto \lambda\, \mathrm{e}^{x/2}$ for arbitrary $\lambda \in \mathbb{C}^*$, hence $G = \mathbb{C}^*$ in this case (this also follows from the fact that $\mathbb{C}(\mathrm{e}^{x/2})$ has transcendence degree 1 over $\mathbb{C}$). Over $k = \mathbb{C}(\mathrm{e}^x)$, the differential Galois group coincides with the classical Galois group of the algebraic extension $\mathbb{C}(\mathrm{e}^x) \subset \mathbb{C}(\mathrm{e}^{x/2})$, that is, $G = \{1, -1\}$. Over $k = \mathbb{C}(\mathrm{e}^{x/2})$, $G = \{1\}$.

In Example 2.25, the differential Galois group is a subgroup of $\mathrm{SL}(2, C)$. More generally, subgroups of $\mathrm{SL}(p, C)$ are called *unimodular groups*. Unimodular differential Galois groups play an important role, for instance in the algorithmic approach of the theory. One criterion is the following.

Proposition 2.26. *The differential Galois group G of (S): $y' = Ay$ over k is a subgroup of $\mathrm{SL}(p, C_k)$ if and only if the scalar differential equation $y' = \mathrm{tr}(A) y$ has a non-trivial solution in k.*

Proof. Let $Z \in \mathrm{GL}(p,K)$ be a fundamental solution of (S) generating the Picard-Vessiot extension K of k for (S). The differential Galois group G is unimodular if and only if $\det\sigma = 1$ for all $\sigma \in G$, that is, if $\det\sigma(Z) = \det Z$ for all $\sigma \in G$. Since σ is a field automorphism, $\det\sigma(Z) = \sigma(\det Z)$, hence $G \subset \mathrm{SL}(p,C)$ if and only if $\sigma(\det Z) = \det Z$ for all $\sigma \in G$, that is, if $\det Z \in k$. It is well known that for any $p \times p$ matrix Z solution of (S), the determinant $\det(Z)$ satisfies the scalar equation $y' = (\mathrm{tr}A)y$. Note that $\mathrm{tr}A \in k$ since A has entries in k, and that the solutions of $y' = (\mathrm{tr}A)y$ belong to k if and only if a non-trivial one does, which ends the proof. □

Corollary 2.27. *Let k be a differential field. The scalar differential equation*

$$y^{(p)} - \sum_{i=0}^{p-1} a_i y^{(i)} = 0, \ a_i \in k$$

has a unimodular differential Galois group over k if and only if the differential equation $y' = a_{p-1}y$ *has a non-trivial solution in k, in particular if* $a_{p-1} = 0$.

This criterion is illustrated on Example 2.25.

2.2 Analytic differential Galois theory

For differential systems over the field of complex rational functions, the differential Galois group has a special structure which can be described in terms of analytic and formal invariants.

2.2.1 *Regular singular systems*

Consider the linear differential system

$$y' = A(x)y, \qquad (S)$$

of order p with coefficients in $\mathbb{C}(x)$. Fix a non-singular $x_0 \in \mathbb{C}$ and a germ of holomorphic fundamental solution Y of (S) at x_0. The field extension $K = k(Y)$ of $k = \mathbb{C}(x)$ generated by the entries of Y is a Picard-Vessiot extension of k for the derivation $y' = A(x)y$ (it produces no new constants). Let Σ be the set of singular points of (S) in $\overline{\mathbb{C}}$, and $U_\Sigma = \overline{\mathbb{C}} \setminus \Sigma$. The following result goes back to Schlesinger[3] [Sch87].

Theorem 2.28 (Schlesinger). *Assume that the system (S) has regular singular points only. Fix $x_0 \in \mathbb{C}$, non-singular, and a germ of holomorphic fundamental solution Y of (S) at x_0. Then the monodromy group*

[3] who stated it in the language of his time (1887).

$$\mathrm{Mon}(\pi_1(U_\Sigma;x_0)) \subset \mathrm{GL}(p,\mathbb{C})$$

with respect to Y is a Zariski dense subgroup of the differential Galois group of (S) over $\mathbb{C}(x)$.

Proof. Consider the Picard-Vessiot extension $K = k(Y)$ of $k = \mathbb{C}(x)$ for (S) and fix a loop γ with initial point x_0 in $U_\Sigma = \mathbb{P}^1_{\mathbb{C}} \setminus \Sigma$. Since analytic continuation along γ leaves rational functions invariant (these are single-valued) and commutes with the derivation (cf. Exercise 1.21 p. 13 and Corollary 1.22 p. 13) it defines a differential k-automorphism of K via its action on the entries of Y, that is, an element of the differential Galois group G of (S) over k, represented by a matrix $\Gamma_\gamma \in \mathrm{GL}(p,\mathbb{C})$ with respect to Y. Let $H = \mathrm{Mon}(\pi_1(U_\Sigma;x_0))$ be the monodromy group of (S) with respect to Y (cf. Section 1.2.3). In view of Proposition 2.17, to prove that $\overline{H} = G$ we need to prove that $K^H = k$, that is, that any element $f \in K$ invariant by H is a rational function. To this end, it is sufficient by Theorem 1.2 p. 5 to prove that f is meromorphic in $\overline{\mathbb{C}}$. By analytic continuation, Y can be extended as a solution $\tilde{Y}$ whose entries are analytic on the universal covering $\widetilde{U_\Sigma}$ of U_Σ with base-point x_0 (see Section 4.3.2, Definition 4.43). An element $f \in K$ is a rational function of x and the entries of Y, that is, $f = f_1 / f_2$ where f_1 and f_2 are polynomial expressions in x and the entries of Y. By analytic continuation f_1 and f_2 can be extended as analytic functions $\tilde{f}_1, \tilde{f}_2$ on $\widetilde{U_\Sigma}$, and f as a meromorphic function $\tilde{f}$ on $\widetilde{U_\Sigma}$. Assume that f is invariant by H. This means that for each singular point $a \in \Sigma$ and elementary loop γ from x_0 around a, the continuation of f along γ, which by definition is f_1^γ / f_2^γ, equals f. Since $\tilde{f}$ is meromorphic on $\widetilde{U_\Sigma}$ and invariant by γ, it defines a meromorphic function in a full punctured disk $\mathring{D}_a$ with center a. As in the proof of Corollary 1.37 p. 20 we can write (the analytic continuation of) each entry of Y, locally at a, as a finite sum

$$\sum_{k,\ell} (x-a)^{\rho_k} (\log(x-a))^{b_\ell} \varphi_{k\ell} \tag{2.1}$$

where ρ_k are the normalized eigenvalues of E (with $E = (\log \Gamma_\gamma)/2\pi\mathrm{i}$), b_ℓ are non-negative integers and $\varphi_{k\ell}$ are meromorphic germs at a. This implies that $\tilde{f}_1$ and $\tilde{f}_2$ are each, near a in the local coordinate $x-a$, of the form

$$\sum_{i=1}^{N} (\log(x-a))^i \left(\sum_{j=1}^{m_i} (x-a)^{\beta_{ij}} \phi_{ij} \right) \tag{2.2}$$

where all β_{ij} have a real part $\Re(\beta_{ij}) \in [0,1[$ and all ϕ_{ij} are meromorphic germs at a for each i,j. The fact that f extends as a meromorphic function on $\mathring{D}_a$, together with the expression (2.2) for $\tilde{f}_1$ and $\tilde{f}_2$, implies (see for instance [Ku93, Theorem 18.2 p.118]) that f is meromorphic at a. To see this, assume $a = 0$ and consider for $\ell = 1,2$ an expression of $\tilde{f}_\ell$ of the form (2.2) locally at 0,

$$\tilde{f}_\ell = \sum_{i=1}^{N} \sum_{j=1}^{m_i} (\log x)^i x^{\beta_{ij}} \phi_{\ell,ij}$$

where β_{ij} are pairwise distinct exponents with $\Re(\beta_{ij}) \in [0,1[$ and $\phi_{\ell,ij}$ are meromorphic at 0. Since $\tilde{f}_2$ is analytic in $\widetilde{U_\Sigma}$ and not identically 0, there is an open subset U in $\widetilde{U_\Sigma}$ on which $\tilde{f}_2 \neq 0$ and $\tilde{f}_1 = \tilde{f}\tilde{f}_2$. Since no two β_{ij} differ by an integer, the functions

$$(\log x)^i x^{\beta_{ij}}$$

are known to be linearly independent over the field of meromorphic germs at 0 (cf. [Ku93]). Since $\tilde{f}_2 \neq 0$ on U, we can fix some $\varphi_{2,ij}$ which is not identically 0 on U. By identification in $\tilde{f}_1 = \tilde{f}\tilde{f}_2$ we get

$$\phi_{1,ij} = \tilde{f}\phi_{2,ij}$$

hence $\tilde{f} = \phi_{1,ij}/\phi_{2,ij}$ in U. By analytic continuation this equality extends to $\widetilde{U_\Sigma}$ and proves that f is meromorphic at 0, as a quotient of meromorphic germs. Since $\tilde{f}$ is meromorphic on $\widetilde{U_\Sigma}$ and at each $a \in \Sigma$, it is meromorphic on $\overline{\mathbb{C}}$, which ends the proof. □

Let us describe the monodromy and differential Galois group on examples of Chapter 1.

Example 2.29. In Examples 1.27 p. 16 and 1.34 p. 18, the system

$$\frac{dy}{dx} = \begin{pmatrix} 1/x & 1 \\ 0 & 0 \end{pmatrix} y$$

has two singular points 0 and ∞. With respect to the fundamental solution

$$y = \begin{pmatrix} x & x\log x \\ 0 & 1 \end{pmatrix}$$

its monodromy group is

$$H = \mathrm{Mon}(\pi_1(U_\Sigma; x_0)) = \left\{ \begin{pmatrix} 1 & 2\pi i k \\ 0 & 1 \end{pmatrix}, \ k \in \mathbb{Z} \right\}$$

which is an infinite subgroup of

$$\left\{ \begin{pmatrix} 1 & \lambda \\ 0 & 1 \end{pmatrix}, \ \lambda \in \mathbb{C} \right\} \simeq (\mathbb{C}, +),$$

hence $\overline{H} \simeq \mathbb{C}$ since proper closed subroups of $\mathbb{C}$ are finite ($\overline{H}$ denotes the Zariski closure of H). Both 0 and ∞ are regular singular (the growth of the fundamental solution is moderate at each, and 0 is moreover Fuchsian). The differential Galois group over $\mathbb{C}(x)$ is therefore $\overline{H} = (\mathbb{C}, +)$ by Theorem 2.28 p. 35.

Example 2.30. In Example 2.24

$$y' = \frac{\alpha}{x} y, \ \ \alpha \in \mathbb{C}$$

both 0 and ∞ are Fuchsian, hence regular singular points. It is easy to see that the monodromy 'matrices' here are $m_0 = \mathrm{e}^{2\pi \mathrm{i}\alpha}$ at 0 and $m_\infty = \mathrm{e}^{-2\pi \mathrm{i}\alpha}$ at ∞. If $\alpha \notin \mathbb{Q}$, then the monodromy group is an infinite subgroup of $\mathbb{C}^*$. This implies that the differential Galois group, which is its Zariski closure, is $\mathbb{C}^*$. If $\alpha = p/q$ where p,q are relatively prime integers then the monodromy group is isomorphic to $\mathbb{Z}/q\mathbb{Z}$, hence Zariski closed, equal to the differential Galois group. We recover the results found on p. 33.

Example 2.31. In Example 1.29 p. 16 the system

$$\frac{dy}{dx} = \begin{pmatrix} 0 & 1 \\ 1/x^2 & -1/x \end{pmatrix} y$$

has two regular singular points and the fundamental solution has rational coefficients. Both the monodromy and the differential Galois groups are trivial in this case (the Picard-Vessiot extension is a trivial extension of $\mathbb{C}(x)$).

Remark 2.32. Various equivalent definitions of regular singular points are being used in the literature. One of them is given by the following characterization. Since the definition is local, we may assume that the singular point is 0 and consider the system over $k = \mathbb{C}(\{x\})$.

Theorem 2.33. *Let (S) be a linear differential system with coefficients in* $k = \mathbb{C}(\{x\})$. *Assume that* 0 *is a singular point of (S). Then* 0 *is regular singular, in the sense of Definition 1.24 p. 14, if and only if (S) is k-equivalent to a system with a Fuchsian singular point at* 0*, that is, if* 0 *is a simple pole of some k-equivalent system.*

We refer for instance to [Sab10, Théorème 1.4.6] for a proof of this result.

In the general case of a system (S) with possibly irregular singular points, the differential Galois group is the Zariski closure in $\mathrm{GL}(p,\mathbb{C})$ of a group generated by further elements, other than the monodromy matrices only. To define these additional generators we need to recall the existence, at a singular point, of a formal fundamental solution of a special form.

2.2.2 *Formal fundamental solutions*

In this section we describe a convenient formula for a formal fundamental solution at a singular point of (S), say 0. This special formula originally appeared in the formal meromorphic classification of linear differential systems. It is useful as well in the (convergent) meromorphic classification of systems as in differential Galois theory and in the study of the generalized Riemann-Hilbert problem.

Assume that $0 \in \Sigma$ is a singular point of the system

$$y' = A(x)y \qquad (S)$$

with coefficients in $\mathbb{C}(x) \subset \mathbb{C}(\{x\})$. The system (S) is now to be considered over the differential field $\mathbb{C}(\{x\})$ of meromorphic germs at 0.

It is known from the general theory of linear ordinary differential systems (cf. [BJL79], [Lod95], [Lod01]) that (S) has a formal fundamental solution of the form

$$\hat{Y} = \hat{\phi}(x)\, x^L\, \mathrm{e}^{Q(1/t)} \tag{2.3}$$

where

(a) $t^\nu = x$ for some $\nu \in \mathbb{N}^*$,
(b) $Q(1/t) = \mathrm{diag}(q_1, \ldots, q_p)$, where $q_i \in \frac{1}{t}\mathbb{C}[\frac{1}{t}]$,
(c) L is a constant $p \times p$ matrix,
(d) $\hat{\phi} \in \mathrm{GL}(p\ \mathbb{C}((x)))$ is a formal meromorphic invertible matrix of power series in the variable x.

The entries of $\hat{\phi}$ are in general *divergent series* which conceal the relevant information about the irregular singular point. To disclose this information and 'reincarnate" the divergent series as convergent meromorphic 'sums' (which still give a solution by formula 2.3), requires various *summability* processes, depending on the complexity and level of irregularity of the singular point. We refer to the second part of this volume for an introduction to these techniques, and to the second volume [Lod16] for an extensive study of summability in relation to differential equations.

Let us reorder the polynomials $q_i = q_i(1/t)$ in such a way that Q which will throughout stand for $Q(1/t)$, reads as a direct sum of scalar blocks

$$Q = \bigoplus q_{i_j} I_{k_j}$$

where I_{k_j} is the $k_j \times k_j$ identity matrix, q_{i_j} are distinct polynomials and $\sum k_j = p$. The smallest possible ν in a formal fundamental solution of this form is usually called the *degree of ramification* of the system.

If it is possible to choose $\nu = 1$, this case is referred to as *without ramification* or *without roots* or *unramified* even if the singular point 0 is a branch-point of the solutions. In this case the so-called *determining polynomials* q_i are indeed actual polynomials in $1/x$. This implies that the matrix L can be chosen in its Jordan form J, where furthermore the Jordan blocks are compatible with the (scalar) block-structure of Q, meaning that they correspond to sub-blocks of the scalar blocks of Q, so that the matrices J and Q commute in this case, as well as the matrices x^L and $\mathrm{e}^{Q(1/t)}$ (this may be of importance in computations).

If the smallest possible ν is > 1, this case is referred to as *with ramification* or *with roots* or *ramified*. The determining polynomials q_i in this case are no longer polynomials in $1/x$, but in $1/t$, whereas the family $\{q_1, \ldots, q_p\}$ is invariant under the action of the Galois group $\Gamma_\nu \simeq \mathbb{Z}/\nu\mathbb{Z}$ of the ramification. More precisely, replacing t by ζt in $Q(1/t)$, with $\zeta = \mathrm{e}^{2\pi \mathrm{i}/\nu}$, induces a permutation of the scalar diagonal blocks of Q. The matrix L may be chosen in such a way that its Jordan form $J = ULU^{-1}$ commutes with Q. However, if we want the matrix J to appear in the formula of the formal solution we have to write

$$\hat{Y} = \hat{\phi}(x)\, U^{-1} x^J U\, \mathrm{e}^{Q(1/t)}. \tag{2.4}$$

The matrices J and Q do commute but they do not commute with U, which implies that the matrices x^L and $\mathrm{e}^{Q(1/t)}$ in (2.3) do not commute either. We refer to Example 2.50 below for an illustration of this fact by the Airy equation. The form of the matrix U is given explicitly in [BJL79] and in [Lod01, p. 239], where it appears as a direct sum of van der Monde matrices built on the full set or a cyclic subset of the ν-th roots of unity $(1, \zeta, \dots, \zeta^{\nu-1})$ tensored by unit matrices of an appropriate size.

It turns out that the system (S_0) satisfied by the formal fundamental solution

$$\hat{Y}_0 = x^L \mathrm{e}^{Q(1/t)}$$

has (convergent) meromorphic coefficients. Its *formal equivalence class* over $\mathbb{C}((x))$, that is, the $\mathbb{C}((x))$- equivalence class of (S_0) in the sense of Definition 2.18, over the field of formal Laurent series, contains the system (S) by construction (with $P = \hat{\phi}$ in the notation of Def. 2.18). The system (S_0) as well as any system in its (convergent) meromorphic class, that is, its $\mathbb{C}(\{x\})$- equivalence class, is called a *normal form* of (S) and the fundamental solution $x^L \mathrm{e}^{Q(1/t)}$ as well as any $\phi(x) x^L \mathrm{e}^{Q(1/t)}$ with convergent ϕ is called a *normal solution*. The pair (J, Q) described in detail above, although non-unique, provides a full set of formal invariants for the systems in the formal class of (S_0). In other words, the entries of J and Q completely determine the equivalence class of (S_0) modulo $C((x))$- equivalence. Note that redundancy may occur in this set of entries. For instance, for a given determining polynomial $q_i(1/t)$ the matrix Q contains all polynomials $q_i(1/\zeta t), \dots, q_i(1/\zeta^{\nu-1} t)$, with the same multiplicity, whereas one only of these polynomials is needed to characterize the full orbit under the action of Γ_ν. The expression of a normal form in terms of a minimal set of formal invariants is given explicitly in [Lod01, Theorem 2.2], by characteristic formulas.

Let us mention here a 'weaker' formula for a formal fundamental solution of (S) which appeared in [Tu55] and leads to a normal form in the formal classification over $C((t))$ rather than $C((x))$. This formula is

$$\hat{Y} = \hat{\psi}(t) x^J \mathrm{e}^{Q(1/t)} \tag{2.5}$$

with Q, J as before and where the entries of $\hat{\psi}(t)$ are now formal Laurent series in t. The matrices Q and J still commute here.

Remark 2.34. By the results of Section 1.2.4 (see also Section 4.1) one can prove that if 0 is a regular singular point, then the formal series $\hat{\phi}$, $\hat{\psi}$ above are convergent. Moreover in this case, $Q = 0$ in formulas (2.3) to (2.5), and this actually characterizes regular singular points (cf. [Lod16, Definition 3.3.2 and Proposition 3.3.3]), whereas the convergence alone of $\hat{\phi}$, $\hat{\psi}$ does not. We refer to ([Lod16, Proposition 3.3.14]) for yet another characterization of regular singular points in terms of the Newton polygon of the system.

2.2.3 *Local differential Galois groups*

Consider a linear differential system

$$y' = A(x)\,y \qquad (S)$$

with coefficients in $k = \mathbb{C}(\{x\})$ and assume that $0 \in \Sigma$ is a singular point of (S). The *local differential Galois group* at 0 is the differential Galois group of any Picard-Vessiot extension of k for (S), since by Theorem 2.5 p. 26 these are isomorphic. We will describe this group for the particular Picard-Vessiot extension $K = k(\hat{Y})$ generated by a *formal* fundamental solution $\hat{Y}$ of the form (2.3):

$$\hat{Y} = \hat{\phi}(x)\,x^L\,\mathrm{e}^{Q(1/t)}.$$

From the results of Section 2.2.2 we know that K can be viewed as a differential subfield of the larger differential field

$$\mathscr{L} = \mathbb{C}((x))(t, \log x, x^{\alpha_1}, \ldots, x^{\alpha_N}, \mathrm{e}^{q_1}, \ldots, \mathrm{e}^{q_p}) \qquad (2.6)$$

for some $N \in \mathbb{N}$ and suitable exponents $\alpha_i \in \mathbb{C}^*$, with $Q(1/t) = \mathrm{diag}(q_1, \ldots, q_p)$ and where t, $\log x$, $x^{\alpha_1}, \ldots, x^{\alpha_N}$, $\mathrm{e}^{q_1}, \ldots \mathrm{e}^{q_p}$ are purely *formal* objects so far. More precisely, formula (2.6) means that the differential extension $\mathscr{L}$ of $\mathbb{C}((x))$ results from successive Picard-Vessiot extensions with respect to the differential equations

(a) $\nu x \partial y = y$
(b) $x\partial^2 y + \partial y = 0$
(c) $x\partial y = \alpha_i y$ for $i = 1, \ldots, N$
(d) $\partial y = q_i^*\, y$ for $i = 1, \ldots, p$

respectively, where $q_i^* = \partial(q_i(1/t))$ for the derivation ∂ defined by (a) (cf. Exercise 2.3 p. 26). It is important here to point out that the field of constants of $\mathscr{L}$ is $\mathbb{C}$ (we leave this as an exercise).

Exercise 2.35. Write q_i^* explicitely.

We are now ready to describe the additional meromorphic invariants needed to generate a dense subgroup of the differential Galois group in the same way as monodromy matrices in the regular singular case.

2.2.3.1 Formal monodromy

The formal monodromy accounts for the action on the formal solution (2.3) of a 'formal turn' around 0, that is, of the change of variable $x \leftarrow x\mathrm{e}^{2\pi\mathrm{i}}$ (and subsequent substitution $t \leftarrow t\,\mathrm{e}^{2\pi\mathrm{i}/\nu}$). It is defined as follows.

Consider the differential $\mathbb{C}((x))$-automorphism $\hat{\mu}$ of $\mathscr{L}$ defined by

(a) $\hat{\mu}(t) = \zeta t$

(b) $\hat{\mu}(x^{\alpha_i}) = e^{2\pi i \alpha_i} x^{\alpha_i}$ for $i = 1, \ldots, N$

(c) $\hat{\mu}(\log x) = \log x + 2\pi i$

(d) $\hat{\mu}(e^{q_i}) = e^{q_{j_i}}$ for $i = 1, \ldots, p$

with $\zeta = e^{2\pi i/\nu}$ and where $q_i \mapsto q_{j_i}$ denotes the permutation of the q_i induced by $t \leftarrow \zeta t$ as explained in Section 2.2.2 (we leave it as an exercise to show that the automorphism thus defined is differential.) Note that $\hat{\mu}$ is *a fortiori* a k-automorphism of $\mathscr{L}$ since $k = \mathbb{C}(\{x\}) \subset \mathbb{C}((x))$. Let $\hat{\mu}(\hat{Y})$ denote the matrix obtained by applying $\hat{\mu}$ to each entry of $\hat{Y}$ in $\mathscr{L}$. Since $\hat{\mu}$ commutes with the derivation of $\mathscr{L}$, the matrix $\hat{\mu}(\hat{Y})$ is again a fundamental solution of (S) and hence there is a matrix $\hat{M} \in \mathrm{GL}(p, \mathbb{C})$ such that

$$\hat{\mu}(\hat{Y}) = \hat{Y}\hat{M}.$$

This shows that $K = k(\hat{Y})$ is invariant under the k-automorphism $\hat{\mu}$ of $\mathscr{L}$. In other words, $\hat{\mu}$ induces a differential k-automorphism of K, that is, an element of the differential Galois group of (S), completely determined by the linear map $\hat{Y} \mapsto \hat{Y}\hat{M}$ on the space of formal solutions of (S). This element of the differential Galois group is by definition the *formal monodromy*, and $\hat{M}$ the formal monodromy matrix of (S) with respect to $\hat{Y}$.

2.2.3.2 Exponential torus

The exponential torus accounts for the possible algebraic dependence of the exponential factors in formula (2.3). To define this specific subgroup of the differential Galois group we proceed as for the formal monodromy, by defining $\mathbb{C}((x))$-automorphisms of $\mathscr{L}$, then restricting them to $k(\hat{Y})$.

Let $\mathscr{T}$ denote the differential Galois group over $\mathbb{C}(t)$ of the differential field $\mathbb{C}(t)(e^{q_1}, \ldots, e^{q_p})$, which is indeed a Picard-Vessiot extension. To describe $\mathscr{T}$ in more detail let $\{p_1, \ldots, p_r\}$ denote a $\mathbb{Z}$-basis of the $\mathbb{Z}$-module $\mathbb{Z}q_1 + \ldots + \mathbb{Z}q_p$. Then

$$\mathbb{C}(t)(e^{q_1}, \ldots, e^{q_p}) = \mathbb{C}(t)(e^{p_1}, \ldots, e^{p_r}).$$

Let φ be a differential $\mathbb{C}(t)$-automorphism of $\mathbb{C}(t)(e^{p_1}, \ldots, e^{p_r})$. Then $\varphi(e^{p_j})$ satisfies for each j the same differential equation as e^{p_j}

$$\frac{d\varphi(e^{p_j})}{dt} = p_j^* \varphi(e^{p_j})$$

since $p_j^* = \frac{d}{dt}(p_j(\frac{1}{t})) \in \mathbb{C}(t)$, and hence $\varphi(e^{p_j}) = \tau_j e^{p_j}$ for some $\tau_j \in \mathbb{C}^*$. Conversely,

$$\varphi(e^{p_j}) = \tau_j e^{p_j}$$

for each j and arbitrary $\tau_j \in \mathbb{C}^*$, defines a $\mathbb{C}(t)$-automorphism φ of $\mathbb{C}(t)(e^{p_1}, \ldots, e^{p_r})$ which commutes with the derivation d/dt since

$$\varphi\left(\frac{d}{dt}\mathrm{e}^{p_j}\right) = \varphi(p_j^*\mathrm{e}^{p_j}) = p_j^*\,\varphi(\mathrm{e}^{p_j}) = p_j^*\,\tau_j\,\mathrm{e}^{p_j} = \frac{d}{dt}(\tau_j\,\mathrm{e}^{p_j}).$$

This proves that $\mathscr{T}$ is isomorphic to $\mathbb{C}^{*r}$, in other words that $\mathscr{T}$ is an r-dimensional torus (see Exercise 2.36 below for a precise description of its torus structure).

We can extend any $\varphi \in \mathscr{T}$ given by $\varphi(\mathrm{e}^{p_j}) = \tau_j\,\mathrm{e}^{p_j}$ for $\tau_j \in \mathbb{C}^*$, $j = 1,\ldots,r$, to a differential $\mathbb{C}((x))$-automorphism $\widetilde{\varphi}$ of $\mathscr{L}$ by letting $\widetilde{\varphi}$ be the identity map on

$$\mathbb{C}((x))(t, \log x, x^{\alpha_1},\ldots,x^{\alpha_N})$$

and

$$\widetilde{\varphi}(\mathrm{e}^{p_j}) = \tau_j\,\mathrm{e}^{p_j}$$

for each j. Since $\varphi(t) = t$, and $\log x$, x^{α_i}, e^{p_j}, for $i = 1,\ldots,p$, $j = 1,\ldots,r$ are algebraically independent over $\mathbb{C}((x))(t)$ and since φ commutes with the derivation $\nu^{-1}t^{1-\nu}d/dt$ extending d/dx, we have defined a differential $\mathbb{C}((x))$-automorphism $\widetilde{\varphi}$ of $\mathscr{L}$. The Picard-Vessiot extension $K = k(\hat{Y})$ of $k = \mathbb{C}(\{x\})$, as a subfield of $\mathscr{L}$, is clearly invariant by $\widetilde{\varphi}$ and by restricting each $\widetilde{\varphi}$ to K we identify $\mathscr{T}$ with a subgroup of the differential Galois group G of (S) over k. As a torus, $\mathscr{T}$ plays an essential role in the algebraic group structure of G via its adjoint action on the Lie algebra of G (cf. Section 2.2.3.4 p. 49).

Exercise 2.36. Write the matrix representing an element of the exponential torus with respect to a formal fundamental solution $\hat{Y}$ of the form (2.3).

2.2.3.3 Stokes matrices

Let us as before fix a formal fundamental solution of the form (2.3)

$$\hat{Y} = \hat{\phi}(x)\,x^L\,\mathrm{e}^{Q(1/t)} \tag{2.7}$$

with $\hat{\phi} \in \mathrm{GL}(p,\mathbb{C}((x)))$, $t^\nu = x$ and $Q = \mathrm{diag}\{q_1,\ldots,q_p)$, of a system (S) with coefficients in $\mathbb{C}(\{x\})$. In its representation in $\mathrm{GL}(p,\mathbb{C})$ via $\hat{Y}$ we are going to define new elements of the local differential Galois group, called the *Stokes matrices*. These fill in a way the gap of information between the formal solution (2.7) and its analytic realizations. With respect to $\hat{Y}$, we may for example define these matrices by means of 'multisums' of the factor $\hat{\phi}$ as follows.

Let $\ell_1,\ldots,\ell_N$ denote the *singular rays* of Q, or (S), also called *singular lines*, ordered counterclockwise around 0 in $\mathbb{C}^*$ (named *anti-Stokes lines* in the second volume [Lod16]). These are, in our definition, the open half-lines issued from 0 in the x-plane, on each of which the module of some $\mathrm{e}^{q_i-q_j}$ decreases maximally as x tends to 0, for any given determination of the argument of x that gives the exponentials $\mathrm{e}^{q_i-q_j}$ an analytic, non-formal meaning. We will indistinctively use the terminology *ray* or *line*, always meaning an open half-line issued from 0.

As before $k = \mathbb{C}(\{x\})$ denotes the field of convergent meromorphic series and $K = k(\hat{Y})$ the Picard-Vessiot extension of k for (S) generated as a field by k and the entries of $\hat{Y}$. Various theories of summability (cf. [Ba94], [Ba00], [MR92], [BBRS91], [Lod94], [Lod16]) provide, on a given *non-singular* ray λ, a so-called *sum* ϕ_λ *of* $\hat{\phi}$ with asymptotic expansion $\hat{\phi}$ as x tends to 0, which together with an analytic interpretation of the normal solution

$$\hat{Y}_0 = x^L \mathrm{e}^{Q(1/t)}$$

yields a fundamental solution

$$Y_\lambda = \phi_\lambda(x) x^L \mathrm{e}^{Q(1/t)}$$

of (S) (cf. [Lod16, Theorems 5.2.5, 7.3.5]) which is abusively called a *sum of* $\hat{Y}$. The sum Y_λ is in particular holomorphic in an open sectorial neighborhood with vertex 0 containing λ. Moreover, the operators which occur in the summation process are *morphisms of differential algebras* between certain differential algebras, all endowed with the derivation d/dx or its extensions. In the particular case where all non-zero $q_i - q_j$ are of the same degree d in $1/t$, the sum along a non-singular line λ (it can for example be obtained via Borel and Laplace transformations, see Chapter 5 below and [Lod16]) is such that $\phi_\lambda(x)$ is (ν/d)-*Gevrey asymptotic* to $\hat{\phi}(x) = \sum_{n\geq 0} a_n x^n$ on some sectorial neighbourhood $\mathscr{V}_\lambda$, sector for short, of opening $> \nu\pi/d$ bisected by λ, meaning that for any proper subsector $\mathscr{W}$ of $\mathscr{V}$ we have

$$|x|^{-n} \left\| \phi_\lambda(x) - \sum_{k=0}^{n-1} a_k x^k \right\| < C_{\mathscr{W}} (n!)^{\frac{\nu}{d}} A_{\mathscr{W}}^n$$

for all $x \in \mathscr{W}$, $n \in \mathbb{N}$ and constants $C_{\mathscr{W}}, A_{\mathscr{W}}$ depending on $\mathscr{W}$ only.

For each singular line ℓ though, one can only define the left-hand limit ϕ_ℓ^- of ϕ_λ as λ tends to ℓ from 'below', that is, $\arg(\lambda)$ tends to $\arg(\ell)^-$, and the right-hand limit ϕ_ℓ^+ of ϕ_λ as λ tends to ℓ from 'above', that is, $\arg(\lambda)$ tends to $\arg(\ell)^+$, and not both sides at once. One can indeed prove that the ϕ_λ, hence the sums Y_λ are analytic continuations of each other as λ moves around 0, *as long as no singular line is to be crossed*, see Lemma 2.38, p. 47 in the case of one level of summability. The limits ϕ_ℓ^- and ϕ_ℓ^+ are defined in a common open sectorial neighbourhood $\mathscr{U}_\ell$ of 0 containing ℓ.

For the definition and existence of sums of the formal solution $\hat{Y}$ we refer to [Lod94, Definition III.2.2, Theorem III.2.8], and to [Lod16, Theorems 5.2.5, 7.3.5] in the second volume, where different theories of summability are presented and compared in the case of one-level summability as well as multisummability.

Our purpose is to define the Stokes matrices as elements of the differential Galois group relative to the Picard-vessiot extension $k \subset k(\hat{Y})$. To achieve this we will again define them as differential automorphisms of some larger differential extension of k containing $K = k(\hat{Y})$. The 'existential problem' here is so to speak the fact that the formal monodromy and exponential torus are *formal* invariants that can be defined from the normal form alone (they can be viewed, as mentionned earlier, as elements of the differential Galois group of (S_0) over $k = \mathbb{C}(\{x\})$ as well as $\mathbb{C}((x))$) whereas the Stokes matrices are *analytic invariants* of (S), defined via analytic functions.

Let $\widetilde{K} = k(\hat{\phi}, \hat{Y}_0)$ denote the extension of k by the entries of $\hat{\phi}$ and

$$\hat{Y}_0 = x^L e^{Q(1/t)}.$$

Then $K = k(\hat{Y}) \subset \widetilde{K}$. With notation as before, fix a singular line ℓ and an *analytic* fundamental solution $Y_{0,\ell}$ in $\mathcal{U}_\ell$ representing ('equal to') the formal solution $\hat{Y}_0 = x^L e^Q$ *for a given determination of* $\arg(x)$ *on* ℓ *and given initial values of the entries of* e^Q (remember that each of these exponentials was so far only defined by a scalar differential equation) and where the ray ℓ itself is now to be considered with its assigned argument. By definition of the sums ϕ_ℓ^- and ϕ_ℓ^+ we know that $\phi_\ell^- Y_{0,\ell}$ and $\phi_\ell^+ Y_{0,\ell}$ are both solutions of (S) on $\mathcal{U}_\ell$. There is therefore an invertible matrix $S_\ell \in \mathrm{GL}(p, \mathbb{C})$ such that [4]

$$\phi_\ell^- \, Y_{0,\ell} = \phi_\ell^+ \, Y_{0,\ell} \, S_\ell. \tag{2.8}$$

To show that S_ℓ belongs to the differential Galois group, let us define two (non-canonical) injective morphisms $u_\ell{}^+$ and $u_\ell{}^-$ of k-algebras from $\widetilde{K} = k(\hat{\phi}, \hat{Y}_0)$ to the field $k(\phi_l^-, \phi_l^+, Y_{0,\ell})$, namely

$$k(\hat{\phi}, \hat{Y}_0) \xrightarrow{u_\ell{}^+} k(\phi_\ell^-, \phi_\ell^+, Y_{0,\ell})$$

which maps $\hat{\phi}$ to its sum ϕ_ℓ^+ and $\hat{Y}_0$ identically to $Y_{0,\ell}$, and

$$k(\hat{\phi}, \hat{Y}_0) \xrightarrow{u_\ell{}^-} k(\phi_\ell^-, \phi_\ell^+, Y_{0,\ell})$$

which maps $\hat{\phi}$ to its sum ϕ_ℓ^- and $\hat{Y}_0$ identically to $Y_{0,\ell}$.

From various properties of the summation operators it follows that $u_\ell{}^+$ and $u_\ell{}^-$ are *differential* k-homomorphisms from the differential field $\widetilde{K}$ to the differential field $k(\phi_\ell^-, \phi_\ell^+, Y_{0,\ell})$. Moreover,

$$u_\ell{}^+(K) = u_\ell{}^-(K)$$

since $\phi_\ell^- \, Y_{0,\ell} = \phi_\ell^+ \, Y_{0,\ell} \, S_\ell$ and $S_\ell \in \mathrm{GL}(\mathbb{C})$. Let

$$\overline{K} = u_\ell{}^+(K) = u_\ell{}^-(K)$$

and let $\overline{u}_\ell{}^+$, $\overline{u}_\ell{}^-$ denote the (bijective) differential k-homomorphisms from K to $\overline{K}$ induced by $u_\ell{}^+$ and $u_\ell{}^-$ respectively.

We are now ready to define the *Stokes automorphism* of the Picard-Vessiot extension $K = k(\hat{Y})$ of (S) over k with respect to ℓ as the *differential* k-automorphism

$$s_\ell = (\overline{u}_\ell{}^+)^{-1} \, \overline{u}_\ell{}^- \tag{2.9}$$

of K, as sketched on the following commutative diagram

[4] In the second volume [Lod16] the Stokes matrices are defined with respect to the clockwise orientation of the unit circle *and* in the reverse way $\phi_\ell^+ \, Y_{0,\ell} = \phi_\ell^- \, Y_{0,\ell} \, S_\ell$ and are therefore equal to ours.

$$\begin{array}{ccc} \widetilde{K} \supset K & \xrightarrow{\quad s_\ell \quad} & K \subset \widetilde{K}\,. \\ & {\scriptstyle \overline{u}_\ell^{\,-}} \searrow \quad \swarrow {\scriptstyle \overline{u}_\ell^{\,+}} & \\ & \overline{K} & \end{array}$$

It is easy to see that S_ℓ is indeed the matrix of s_ℓ with respect to the fundamental solution $\hat{Y}$ and that S_ℓ belongs as expected to the differential Galois group of (S) over $k = \mathbb{C}(\{x\})$. It is by definition the *Stokes matrix* of (S) with respect to $\hat{Y}$ and the singular line ℓ, *for the given determination of the argument.*

An important property of the Stokes matrices, which is useful in particular for the computation of local differential Galois groups, is the following.

Proposition 2.37. *The Stokes matrices are unipotent.*

Proof. Fix a formal fundamental solution of (S) of the form (2.4)

$$\hat{Y} = \hat{\phi}(x)U^{-1}x^J U\, \mathrm{e}^{Q(1/t)}. \tag{2.10}$$

We will prove the result in the particular case of *one level* of summability (cf. [Lod94], and [Lod16, Definition 3.3.4]). In other words, we assume that all non-zero polynomials $q_i - q_j$ for $i \neq j$ have the same degree d in $1/t$ in the matrix $Q = \mathrm{diag}(q_1, \ldots, q_p)$ of (2.10). Let S denote the Stokes matrix with respect to $\hat{Y}$ on a given singular line ℓ, for a given determination of $\arg(x)$. For any *non-singular* ray λ there is a unique invertible matrix ϕ_λ, holomorphic in a punctured sectorial neighborhood $\mathscr{V}_\lambda$ of 0 of opening $> \pi\nu/d$, such that $\hat{\phi}(x)$ is the asymptotic expansion of $\phi_\lambda(x)$ as x tends to 0 in $\mathscr{V}_\lambda$, and such that

$$Y_\lambda = \phi_\lambda(x)U^{-1}x^J U\, \mathrm{e}^{Q(1/t)}$$

is a solution of (S) for the given determination of $\arg(x)$ (which gives a meaning to each factor). The matrices ϕ_λ and Y_λ are the sums of $\hat{\phi}$ and $\hat{Y}$ respectively along λ (cf. [Lod16, Theorem 5.2.5]) and we will abusively say that $\hat{Y}$ is the asymptotic expansion of Y at 0. Let ℓ_-, ℓ_+ such that $\arg(\ell_-) < \arg(\ell) < \arg(\ell_+)$ in the counterclockwise order, be non-singular rays issued from 0, close to the singular ray ℓ, and let ϕ_-, Y_-, ϕ_+, Y_+ denote the sums of $\hat{\phi}$ and $\hat{Y}$ along ℓ_- and ℓ_+ respectively. In general, a $p \times p$ matrix function $\Phi(x)$ is said to be *tangent to the identity* in a sector $\mathscr{U}$ and we write this $\Phi \approx I$, if $\Phi(x) = I + \Psi(x)$ where $\Psi(x)$ has the asymptotic expansion 0 in $\mathscr{U}$. Assume that

$$\arg(\ell_+) - \arg(\ell) = \arg(\ell) - \arg(\ell_-) = \varepsilon > 0.$$

For small enough ε the sums Y_- are, as ε varies, analytic continuations of each other by Lemma 2.38 below, and so are the sums Y_+. This allows us, as ε tends to 0, to consider by analytic continuation the sums Y_- and Y_+ of $\hat{Y}$ in a punctured sectorial neighborhood $\mathscr{U}_\ell$ of 0 bisected by ℓ, of opening $\pi\nu/d$. By definition of the Stokes matrix S along ℓ we have $Y_- = Y_+ S$ on $\mathscr{U}_\ell$ and hence

$$\phi_- U^{-1} x^J U \mathrm{e}^{Q(\frac{1}{t})} = \phi_+ U^{-1} x^J U \mathrm{e}^{Q(\frac{1}{t})} S$$

for the given determination of $\arg(x)$, which implies that

$$\mathrm{e}^Q S \mathrm{e}^{-Q} = (U^{-1} x^J U)^{-1} \; \phi_+^{-1} \phi_- \; (U^{-1} x^J U) \approx I$$

holds on $\mathscr{U}_\ell$. The (i,j) entry of $\mathrm{e}^Q S \mathrm{e}^{-Q}$, with $S = (s_{ij})$, is

$$s_{ij} \mathrm{e}^{q_i - q_j}.$$

Let us, with respect to ℓ, order the polynomials q_i in such a way that

$$q_i \prec q_j$$

whenever $\Re(q_i - q_j) \le 0$ as x gets close to 0 on ℓ (with $x = t^\nu$), and let us reorder the columns of $\hat{Y}$ so that

$$q_1 \prec q_2 \prec \ldots \prec q_p.$$

If $i > j$, the exponential $\mathrm{e}^{q_i - q_j}$ is oscillating or tends to ∞ as x tends to 0 on ℓ. From

$$\mathrm{e}^Q S \mathrm{e}^{-Q} \approx I \tag{2.11}$$

on $\mathscr{U}_\ell$ it follows that $s_{ij} \mathrm{e}^{q_i - q_j}$ has an asymptotic expansion equal to 0 as x (hence t) tends to 0 on ℓ, which is possible only if $s_{ij} = 0$. From (2.11) we also deduce that $s_{ii} = 1$ for all i, which proves that S is upper-triangular and unipotent. We moreover see that for any $i < j$, if $s_{ij} \neq 0$ then ℓ is a ray on which the module of $\mathrm{e}^{q_i - q_j}$ decreases maximally as x tends to 0. Indeed, (2.11) is valid in the sector $\mathscr{U}_\ell$ of opening $\pi\nu/d$. This implies that $\mathrm{e}^{q_i - q_j}$ is 'flat', meaning that it has an asymptotic expansion equal to 0 in $\mathscr{U}_\ell$ as x tends to 0. Since the successive contiguous open sectors in which (the module of) $\mathrm{e}^{q_i - q_j}$ is alternately decreasing and increasing as x tends to 0, have opening $\pi\nu/d$ (note that $q_i - q_j$ is equivalent to some $ax^{-d/\nu}$, $a \in \mathbb{C}$) the sector $\mathscr{U}_\ell$ must be a decrescence sector, and $\mathrm{e}^{q_i - q_j}$ maximally decreasing on its bisector ℓ. For a full proof in the general 'multilevel' case we refer to [Lod16, Proposition 3.5.8, Theorem 3.5.14]. □

In the following lemma and its proof, we will for convenience identify any ray (half-line) λ from 0 with its argument $\arg(\lambda)$ for a given determination of arg, say $\arg \in [0, 2\pi[$. Any sectorial neighbourhood $\mathscr{V}$ of 0, or sector for short, is likewise identified with an interval $]a,b[\subset [0,2\pi[$ and its opening denoted $|\mathscr{V}| = b - a$, where a and b are its edges.

Lemma 2.38. *With notation as before, assume in* (2.10) *that all non-zero polynomials $q_i - q_j$, for $i \neq j$, are of the same degree d in $1/t$. Let $\theta_1, \theta_2 \in [0, 2\pi[$, $\theta_1 < \theta_2$, be non-singular rays for* (2.10) *such that $[\theta_1, \theta_2]$ contains no singular ray. Then the sums Y_1 and Y_2 of $\hat{Y}$ along θ_1 and θ_2 respectively are analytic continuations of each other.*

Proof. Let us prove the result in the particular 'unramified' case where $\nu = 1$. The sums Y_1 and Y_2 are defined on sectors $\mathscr{V}_1$ and $\mathscr{V}_2$ of opening $> \pi/d$ bisected by θ_1

and θ_2 respectively and we may assume that

$$|\mathscr{V}_1| = |\mathscr{V}_2| = \frac{\pi}{d} + \varepsilon$$

for some $\varepsilon > 0$. The sector $\mathscr{V} = \mathscr{V}_1 \cap \mathscr{V}_2$ is bisected by $\tau = (\theta_1 + \theta_2)/2$ and its opening is

$$|\mathscr{V}| = \frac{\pi}{d} + \varepsilon - (\theta_2 - \theta_1).$$

Let us prove that $\mathscr{V}$ is not contained in any *singular sector*, by which we mean a sector of opening π/d bisected by a singular line. Let $\eta = \theta_2 - \theta_1 - \varepsilon$. The result is clearly true if $\eta < 0$. Suppose, if $\eta \geq 0$, that $\mathscr{V} \subset \mathscr{S}$, where $\mathscr{S}$ is a singular sector bisected by the singular ray $\ell \in [0, 2\pi[$. Since $|\mathscr{S}| = \pi/d$ and $|\mathscr{V}| = \pi/d - \eta$, we have

$$|\ell - \tau| \leq \frac{\eta}{2},$$

which implies that

$$\varepsilon - (\theta_2 - \theta_1) \leq 2\ell - \theta_1 - \theta_2 \leq (\theta_2 - \theta_1) - \varepsilon, \tag{2.12}$$

that is,

$$\theta_1 + \frac{\varepsilon}{2} \leq \ell \leq \theta_2 - \frac{\varepsilon}{2}$$

hence

$$\theta_1 < \ell < \theta_2$$

which contradicts our asumption and proves the statement. Since Y_1 and Y_2 are both fundamental solutions of (S) there is a constant invertible matrix C such that

$$Y_1 = Y_2 C \quad \text{in} \quad \mathscr{V} = \mathscr{V}_1 \cap \mathscr{V}_2$$

and since Y_1 and Y_2 are both asymptotic to $\hat{Y}$, we easily see that

$$\phi_2^{-1}\phi_1 = x^L \, \mathrm{e}^Q C \mathrm{e}^{-Q} \, x^{-L}$$

holds in $\mathscr{V}$ with

$$\mathrm{e}^Q C \mathrm{e}^{-Q} \approx I.$$

(Recall that $\approx$ means 'asymptotic to'). This, as in the proof of Proposition 2.37, implies that in $C = (c_{ij})$ we have $c_{ii} = 1$ for each i and $c_{ij}\mathrm{e}^{q_i - q_j} \approx 0$ for $i \neq j$, that is,

$$\lim_{x \to 0} c_{ij} \mathrm{e}^{q_i - q_j} = 0 \tag{2.13}$$

in $\mathscr{V}$. From the fact that $\mathscr{V}$ is not contained in any singular sector it follows that for any fixed $i \neq j$, $\mathscr{V}$ contains a nonempty sector $\mathscr{W}$ on which $\mathrm{e}^{q_i - q_j}$ is unbounded, hence $c_{ij} = 0$ by considering the limit (2.13) in $\mathscr{W}$. We have thus proved that $C = I$, that is, $Y_1 = Y_2$ in $\mathscr{V}$, which implies that Y_1 and Y_2 are analytic continuations of each other. We leave it as an exercise to complete the proof in the case $\nu \neq 1$. □

A useful formula relating the Stokes matrices to the monodromy, formal and topological, is given by the following exercise.

Exercise 2.39. With notation as before, prove the *cyclic relation*

$$\hat{M}\,S_{\ell_N}\ldots S_{\ell_1} = M$$

where M is the topological monodromy matrix around 0 with respect to a sum Y of $\widehat{Y}$, with a determination of $\arg(x)$ such that $0 \leq \arg(\ell_1) < \ldots < \arg(\ell_N) < 2\pi$.

The following exercise shows why, in the definition of the Stokes matrices, it is important to fix a determination of the argument.

Exercise 2.40. Assume 0 is a singular point of (S) and let ℓ be a singular ray. Let $\tilde{\ell}$ denote this ray for a given determination of $\arg(x)$, in other words, we consider $\tilde{\ell}$ on the Riemann surface $\widetilde{\mathbb{C}}$ of the logarithm. Let $\tilde{\ell}'$ represent ℓ for another determination of the logarithm, say $\arg(\tilde{\ell}') = \arg(\tilde{\ell}) + 2\pi$. Show that

$$S_{\tilde{\ell}'} = \hat{M}\,S_{\tilde{\ell}}\hat{M}^{-1}$$

where $S_{\tilde{\ell}}$ and $S_{\tilde{\ell}'}$ are the corresponding Stokes matrices and $\hat{M}$ the formal monodromy with respect to a given formal solution of the form (2.3).

2.2.3.4 Infinitesimal Stokes matrices

The fact that the Stokes matrices are unipotent has important applications. Consider a differential system with a fundamental solution (2.10), and let S_ℓ as before denote the Stokes matrix with respect to a singular line ℓ, for a given determination of the argument. Since S_ℓ is a unipotent element of the differential Galois group G of (S) over $k = \mathbb{C}(\{x\})$, it is the exponential

$$S_\ell = \exp \sigma_\ell \tag{2.14}$$

of a unique, nilpotent 'infinitesimal' Stokes matrix σ_ℓ belonging to the Lie algebra $\mathfrak{G} = \mathrm{Lie}_{\mathbb{C}}(G)$ of G. We will see how we can decompose σ_ℓ in $\mathfrak{G}$ under the action of the exponential torus $\mathscr{T}$.

Let us first recall some basic facts about group representations. A *representation* of a group Γ on a finite-dimensional $\mathbb{C}$-vector space V is a group homomorphism

$$\alpha : \Gamma \to \mathrm{GL}(V)$$

from Γ to the group of linear automorphisms of V. A *character* of Γ over $\mathbb{C}$ is a one-dimensional representation of Γ, that is, a group homomorphism

$$\chi : \Gamma \to \mathrm{GL}(\mathbb{C}) = \mathbb{C}^*.$$

For a given representation α of Γ on V, a character χ is said to be a *weight* if there is a non-zero element $v \in V$, called a *weight vector* for χ, such that

$$\alpha(\gamma)v = \chi(\gamma)v$$

for all $\gamma \in \Gamma$, that is, v is an eigenvector of $\alpha(\gamma)$ with eigenvalue $\chi(\gamma)$. The corresponding *weight space* V_χ is the subspace of V spanned by the weight vectors for the weight χ.

Exercise 2.41. Let α be a representation of the group Γ over $\mathbb{C}$ (or any other field). Show that the weight spaces are linearly independent.

An important example is the *adjoint representation* of a linear algebraic group H on its Lie algebra $\mathfrak{H}$. If $H \subset \mathrm{GL}(p,\mathbb{C})$, it is the restriction

$$\mathrm{Ad} : H \to \mathrm{GL}\,(\mathfrak{H})$$

of the representation

$$\mathrm{Ad} : \mathrm{GL}(p,\mathbb{C}) \to \mathrm{GL}\,(\mathfrak{gl}(p,\mathbb{C}))$$

of $\mathrm{GL}(p,\mathbb{C})$ on its Lie algebra $\mathfrak{gl}(p,\mathbb{C})$ defined by

$$\mathrm{Ad}(\gamma)(g) = \gamma g \gamma^{-1}$$

for each $\gamma \in \mathrm{GL}(p,\mathbb{C})$ and $g \in \mathfrak{gl}(p,\mathbb{C})$. To learn more about the adjoint representation, see for instance [Bor91, 3.13], [Var84, 2.13]. Since we are working over $\mathbb{C}$, a more analytic way to think of Ad is to consider the action

$$i : \mathrm{GL}(p,\mathbb{C}) \to \mathrm{Aut}(\mathrm{GL}(p,\mathbb{C}))$$

of $\mathrm{GL}(p,\mathbb{C})$ on itself by inner automorphisms given by $i_\gamma(\xi) = \gamma \xi \gamma^{-1}$ for each $\gamma \in \mathrm{GL}(p,\mathbb{C})$ and $\xi \in \mathrm{GL}(p,\mathbb{C})$. The map i_γ is an (analytic) morphism of analytic varieties and for each $\gamma \in \mathrm{GL}(p,\mathbb{C})$, $\mathrm{Ad}(\gamma)$ is the differential $(di_\gamma)_I$ of this map at $I \in \mathrm{GL}(p,\mathbb{C})$, that is, an automorphism of the tangent space $\mathfrak{gl}(p,\mathbb{C}))$ of $\mathrm{GL}(p,\mathbb{C})$ at its identity element I. Over general fields, $\mathrm{Ad}(\gamma)$ is the differential at I of the morphism of algebraic varieties $i_\gamma : G \to G$, in the sense of algebraic geometry.

The restriction of Ad to the differential Galois group G yields the adjoint representation

$$\mathrm{Ad} : G \to \mathrm{GL}(\mathfrak{G})$$

of G on its Lie algebra $\mathfrak{G}$ and since the exponential torus $\mathscr{T}$ is a linear algebraic subgroup of G it induces a representation

$$\mathrm{Ad} : \mathscr{T} \to \mathrm{GL}(\mathfrak{gl}(p,\mathbb{C})))$$

of $\mathscr{T}$ on $\mathfrak{gl}(p,\mathbb{C})$ as well as a representation

$$\mathrm{Ad} : \mathscr{T} \to \mathrm{GL}(\mathfrak{G})$$

of $\mathscr{T}$ on $\mathfrak{G}$, defined by $\mathrm{Ad}(\tau)(g) = \tau g \tau^{-1}$ for each $\tau \in \mathscr{T}$ and $g \in \mathfrak{gl}(p,\mathbb{C})$.

Let $\{p_1, \dots, p_r\}$ be a $\mathbb{Z}$-basis of the $\mathbb{Z}$-module $\mathbb{Z}q_1 + \ldots + \mathbb{Z}q_p$, generated by the entries of Q. Each exponential e^{q_i} can be written

$$e^{q_i} = m_i\left(e^{p_1}, \dots, e^{p_r}\right)$$

where m_i is a monomial in r variables. It follows from the definition of the exponential torus in Section 2.2.3.2, see also Exercise 2.36 p. 43, that $\mathscr{T}$ is isomorphic to an r-dimensional torus via the isomorphism $(\mathbb{C}^*)^r \to \mathscr{T}$ defined by

$$\tau = (\tau_1, \dots, \tau_r) \in (\mathbb{C}^*)^r \longmapsto T_\tau = \mathrm{diag}(m_1(\tau), \dots, m_p(\tau)) \in \mathscr{T}. \tag{2.15}$$

Let $T_\tau \in \mathscr{T}$ denote a generic element of $\mathscr{T}$, that is, we consider $\tau = (\tau_1, \dots, \tau_r)$ as the r-tuple of coordinate functions on the affine group $\mathscr{T} \simeq (\mathbb{C}^*)^r$. Note that any monic Laurent monomial $m(\tau)$ (i.e. a monomial with exponents $\in \mathbb{Z}$) is a character of $\mathscr{T} \simeq (\mathbb{C}^*)^r$. Note also that for $\tau^{-1} := (\tau_1^{-1}, \dots, \tau_p^{-1})$ the monomial $m(\tau^{-1})$ is the inverse character of $m(\tau)$ and that

$$T_{\tau^{-1}} = (T_\tau)^{-1}, \quad \mathrm{Ad}\,(T_{\tau^{-1}}) = \mathrm{Ad}\left((T_\tau)^{-1}\right) = (\mathrm{Ad}\,(T_\tau))^{-1}.$$

For all $g = (g_{ij}) \in \mathfrak{gl}(p, \mathbb{C})$ we have

$$\mathrm{Ad}\,(T_\tau)\,(g) = \left(m_i(\tau)\,m_j(\tau)^{-1}\,g_{ij}\right)$$

and in particular

$$\mathrm{Ad}\,(T_\tau)\,(E_{ij}) = m_i(\tau)\,m_j(\tau)^{-1}\,E_{ij},$$

where $\{E_{ij}\}$ is the canonical basis of $\mathfrak{gl}(p, \mathbb{C})$. In other words, E_{ij} is a weight vector for the weight $m_i(\tau)m_j(\tau)^{-1}$. Let $\mathscr{M}$ denote the set of pairwise distinct Laurent monomials of the form $m_i(\tau)m_j(\tau)^{-1}$, $1 \leq i, j \leq p$. It is easy to see that $\mathrm{Ad}\,(T_\tau)(g)$, for $g \in \mathfrak{G}$, has a unique decomposition

$$\mathrm{Ad}\,(T_\tau)(g) = \sum_{w \in \mathscr{M}} w(\tau)\,g_w$$

which, after applying $\mathrm{Ad}\,(T_{\tau^{-1}})$, gives a decomposition

$$g = \sum_{w \in \mathscr{M}} g_w \tag{2.16}$$

where each non-zero $g_w \in \mathfrak{gl}(p, \mathbb{C})_w$ is a weight vector of Ad in $\mathfrak{gl}(p, \mathbb{C})$ for the weight $w \in \mathscr{M}$. Note that the uniqueness of the decomposition also follows from Exercise 2.41. Let $\mathscr{W}$ denote the set of $w \in \mathscr{M}$ that are weights of the adjoint representation of $\mathscr{T}$ on $\mathfrak{G}$.

Lemma 2.42. *Each $g \in \mathfrak{G}$ has a unique decomposition*

$$g = \sum_{w \in \mathscr{W}} g_w$$

with $g_w \in \mathfrak{G}_w$ for $w \in \mathscr{W}$.

Proof. Let $\mathscr{M}_g$, for a given $g \in \mathfrak{G}$, denote the subset of elements $w \in \mathscr{M}$ such that $g_w \neq 0$ in the decomposition (2.16). Let us show that $g_w \in \mathfrak{G}$ for each $w \in \mathscr{M}_g$. Let

$\mathscr{M}_g = \{w_1, \ldots, w_s\}$. We have

$$\sum_{j=1}^{s} w_j(\tau)\, g_{w_j} = \mathrm{Ad}(T_\tau)(g).$$

Since the w_j are pairwise distinct, hence $\mathbb{C}$-linearly independent, we can choose weight vectors $v_1, \ldots, v_s \in \mathbb{C}^{*r}$ such that

$$\det(w_j(v_i)) \neq 0.$$

This proves that for the given $g \in \mathfrak{G}$, the system of s linear equations

$$\sum_{j=1}^{s} w_j(v_i)\, g_{w_j} = \mathrm{Ad}(T_{v_i})(g) \in \mathfrak{G}$$

for $i = 1, \ldots, s$ has a unique solution $(g_{w_1}, \ldots, g_{w_s})$ in $\mathfrak{G}^s$, hence that $g_{w_j} \in \mathfrak{G}_{w_j}$ for each j. We have thus proved that $\mathrm{Ad}(T_\tau)(g) = \sum_{w \in \mathscr{W}} w(\tau)\, g_w$ with $g_w \in \mathfrak{G}_w$ for each $w \in \mathscr{W}$, and we apply $\mathrm{Ad}(T_{\tau^{-1}})$ to get the result. □

Let us choose g to be an infinitesimal Stokes matrix σ_ℓ defined by (2.14), where ℓ refers to a singular line of the differential equation, that is, a ray ℓ on which one at least of the $\mathrm{e}^{q_i - q_j}$ decreases maximally. Lemma 2.42 provides a unique decomposition

$$\sigma_\ell = \sum_{w \in \mathscr{W}} \dot{\Delta}_{\ell,w} \tag{2.17}$$

with the notation $\dot{\Delta}_{\ell,w}$ for $g_w = (\sigma_\ell)_w \in \mathfrak{G}_w$. The weights w are of the form

$$w = m_i m_j^{-1}$$

with m_i, m_j as in (2.15), and the $w = m_i m_j^{-1}$ a priori occurring in (2.17) correspond to those $\mathrm{e}^{q_i - q_j}$ which decrease maximally on ℓ as x tends to 0. Generically, $\dot{\Delta}_{\ell,w} \neq 0$ for ℓ and w related in this way. However, the vanishing of terms $\dot{\Delta}_{\ell,w}$ in (2.17) provides useful information on the solutions, in particular on their algebraic properties, as we will illustrate below on the example of a generalized hypergeometric equation.

These elements $\dot{\Delta}_{\ell,w} \in \mathfrak{G}$ actually have an interpretation as *alien derivations* in resurgence theory (cf. Section 6.12.2 where σ_ℓ would correspond to Δ_ℓ there). They correspond to elements of a 'resurgence algebra' introduced by Écalle [Éc81] in a very general setting and used by Ramis [Ra88] in his Tannakian approach of analytic differential Galois theory. Ramis actually defined the local differential Galois groups as representations of a so-called 'wild fundamental group' (π_1 sauvage). In the case of a differential system (S) with the single level one (of summability) we refer the reader to Section 6.12.2 to see how the Stokes matrices, as elements of the differential Galois group G, are the restriction to the formal solution space of (S) of the *symbolic Stokes automorphisms* defined on a much larger space. We also refer to [MaRa89], [LR11] for further results relating Stokes matrices to alien calculus.

Note that the decomposition (2.17) is in particular useful to compute the local differential Galois group. Indeed, with a view to applying Ramis's theorem 2.47, it provides additional generators of the Lie algebra $\mathfrak{G}$, hence of the identity component G^0 of the differential Galois group, by 'splitting' in generic cases the infinitesimal Stokes matrices into further elements of $\mathfrak{G}$ (cf. [Mi96]) that are weight vectors for the adjoint action of $\mathscr{T}$.

Example 2.43. Let us illustrate the adjoint action of the exponential torus with the family of generalized confluent hypergeometric equations of order 7

$$D_{7,1}(y) = 0$$

where $D_{7,1}$ is the linear differential operator

$$D_{7,1} = z(\partial + \mu) - \prod_{j=1}^{7} (\partial + \nu_j - 1) \tag{2.18}$$

with $\partial = zd/dz$. This is a so-called Hamburger differential equation, with an irregular singular point at infinity and a regular singular point at 0. The global differential Galois group G of this equation over $\mathbb{C}(z)$ (cf. Section2.2.4) coincides in this case with the local differential Galois group at infinity, which can be computed using Ramis's theorem 2.47 p. 58. We consider a particular formal fundamental solution of this equation of the form (2.5) (cf. [Mi96, (2.3) p. 372])

$$\hat{Y} = \hat{\psi}(t)\, z^J\, e^{Q(t)}$$

with $t^6 = z$. Let $\zeta = e^{\pi i/3}$. The matrices Q and J in this example are

$$Q(t) = \mathrm{diag}(q_1, \dots, q_7) = -6t\, \mathrm{diag}\left(0, \zeta^{-2}, \zeta^{-1}, 1, \zeta, \zeta^2, -1\right)$$

$$J = \mathrm{diag}(-\mu, 0, \dots, 0) \tag{2.19}$$

with $\zeta = e^{\pi i/3}$. Note that the entries of Q are polynomials in t, not in $1/t$ since the irregular singular point is at infinity. At infinity, the singular lines of this equation with respect to $\hat{Y}$ are $\{\arg(z) = 0\}$ and $\{\arg(z) = \pi\}$ modulo 2π. Let us fix the determination $\arg(z) \in]-\pi, \pi]$ for the argument, and let

$$p_1(t) = -6\zeta^{-2} t, \quad p_2(t) = -6\zeta^{-1} t.$$

Since $\zeta^2 - \zeta + 1 = 0$, and ζ^{-1}, ζ^{-2} are $\mathbb{Z}$-linearly independent, $\{p_1, p_2\}$ is a basis of the $\mathbb{Z}$-module $\sum_{j=1}^{7} \mathbb{Z} q_j$ and

$$Q = \mathrm{diag}(0, p_1, p_2, p_2 - p_1, -p_1, -p_2, p_1 - p_2).$$

The exponential torus (cf. solution of Exercise 2.36 p. 68) is therefore isomorphic to $(\mathbb{C}^*)^2$ and equal to

$$\mathscr{T} = \left\{ T_\tau = \operatorname{diag}(1,\, \tau_1,\, \tau_2,\, \tau_1^{-1}\tau_2,\, \tau_1^{-1},\, \tau_2^{-1},\, \tau_1\tau_2^{-1}),\ \tau = (\tau_1, \tau_2) \in (\mathbb{C}^*)^2 \right\}.$$

The Stokes matrices at infinity (cf. [Mi91, p. 174]) relative to the singular lines $arg(z) = 0$ and $\arg(z) = \pi$ are the unipotent matrices

$$S_0 = I + N_0$$

$$S_\pi = I + N_\pi$$

respectively, with

$$N_0 = \alpha E_{4,1} + \beta E_{1,7} + \gamma E_{3,2} + \delta E_{5,6} + \eta E_{4,7},$$

$$N_\pi = c E_{3,7} + d E_{4,6},$$

where $\{E_{ij}\}$ as before denotes the canonical basis of $\mathfrak{gl}(7,\mathbb{C})$ and the coefficients are as follows. Let

$$b = \mathrm{e}^{-2\pi \mathrm{i}\mu}, \qquad \lambda = \frac{7}{2} + \mu - \sum_{j=1}^{7} \nu_j$$

and let e_r (resp. e'_r), $1 \le r \le 7$, be the elementary symmetric functions in $(e^{-2\pi \mathrm{i}\nu_j})_{1\le j\le 7}$ (resp. $(e^{2\pi \mathrm{i}\nu_j})_{1\le j\le 7}$). Then

$$\alpha = \frac{2\pi \mathrm{i}}{\prod_{j=1}^{7} \Gamma(1+\mu-\nu_j)}, \qquad \beta = \frac{(2\pi)^6 \mathrm{i}}{\prod_{j=1}^{7} \Gamma(\nu_j-\mu)}$$

and

$$\begin{aligned}
\gamma &= \zeta^{\lambda}\left(b^{-1} - e'_1\right),\\
\delta &= -\zeta^{-\lambda}\,(b - e_1),\\
c &= \zeta^{2\lambda}\left(e'_2 - b^{-1} e'_1 + b^{-2}\right),\\
d &= -\zeta^{-2\lambda}\left(e_2 - b e_1 + b^2\right),\\
\eta &= -\zeta^{-3\lambda}\left(b^3 - b^2 e_1 + b e_2 - e_3\right).
\end{aligned}$$

Remark 2.44. We know from Section 2.2.3.3 that the Stokes matrices are defined for a given determination of the argument, here $\arg(z) \in]-\pi, \pi]$, and (cf. Exercise 2.40) that the Stokes matrices for other determinations of the argument are obtained by conjugating the first by a power of the formal monodromy $\hat{M}$. In our example, we get in this way all Stokes matrices as conjugates of S_0 and S_π by powers $\hat{M}^k$, $k \in \mathbb{N}^*$, of the formal monodromy

$$\hat{M} = \operatorname{diag}(\mathrm{e}^{-2\pi\mu \mathrm{i}}, \zeta^{\lambda} P)$$

where P is the 6×6 permutation matrix (p_{ij}) such that $p_{ij}=1$ if $i\equiv j+1$ modulo 6, $p_{ij}=0$ else.

Moreover, it is equivalent to attach a Stokes matrix to each singular z-line and given determination of the argument, or to attach a Stokes matrix to each 'singular line' in the covering of the z-plane by the t-plane given by $t^6=z$, that is, to each t-line projecting on a singular z-line. Let us denote a Stokes matrix by $\mathbf{s}_\lambda$ whenever we consider it attached to a singular line λ in the t-plane.

We have in particular

$$S_0=\mathbf{s}_0,\qquad S_\pi=\mathbf{s}_{\pi/6}.$$

With this notation, the full set of Stokes matrices consists of

$$\mathbf{s}_{k\pi/3}=\hat{M}^k\,\mathbf{s}_0\,\hat{M}^{-k}\quad\text{and}\quad \mathbf{s}_{(2k+1)\pi/6}=\hat{M}^k\,\mathbf{s}_{\pi/6}\,\hat{M}^{-k},\quad\text{for all }\ k\in\mathbb{Z}.$$

Note that the possible non-zero entries off the diagonal in $\mathbf{s}_0$ (resp. $\mathbf{s}_{\pi/6}$) are (i,j)-entries such that $(q_i(t)-q_j(t))$ decreases maximally on $\arg(t)=0$ (resp. on $\arg(t)=\pi/6$), and we have a similar description for all Stokes matrices. Taking the logarithm of $\mathbf{s}_0$ and $\mathbf{s}_{\pi/6}$ yields the infinitesimal Stokes matrices

$$\begin{aligned}\sigma_0 &=\log(\mathbf{s}_0)=\alpha E_{4,1}+\beta E_{1,7}+\gamma E_{3,2}+\delta E_{5,6}+\left(\eta-\frac{\alpha\beta}{2}\right)E_{4,7},\\ \sigma_{\frac{\pi}{6}} &=\log(\mathbf{s}_{\frac{\pi}{6}})=c\,E_{3,7}+d\,E_{4,6}\end{aligned}$$

respectively. Let $\rho=\eta-\alpha\beta/2$. The adjoint action of $\mathscr{T}$ on σ_0 and $\sigma_{\pi/6}$ gives

$$\begin{aligned}\mathrm{Ad}(T_\tau)(\sigma_0) &=\tau_1^{-1}\tau_2\,(\alpha E_{4,1}+\beta E_{1,7}+\gamma E_{32}+\delta E_{5,6})+\tau_1^{-2}\tau_2^2\,(\rho E_{4,7}),\\ \mathrm{Ad}(T_\tau)(\sigma_{\frac{\pi}{6}}) &=\tau_1^{-1}\tau_2^2\,\sigma_{\frac{\pi}{6}}.\end{aligned}$$

Let us write the weights $w\in\mathscr{W}$ as

$$w(a,b)=\tau_1^a\,\tau_2^b,$$

where $a,b\in\{-2,-1,0,1,2\}$. The weights occurring in the adjoint action on σ_0 and $\sigma_{\pi/6}$ are

$$w(-1,1)=\tau_1^{-1}\tau_2,\quad w(-1,2)=\tau_1^{-1}\tau_2^2,\quad w(-2,2)=\tau_1^{-2}\tau_2^2$$

and the corresponding decomposition in weight vectors is

$$\begin{aligned}\sigma_0 &=\dot{\Delta}_{0,w(-1,1)}+\dot{\Delta}_{0,w(-2,2)}\\ \sigma_{\frac{\pi}{6}} &=\dot{\Delta}_{\frac{\pi}{6},w(-1,2)}\end{aligned}\tag{2.20}$$

which provides new elements

$$\dot{\Delta}_{0,w(-1,1)} = \alpha E_{4,1} + \beta E_{1,7} + \gamma E_{3,2} + \delta E_{5,6}$$
$$\dot{\Delta}_{0,w(-2,2)} = \rho E_{4,7}$$

of the Lie algebra $\mathfrak{G}$ of G, as long as $\rho \neq 0$. The adjoint action of $\mathscr{T}$ on all infinitesimal Stokes matrices will in this way, after conjugation by the formal monodromy, together provide twelve additional elements of the Lie algebra, whenever $\rho \neq 0$.

To describe these elements in the language of resurgence theory, note that if we express the equation in the variable t, we are in a case of 1-summability of the formal fundamental solution since all non-zero polynomials $q_i - q_j$, with q_i given by (2.19), are of degree 1 (cf. Volume 2, Thm 5.2.5 and Cor. 5.2.7).

Let us write each $\mathrm{e}^{q_i - q_j}$ as

$$\mathrm{e}^{q_i - q_j} = \mathrm{e}^{-a_{ij}t}$$

The complex numbers a_{ij} correspond to the singular points in the so-called *Borel plane* of the (Borel transform of) the differential equation $D_{7,1}$. With notation from Section 6.12.2, we are concerned here with the lattice Ω generated over $\mathbb{Z}$ by

$$\{0, \zeta^{-2}, \zeta^{-1}, 1, \zeta, \zeta^2, -1\},$$

that is, by 1 and ζ for instance. If ℓ is a singular t-line and w some weight $m_i m_j^{-1}$ with i, j such that

$$q_i - q_j = -a_{ij}t$$

decreases maximally on ℓ, the element $\dot{\Delta}_{\ell,w}$ can now be interpreted, with notation from Section 6.12.1, as the *alien derivation*

$$\dot{\Delta}_{\omega} \quad \text{with} \quad \omega = a_{ij} \in \Omega.$$

Note that $\arg(\omega) = \arg(a_{ij}) = -\arg(\ell)$. With this interpretation we can rewrite the additional elements of the decomposition (2.20) as

$$\dot{\Delta}_6 = \dot{\Delta}_{0,w(-1,1)}, \quad \dot{\Delta}_{12} = \dot{\Delta}_{0,w(-2,2)}$$

with 6, $12 \in \Omega$ in the direction 0, and

$$\dot{\Delta}_{12-6\zeta} = \dot{\Delta}_{\frac{\pi}{6},w(-1,2)}$$

with $12 - 6\zeta \in \Omega$ in the direction $-\pi/6 = \arg(2 - \zeta)$. For instance, the decomposition of all infinitesimal Stokes matrices corresponding to the particular determination in $]-\pi, \pi]$ of the argument, will in this way yield eighteen possible different alien derivations, two for each line $\{\arg(\lambda) = 0 \text{ mod. } \pi/3\}$:

$$\dot{\Delta}_6, \quad \dot{\Delta}_{12}, \qquad \dot{\Delta}_{6\zeta}, \quad \dot{\Delta}_{12\zeta}, \qquad \dot{\Delta}_{6\zeta-6}, \quad \dot{\Delta}_{12\zeta-12},$$

$$\dot{\Delta}_{-6}, \quad \dot{\Delta}_{-12}, \qquad \dot{\Delta}_{-6\zeta}, \quad \dot{\Delta}_{-12\zeta}, \qquad \dot{\Delta}_{6-6\zeta}, \quad \dot{\Delta}_{12-12\zeta},$$

and one for each line $\{\arg(\lambda) = -\pi/6 \bmod. \pi/3\}$:

$$\dot{\Delta}_{12-6\zeta}, \quad \dot{\Delta}_{6\zeta+6}, \qquad \dot{\Delta}_{12\zeta-6}, \quad \dot{\Delta}_{6\zeta-12}, \qquad \dot{\Delta}_{-6\zeta-6}, \quad \dot{\Delta}_{6-12\zeta}.$$

obtained after conjugating the first three by the formal monodromy. Iterating this conjugation, we get in the generic case further alien derivations (infinitely many if $\hat{M}$ is not of finite order).

Example 2.45. Consider the particular case

$$D_{7,1} = z\left(\partial + \frac{1}{2}\right) - \partial\,(\partial-1)^4\left(\partial - \frac{1}{3}\right)\left(\partial - \frac{2}{3}\right),$$

where $\rho = \eta - \alpha\beta/2 = -1$ since in this case

$$\alpha = \frac{96\mathrm{i}}{\pi^2\sqrt{\pi}}, \qquad \beta = -\frac{\pi^2\sqrt{\pi}\,\mathrm{i}}{3},$$

$$\gamma = 5\mathrm{e}^{-2\pi\mathrm{i}/3}, \qquad \delta = -5, \qquad \eta = 15, \qquad \lambda = 2.$$

By [Mi96, Theorem 7.3] we know that the differential Galois group G of this equation is rather 'large', equal to $\mathrm{SL}(7,\mathbb{C})$. As appears in the computation of this group using Ramis's theorem 2.47, this is indeed due to the additional generators listed above. Note that the formal monodromy

$$\hat{M} = \mathrm{diag}(\mathrm{e}^{-\pi\mathrm{i}}, \mathrm{e}^{2\pi\mathrm{i}/3}P)$$

is in this case of order 6, that is, we get no further alien derivations by conjugation by $\hat{M}$, than the eighteen listed above. The most generic example has $G = \mathrm{GL}(7,\mathbb{C})$.

Example 2.46. Consider a more 'degenerate' equation:

$$D_{7,1} = z\left(\partial + \frac{1}{14}\right) - \prod_{r=0}^{6}\left(\partial - \frac{r}{7}\right)$$

in the same family. In this particular case we have

$$\eta = -1, \qquad \alpha = (2\pi)^{-5/2}\mathrm{i}, \qquad \beta = -2/\alpha, \qquad c = -d = 1$$

and the adjoint action of the exponential torus provides no new generators, since in this case $\rho = 0$, hence

$$\sigma_0 = -2\alpha^{-1}E_{1,7} + \alpha E_{4,1} - E_{5,6} + E_{3,2}, \qquad \sigma_{\frac{\pi}{6}} = E_{3,7} - E_{4,6}$$

and

$$\mathrm{Ad}(T_\tau)(\sigma_0) = w(-1,1)\,\sigma_0, \qquad \mathrm{Ad}(T_\tau)(\sigma_{\frac{\pi}{6}}) = w(-1,2)\,\sigma_{\frac{\pi}{6}},$$

with similar formulas for all infinitesimal Stokes matrices. In other words, the alien derivations (more precisely their restriction to the formal solution space of $D_{7,1}$)

$$\dot{\Delta}_{12}, \quad \dot{\Delta}_{12\zeta}, \quad \dot{\Delta}_{12\zeta-12}, \quad \dot{\Delta}_{-12}, \quad \dot{\Delta}_{-12\zeta}, \quad \dot{\Delta}_{12-12\zeta}$$

(one in each direction $k\pi/3$, $k = 0,\ldots,5$) vanish here and this accounts for more symmetries on the solution space of the differential equation D_{71}, than in the previous case. In particular, one shows that the Lie algebra of the subgroup generated by the exponential torus and the infinitesimal Stokes matrices leaves some bilinear symmetric form invariant and hence is a subalgebra of $\mathfrak{so}(7,\mathbb{C})$. We know by [Mi96, Proposition 7.6] that the differential Galois group of this equation is precisely

$$G = \mathrm{G}_2 \times \mathbb{Z}/7\mathbb{Z}.$$

2.2.3.5 Ramis's theorem

The following theorem is due to Ramis (cf. [Ra85a], [Ra85b], see also [Lod94] and [PSi01]). It generalizes Schlesinger's theorem to the case of irregular singular points.

Theorem 2.47 (Ramis). *With notation as before, the local differential Galois group* $\mathrm{Gal}_0((S))$ *is the Zariski closure in* $\mathrm{GL}(p,\mathbb{C})$ *of the subgroup generated by the formal monodromy* $\hat{M}$*, the exponential torus* $\mathscr{T}$ *and the Stokes matrices* $(S_{\ell_j})_{1\le j\le N}$ *of (S) at 0.*

Ramis's original proof (cf. [Ra85a], [Ra85b]) is essentially analytic, based on the theory of multisummability, whereas the proof given in [Lod94] is specifically based on Tannakian categories and does not require any theory of summability, nor does it to define the Stokes automorphisms. Another proof is given in [PSi01] and is closer to the original proof, involving multisummability. In [IK90], a slightly different definition of the Stokes matrices also meets the conclusions of Ramis's theorem.

Exercise 2.48. Show that Theorem 2.47 holds independently of the determination of $\arg(x)$ that was fixed for each singular ray ℓ_j to define the Stokes matrix S_{ℓ_j}.

Remark 2.49. If 0 is a regular singular point, then $\hat{\phi}$ is convergent at 0 (cf. Remark 2.34 p. 40). There is no Stokes phenomenon in this case and the formal monodromy $\hat{M}$ is nothing but the usual monodromy M. The local differential Galois group in this case is the closure of the monodromy group, as in the global case over $\mathbb{C}(x)$ for regular singular systems.

Let us illustrate Ramis's theorem on a classical example.

Example 2.50. The *Airy equation* $y'' = xy$ has a single singular point, at infinity. Its differential Galois group over $\mathbb{C}(x)$ is therefore equal to its local differential Galois group at infinity, see Section 2.2.4 below. Note that it is a subgroup of $\mathrm{SL}(2,\mathbb{C})$ since there is no term y'. The Airy equation (cf. ([Wa87], p. 131) is known to have the basis of formal solutions

$$\begin{cases} \tilde{y}_1 = \frac{1}{2\sqrt{\pi}}\, x^{-\frac{1}{4}}\, \mathrm{e}^{-\frac{2}{3}x^{\frac{3}{2}}}\ \sum_{n\geq 0} (-1)^n a_n x^{-\frac{3n}{2}} \\ \tilde{y}_2 = -\frac{i}{2\sqrt{\pi}}\, x^{-\frac{1}{4}}\, \mathrm{e}^{\frac{2}{3}x^{\frac{3}{2}}}\ \sum_{n\geq 0} \quad a_n x^{-\frac{3n}{2}} \end{cases}$$

in the 'weak' form (2.5) where in particular $\tilde{y}_1$ is the asymptotic expansion of the Airy integral $Ai(x)$. We refer to Section 6.14 of this volume for a study of the Airy equation from the resurgence point of view (see also Example 9 in [Lod16]). To apply the definitions given in this section, let us perform the change of variable $x \leftarrow 1/x$ which transforms the initial equation into

$$x^5 y'' + 2x^4 y' - y = 0$$

with a single singular point at 0. The companion system of this equation has a solution of the form (2.4) (cf. discussion p. 39) with a formal, non-fractional, power series in x (cf. [Lod95], p. 128)

$$\hat{Y} = \hat{F}(x)\, x^J\, U\, \mathrm{e}^{Q(\frac{1}{t})} \tag{2.21}$$

where $t^2 = x$ and

$$J = \begin{pmatrix} \frac{1}{4} & 0 \\ 0 & \frac{3}{4} \end{pmatrix}, \quad U = \begin{pmatrix} 1 & 1 \\ 1 & -1 \end{pmatrix}, \quad Q(\frac{1}{t}) = \begin{pmatrix} -\frac{2}{3t^3} & 0 \\ 0 & \frac{2}{3t^3} \end{pmatrix},$$

$$\hat{F} = \begin{bmatrix} \sum\limits_{n\geq 0} a_{2n} x^{3n} & \sum\limits_{n\geq 0} -a_{2n+1} x^{3n+1} \\ \sum\limits_{n\geq 0} \Big((3n - \frac{1}{4})a_{2n} - a_{2n+1}\Big) x^{3n-1} & \sum\limits_{n\geq 0} \Big(a_{2n} - (3n - \frac{5}{4})a_{2n-1}\Big) x^{3(n-1)} \end{bmatrix}$$

with coefficients $a_{-1} = 0$ and

$$a_n = \frac{1}{2\pi} \Big(\frac{3}{4}\Big)^n \frac{1}{n!}\, \Gamma\Big(n + \frac{5}{6}\Big)\, \Gamma\Big(n + \frac{1}{6}\Big)$$

for $n \geq 0$. Note that J and Q in (2.21) commute. To read this solution in the concise form (2.3) p. 39, let $\hat{\phi} = \hat{F}U$ and $L = U^{-1}JU$. Then

$$\hat{Y} = \hat{\phi}(x)\, x^L\, \mathrm{e}^Q,$$

where we easily see that

$$6t^3 LQ = \begin{pmatrix} -2 & -1 \\ 1 & 2 \end{pmatrix}, \quad 6t^3 QL = \begin{pmatrix} -2 & 1 \\ -1 & 2 \end{pmatrix}$$

that is, L and Q do *not* commute any longer, as was pointed out in the discussion of Section 2.2.2 p. 38. If x is formally replaced by $x\mathrm{e}^{2\pi \mathrm{i}}$ in (2.21) the formal solution $\hat{Y}$ is changed into

$$\hat{Z} = \hat{F}\,x^J\,\mathrm{e}^{2\pi \mathrm{i} J}\,U\,\mathrm{e}^{-Q}$$

An easy calculation of the formal monodromy gives

$$\hat{M} = \hat{Y}^{-1}\hat{Z} = \begin{pmatrix} 0 & \mathrm{i} \\ \mathrm{i} & 0 \end{pmatrix}.$$

Note that the 'topological monodromy' here is trivial since on the punctured Riemann sphere $\mathbb{P}^1_{\mathbb{C}} \setminus \{0\}$ any loop enclosing 0 is homotopically trivial. But the cyclic relation (cf. Exercise 2.39 p. 49) tells us that some Stokes phenomenon is occurring, that is, 0 is an *irregular* singular point and the product of the Stokes matrices is $\hat{M}$. Fix a determination of $\arg(x)$, say $\arg(x) \in [0, 2\pi[$. With respect to $Q = \mathrm{diag}(q_1, q_2)$ the singular rays in $[0, 2\pi[$, along which

$$\mathrm{e}^{q_1 - q_2} = \mathrm{e}^{-\frac{4}{3t^3}} = \mathrm{e}^{-\frac{4}{3}x^{-\frac{3}{2}}} \quad \text{and} \quad \mathrm{e}^{q_2 - q_1} = \mathrm{e}^{\frac{4}{3t^3}} = \mathrm{e}^{\frac{4}{3}x^{-\frac{3}{2}}}$$

decrease maximally (in module) as x tends to 0, are the rays ℓ_1, ℓ_2, ℓ_3 with argument $0, \frac{2\pi}{3}, \frac{4\pi}{3}$ respectively. The corresponding Stokes matrices are of the form

$$S_{\ell_1} = \begin{pmatrix} 1 & a \\ 0 & 1 \end{pmatrix}, \quad S_{\ell_2} = \begin{pmatrix} 1 & 0 \\ b & 1 \end{pmatrix}, \quad S_{\ell_3} = \begin{pmatrix} 1 & c \\ 0 & 1 \end{pmatrix}$$

with $abc \neq 0$. This follows from the proof of Proposition 2.37 p. 46 since, with notation from this proof, it is easy to show that $q_1 \prec q_2$ on $\tilde{\ell}_1$ and $\tilde{\ell}_3$, and $q_2 \prec q_1$ on $\tilde{\ell}_2$, which implies that the Stokes matrices have the announced form. Moreover, it is easy to deduce from this form of the Stokes matrices and their cyclic relation

$$\hat{M}\,S_{\ell_3}\,S_{\ell_2}\,S_{\ell_1} = M$$

with $\hat{M} = \begin{pmatrix} 0 & \mathrm{i} \\ \mathrm{i} & 0 \end{pmatrix}$ and $M = I$, that all three S_{ℓ_j} are $\neq I$, that is, $abc \neq 0$. (We actually have $a, b, c = -\mathrm{i}$, see [Lod16, Exercise 9] and footnote p. 45, but these precise values are not needed here.)

The *exponential torus* is easily seen to be

$$\mathscr{T} = \left\{ \begin{pmatrix} \lambda & 0 \\ 0 & \lambda^{-1} \end{pmatrix}, \lambda \in \mathbb{C}^* \right\}.$$

There are four types of subgroups of $\mathrm{SL}(2, \mathbb{C})$ (cf. [Kov86])[5]:

(a) reducible subgroups (there is a line of $\mathbb{C}^2$ invariant by the elements of the group),

(b) irreducible, imprimitive groups (there is no invariant line, but a pair of two lines that are exchanged by the elements of the group),

[5] In [Kov86], Kovacic's algorithm for solving second order linear differential equations was based on this classification of the subgroups of $\mathrm{SL}(2, \mathbb{C})$. See also [DL92] for a reworking of the algorithm.

(c) finite subgroups,
(d) $\mathrm{SL}(2,\mathbb{C})$.

It is easy, from the form of $\hat{M}$ and the Stokes matrices, to exclude cases (a) and (b), as well as case (c) since $\mathscr{T}$ is infinite. The differential Galois group over $\mathbb{C}(x)$ of the Airy equation is therefore $\mathrm{SL}(2,\mathbb{C})$.

Remark 2.51. For a long time, the Airy equation was the only known example whose differential Galois group had been explicitly calculated (cf. [Kap57]). The Airy equation is a perfect textbook example which appears in many areas of mathematics and physics. In [Au01, p. 142] for instance, it is used to illustrate the Morales-Ramis theorem (cf. [MoRa01]) for integrable systems. This theorem states that if a Hamiltonian system is integrable (meaning, roughy speaking, that it has 'enough' first integrals) then the differential Galois group of its first variational equation (which is a linear differential system) has an abelian identity component. The fact that the Airy equation has a non-abelian (connected) differential Galois group, namely $\mathrm{SL}(2,\mathbb{C})$, was used to show that some Hénon-Heiles Hamiltonian systems, appearing in cosmology, are not integrable since their first normal variational equation leads to the Airy equation (cf. [Au01]).

In the general case, it is important to note that the Stokes phenomenon actually appears with the factor $\hat{\phi}$ of the formal solution (2.3). With notation of Section 2.2.2, consider the normal form (S_0) of the system (S), that is, the linear differential system satisfied by the fundamental solution

$$\hat{Y}_0 = x^L \mathrm{e}^{Q(1/t)}.$$

Since $\hat{\phi} = I$, there is no problem of summation here, and hence no Stokes matrices. By Ramis's theorem, the differential Galois group of (S_0), both over $\mathbb{C}(\{x\})$ and $\mathbb{C}((x))$, is topologically generated (i.e. the Zariski closure of the subgroup generated) by the exponential torus and the formal monodromy, which are the same for (S_0) and (S). The differential Galois group $\hat{G}$ of (S_0) over $\mathbb{C}(\{x\})$ thus appears as a subgroup of the differential Galois group G of (S) over $\mathbb{C}(\{x\})$.

2.2.4 The global differential Galois group

The differential Galois group over $\mathbb{C}(x)$ is related to the local differential Galois groups in a natural way. To see this, consider a system (S) with coefficients in $\mathbb{C}(x)$. Let Σ be the set of its singular points and $\mathscr{M}_a$, for $a \in \Sigma$, the field of germs of meromorphic functions at a.

Proposition 2.52. *The global differential Galois group* $\mathrm{Gal}_{\mathbb{C}(x)}((S))$ *is, in a sense to be defined, topologically generated by the local differential Galois groups* $\mathrm{Gal}_{\mathscr{M}_a}((S))$ *where* $a \in \Sigma$.

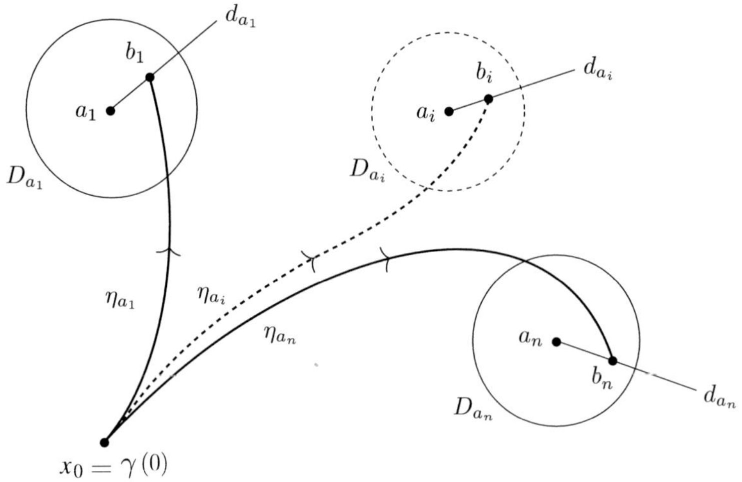

Fig. 2.1: Relation between local and global differential Galois groups

Proof. Consider a germ of holomorphic fundamental solution Z of (S) in the neighborhood of a fixed non-singular point $x_0 \in U_\Sigma$, and for each $a \in \Sigma$ fix moreover (see Fig. 2.1)

(a) an open disk D_a with center a, such that $D_a \cap \Sigma = \{a\}$
(b) some $b \in D_a$, $b \neq a$, on a ray d_a from a in D_a
(c) some path η_a from x_0 to b in U_Σ
(d) the open sector $S_a = \{x \in D_a, x \neq a \mid \arg(x-a) - \arg(d_a) \in [0, 2\pi[\}$.

With these data, we can identify each $\mathrm{Gal}_{\mathcal{M}_a}((S))$, $a \in \Sigma$, with a subgroup of $G = \mathrm{Gal}_{\mathbb{C}(x)}((S))$ in the following way. The analytic continuation Z^{η_a} of Z along η_a yields a fundamental solution of (S) in S_a, with respect to which we define $\mathrm{Gal}_{\mathcal{M}_a}((S))$. Then we construct a one-to-one map

$$\phi_a : \mathrm{Gal}_{\mathcal{M}_a}((S)) \to G$$

as follows. For any $\sigma \in \mathrm{Gal}_{\mathcal{M}_a}((S))$, we define $\phi_a(\sigma)$ by

$$\phi_a(\sigma)(Z) = \left(\sigma\left(Z^{\eta_a}\right)\right)^{\eta_a^{-1}}$$

(we 'conjugate' σ by forth and back analytic continuation along η_a). Let

$$G_a = \phi_a\Big(\mathrm{Gal}_{\mathcal{M}_a}((S))\Big) \subset G.$$

With this definition, an element f of the Picard-Vessiot extension $\mathbb{C}(x)(Z)$ of $\mathbb{C}(x)$ is invariant by G_a if and only if $f^{\eta_a} \in \mathcal{M}_a(Z^{\eta_a})$ is invariant by $\mathrm{Gal}_{\mathcal{M}_a}((S))$, that is, if $f^{\eta_a} \in \mathcal{M}_a = \mathbb{C}(\{x-a\})$ by Theorem 2.15 p. 31. Let H be the subgroup of G

generated by all G_a, $a \in \Sigma$. An element $f \in \mathbb{C}(x)(Z)$ is invariant by H if and only it is invariant by each G_a, that is, if for each $a \in \Sigma$ the function f^{η_a} is meromorphic at a, which by Theorem 1.2 p. 5 is equivalent to $f \in \mathbb{C}(x)$. We have thus proved that G is the Zariski closure of H, by Proposition 2.17 p. 31. □

In this section we have presented various tools to compute the monodromy group or the differential Galois group, although such computation, referred to as the *direct problem*, is in general difficult, even with computer algebra. The main problem is actually to find *connection formulas* relating local solutions at different singular points, and this is a highly transcendental problem.

Let us mention here an important result of model theory by Hrushovski [Hru02] who gave a general algorithm for computing the differential Galois group over any field $C(x)$ of rational functions. This algorithm has recently been reworked in [Fe15] and [Re14].

Note that in particular cases over $\mathbb{C}(x)$, for instance linear differential equations with two singular points, one regular singular and one irregular, also named *Hamburger equations*, the global differential Galois group coincides with the local group at the irregular singular point, and Ramis's theorem 2.47 applies. It was for instance applied to generalized confluent hypergeometric equations in [Mi91], [Mi96] and [DM89], where the existence of convenient integral formulas for the solutions led to an explicit calculation of the Stokes matrices, like those we considered in Examples 2.43 to 2.46 p. 53. Related to Example 2.46 via the change of variable $x = (t/7)^7$, the differential equation

$$7y^{(7)} + ty' + y = 0$$

is for instance shown, in this way, to have the differential Galois group G_2. Note that (the Lie algebra of) the differential Galois group of all generalized confluent hypergeometric equations had been originally determined in [Kat87] by purely algebraic methods using D-modules.

For references and results about the direct problem, see [Sin99], [PSi01], [Sin09] and bibliography therein.

As was mentionned after Example 2.50 p. 58, the computation of the differential Galois group, at least of its identity component, is used to determine whether Hamiltonian systems are (non-) integrable, using the Morales-Ramis theorem (cf. [MoRa01], [Au01]). In [CG15] for example, it was recently shown that some higher Painlevé equations are non-integrable, by reducing the problem to computing the differential Galois group of some confluent generalized hypergeometric equations. In [CG15(2)] the same authors gave criteria of integrability for the stationary solutions in Bose–Fermi mixtures, using a similar method.

To end this chapter, let us mention analogues of the Picard-Vessiot theory for other types of equations. The first extension was to difference equations in [PSi97], then more specifically to q-difference equations in [Saul03], [RaSa15]. A more general theory for difference-differential equations was developed in [HS08], then in [DHW14]. Let us also mention a Hopf algebra approach of differential Galois theory in [AMT09]. For parameterized families of linear differential equations a specific parameterized Picard-Vessiot theory was established in [CaSi07]. This theory and its applications are naturally related to monodromy, in particular to isomonodromy and monodromy evolving deformations (cf. [MS12], [MS13], [Mi15]). Stronger conditions on the field of constants (of the base-field) are required here, depending for instance on the number of parameters. In [CaSi07], the field of constants is required to be 'differentially closed', whereas in [GGO11] it is assumed to be 'existentially closed' and in [Wi12] it needs only be algebraically closed in the particular case of a single parameter.

2.3 Solutions to exercises of Chapter 2

Exercise 2.3 p. 26 It follows from $t^n = x$ that the unique derivation ∂ extending the derivation d/dx on the field extension must in each case satisfy

$$\partial(t) = \frac{1}{nt^{n-1}}$$

which completely determines ∂ by the usual derivation rules.

Exercise 2.7 p. 27 The differential field $L = \mathbb{C}(x, e^{\alpha_1 x}, \ldots, e^{\alpha_m x})$ is a Picard-Vessiot extension

(a) of $\mathbb{C}(x)$ for the linear differential system $\partial y = Ay$ where

$$A = \mathrm{diag}(\alpha_1, \ldots, \alpha_m);$$

a fundamental solution of this sytem is

$$Y_1 = \mathrm{diag}(\mathrm{e}^{\alpha_1 x}, \ldots, \mathrm{e}^{\alpha_m x}).$$

(b) of $\mathbb{C}$ for the differential system $\partial y = By$ where

$$B = \begin{pmatrix} C & 0 \\ 0 & A \end{pmatrix}$$

with $C = \begin{pmatrix} 0 & 1 \\ 0 & 0 \end{pmatrix}$ and A as in (a); A fundamental solution of this system is

$$Y_3 = \begin{pmatrix} Y_2 & 0 \\ 0 & Y_1 \end{pmatrix}$$

with $Y_2 = \begin{pmatrix} 1 & x \\ 0 & 1 \end{pmatrix}$ and Y_1 as in (a).

This example in particular illustrates the fact that if a given differential field extension is a Picard-Vessiot extension, it can be viewed as such in many ways. Here, for instance, the order of the system depends on the number of exponentials $e^{\alpha_i x}$ generating the field extension together with x, and there is an infinite choice of such generators, and of their number m.

Exercise 2.8 p. 27 L-linear independence clearly implies C-linear independence. The converse is proved by induction on the number ℓ of solutions. It holds for $\ell = 1$, since linear independence (over any field) is here equivalent to $y_1 \neq 0$. Assume $\ell > 1$ and the result holds for any $\ell - 1$ solutions. Let $y_1, \ldots, y_\ell$ be C-linearly independent solutions and assume

$$\sum_{i=1}^{\ell} \lambda_i y_i = 0$$

for some $\lambda_i \in L$, $i = 1, \ldots, \ell$. If $\lambda_i = 0$ for some i, then the solutions y_j for $j \neq i$ are C-linearly, hence L-linearly independent by assumption, which implies that $\lambda_j = 0$ for all $j \neq i$, hence for all j. If all $\lambda_i \neq 0$, we may assume $\lambda_1 = 1$, hence

$$y_1 + \lambda_2 y_2 + \ldots + \lambda_\ell y_\ell = 0.$$

Applying the derivation ∂ we get

$$A\,(y_1 + \lambda_2 y_2 + \ldots \lambda_\ell y_\ell) + \partial(\lambda_2)\,y_2 + \ldots + \partial(\lambda_\ell)\,y_\ell = 0,$$

hence

$$\partial(\lambda_2)\,y_2 + \ldots + \partial(\lambda_\ell)\,y_\ell = 0.$$

Since $y_2, \ldots, y_\ell$ are C-linearly, hence L-linearly independent by assumption, this implies that $\partial(\lambda_2) = \ldots = \partial(\lambda_\ell) = 0$, that is, all $\lambda_i \in C$, which contradicts the C-linear independence of $y_1, \ldots, y_\ell$ since $\lambda_1 = 1$.

Exercise 2.11 p. 30 The differential ideal J is defined by $J = M\,L[Y_{ij}, -]$ where L is the fraction-field of R/M, R is the differential field $k[Y_{ij}, -]$ and M a given maximal differential ideal of R. New indeterminates $Z = (Z_{ij})$ are defined by $Y = yZ$, where y is the image of $Y = (Y_{ij})$ in $R/M \subset L$. Let

$$I = J \cap C[Z_{ij}, -]$$

in $L[Y_{ij}, -] \cong L[Z_{ij}, -]$. Consider the differential rings $\mathscr{A} = C[Z_{ij}, -]$ and $\mathscr{A}_L = \mathscr{A} \otimes_C L = L[Z_{ij}, -]$. Note that $\partial(Z) = 0$ since $\partial(Y) = AY$ and $\partial y = Ay$. Any basis

$$\mathscr{E} = \{e_\beta\}_{\beta \in \mathscr{B}}$$

of the C-vector space $\mathscr{A}$ is a basis of the L-vector space $\mathscr{A}_L$ and $\partial(e_\beta) = 0$ for all $\beta \in \mathscr{B}$ since $\partial Z = 0$. Let $u = \sum \lambda_\beta\, e_\beta$, with $\lambda_\beta \in L$ (all $\lambda_\beta = 0$ but a finite number) be a non-zero element of $J \subset \mathscr{A}_L$. Let us prove that $u \in I\mathscr{A}_L$ by induction on the number $l(u)$ of non-zero coefficients λ_β. If $l(u) = 1$ then $u = \lambda e_{\beta_0}$ with $\lambda \neq 0$, and $u \in J$ if and only if $e_{\beta_0} \in J$, hence $u \in I\mathscr{A}_L$. Assume $l(u) > 1$. We may assume that $\lambda_{\beta_1} = 1$ for some β_1 and that $\lambda_{\beta_2} \in L \setminus C$ for some β_2 (note that if all $\lambda_\beta \in C$, then

$u \in C[Z_{ij}, -]$ as announced). Since $\lambda_{\beta_1} = 1$ we have $l(\partial u) < l(u)$ and $\partial u \in J$ since J is a differential ideal. The same holds for the element $\lambda_{\beta_2}^{-1} u$. We have

$$\partial(\lambda_{\beta_2}^{-1} u) = \partial(\lambda_{\beta_2}^{-1}) u + \lambda_{\beta_2}^{-1} \partial u$$

By the induction hypothesis, the elements $\partial(\lambda_{\beta_2}^{-1} u)$ and $\lambda_{\beta_2}^{-1} \partial u$, which have length less than $l(u)$ both belong to $I\mathscr{A}_L$, an so does $\partial(\lambda_{\beta_2}^{-1})u$. We have $\partial(\lambda_{\beta_2}^{-1}) \neq 0$ since $\lambda_{\beta_2} \notin C$, hence $u \in C[Z_{ij}, -]$, which ends the proof.

Exercise 2.13 p. 30 If $\{\beta_1, \ldots, \beta_r\}$ denotes a basis of the $\mathbb{Z}$-module

$$\sum_{i=1}^{m} \mathbb{Z}\alpha_i$$

then

$$K = \mathbb{C}(x, e^{\alpha_1 x}, \ldots, e^{\alpha_m x}) = \mathbb{C}(x, e^{\beta_1 x}, \ldots, e^{\beta_r x})$$

and K is (for instance) a Picard-Vessiot extension of $\mathbb{C}$ for the differential system

$$\partial Y = AY = \begin{pmatrix} A_1 & 1 \\ 0 & A_2 \end{pmatrix}$$

where

$$A_1 = \begin{pmatrix} 0 & 1 \\ 0 & 0 \end{pmatrix}$$

and $A_2 = \mathrm{diag}(\beta_1, \ldots, \beta_r)$. The block-diagonal form of A implies that the differential Galois group of K over $\mathbb{C}$ is the direct product of the differential Galois groups over $\mathbb{C}$ of $\partial Y = A_1 Y$ and $\partial Y = A_2 Y$ respectively. Since by assumption the elements $e^{\beta_1 x}, \ldots, e^{\beta_r x}$ are algebraically independent over $\mathbb{C}$ (see Exercise 2.14 below) the only $\mathbb{C}$-differential automorphisms of $\mathbb{C}(e^{\beta_1 x}, \ldots, e^{\beta_r x})$, which is the Picard-Vessiot extension of $\mathbb{C}$ for $\partial Y = A_2 Y$, are those ϕ given by

$$\phi(e^{\beta_i x}) = \lambda_i e^{\beta_i x},\ i = 1, \ldots, r$$

for arbitraty $\lambda_i \in \mathbb{C}^*$. The differential Galois group of $\partial Y = A_2 Y$ is therefore a torus $(\mathbb{C}^*)^r$. It is easy to see that the only differential $\mathbb{C}$-automorphisms of the Picard-Vessiot extension $\mathbb{C}(x)$ of $\mathbb{C}$ for $\partial Y = A_1 Y$ are of the form $x \mapsto x + \lambda$, for arbitrary $\lambda \in \mathbb{C}$ (equivalently, given by matrices $\begin{pmatrix} 0 & \lambda \\ 0 & 0 \end{pmatrix}$ with respect to the fundamental solution $\begin{pmatrix} 1 & x \\ 0 & 1 \end{pmatrix}$ of the system $\partial Y = A_1 Y$). This shows that the differential Galois group of K has the expected form $\mathbb{C} \times (\mathbb{C}^*)^r$.

Note that the differential Galois group of a Picard-Vessiot extension only depends on the differential fields in the extension, not on the specific Picard-Vessiot structure of this extension, which is not unique as seen in Exercise 2.13.

Exercise 2.14 p. 30 Assume $\beta_1,\ldots,\beta_r$ are $\mathbb{Z}$-linearly independent complex numbers. For a monomial $m=\prod_{j=1}^{r} X_j^{\tau_j}$ in r variables, write $\ell_m=\sum_{j=1}^{r}\tau_j X_j$. The map

$$\ell : m \mapsto \ell_m$$

from the monoid of monic monomials (including the monomial 1) to the additive group of polynomials of degree 1 in r variables, is a one-to-one morphism. In particular,

$$m(\mathrm{e}^{\beta_1 x},\ldots,\mathrm{e}^{\beta_r x}) = \mathrm{e}^{\ell_m(\beta_1,\ldots,\beta_r)x}$$

for any monic monomial m. Assume that $\mathrm{e}^{\beta_1 x},\ldots,\mathrm{e}^{\beta_r x}$ are algebraically dependent over $\mathbb{C}$. Then

$$P(\mathrm{e}^{\beta_1 x},\ldots,\mathrm{e}^{\beta_r x}) = 0.$$

for some non-trivial polynomial

$$P=\sum_{k=1}^{s}\lambda_k m_k$$

in r variables, where $\lambda_k \in \mathbb{C}^*$ and m_k are s distinct monic monomials, $s>1$. Differentiating this relation $s-1$ times with respect to x yields a linear system satisfied by the non-trivial s-tuple

$$\left(\lambda_1\,\mathrm{e}^{\ell_{m_1}(\beta_1,\ldots,\beta_r)x},\ldots,\lambda_s\,\mathrm{e}^{\ell_{m_s}(\beta_1,\ldots,\beta_r)x}\right).$$

The determinant of this system, which is a Vandermonde determinant, is equal to

$$\prod_{1\le i<i\le s}(\gamma_i-\gamma_j)$$

with $\gamma_k=\ell_{m_k}(\beta_1,\ldots,\beta_r)$ for $k=1,\ldots,s$. Since the system has a non-trivial solution, this determinant is equal to zero, that is,

$$(\ell_{m_i}-\ell_{m_j})(\beta_1,\ldots,\beta_r)=0 \tag{2.22}$$

for some $1\le i<j\le s$. As the map ℓ is one-to-one, (2.22) is a non-trivial linear relation over $\mathbb{Z}$, which contradicts the linear independence of $\beta_1,\ldots,\beta_r$.

Exercise 2.19 p. 32 Let $\partial y = A_1 y$ and $\partial y = A_2 y$ be two k-equivalent differential systems (of order n) and Y_1 some fundamental solution of $\partial y = A_1 y$. Then, by definition of k-equivalence, PY_1 is a fundamental solution of $\partial y = A_2 y$ for some invertible matrix $P\in \mathrm{GL}(n,k)$. The Picard-Vessiot extensions $k(Y_1)$ and $k(PY_1)$ of k for the systems of $\partial y = A_1 y$ and $\partial y = A_2 y$ respectively are equal since the entries of P belong to k.

Exercise 2.22 p. 32 The differential field $K=\mathbb{C}(x,\mathrm{e}^{x^2})$ is a Picard-Vessiot extension of $\mathbb{C}(x)$ for the scalar linear differential equation $y''-2xy=0$. Let us show that it is not a Picard-Vessiot extension of $\mathbb{C}$. Given a scalar linear differential equation with constant coefficients

$$\sum_{i=0}^{n} a_i y^{(i)} = 0, \quad a_i \in \mathbb{C} \tag{E}$$

we know that there are constants $\alpha_1, \ldots, \alpha_m \in \mathbb{C}$ such that any solution of (E) is of the form

$$y = \sum_{j=0}^{m} P_j \, \mathrm{e}^{\alpha_j x}$$

where the P_j are polynomials in x. Let $\{y_1, \ldots, y_n\}$ be a basis of the solution space of E. The Picard-Vessiot extension of $\mathbb{C}$ for (E) is generated by a fundamental solution of the companion system, that is, by the y_i and their successive ℓ-th derivatives, for any $\ell \in \mathbb{N}$. Using some growth argument, one easily sees that e^{x^2} cannot be written as a rational function of the ℓ-th derivatives $y_i^{(\ell)}$, $\ell \in \mathbb{N}$.

Exercise 2.35 p. 41 An easy calculation using Exercise 2.3 shows that

$$q_i^*(t) = -\frac{1}{\nu x t} q_i'(\frac{1}{t})$$

for each $i = 1, \ldots, p$, where q_i' is the formal derivative of the polynomial q_i.

Exercise 2.36 p. 43 Consider as in (2.3) p. 39 a formal fundamental solution

$$\hat{Y} = \hat{\phi}(x) x^L \mathrm{e}^{Q(\frac{1}{t})}$$

with $Q = \mathrm{diag}(q_1, \ldots, q_p)$, $q_i \in \frac{1}{t}\mathbb{C}[\frac{1}{t}]$. Choose a basis $\{p_1, \ldots, p_r\}$ of the $\mathbb{Z}$-module $\mathbb{Z}q_1 + \ldots + \mathbb{Z}q_p$ and let $\tau \in \mathscr{T}$ be an element of the exponential torus, defined by $\tau(\mathrm{e}^{p_j}) = \tau_j \mathrm{e}^{p_j}$ for $j = 1, \ldots, r$, with $\tau_j \in \mathbb{C}^*$. Each e^{q_i} can be written in a unique way as

$$\mathrm{e}^{q_i} = m_i\,(\mathrm{e}^{p_1}, \ldots, \mathrm{e}^{p_r})$$

where m_i is a monomial $m_i(X_1, \ldots, X_r) = X_1^{\alpha_{i,1}} \ldots X_r^{\alpha_{i,r}}$. The automorphism τ preserves $\hat{\phi}$ and x^L, and

$$\tau(\mathrm{e}^{q_i}) = m_i\,(\tau_1, \ldots, \tau_r)\mathrm{e}^{q_i}$$

for each $i = 1, \ldots, p$. Let $\tau(\hat{Y})$ be the formal solution obained by applying τ to the entries of $\hat{Y}$. The matrix T_τ of τ with respect to $\hat{Y}$, given by $\tau(\hat{Y}) = \hat{Y}\, T_\tau$ is

$$T_\tau = \mathrm{diag}\,(\nu_1, \ldots, \nu_p),$$

where $\nu_i = m_i\,(\tau_1, \ldots, \tau_r) = \tau_1^{\alpha_{i,1}} \ldots \tau_r^{\alpha_{i,r}}$ for $i = 1, \ldots, p$. In Section 2.2.3.2 we have seen that, conversely, any $\underline{\tau} = (\tau_1, \ldots, \tau_r) \in \mathbb{C}^{*r}$ defines an element τ of the exponential torus by

$$\tau(\mathrm{e}^{p_j}) = \tau_j \, \mathrm{e}^{p_j}, \quad j = 1, \ldots, r.$$

This shows that the exponential torus is

$$\mathscr{T} = \left\{\mathrm{diag}\,((m_1(\underline{\tau}), \ldots, m_p(\underline{\tau})), \quad \underline{\tau} \in \mathbb{C}^{*r}\right\}$$

and that it is isomorphic to $(\mathbb{C}^*)^r$ via

$$\underline{\tau} \mapsto (m_1(\underline{\tau}), \ldots, m_p(\underline{\tau}).$$

Exercise 2.39 p. 49 We keep notation and terminology from Section 2.2.3, in particular from the proof of Proposition 2.37 p. 46. Let us fix a determination $0 \leq \arg(x) < 2\pi$ of the argument and let $\ell_1, \ldots, \ell_N$, with

$$0 < \arg(\ell_1) < \ldots < \arg(\ell_N) < 2\pi$$

be the singular rays of (S) with respect to a given formal solution

$$\hat{Y} = \hat{\phi}(x) x^L \mathrm{e}^{Q(\frac{1}{t})}$$

(we may for simplicity assume that $\mathbb{R}_+$ is not a singular ray). Fix a ray λ such that $0 < \arg(\lambda) < \arg(\ell_1)$ and let $Y = \phi_\lambda x^L \mathrm{e}^Q$ be the solution of (S), also called the *sum* of $\hat{Y}$ on λ for this determination of $\arg(x)$ and given meaning of e^Q, such that ϕ_λ is the *sum* of $\hat{\phi}$ with respect to λ, as explained in Section 2.2.3 and in the proof of Proposition 2.37 p. 46. Let us analytically continue Y along an elementary loop γ around 0. Let $c_{\ell_1}(Y)$ denote the analytic continuation of Y (along γ) across the first singular line ℓ_1, but still in $\arg(x) < \arg(\ell_2)$. By definition of the Stokes matrices (cf. (2.8) p. 45), the sum Y of $\hat{Y}$ is continued across ℓ_1 as

$$c_{\ell_1}(Y) = Y_{\ell_1}^+ S_{\ell_1}$$

where $Y_{\ell_1}^+$ is the sum of $\hat{Y}$ along any l such that $\arg(\ell_1) < \arg(l) < \arg(\ell_2)$. In other words, the sum of $\hat{Y}$ is now

$$c_{\ell_1}(Y) S_{\ell_1}^{-1}$$

As we cross ℓ_2, $\hat{Y}$ will similarly, and by $\mathbb{C}$-linearity, have along any ray l such that $\arg(\ell_2) < \arg(l) < \arg(\ell_3)$ the sum

$$c_{\ell_2}\left(c_{\ell_1}(Y)\right) S_{\ell_1}^{-1} S_{\ell_2}^{-1}$$

where $c_{\ell_2}\left(c_{\ell_1}(Y)\right)$ is the resulting analytic continuation of Y across ℓ_1 and ℓ_2 successively, and so on. Once we have crossed the last singular ray ℓ_N in $\{\arg(x) \in]0, 2\pi[\}$, $\hat{Y}$ has along any l such that $\arg(\ell_N) < \arg(l) < 2\pi$ the sum

$$Z = c_{\ell_N}\left(\ldots(c_{\ell_1}(Y)\right) S_{\ell_1}^{-1} \ldots S_{\ell_N}^{-1}.$$

As we cross $\mathbb{R}_+$ though, the determination of $\arg(x)$ changes: $\hat{Y}$ is changed into $\hat{Y}\hat{M}$ (see Section 2.2.3.1 p. 41). On the other hand, back to λ after analytic continuation along the full loop γ, the continuation of Y is

$$c_{\ell_N}\left(\ldots(c_{\ell_1}(Y)\right) = YM$$

where M is the topological monodromy. This implies that the continuation $c_{\mathbb{R}_+}(Z)$ of Z across $\mathbb{R}_+$ is now asymptotic to $\hat{Y}\hat{M}$, that is, $\hat{Y}$ has along λ the sum

$$c_{\mathbb{R}_+}(Z)\hat{M}^{-1} = YMS_{\ell_1}^{-1}\dots S_{\ell_N}^{-1}\hat{M}^{-1}.$$

By the uniqueness of the sum of $\hat{Y}$ along λ, this gives the announced relation

$$\hat{M}S_{\ell_N}\dots S_{\ell_1} = M.$$

Exercise 2.40 p. 49 Fix a formal fundamental solution of the form (2.3) p. 39

$$\hat{Y} = \hat{\phi}\hat{Y}_0.$$

Let $\tilde{\ell}$ for a given determination of $\arg(x)$, and $\tilde{\ell}'$ for another determination of $\arg(x)$ both represent the singular line ℓ, with $\arg(\tilde{\ell}') = \arg(\tilde{\ell}) + 2\pi$. In the definition of the Stokes matrices, the formal solution $\hat{Y}_0$ of the normal form (S_0) is given an analytic meaning as functions $Y_{0,\tilde{\ell}}$ and $Y_{0,\tilde{\ell}'}$ respectively, for these determinations of $\arg(x)$. If $\arg(\tilde{\ell}') = \arg(\tilde{\ell}) + 2\pi$ it is easy to see that $Y_{0,\tilde{\ell}'} = Y_{0,\tilde{\ell}}\hat{M}$ where $\hat{M}$ is the formal monodromy with respect to this formal solution, hence

$$k(\phi_l^-,\phi_l^+,Y_{0,\tilde{\ell}}) = k(\phi_l^-,\phi_l^+,Y_{0,\tilde{\ell}'})$$

and we have

$$u_{\tilde{\ell}'}{}^+(\hat{\phi}\hat{Y}_0) = \phi^+Y_{0,\tilde{\ell}'} = \phi^+Y_{0,\tilde{\ell}}\hat{M} = u_{\tilde{\ell}}{}^+(\hat{\phi}\hat{Y}_0)\hat{M},$$

$$u_{\tilde{\ell}'}{}^-(\hat{\phi}\hat{Y}_0) = \phi^-Y_{0,\tilde{\ell}'} = \phi^-Y_{0,\tilde{\ell}}\hat{M} = u_{\tilde{\ell}}{}^-(\hat{\phi}\hat{Y}_0)\hat{M}.$$

The Stokes matrices are defined by

$$(\overline{u}_{\tilde{\ell}}{}^+)^{-1}\,\overline{u}_{\tilde{\ell}}{}^-(\hat{\phi}Y_{0,\tilde{\ell}}) = \hat{\phi}Y_{0,\tilde{\ell}}S_{\tilde{\ell}}$$

$$(\overline{u}_{\tilde{\ell}'}{}^+)^{-1}\,\overline{u}_{\tilde{\ell}'}{}^-(\hat{\phi}Y_{0,\tilde{\ell}'}) = \hat{\phi}Y_{0,\tilde{\ell}'}S_{\tilde{\ell}'}$$

from which we easily conclude that

$$S_{\tilde{\ell}'} = \hat{M}S_{\tilde{\ell}}\hat{M}^{-1}.$$

Exercise 2.41 p. 50 Assume that the weight spaces V_χ are linearly dependent and choose $n \in \mathbb{N}$, necessarily > 1, minimal such that

$$v_1 + \dots + v_n = 0$$

where each $v_i \in V_{\chi_i}$ is a weight vector and $\chi_1,\dots,\chi_n$ are pairwise distinct weights. Since $\chi_1 \neq \chi_2$ we can choose $\gamma \in \Gamma$ such that $\chi_1(\gamma) \neq \chi_2(\gamma)$. The action of γ on V gives

$$\chi_1(\gamma)v_1 + \ldots + \chi_n(\gamma)v_n = \alpha(\gamma)(0) = 0.$$

If we substract $\chi_1(\gamma)^{-1}$ times this equality from the first we get an non-trivial relation

$$(\chi_2(\gamma)(\chi_1(\gamma))^{-1} - 1)v_2 + (\ldots + \chi_n(\gamma)(\chi_1(\gamma))^{-1} - 1)v_n = 0$$

with strictly less terms than n, which contradicts the minimality of n.

Exercise 2.48 p. 58 In Exercise 2.40 we proved that for a given singular line ℓ, if we change the determination of $\arg(x)$ and consider the corresponding lines $\widetilde{\ell}$ and $\widetilde{\ell}'$ representing ℓ, the Stokes matrices $S_{\widetilde{\ell}}$ and $S_{\widetilde{\ell}'}$ are conjugate of each other by some power of the formal monodromy $\hat{M}$. The conclusion of the theorem therefore holds independently of the argument attributed to each singular line.

References

AMT09. Amano, K., Masuoka, A., Takeuchi, M.: Hopf algebraic approach to Picard-Vessiot theory. In: Handbook of Algebra vol. 6, 127-171, Elsevier/North Holland (2009)

Au01. Audin, M.: Les systèmes hamiltoniens et leur intégrabilité. Cours spécialisés, Société Mathématique de France et EDP-Sciences (2001)

Ba94. Balser, W.: From Divergent Power Series to Analytic Functions. Lect. Notes Math. vol. 1582, Springer (1994)

Ba00. Balser, W.: Formal Power Series and Linear Systems of Meromorphic Ordinary Diffenrential Equations. Universitext, Springer (2000)

BJL79. Balser, W., Jurkat, W. B., Lutz, D. A.: A general theory of invariants for meromorphic differential equations. Part I: Formal invariants. Part II: Proper invariants. Funcialaj. Ekvacioj **22**, 197–221, 257–283 (1979)

BBRS91. Balser, W., Braaksma, B. L. J., Ramis, J.-P., Sibuya, Y.: Multisummability of formal power series solutions of linear ordinary differential equations. Asymptotic Anal. **5-1**, 27–45 (1991)

BCL03. Barkatou, M., Chyzak, F., Loday-Richaud, M.: Remarques algorithmiques liées au rang d'un opérateur différentiel linéaire. In: Fauvet, F., Mitschi, C. (eds), From Combinatorics to Dynamical Systems, Journées de Calcul Formel, Strasbourg 2002. Irma Lectures in Mathematics and Theoretical Physics **3**, 87–129 De Gruyter (2003)

Bor91. Borel, A.: Linear Algebraic Groups. Second edition. Graduate Texts in mathematics, Springer (1991)

CaSi07. Cassidy, Ph. J., Singer, M. F.: Galois theory of parameterized differential equations and linear differential algebraic groups. In: Bertrand et al. (eds): Differential equations and quantum groups, IRMA Lect. Math. Theor. Phys. vol. 9, EMS (2007)

CG15. Christov, O., Georgiev, G.: Non-integrability of some higher-order Painlevé equations in the sense of Liouville. SIGMA **11** (2015). ArXiv.1412.2367

CG15(2). Christov, O., Georgiev, G.: On the integrability of a system describing the stationary solutions in BoseFermi mixtures. ArXiv.1503.08171 (2015)

Cop34. Cope, F.: Formal Solutions of irregular linear differential equations I. Am. J. Math. **56**, 411–437 (1934)

Cop36. Cope, F.: Formal Solutions of irregular linear differential equations II. Am. J. Math. **58**, 130–140 (1936)

CH11. Crespo, T., Z. Hajto, Z.: Algebraic Groups and Differential Galois Theory. Graduate Studies in Mathematics, vol. 122, American Mathematical Society (2011)

CR62. Curtis, C., Reiner, I.: Representation Theory of Finite Groups and Associative Algebras. Wiley Interscience, New York (1962)

DHW14. Di Vizio, L., Hardouin, C., Wibmer, M.: Difference Galois theory of linear difference equations. Adv. Math. **260**, 1–58 (2014)

DL92. Duval, A., Loday-Richaud, M.: Kovacic's algorithm and its application to some families of special functions. Applicable Algebra in Engineering, Communication and Computing, **38-3**, 211–246 (1992)

DM89. Duval, A., Mitschi, C.: Matrices de Stokes et groupe de Galois des équations hypergéométriques confluentes généralisées. Pacific J. Math. **38**, 25–56 (1989)

Éc81. Écalle, J.: Les fonctions résurgentes, tome I : Les algèbres de fonctions résurgentes, vol. 81-05. Publ. Math. Orsay (1981)

Fe15. Feng, R.: Hrushovski's algorithm for computing the Galois group of a linear differential equation. Advances in Applied Mathematics **65**, 1–37 (2015)

GGO11. Gillet, H., Gorchinskyi, Ovchinnikov, A.: Parametrized Picard-Vessiot extensions and Atiyah extensions. Adv. Math. **238**, 322–411 (2013)

HS08. Hardouin, C., Singer, M. F.: Differential Galois theory of lincar difference equations. Math. Ann. **3**42, 333–377 (2008). Erratum: Math. Ann. **350**, 243–244 (2011)

Hru02. Hrushovski, E.: Computing the Galois group of a linear differential equation. Differential Galois Theory. Warszawa: Institute of Mathematics, Polish Academy of Sciences. Banach Center Publications vol. 58, 97–138 (2002)

Hu75. Humphreys, J. E.: Linear algebraic groups. Springer (1975)

IK90. Ilyashenko, Yu., A. Khovansky, A.: Galois groups, Stokes multipliers and Ramis' theorem. Funct. Anal. Appl. **24-4**, 286–296 (1990)

Ja37. Jacobson, N.: Pseudo-linear Transformations. Annals of Math. **38** (1937)

Kap57. Kaplansky, I.: An Introducation to Differential Algebra. First edition. Hermann, Paris (1957)

Kap76. Kaplansky, I.: An Introducation to Differential Algebra. Second edition. Hermann, Paris (1976)

Kat87. Katz, N.: On the calculation of some differential Galois groups. Invent. Math. **87**, 13–61 (1987)

Kat95. Katz, N.: Rigid local systems. Annals of Mathematics Studies, vol. 139 , Princeton University Press (1995)

Kol76. Kolchin, E.: Differential Algebra and Algebraic Groups. Academic Press, New York (1976)

Kov86. Kovacic, J.: An algorithm for solving second order linar homogeneous differential equations. J. Symb. Comput. **2**, 3–43 (1986)

Ku93. Kuga, M.: Galois' Dream : Group Theory and Differential Equations. Birkhäuser (1993)

Lod94. Loday-Richaud, M.: Stokes phenomenon, multisummability and differential Galois groups. Ann. Inst. Fourier **44-3**, 849–906 (1994)

Lod95. Loday-Richaud, M.: Solutions formelle des systèmes différentiels linéaires méromorphes et sommation. Expositiones Mathematicae **1**3, 116–162 (1995)

Lod01. Loday-Richaud, M.: Rank reduction, normal forms and Stokes matrices. Expositiones Mathematicae **19**, 229–250 (2001)

Lod16. Loday-Richaud, M.: Divergent series, Summability and Resurgence II. Simple and multiple summability. Lect. Notes Math., vol. 2154, Springer (2016)

LR11. Loday-Richaud, M., Remy, P.: Resurgence, Stokes phenomenon and alien derivatives for level-one linear differential systems. J. Differential Equations **250-3**, 1591–1630 (2011)

Mag94. Magid, A.: Lectures on Differential Galois Theory. Second Edition. University Lecture Series, American Mathematical Society (1994)

MR92. Malgrange, B., Ramis, J.-P.: Fonctions multisommables. Ann. Inst. Fourier **42-1-2**, 353–368 (1992)

MaRa89. Martinet, J., Ramis, J.-P.: Théorie de Galois différentielle et resommation. In: Tournier, E. (ed): Computer Algebra and Differential Equations, 117–214, Academic Press (1989)

Mi91. Mitschi, C.: Differential Galois groups and G-functions. In: Singer, M. F. (ed): Differential Equations and Computer Algebra, 149–180, Academic Press (1991)

Mi96. Mitschi, C.: Differential Galois groups of confluent generalized hypergeometric equations: an approach using Stokes multipliers. Pacific J. Math. **176-2**, 365-405 (1996)

Mi15. Mitschi, C.: Some applications of the parameterized Picard-Vessiot theory. Izvestiya: Mathematics **80-1** (2016), to appear.

MS12. Mitschi, C., Singer, M. F.: Monodromy groups of parameterized linear differential equations. Bull. London Math. Soc. **44-5**, 913–930 (2012)

MS13. Mitschi, C., Singer, M. F.: Projective isomonodromy and Galois groups Proceedings of the Am. Math. Soc. **141-2**, 605-617 (2013)

MoRa01. Morales-Ruiz, J. J., Ramis, J.-P.: Galoisian obstructions to integrability of Hamiltonian systems I & II. Methods Appl. Anal. **8-1**, 33–111 (2001)

PSi97. van der Put, M., Singer, M. F.: Galois Theory of Difference Equations. Lect. Notes Math. vol. 1666, Springer (1997)

PSi01. van der Put, M., Singer, M. F.: Galois Theory of Linear Differential Equations. Grundlehren des mathematischen Wissenschaft vol. 328, Springer (2001)

Ra85a. Ramis, J.-P.: Filtration Gevrey sur le groupe de Picard-Vessiot d'une équation différentielle irrégulière. Informes de Matematica. Preprint IMPA Series A-045/85 (1985)

Ra85b. Ramis, J.-P.: Phénomène de Stokes et filtration Gevrey sur le groupe de Picard-Vessiot. C. R. Acad. Sci. Paris **301**, 165–167 (1985)

Ra88. Ramis, J.-P.: Irregular connections, savage π_1 and confluence. Proceedings of a Conference at Katata, Japan, 1987. Taniguchi Foundation (1988)

RaSa15. Ramis, J.-P., Sauloy, J.: The q-analogue of the wild fundamental group and the inverse problem of the Galois thoery of q-difference equations Ann. Sci. Éc. Norm. Supér. (4) **48-1**, 171–226 (2015)

Re14. Rettstadt, D.: On the computation of the differential Galois group. Ph.D. thesis, RWTH Aachen University (2014)

Sab10. Sabbah, C.: Équations Différentielles Linéaires Algébriques. Cours M2, Strasbourg (2010). Available at: http://www.math.polytechnique.fr/cmat/sabbah/livres.html

Saul03. Sauloy, J.: Galois theory of Fuchsian q-difference equations Ann. Sci. Éc. Norm. Supér. **36**, 925-968 (2003)

Sch87. Schlesinger, L.: Handbuch der Theorie der linearen Differentialgleichungen. Teubner, Leipzig (1887)

Sin99. Singer, M. F.: Direct and inverse problems in differential Galois theory. In: Bass, H., Buium, A., Cassidy, P. (eds): Selected Works of Ellis Kolchin, 527–554 American Mathematical Society (1999) Available at Michael F. Singer's Webpage, North Carolina University.

Sin09. Singer, M. F.: Introduction to the Galois Theory of Linear Differential Equations. In: MacCallum, M. A. H., Mikhalov, A. V. (eds): Algebraic Theory of Differential Equations.. London Mathematical Society Lect. Note Series vol. 357, 1–82, Cambridge University Press (2009)

Tu55. Turrittin, H.: Convergent solutions of ordinary differential equations in the neighborhood of an irregular singular point. Acta. Math. **93**, 27–66 (1955)

Var84. Varadarajan, V.S.: Lie groups, Lie algebras, and their representations Springer-Verlag (1984)

Wa87. Wasov, W.: Asymptotic Expansions for Ordinary Differential Equations. Dover Publications, Inc., New York (1987)

Wi12. Wibmer, M.: Existence of ∂-parameterized Picard-Vessiot extensions over fields with algebraically closed constants. J. Algebra **361**, 163–171 (2012)

Chapter 3
Inverse Problems

We are now able to state the *inverse problems* of characterizing those groups that can be realized as the monodromy group or the differential Galois group of some differential system, although an effective construction of such systems remains a difficult problem.

3.1 The Riemann-Hilbert problem

Given $\Sigma = \{a_1, \dots; a_n\} \subset \overline{\mathbb{C}}$, $x_0 \in U_\Sigma$ and a representation

$$\chi : \pi_1(U_\Sigma; x_0) \longrightarrow \mathrm{GL}(p, \mathbb{C})$$

one may ask:

1. whether there exists a differential system (S) with χ as its monodromy representation,
2. whether there exist such (S) with regular singular points only,
3. whether there exist such (S) with singular points all Fuchsian but one,
4. whether there exist such (S) with Fuchsian singular points only.

The answer is *yes* to Questions 1 to 3, *no* in general to Question 4 which is today commonly referred to as the classical *Riemann-Hilbert problem* (RH problem for short). It is a rare example of a problem that had been considered as solved since 1908 (by Plemelj, [Ple64]) until Treibich Kohn [Tr83] in 1979 discovered an error in Plemelj's proof, and Bolibrukh [Bo90] eventually produced the first counterexample in 1989. The problem still remains open of finding necessary and sufficient conditions for a representation χ to be the monodromy representation of a Fuchsian system (S), that is, of a system with only Fuchsian singular points. Significant progress on this problem has been achieved. If χ is irreducible, Bolibrukh

C. Mitschi, D. Sauzin, *Divergent Series, Summability and Resurgence I*,
Lecture Notes in Mathematics 2153, DOI 10.1007/978-3-319-28736-2_3

[Bo92] and Kostov [Kos92] independently proved that such systems exist. For systems of order two a complete characterization was given by Dekkers [Dek79], and for $p = 3$ and 4 by Bolibrukh [Bo99] and Gladishev [Gla00]. If one of the monodromy matrices is diagonalizable, then the RH problem can be solved. This was actually the missing asumption in Plemelj's proof, corrected by Treibich Kohn. The special case of reducible representations χ has been studied by Malek [Ma02] who in particular showed how to produce new families of reducible counterexamples. We also refer to [GP08] for a survey on the RH problem. The RH problem will be presented in detail in Chapter 4.

3.2 The generalized Riemann-Hilbert problem

The classical Riemann-Hilbert problem can be generalized to systems with irregular singular points. If the location, or the number only of singular points is prescribed but not their type, we just saw that the inverse monodromy problem has a solution, for instance with regular singular points, and even with singular points all Fuchsian but one. The problem of characterizing those representations χ that are realizable as monodromy representations turns out to be difficult if one moreover imposes a minimal Poincaré rank on the singular points, which means being Fuchsian in the case of regular singular points. In the classical RH problem, both the growth rate (moderate) and the Poincaré rank (minimal) are prescribed. The *generalized Riemann-Hilbert problem* (GRH problem for short) extends these requirements to irregular singular points. It starts with the following data, called *generalized monodromy data*:

- A finite subset Σ of $\mathbb{P}^1_{\mathbb{C}}$, $x_0 \in U_\Sigma$ and a representation

$$\chi : \pi_1(U_\Sigma; x_0) \longrightarrow \mathrm{GL}(p, \mathbb{C})$$

- For each $a \in \Sigma$, a local linear differential system (S_a) with coefficients in $\mathcal{M}_a$.

The GRH problem asks under which conditions there exists a system (S) with χ as its monodromy representation, which locally at each $a \in \Sigma$ is meromorphically equivalent to (S_a) and whose Poincaré rank at each $a \in \Sigma$ is minimal (in the class of meromorphically equivalent systems at a). If the given local systems are regular singular, the problem is the classical RH problem. If Σ has two elements, say 0 and ∞, and one of the two minimal Poincaré ranks is 0, the problem is known as the *Birkhoff standard form problem*, which is still open (cf. [JLP76], [Ba89], [BB97], [Sab02]). As in the classical case, the irreducibility of χ, together with the condition that one at least of the singular points is 'without roots' (unramified) is a sufficient condition for the solvability of the GRH problem (cf. [BMM06]). The problem always has a solution if at one of the singular points the Poincaré rank is not required to be minimal. Gontsov and Vyugin [GV09] proved that one can in fact, in this case, impose an upper bound on the Poincaré rank so that the problem has a solution.

In Section 2.2 we defined the formal monodromy, exponential torus and Stokes matrices at an irregular singular point. Together with the representation of the monodromy, these are also referred to as 'generalized monodromy data' since they in many ways generalize the topological monodromy: they play the same role in Ramis's density theorem (cf. Theorem 2.47 p. 58) which describes the local analytic differential Galois groups as the Zariski closure of a subgroup generated by these data, as the usual monodromy in Schlesinger's theorem which states that in the regular singular case, the differential Galois group over $\mathbb{C}(x)$ is the Zariski closure of the monodromy group. Indeed, the formal monodromy and exponential torus together with the Stokes matrices completely determine the meromorphic equivalence class of a local differential system at an irregular singular point. This is based on results by Malgrange, Ramis and Sibuya (cf. [Mal79], [Mal83], [RS89], [Sib90], see also [Lod94]).

3.3 Related problems

The Riemann-Hilbert problem, classical and generalized, is closely related to other questions about monodromy. It has been and still is extensively studied in symplectic and Hamiltonian geometry in relation to Painlevé equations and isomonodromic deformations, from the point of view of moduli spaces for instance (cf. [Boa01], [IIS06a], [IIS06b], [PSa09], [Boa11], [IS13], [BS13], [Boa14], [HY14]) and in relation to quiver varieties, in the Fuchsian case (cf. [CB03] as well as non-Fuchsian (cf. [Boa08], [Boa12], [HY14]). To learn more about Painlevé equations and their relation to isomonodromy, we also refer the reader to the third volume [De16, Section 1.3.3 and Chapter 2] of this book, where the First Painlevé equation is studied in detail.

The Riemann-Hilbert problem has over the years extended to new problems, far beyond the material presented here. As explained in Chapter 4, solutions of the Riemann-Hilbert problem (classical or generalized) consist, given a monodromy representation or generalized monodromy data, in realizing local differential systems with these data, then patching them together into a global differential system with these data, via a trivializing fiber bundle and connection. As we have seen in Chapter 2, a local differential Galois group and all its constituents, in particular the monodromy matrix and the Stokes matrices, are defined up to conjugation only in $\mathrm{GL}(p,\mathbb{C})$.

In [Kat95] Katz introduced the notion of *rigid local systems* to characterize those local systems (cf. [Del70] for a precise definition), roughly speaking those simultaneous data of local differential systems at the prescribed singular points, which completely determine the global system, in the same way as Riemann would characterize the hypergeometric equation from its monodromy.

This involves the purely algebraic *Deligne-Simpson problem*, DS problem for short, strongly related to the Riemann-Hilbert problem: to give necessary and sufficient conditions on the choice of $n+1$ conjugacy classes $C_j \subset \mathrm{GL}(p,\mathbb{C})$ so that there exist irreducible $n+1$-tuples of matrices $M_j \in C_j$ satisfying the relation

$$M_0 \dots M_n = I,$$

'irreducible' meaning here that the M_j have no common proper invariant subspace. The problem was first stated by Simpson [Simp91], who gave a necessary and sufficient condition for the existence of solutions under some conditions on the conjugacy classes. It was given its name by Kostov who has dealt with this problem in a number of papers (cf. [Kos04] for a general survey and references). Kostov also introduced an additive analogue of the problem, closer to the RH problem and called the *additive DS problem*, of finding conjugacy classes of matrices $A_0, \dots A_n \in \mathfrak{gl}(p, \mathbb{C})$ with

$$A_0 + \dots + A_n = 0,$$

realizing a *Fuchsian* system

$$\frac{dY}{dx} = \sum_{i=1}^{n} \frac{A_i}{x - a_i} Y$$

where A_0 is the residue at infinity. Kostov gave necessary and sufficient conditions for the existence of solutions to both the original and additive DS problems under some generic conditions on conjugacy classes. The additive problem was completely settled by Crawley-Boevey [CB03] who initiated the study of these problems from the *quiver theory* point of view and also gave a sufficient condition (cf. joint work [CBS06]) weaker than previously available ones, for a solution of the original DS problem.

A generalization of the DS problem to the unramified irregular singular case can be found in [Boa14], and a generalization of the additive DS problem to the unramified irregular singular case was considered in [Boa08] in several cases, and finally solved in[Hir13].

In [Kat95], Katz also introduced the notion of *middle convolution* which was later reformulated into an algebraic algorithm and applied to inverse Galois problems (cf. [Vö01], [DR00], [DR07]). In simple words, let us say that the middle convolution algorithm is a method by which one can for instance construct any irreducible rigid local system on the punctured complex projective line, by a repeated application of the algorithm starting from a local system of rank one. Middle convolution has provided interesting results not only for rigid systems but, more generally, to solve modern versions of the Riemann-Hilbert problem, classical and generalized, and related connection problems, as well as inverse Galois problems, classical and differential. To learn about middle convolution and its developments, we refer the reader to [Osh12], [Fi06], [HF07], [BF15b], [BF15c]. There has been much progress lately in the subject, in particular in Japan where, following previous generalization in [Ar10], different authors have extended middle convolution to irregular systems ([Kaw10], [Ta11], [Ya11]), holonomic systems [Hara12] and q-difference equations [SY14].

3.4 The inverse problem in differential Galois theory

Let G be a linear algebraic group over an algebraically closed field C of characteristic 0, and k a differential field with C as field of constants. The inverse problem asks whether G can be realized as the differential Galois group of some system (S) with coefficients in k. Let us show on an example that, depending on the ground field, not any linear algebraic group is realizable. Over the differential field $(k=\mathbb{C}, \partial=0)$ for instance, we have the following characterization.

Proposition 3.1. *A linear algebraic group G over $\mathbb{C}$ is realizable as a differential Galois group over $\mathbb{C}$ if and only if it is a torus $(\mathbb{C}^*)^\ell$ or a product $\mathbb{C}\times(\mathbb{C}^*)^\ell$.*

Proof. Let us first show that G is realizable over $\mathbb{C}$ if and only it is a quotient of some product $\mathbb{C}\times(\mathbb{C}^*)^m$. Assume G is realizable, say for a scalar linear differential equation with constant coefficients. Then, as is well known, there are non-zero constants $\lambda_1,\ldots,\lambda_m\in\mathbb{C}$ such that any solution of the equation can be written as

$$y = P_0 + \sum_{j=1}^{m} P_j \mathrm{e}^{\lambda_j x}$$

with polynomials $P_0, P_1, \ldots, P_m$. This shows that the Picard-Vessiot extension over $\mathbb{C}$ is a differential subfield of

$$K = \mathbb{C}(x, \mathrm{e}^{\lambda_1 x}, \ldots, \mathrm{e}^{\lambda_m x})$$

which is itself a Picard-Vessiot extension of $\mathbb{C}$ (cf. Exercise 2.13 p. 30) with a differential Galois group of the form $\mathbb{C}\times(\mathbb{C}^*)^r$, where r is the rank of the $\mathbb{Z}$-module $\sum_{j=1}^m \mathbb{Z}\lambda_j$. The differential Galois correspondence (cf. Theorem 2.15 p. 31) then tells us that G, the differential Galois group of the equation, is a quotient of $\mathbb{C}\times(\mathbb{C}^*)^r$.

Exercise 3.2. Prove that a linear algebraic group which is a quotient of $\mathbb{C}\times(\mathbb{C}^*)^r$ is isomorphic to $(\mathbb{C}^*)^\ell$ or $\mathbb{C}\times(\mathbb{C}^*)^\ell$ for some $\ell\le r$.

The solution of the inverse problem over $\mathbb{C}$ now follows easily. Choose $\mathbb{Z}$-linearly independent constants $\lambda_1,\ldots,\lambda_\ell\in\mathbb{C}$. Then $(\mathbb{C}^*)^\ell$ is the differential Galois group of $Y' = AY$ where $A=\mathrm{diag}(\lambda_1,\ldots,\lambda_\ell)$, and $\mathbb{C}\times(\mathbb{C}^*)^\ell$ is the differential Galois group of $Y' = BY$ where

$$B = \begin{pmatrix} C & 0 \\ 0 & A \end{pmatrix}, \quad \text{with} \quad C = \begin{pmatrix} 0 & 1 \\ 0 & 0 \end{pmatrix}.$$

□

Over $\mathbb{C}(\{x\})$, the inverse problem has a solution if and only if G has a so-called *local Galois structure* (cf. Ramis [Ra94], see also [MS96a]), namely if $G/L(G)$ is cyclic, where $L(G)$ is the subgroup of G generated by its tori. The group $\mathrm{SL}(n,\mathbb{C})$ for instance can be realized as a local differential Galois group whereas $\mathbb{C}^n$ cannot, for $n\ge 2$.

Over $\mathbb{C}(x)$ the problem always has a solution. This was proved by Tretkoff and Tretkoff [TT79]. An outline of their proof is presented below. It is based on the Riemann-Hilbert problem (Section 3.1, Question 1).

Over $C(x)$, where C is any algebraically closed field of characteristic zero, the problem was completely solved by Hartmann [Hart05]. Constructive solutions had previously been given by Singer and the author ([MS96b], [MS02])) for connected algebraic groups and large families of non-connected groups, in particular those groups whose identity component is solvable, and later in joint work [CMS05] with Cook, when the identity component is a finite product of simple algebraic groups of the types A_l, C_l, D_l, E_6 and E_7.

For more references, see [Sin99].

3.5 The differential Galois inverse problem over $\mathbb{C}(x)$

To conclude this review of inverse problems we will sketch the solution by Tretkoff and Tretkoff [TT79] of the differential inverse Galois problem over $\mathbb{C}(x)$, based on the weak form of the Riemann-Hilbert problem. Their solution essentially uses the following result.

Theorem 3.3. *Any algebraic group $G \subset \mathrm{GL}(p,k)$, where k is an algebraically closed field of characteristic 0, is the Zariski-closure of a finitely generated subgroup.*

The proof of this theorem is based on two lemmas which hold over any algebraically closed field k of characteristic 0. By a *periodic group* one means a group in which all elements are of finite order.

Lemma 3.4. *Any periodic algebraic group $G \subset \mathrm{GL}(p,k)$ is finite.*

To prove the lemma, we need in particular the following results.

Exercise 3.5. Show that any closed subgroup of finite index of G contains G^0.

Exercise 3.6. Show that any set M of commuting diagonalizable $n \times n$ matrices over k is (simultaneously) diagonalizable.

Proof (of Lemma 3.4). By a theorem of Schur (cf. [CR62, Theorem 36.14]) we know that G, as a periodic group, must contain an abelian normal subroup G_1 of finite index. The identity component G^0 of G is a normal subgroup too of finite index, hence G_1 contains G^0 by Exercise 3.5, and since G_1 is abelian, so is G^0. Note that all elements of G are diagonalizable since they are of finite order (their minimal polynomial has simple roots only, where all of them are roots of unity). Since G^0 is abelian, its elements are simultaneously diagonalizable by Exercise 3.6, that is, G^0 is a *diagonal group*. It is isomorphic to a closed subgroup of the group $\mathrm{D}(p,k)$ of diagonal matrices, and therefore isomorphic to $(k^*)^r$ for some $r \in \mathbb{N}$ (cf. [Hu75], p. 104 : the proof of this result uses the characters of the group). Since k is of characteristic zero, $(k^*)^r$ can be periodic only if $r = 0$, which ends the proof. □

Lemma 3.7. *Any algebraic group* $G \subset \mathrm{GL}(p,k)$ *of dimension* ≥ 1 *contains a finitely generated subgroup* H *whose Zariski-closure has dimension* ≥ 1.

Proof. Assume that the conclusion doesn't hold, that is, the closure of any finitely generated subgroup of G is of dimension 0, hence finite. In particular, the closure of the subgroup generated by any element of G is finite. The algebraic group G is therefore periodic, hence of dimension 0 since by Lemma 3.4 it is finite. This contradicts the assumption, and proves the lemma. □

Theorem 3.3 is proved by induction, and easily restricts to the connected case via the following result.

Exercise 3.8. Show that if Theorem 3.3 holds for *connected* linear algebraic groups over k, then it holds for any algebraic group $G \subset \mathrm{GL}(p,k)$.

Proof (**of Theorem 3.3**). By Exercise 3.8 we may assume that G is connected and proceed by induction on its Zariski dimension. If $\dim G = 0$, then G is trivial (since it is finite and connected) and the result holds. If $\dim G = 1$, then by Lemma 3.7 , G has a finitely generated subgroup H such that $\dim \overline{H} \geq 1$, hence $\dim \overline{H} = 1$ and $\overline{H} = G$ since G is connected. Assume that $\dim G > 1$ and select a *proper*, connected, subgroup H of G of maximal dimension. Note that $\dim H < \dim G$ since G is connected.

If H is normal in G then by induction there are finitely generated, Zariski-dense subgroups R of H and S of G/H respectively, where $R = \langle h_1, \ldots, h_r \rangle$ and $S = \langle g_1 H, \ldots, g_s H \rangle$ (S consists of cosets). Let us prove that $G = \overline{L}$, where

$$L = \langle h_1, \ldots, h_r, g_1, \ldots, g_s \rangle.$$

Let $\pi : G \to G/H$ denote the quotient-map. Since R is a subgroup of L, then $H = \overline{R}$ is a subgroup of $\overline{L}$, which implies that $\overline{L} = \pi^{-1}(\pi(\overline{L}))$ on one hand, and S as well as its closure $\overline{S} = G/H$ are subgroups of $\overline{L}/H = \pi(\overline{L})$ on the other hand. This implies that $\overline{L}/H = G/H$, that is, $\pi(\overline{L}) = \pi(G)$, hence

$$\overline{L} = \pi^{-1}(\pi(\overline{L})) = \pi^{-1}(\pi(G)) = G$$

which proves the result.

If H is *not* normal in G, then $H \neq gHg^{-1}$ for some $g \in G$. By the induction hypothesis there is a finitely generated subgroup R of H such that $\overline{R} = H$. Consider the subgroup $L = \langle R, g \rangle$ of G generated by by R and g. Then both H and gHg^{-1} are closed subgroups of $\overline{L}$, hence of $\overline{L}^0$ since they are connected. If

$$\dim \overline{L}^0 = \dim G$$

then $\overline{L}^0 = G$ since G is connected, hence $G = \overline{L}$ since $\overline{L}^0 \subset \overline{L} \subset G$. If

$$\dim \overline{L}^0 < \dim G$$

then $\dim \overline{L}^0 = \dim H$ by the maximality of $\dim H$, hence $\overline{L}^0 = H$ since H is connected, and $\overline{L}^0 = gHg^{-1}$ for the same reason. This contradicts $H \neq gHg^{-1}$. □

We are now ready to give the proof of Tretkoff and Tretkoff's solution of the RH problem.

Theorem 3.9 (Tretkoff–Tretkoff). *The differential Galois inverse problem has a solution over* $\mathbb{C}(x)$.

Proof. Assume that the linear algebraic group G is topologically generated by $g_1, \dots, g_{n-1}$, that is, $G = \overline{H}$ where $H = \langle g_1, \dots, g_{n-1} \rangle$, then choose any subset $\Sigma = \{a_1, \dots, a_n\}$ of $\mathbb{C}$ of pairwise distinct points, and a base-point $x_0 \in U_\Sigma$. Consider the representation

$$\chi : \pi_1(U_\Sigma; x_0) \longrightarrow \mathrm{GL}(p, \mathbb{C})$$

defined by

$$\chi([\gamma_i]) = g_i,\ i = 1, \dots, n-1, \quad \chi([\gamma_n]) = (g_1 \dots g_n)^{-1}$$

where γ_i is an elementary loop around a_i issued from x_0. We have seen that χ can be realized as the monodromy representation of some regular singular system(S) (the positive answer to Question 1.) and by Schlesinger's Theorem 2.28 p. 35, we know that the differential Galois group of (S) is $\overline{H} = G$. If we moreover require the system to be Fuchsian, then we may just choose an additional a_{n+1} with $\chi([\gamma_{n+1}]) = \mathrm{I}$ and apply Theorem 4.49 p. 112. □

3.6 Solutions to exercises of Chapter 3

Exercise 3.2 p. 79 Let G be a linear algebraic group and assume that G is a quotient Γ/H of $\Gamma = \mathbb{C} \times (\mathbb{C}^*)^r$ by a closed subgroup H. We know that the only algebraic subgroups of $\mathbb{C}$ are (0) and $\mathbb{C}$. Note that $\mathbb{C}$ has indeed no elements of finite order and the Zariski closure of any infinite subset of an algebraic variety is of dimension > 0. Since $\mathbb{C}$ is connected, the property follows from elementary facts about dimension and subvarieties (see ([Hu75, Section 3]).

Assume that $H \subset (0) \times (\mathbb{C}^*)^r$. Then G is isomorphic to $\mathbb{C} \times \big((\mathbb{C}^*)^r / K\big)$, where $H = (0) \times K$. The image of a torus $(\mathbb{C}^*)^r$ by a morphism of algebraic groups is again a torus $(\mathbb{C}^*)^l$, $l \leq r$. To see this, note that $(\mathbb{C}^*)^r$ is an affine $\mathbb{C}$-group (affine as a variety over $\mathbb{C}$) and that its quotient by a closed subgroup is again an affine $\mathbb{C}$-group, hence a linear algebraic group (cf. [Bor91, Proposition 1.10, Theorem 6.8]). Moreover, a morphism of algebraic groups is compatible with the multiplicative Jordan decomposition of an element (as the unique product of its commuting semisimple and unipotent factors) and preserves in particular the semisimplicity of an element (cf. [Bor91, Section 4.5]). Since the linear algebraic group $(\mathbb{C}^*)^r$ can be viewed as a connected set of commuting diagonalizable matrices, that is, of commuting semisimple elements, its quotient by K also consists of commuting semisimple elements. It follows from Exercise 3.6 below that $(\mathbb{C}^*)^r/K$ thus can be viewed as a group of simultaneously diagonalizable matrices and since it is connected, it is isomorphic to a torus $(\mathbb{C}^*)^l$ of dimension $l \leq r$, and G to the product $\mathbb{C} \times (\mathbb{C}^*)^l$.

If H is not a subgroup of $(0) \times (\mathbb{C}^*)^r$, then it projects surjectively onto the first factor $\mathbb{C}$ of Γ. Let $\pi : \Gamma \to G = \Gamma/H$ be the quotient morphism. Note that any element $(a,b) \in \Gamma = \mathbb{C} \times (\mathbb{C}^*)^r$ can be written as $(a,b) = (0, bb_1^{-1}).(a,b_1)$ for any $(a,b_1) \in H$, and such elements (a,b_1) exist for each $a \in \mathbb{C}$ by the asumption on H. This implies that $\pi(\Gamma) = \pi((0) \times (\mathbb{C}^*)^r)$ and that G is isomorphic to a quotient $(\mathbb{C}^*)^l$ of $(\mathbb{C}^*)^r$.

Exercise 3.5 p. 80 Let H be a closed subgroup of finite index of the linear algebraic group G. This means that there is a partition

$$G = \bigcup_{i=1}^{l} g_i H$$

of G in finitely many H-cosets represented by $g_1, \ldots, g_l \in G$, where $g_1 = e$ represents the coset $H = eH$. Since H is closed in G, so are all cosets, since multiplication in G by any fixed element $g \in G$ is a homeomorphism. Each coset $g_i H$ is therefore open in G (as the complement in G of a finite union of closed sets). The restriction of this partition to G^0 induces a partition of G^0 into subsets which are both open and closed. Since $G^0 \cap H \neq \emptyset$, the connectedness of G^0 imposes that $G^0 = G^0 \cap H$, that is, $G^0 \subset H$.

Exercise 3.6 p. 80 Let M be a commuting set of diagonalizable $n \times n$ matrices over k, which we identify with the endomorphisms of k^n they represent in the canonical basis of k^n. For any $u \in M$ and $a \in k$, the subspace $W = \ker(u - a\mathrm{I})$ of k^n is stable by any $v \in M$ since v commutes with u. Assume that M does not consist of diagonal matrices only. Then, since k is algebraically closed, we can find $u \in M$ and $a \in k$ such that $W = \ker(u - a\mathrm{I})$ is a proper eigenspace of k^n, and proceed by induction on $n > 1$ (the base case $n = 1$ is clear). Since u is diagonalizable, $k^n = W \oplus W'$ is the direct sum of W and the direct sum $W' \neq (0)$ of the other eigenspaces of u, and since the latter are stable by M, so are both W and W'. Let N (resp. N') be the set of elements of M restricted to W (resp. W'). By the inductive hypothesis the result holds for both N and N', that is, there is a basis B of W (resp. B' of W') such that B (resp. B') diagonalizes all $v \in N$ (resp. $v \in N'$), which implies that $B \cup B'$ diagonalizes all $v \in M$. If we assumed the elements of M to be only commuting, then one could similarly prove that they are simultaneously trigonalizable (cf. [Bor91, Section 4.6]).

Exercise 3.8 p. 81 Assume that the identity component G^0 of some linear algebraic group G is the closure of a finitely generated subgroup $H = \langle h_1, \ldots, h_r \rangle$ of G^0. Since G^0 is a subgroup of finite index in G, we can choose representatives $g_1, \ldots, g_s$ of the cosets $\{\overline{g_1}, \ldots, \overline{g_s}\} = G/G^0$. We have

$$G=\bigcup_{i=1}^{s} g_i G^0.$$

Let $L=\langle h_1,\dots,h_r,g_1,\dots,g_r\rangle$. Then $G^0=\overline{H}\subset\overline{L}$ since $H\subset L$, and since $\overline{L}$ is a group, we have $g_iG^0\subset\overline{L}$ for all i, hence $G=\overline{L}$.

References

Ar10. Arinkin, D.: Rigid irregular connections on $\mathbb{P}^1$. Compositio Math. **146**, 1323–1338 (2010)

Ba89. Balser, W.: Meromorphic transformation to Birkhoff standard form in dimension three. J. Fac. Sci. Univ. Tokyo Sect. IA Math. **36-2**, 233–246 (1989)

BB97. Balser, W., Bolibrukh, A. A.: Transformation of reducible equations to Birkhoff standard form. Ulmer Seminare, 73–81 (1997)

BF15a. Bibilo, Yu., Filipuk, G.: Non-Schlesinger isomonodromic deformations of Fuchsian systems and middle convolution. SIGMA (Symmetry Integrability Geom. Methods Appl.) **11**, Paper 023 (2015)

BF15b. Bibilo, Yu., Filipuk, G.: Middle convolution and non-Schlesinger deformations. Proc. Japan Acad. Ser A, Math. Sci. **91-5**, 66–69 (2015)

BF15c. Bibilo, Yu., Filipuk, G.: Constructive solutions to the Riemann-Hilbert problem and middle convolution. . To appear in Dynam. Control Sysems (2015) DOI 10.1007/s10883-015-9306-3; ArXiv: 1509.05202 (2015)

Boa01. Boalch, P.: Symplectic manifolds and isomonodromic deformations. Adv. Math. **163-2**, 137–205 (2001)

Boa08. Boalch, P.: Irregular connections and Kac-Moody root systems ArXiv:0806.1050 (2008)

Boa11. Boalch, P.: Riemann-Hilbert for tame complex parahoric connections. Transformation groups **16-1**, 27–50 (2011)

Boa12. Boalch, P.: Simply-laced isomonodromy systems. Publ. Math. IHES **116-1**, 1–68 (2012)

Boa14. Boalch, P.: Geometry and braiding of Stokes data; Fission and wild character varieties. Annals of Math. **179**, 301–365 (2014)

Bo90. Bolibrukh, A. A.: The Rieman-Hilbert problem. Russian Math. Surveys **45-2**, 1–47 (1990)

Bo92. Bolibrukh, A. A.: On sufficient conditions for the positive solvability of the Riemann-Hilbert problem. Math. Notes. Acad. Sci. USSR **51-1**, 110–117 (1992)

Bo99. Bolibrukh, A. A.: On sufficient conditions for the existence of a Fuchsian equation with prescribed monodromy. J. Dynam. Control Systems **5**-4, 453–472 (1999)

BIK04. Bolibrukh, A. A., Its, A. R., Kapaev, A. A.: On the Riemann-Hilbert inverse monodromy problem and the Painlevé equations. Algebra i Analiz **16-1**, 121–162 (2004)

BMM06. Bolibrukh, A. A., Malek, S., Mitschi, C.: On the generalized Riemann-Hilbert problem with irregular singularities. Expo. Math. **24-3**, 235–272 (2006)

Bor91. Borel, A.: Linear Algebraic Groups. Second edition. Graduate Texts in mathematics, Springer (1991)

BS13. Bremer, C. L., Sage, D. S.: Moduli spaces of irregular singular connections. Int. Math. Res. Not. IMRN **8**, 188–1872 (2013)

CMS05. Cook, W., Mitschi, C., Singer, M. F.: On the Constructive Inverse Problem in Differential Galois Theory. Comm. in Algebra **33-10**, 3639–3665 (2005)

CB03. Crawley-Boevey, W.,: On matrices in prescribed conjugacy classes with no common invariant subspace and sum zero. Duke Math. J. **118-2**, 339–352 (2003)

CBS06. Crawley-Boevey, W., Shaw, P.: Multiplicative preprojective algebras, middle convolution and the Deligne-Simpson problem. Adv. Math. **201**, 180–208 (2006)

CR62. Curtis, C., Reiner, I. : Representation Theory of Finite Groups and Associative Algebras. Wiley Interscience, New York (1962)

Dek79. Dekkers, W.: The matrix of a connection having regular singularities on a vector bundle of rank 2 on $\mathbb{P}^1_{\mathbb{C}}$. In: Gérard, R., Ramis, J.-P. (eds): Équations différentielles et systèmes de Pfaff dans le champ complexe. Lect. Notes Math. vol. 712, 33–43, Springer (1979)

De16. Delabaere, E.: Divergent seires, Summability and Resurgence III. Resurgent methods and the First Painlevé equation. Lect. Notes Math. vol. 2155, Springer (2016)

Del70. Deligne, P.: Équations Différentielles à Points Singuliers Réguliers. Lect. Notes Math. vol. 163, Springer (1970)

DR00. Dettweiler, M., Reiter, S.: An algorithm of Katz and its applications to the inverse Galois problems. J. Symbolic Comput. **30**, 761–796 (2000)

DR07. Dettweiler, M., Reiter, S.: Middle convolution of Fuchsian systems and the convolution of rigid differential systems. J. Algebra **318**, 1–24 (2007)

Fi06. Filipuk, G.: On the middle convolution and birational symmetries of the sixth Painlevé equation. Kumamoto J. Math. **19**, 15–23 (2006)

Gla00. Gladyshev, A. I.: On the Riemann-Hilbert problem in dimension 4. J. Dynam. Control Sysems **6-2**, 219–264 (2000)

GP08. Gontsov, R. R., Poberezhnyi, V. A.: Various versions of the Riemann-Hilbert problem for linear differential equations. Russian math. Surveys **63-4**, 603–639 (2008)

GV09. Gontsov, R. R., Vyugin, I. V.: Some addition to the generalized Riemann-Hilbert problem. Ann. Fac. Sci. Toulouse Math. (6) **18-3**, 527–542 (2009)

Hara12. Haraoka Y.: Middle convolution for completely integrable systems with logarithmic singularities along hyperplane arrangements. In Arrangements of Hyperplanes (Sapporo 2009), Adv. Stud. Pure Math. **62**, Math. Soc. Japan, Tokyo, 109–136 (2012)

HF07. Haraoka Y., Filipuk G.: Middle convolution and deformation for Fuchsian systems. J. Lond. Math. Soc. **76**, 438–450 (2007)

Hart05. Hartmann, J.: On the inverse problem in differential Galois theory. J. für die reine und angewandte Mathematik **586**, 21–44 (2005)

Hir13. Hiroe, K.: Linear differential equations on the Riemann sphere and representations of quivers. ArXiv:1307.7438 (2013)

HY14. Hiroe, K., Yamakawa, D.: Moduli spaces of meromorphic connections and quiver varieties Adv. Math. **266**, 120–151 (2014)

Hu75. Humphreys, J. E.: Linear algebraic groups. Springer (1975)

IIS06a. Inaba, M., Iwasaki, K., Saito, M.-H.: Moduli of stable parabolic connections, Riemann-Hilbert correspondence and geometry of Painlevé equations of type VI, part I. Publications of the R.I.M.S., University of Kyoto, **42**, 987–1089 (2006)

IIS06b. Inaba, M., Iwasaki, K., Saito, M.-H.: Moduli of stable parabolic connections, Riemann-Hilbert correspondence and geometry of Painlevé equations of type VI, part II. In Moduli Spaces and Arithmetic Geometry, Kyoto 2004. Adv. Stud. Pure Math. **45**, 378–432 (2006)

IS13. Inaba, M., Saito, M.-H.: Moduli of unramified irregular singular parabolic connections on a smooth projective curve. Kyoto Journal of Mathematics **53-2**, 433–482 (2013)

JLP76. Jurkat, W. B., Lutz, D. A., Peyerimhoff, A.: Birkhoff invariants and effective calculations for meromorphic linear differential equations. I. J. Math. Anal. and Appl. **53**, 438–470 (1976)

Kat95. Katz, N.: Rigid local systems. Annals of Mathematics Studies, vol. 139 , Princeton University Press (1995)

Kaw10. Kawakami H., Generalized Okubo systems and the middle convolution. Int. Math. Res. Not. **17**, 3394–3421 (2010)

Kos92. Kostov, V. P.: Fuchsian linear systems on $\mathbb{CP}^1$ and the Riemann-Hilbert problem. C. R. Acad. Sci. Paris Sr. I Math. **315-2**, 143–148 (1992)

Kos04. Kostov, V. P.: The Deligne-Simpson problem – a survey. J. Algebra **281**, 83–108 (2004)

Ma02. Malek, S.: Fuchsian systems with reducible monodromy are meromorphically equivalent to Fuchsian systems. Proc. Steklov Inst. Math. **3-326**, 468–477 (2002)

Mal79. Malgrange, B.: Remarques sur les équations différentielles à points singuliers irréguliers. In: Gérard, R., J.-P. Ramis, J.-P. (eds): Équations différentielles et systèmes de Pfaff dans le champ complexe. Lect. Notes Math. vol. 712, 77–86, Springer (1979)

Mal83. Malgrange, B.: La classification des connexions irrégulières à une variable. In: Boutet de Montvel, L., Douady, A., Verdier, J.-L. (eds): Séminaire ENS, Mathématiques et Physique. Progress in Math. **37**, 381–399, Birkaüser (1983)

MS96a. Mitschi, C., Singer, M. F.: On Ramis's solution f the local inverse problem of differential Galois theory. J. Pure Appl. Algebra **110**, 185–194 (1996)

MS96b. Mitschi, C., Singer, M. F.: Connected Linear Groups as Differential Galois Groups. Journal of Algebra **184**, 333–361 (1996)

MS02. Mitschi, C., Singer, M. F.: Solvable-by-Finite Groups as Differential Galois Groups. Ann. Fac. Sci. Toulouse, Math. **110-6**, 185–194 (2002)

Osh12. Oshima, T.: Fractional calculus of the Weyl algebra and Fuchsuan differential equations. Memoirs vol. 28. Mathematical Society of Japan, Tokyo (2012))

Ple64. Plemelj, J.: Problems in the sense of Riemann and Klein. Tracts in Mathematics vol. 16. Interscience Publishers (1964)

PSa09. van der Put, M., Saito, M.-H.: Moduli spaces for linear differential equations and the Painlevé equations. Annales de l'institut Fourier **59-7**, 2611–2667 (2009)

Ra94. Ramis, J.-P.: About the solution of some inverse problems in differential Galois theory by Hamburger equations. In: Elworthy, Everett and Lee (eds): Differential Equations, Dynamical Systems and Control Science. Lect. Notes in Pure and Applied Math. vol. 152, 277–300, Marcel Dekker, New York (1994)

Sab02. Sabbah, C.: Déformations isomonodromiques et variétés deFrobenius. Savoirs Actuels. EDP Sciences, Les Ulis. CNRS Editions, Paris (2002)

SY14. Sakai, H., Yamaguchi, M.: Spectral types of linear q-difference equations and q-analo of middle convolution. ArXiv. 1410.3674 (2014)

Simp91. Simpson, C.: Products of Matrices, Differential geometry, global analysis, and topology (Halifax, NS, 1990). Canadian Math. Soc. Conf. Proc. **12** Amer. Math. Soc., Providence, RI, 157–185 (1991)

Sin99. Singer, M. F.: Direct and inverse problems in differential Galois theory. In: Bass, H., Buium, A., Cassidy, P. (eds): Selected Works of Ellis Kolchin, 527–554 American Mahematical Society (1999) Available at Michael F. Singer's Webpage, North Carolina University.

Ta11. Takemura K.: Introduction to middle convolution for differential equations with irregular singularities. In New Trends in Quantum Integrable Systems. World Sci. Publ., Hackensack, NJ, 393–420 (2011)

Tr83. Treibich-Kohn, A.: Un résultat de Plemelj. In: Boutet de Montvel, L., Douady, A., Verdier, J.-L. (eds): Séminaire ENS, Mathématiques et Physique. Progress in Math. **37**, 307–312, Birkaüser (1983)

TT79. Tretkoff, C., Tretkoff, M.: Solution of he inverse problem of differential Galois theory in the classical case. Amer. J. Math. **101** 1327–1332 (1979)

Vö01. Völklein, H.: The braid group and linear rigidity. Geom. Dedicata **84, 1-3** 135–150 (2001)

Ya11. Yamakawa, D.: Middle convolution and Harnad duality. Math. Ann. **349-1** 215–262 (2011)

Chapter 4
The Riemann-Hilbert problem

4.1 Levelt's theory for regular singular points

In Section 2.2 we have seen how important it is, at an irregular singular point, to use an appropriate formal fundamental solution to define generalized monodromy data.

In his books and articles [AB94], [Bo90], [Bo92], [Bo95b], [Bo02], [BIK04] about the classical Riemann-Hilbert problem, Bolibrukh used in the same way a well-appropriate form of fundamental solution at a regular singular point. This form is due to Levelt [Le61] and based on the Levelt filtration of the solution space, which we describe now.

Let us throughout this section fix a system

$$\frac{dy}{dx} = A(x)\,y \tag{S}$$

with coefficients in $\mathbb{C}(x)$. By Lemma 1.35 p. 19, we know that we can write any fundamental solution, locally at a singular point $a \in \Sigma$, say $a = 0$, as

$$Y(x) = M(x)\,x^E \tag{4.1}$$

where $M(x)$ is single-valued in the neighborhood of 0, analytic outside 0, and

$$E = \frac{1}{2\pi \mathrm{i}} \log G \tag{4.2}$$

where (using Bolibrukh's notation) G now denotes the monodromy matrix with respect to $Y(x)$ and an elementary loop around 0. Throughout this section we assume that all $a \in \Sigma$ are *regular singular*, i.e. that $M(x)$ is meromorphic at 0 by Corollary 1.37 p. 20.

C. Mitschi, D. Sauzin, *Divergent Series, Summability and Resurgence I*,
Lecture Notes in Mathematics 2153, DOI 10.1007/978-3-319-28736-2_4

4.1.1 The Levelt filtration

We know from Exercise 1.36 p. 20 that the entries of x^E in formula (4.1) are of the form

$$(x^E)_{ij} = \sum_{\ell \geq 1} x^{\rho_\ell} P_{ij\ell}(\log x)$$

where all $P_{ij\ell}$ are polynomials of degree $\leq p-1$ and ρ_ℓ are the eigenvalues of E. The entries of the fundamental solution $Y = (y_{ij})$ of (4.1) are therefore finite sums

$$y_{ij}(x) = \sum_{k,\ell} x^{\rho_k} (\log x)^{b_\ell} f_{k\ell}(x)$$

where ρ_k, $k = 1, \ldots, s$ are the normalized distinct eigenvalues of E, $0 \leq \Re(\rho_k) < 1$, b_ℓ are non-negative integers $< p-1$ and $f_{k\ell}(x)$ are meromorphic germs at 0 (since $M(x)$ is meromorphic) depending on i, j.

Let $\mathscr{X}$ denote the space of (germs of) holomorphic solutions of (S) in the (germ of) sectorial neighborhood $V = \{x \neq 0 \mid \arg x \in [0, 2\pi[\}$. The components of any $y \in \mathscr{X}$ belong to the differential field extension

$$K = \mathbb{C}(\{x\})(x^{\rho_1}, \ldots, x^{\rho_s}, \log x\}$$

of $\mathbb{C}(\{x\})$.

Definition 4.1. Let v be the valuation on K defined by

(a) $v(0) = \infty$,
(b) $v(f(x)) = d$ if $f(x) = \sum_{i \geq d} a_i x^i \in \mathbb{C}(\{x\})$, $a_d \neq 0$,
(c) $v(f(x)x^{\rho_k}) = v(f(x)x^{\rho_k}(\log x)^m) = v(f(x))$ for all $f \in \mathbb{C}(\{x\})$, $m \in \mathbb{Z}$ and $k = 1, \ldots, s$.

The *Levelt valuation* $v : \mathscr{X} \to \mathbb{Z} \cup \{0\}$ is defined by

$$v(y) = \min_{1 \leq j \leq p} v(y_j) \text{ for } y = (y_1, \ldots, y_p)^T \in \mathscr{X}.$$

(We keep the notation v for convenience.)

Exercise 4.2. Show that Definition 4.1 is equivalent to

$$v(y) = \sup\{k \in \mathbb{Z} \mid \lim_{x \to 0} y(x) x^{-\lambda} = 0 \text{ for all } \lambda < k.\} \tag{4.3}$$

We can extend the Levelt valuation to matrices $M = (\mu_{ij})$ with entries in K, by

$$v(M) = \min_{ij}\{v(\mu_{ij})\}.$$

We can think of v as the 'entire' growth order. For instance

$$v(x^{\frac{5}{2}}(\log x)^2) = v(x^2 x^{\frac{1}{2}}) = 2, \quad v(\frac{1}{x}\log x) = -1.$$

The Levelt valuation has the following properties.

Proposition 4.3. *The Levelt valuation* $v : \mathscr{X} \longrightarrow \mathbb{Z} \cup \{\infty\}$ *satisfies*

(a) $v(y^1 + y^2) \geq \min\{v(y^1), v(y^2)\}$ *for any* $y^1, y^2 \in \mathscr{X}$,

(b) $v(cy) = v(y)$ *for any* $c \in \mathbb{C}^*, y \in \mathscr{X}$,

(c) $v(y^\gamma) = v(y)$ *for any* $y \in \mathscr{X}$ *and any loop* γ *in* $\overline{V}$ *around* 0 *from* $x_0 \in V$ *(where* $\overline{V}$ *denotes the closure of* V*).*

Proof. The proof of (a) and (b) is immediate. To prove (c) let y_j be a component of $\underline{y} \in \mathscr{X}$. We have

$$y_j = \sum_{k,\ell} x^{\rho_k} (\log x)^{b_\ell} f_{k\ell}(x)$$

hence

$$y_j^\gamma = \sum_{k,\ell} x^{\rho_k} e^{2\pi i \rho_k} (\log x + 2\pi i)^{b_\ell} f_{kl}(x),$$

which implies that $v(y_j^\gamma) \geq v(y_j)$. We get in the same way that $v(y_j) \geq v(y_j^\gamma)$ by analytically continuing y_j^γ along γ^{-1}. □

Corollary 4.4. *The set* $v(\mathscr{X})$ *is finite.*

Proof. We may label the elements v_k of $v(\mathscr{X}) \cap \mathbb{Z}$ in the decreasing order $v_k \geq v_{k+1}$ and define $\mathscr{X}_k = \{y \in \mathscr{X}, v(y) \geq v_k\}$ for each v_k. It easily follows from Proposition 4.3 that $\mathscr{X}_k$ is a $\mathbb{C}$-linear subspace of $\mathscr{X}$ and that $\mathscr{X}_k \subset \mathscr{X}_{k+1}$. Since $\mathscr{X}$ is finite-dimensional there are finitely many $\mathscr{X}_k$ and v_k only. □

Levelt [Le61] used this valuation to construct special fundamental solutions that became essential tools in Bolibrukh's investigation and proofs about the Riemann-Hilbert problem.

Definition 4.5. With the same notation, let us define

(a) the *Levelt filtration* of $\mathscr{X}$ as the finite sequence $\{0\} \subset \mathscr{X}_1 \subset \ldots \subset \mathscr{X}_m = \mathscr{X}$

(b) the *multiplicity* of $v_k \in v(\mathscr{X})$ as $\kappa_k = \dim_{\mathbb{C}} \mathscr{X}_k / \mathscr{X}_{k-1}$

(c) a *Levelt basis* or *Levelt fundamental solution* as a basis of $\mathscr{X}$ of the form

$$\mathscr{B} = \mathscr{B}_1 \cup \ldots \cup \mathscr{B}_m$$

where $\mathscr{B}_1 \cup \ldots \cup \mathscr{B}_k$ is for each k a basis of $\mathscr{X}_k$, and such that for an elementary loop γ around 0 the monodromy matrix $\mathrm{Mon}(\gamma)$ is an upper triangular matrix with respect to $\mathscr{B}$.

To construct a Levelt basis one proceeds as follows. In short, let $\mathscr{B}_1$ be a basis of $\mathscr{X}_1$ such that $G = \mathrm{Mon}(\gamma)$ is in Jordan form. Since the quotient space $\mathscr{X}_2/\mathscr{X}_1$ is stable by G, we can lift a Jordan basis for G from $\mathscr{X}_2/\mathscr{X}_1$ to $\mathscr{X}_2$, and so on. Iterating the procedure, one recursively constructs a basis $\mathscr{B}$ with the required properties. The construction is of course not unique, but one can refine the definition to construct a special Levelt basis that is unique, in a sense that we shall not develop here.

Let us fix a Levelt basis $\mathscr{B} = \{e_1, \ldots, e_p\}$ of $\mathscr{X}$, that is, a fundamental solution Y of (S) with columns $e_1, \ldots, e_p$ satisfying the conditions of Definition 4.5. Let

$$B = \operatorname{diag}(\varphi_1, \ldots, \varphi_p)$$

where $\varphi_j = v(e_j)$ for $j = 1, \ldots, p$. Then $\varphi_1 \geq \varphi_2 \geq \ldots \geq \varphi_p$. Let as before

$$E = \frac{1}{2\pi \mathrm{i}} \log G$$

with normalized eigenvalues ρ_k.

Lemma 4.6. *With this notation,*

$$v(x^B x^E x^{-B}) = 0.$$

Proof. The matrix

$$x^B x^E x^{-B} = (a_{i,j})_{0 \leq i,j \leq p}$$

is upper triangular. For $i \leq j$,

$$a_{i,j} = x^{\varphi_i - \varphi_j} \sum_{k,\ell} x^{\rho_k} P_{ij\ell}(\log x),$$

where $\varphi_i \geq \varphi_j$ and $P_{ij\ell}$ are polynomials of degree $\leq p-1$. The non-diagonal elements have a non-negative valuation, whereas the diagonal elements have valuation 0. □

The following result gives another characterization of Fuchsian singular points.

Theorem 4.7 (Levelt). *With notation as before:*

(a) If 0 is regular singular then any Levelt fundamental solution is of the form

$$Y = U(x) x^B x^E \tag{4.4}$$

where U is holomorphic at 0.

(b) The regular singular point 0 is Fuchsian if and only if $U(0)$ is invertible.

Proof. The matrix $U = Y x^{-E} x^{-B}$ is meromorphic since $Y = M(x) x^E$ by (4.1), where $M(x)$ is meromorphic. Let $r = \max_k \{\Re \rho_k\}$. To prove (a) it is sufficient to prove that

$$\lim_{x \to 0} U(x) x^{\alpha} = 0$$

for some $\alpha \in [0, 1[$. Let $\alpha = r + 2\varepsilon \in [r, 1[$. Then

$$U x^{\alpha} = U x^{r+2\varepsilon} = Y x^{-E} x^{-B} x^{r+2\varepsilon} = Y x^{-B+\varepsilon I} x^B x^{-E+rI} x^{-B} x^{\varepsilon I}.$$

Let $F_1 = Y x^{-B+\varepsilon I}$ and $F_2 = x^B x^{-E+rI} x^{-B} x^{\varepsilon I}$. Then

$$\lim_{x \to 0} F_1(x) = \lim_{x \to 0} F_2(x) = 0.$$

For F_1 this follows from

$$F_1 = \left(\frac{e_1(x)}{x^{\nu(e_1)-\varepsilon}}, \ldots, \frac{e_p(x)}{x^{\nu(e_p)-\varepsilon}} \right)$$

and the definition of the Levelt valuation. For F_2 let $\tilde{E} = -E + rI$. The entries of $x^{\tilde{E}}$ are of the form

$$(x^{\tilde{E}})_{ij} = \sum_{k,\ell} x^{r-\rho_k} P_{ij\ell}(\log x)$$

where $\Re(r - \rho_k) \geq 0$ and $\nu(x^{\tilde{E}}) = 0$. As in the proof of Lemma 4.6 p. 90 it is easy to see that $\nu(x^B x^{\tilde{E}} x^{-B}) = 0$. The proof of (b), which is long and difficult, can be found in Levelt's thesis [Le61]. □

Example 4.8. A Levelt fundamental solution of the differential system

$$\frac{dy}{dx} = \begin{pmatrix} 1/x & 1 \\ 0 & 0 \end{pmatrix} y$$

in the neighborhood of the regular singular point 0 is

$$Y = \begin{pmatrix} x & x \log x \\ 0 & 1 \end{pmatrix} = \begin{pmatrix} 1 & 0 \\ 0 & 1 \end{pmatrix} x^{\begin{pmatrix} 1 & 0 \\ 0 & 0 \end{pmatrix}} x^{\begin{pmatrix} 0 & 1 \\ 0 & 0 \end{pmatrix}}$$

which is of the form $U x^B x^E$. As expected, $U = U(0) = I$ is invertible since 0 is Fuchsian.

Example 4.9. A Levelt fundamental solution of

$$\frac{dy}{dx} = \begin{pmatrix} 0 & 1 \\ 1/x^2 & -1/x \end{pmatrix} y$$

at the regular singular point 0 is

$$Y = \begin{pmatrix} x & 1/x \\ 1 & -1/x^2 \end{pmatrix} = \begin{pmatrix} x & x \\ 1 & -1 \end{pmatrix} x^{\begin{pmatrix} 0 & 0 \\ 0 & -2 \end{pmatrix}} x^{\begin{pmatrix} 0 & 0 \\ 0 & 0 \end{pmatrix}}$$

which is of the form $U x^B x^E$. As expected $U(0)$ here is not invertible since 0 is not Fuchsian.

4.1.2 The Fuchs relation

Let $Y = Ux^Bx^E$ be a Levelt solution of (S) at 0, where $B = \text{diag}(\varphi_1,\dots,\varphi_p)$ and E is an upper triangular matrix with normalized diagonal entries

$$\rho_1,\dots,\rho_p, \quad 0 \le \Re(\rho_j) < 1$$

for $j = 1,\dots,p$.

Definition 4.10. The complex numbers

$$\beta_j = \varphi_j + \rho_j, \quad j = 1,\dots,p$$

are by definition the *Levelt exponents* of the system (S) at 0.

If the singular point is Fuchsian the exponents can be read directly from the system.

Proposition 4.11. *If* 0 *is a Fuchsian singular point of* (S)*, that is,*

$$A(x) = \frac{A_0(x)}{x}$$

where A_0 *is holomorphic at* 0 *and* $A_0(0) \neq 0$*, then the Levelt exponents are the eigenvalues of* $A_0(0)$.

Proof. Let Y be a Level fundamental solution at 0. Then by Formula (4.7) p. 90

$$A(x) = Y'Y^{-1} = U'U^{-1} + \frac{U}{x}(B + x^BEx^{-B})U^{-1}$$

where $U'U^{-1}$ and x^BEx^{-B} are clearly holomorphic. The diagonal of

$$L(x) = B + x^BEx^{-B}$$

is $\text{diag}(\beta_1,\dots,\beta_n)$. The limit as x tends to 0 of $xA(x)$ is

$$A_0(0) = U(0)L(0)U(0)^{-1},$$

which has the same eigenvalues β_j as $L(0)$. □

The following result is known as the *Fuchs relation* on the exponents.

Theorem 4.12 (Fuchs relation). *Let* $\Sigma = \{a_1,\dots,a_n\}$ *be the set of singular points, all assumed to be regular singular, of the system* (S) *and let* β_{ij}*,* $j = 1,\dots,p$ *be its Level exponents at* a_i*,* $i = 1,\dots,n$*. Then the following inequalty holds*

$$\sum_{i=1}^{n}\sum_{j=1}^{p}\beta_{ij} \le 0. \tag{4.5}$$

Proof. Let Y_i denote Levelt fundamental solutions at $a_i \in \Sigma$. In local coordinates, we can write each such solution as

$$Y_i = U_i\,(x-a_i)^{B_i}\,(x-a_i)^{E_i},$$

hence by the formula $\det(\exp) = \exp(\mathrm{tr})$ where tr denotes the trace,

$$\det Y_i = (\det U_i)(x-a_i)^{\mathrm{tr}(B_i)+\mathrm{tr}(E_i)}$$

Let us write $\det(U_i(x))$, which is holomorphic at a_i, as

$$\det(U_i(x)) = (x-a_i)^{b_i}\,h_i(x)$$

where $b_i \in \mathbb{N}$ and h_i is holomorphic at a_i with $h_i(a_i) \neq 0$. Let

$$s_i = \mathrm{tr}(B_i) + \mathrm{tr}(E_i) = \sum_{j=1}^{p} \beta_{ij}.$$

Then

$$\det Y_i = h_i(x)(x-a_i)^{b_i+s_i}$$

and since $Y_i' = A\,Y_i$ we know that $\det Y_i$ is a solution of the (scalar) differential equation $\omega' = (\mathrm{tr}(A))\,\omega$, hence

$$\frac{(\det Y_i)'}{\det Y_i} = \frac{b_i+s_i}{x-a_i} + \frac{h_i'}{h_i} = \mathrm{tr}(A)$$

at each a_i, $i = 1,\ldots,n$. If we take the sum of the residues of (the rational function) $\mathrm{tr}(A)$ at all a_i we have

$$\sum_{i=1}^{n}(b_i+s_i) = 0,$$

hence

$$\sum_{i=1}^{n} s_i = -\sum_{i=1}^{b} b_i \leq 0$$

which ends the proof. □

The following result is a characterization of Fuchsian systems by means of their exponents.

Corollary 4.13 (Levelt). *A regular singular system (S) is Fuchsian if and only if its exponents satisfy the relation*

$$\sum_{i=1}^{n}\sum_{j=1}^{p} \beta_{ij} = 0.$$

Proof. With notation from the proof of Theorem 4.12, we know that

$$\sum_i s_i = -\sum_i b_i,$$

hence that

$$\sum_{j=1}^{p} \beta_{ij} = \sum_i s_i = 0$$

if and only if $b_i = 0$ for all $i = 1, \ldots, n$, that is, if and only if $\det(U_i(a_i)) \neq 0$ for each i, which by Theorem 4.7 is equivalent to a_i being Fuchsian. □

Remark 4.14. The definition of Levelt exponents was extended by Corel [Cor04] to irregular singular points, and by Bolibrukh to any Kähler variety [Bo02]. See also [Go04] for refined Fuchs inequalities.

4.2 Vector bundles and connections

This section presents elementary facts about vector bundles and is particularly aimed at readers who are not familiar with this language. It paves the way to Bolibrukh's description of the Riemann-Hilbert problem, which will be treated in Section 4.3.

All varieties are complex analytic varieties, with the corresponding morphisms (cf. [Wh72, Chapter 2], [ACMS53], [Na85]; see also Definition 1.1 p. 4 in this volume, replacing $\mathbb{C}$ by $\mathbb{C}^r$ where r is the dimension of the variety). Isomorphisms are bijective maps f such that both f and f^{-1} are morphisms of analytic varieties. Throughout this section B denotes a connected, one-dimensional analytic variety. Any local property on B will be translated in terms of a local complex coordinate, using an atlas on B.

4.2.1 Vector bundles

The simplest example of a vector bundle is the *trivial line bundle* given by the product variety $B \times \mathbb{C}$. More generally, vector bundles (of finite rank) are defined as follows.

Definition 4.15. A *rank p holomorphic vector bundle over B* is a variety F with a surjective morphism $\pi : F \to B$ such that

(a) for each $x \in B$ the fiber $\pi^{-1}(\{x\})$ is a p-dimensional vector space over $\mathbb{C}$,

(b) for each $x \in B$ there is an open neighborhood $U \subset B$ of x and a fiber-preserving isomorphism f_U, that is, commuting with π and the first projection p_1

$$\begin{array}{ccc} \pi^{-1}(U) & \xrightarrow[\sim]{f_U} & U \times \mathbb{C}^p \\ & \searrow^{\pi} \quad \swarrow_{p_1} & \\ & U & \end{array}$$

such that

(c) for each $y \in U$ the restriction of f_U to $\pi^{-1}(\{y\})$ is an isomorphism of $\mathbb{C}$-vector spaces.

By a *vector bundle* we will always mean a holomorphic vector bundle of finite rank, the basic example of which is the *trivial* vector bundle $B \times \mathbb{C}^p$ where π is the projection on B. In this section we will recall definitions and results, the proofs of which can be found in any basic reference book on vector bundles, for instance [St99].

Definition 4.16. Let F be a vector bundle over B.

(a) A *trivializing covering* $(U_i, f_i)_{i\in I}$ *of B for F* is a covering $\mathscr{U} = (U_i)_{i\in I}$ of B together with isomorphisms

$$f_i : \pi^{-1}(U_i) \longrightarrow U_i \times \mathbb{C}^p$$

such that each U_i with $f_{U_i} = f_i$ satisfies parts (b) and (c) of Definition 4.15, for all $i \in I$

(b) A *trivializing atlas on B for F* is a trivializing covering $(U_i, f_i)_{i\in I}$ of B for F such that the covering $\mathscr{U} = (U_i)_{i\in I}$ of B corresponds to an atlas $(U_i, z_i)_{i\in I}$ of the analytic variety B (cf. Definition 1.1 p. 4).

Note that a trivializing atlas (on B) for F does indeed provide an atlas for F, namely $(\pi^{-1}(U_i), \zeta_i \circ f_i)_{i\in I}$, with

$$\zeta_i \circ f_i : \pi^{-1}(U_i) \xrightarrow{f_i} U_i \times \mathbb{C}^p \xrightarrow{\zeta_i} \tilde{U}_i \times \mathbb{C}^p \subset \mathbb{C}^{p+1}$$

with $\zeta_i = z_i \times \mathrm{id}$, where $z_i : U_i \to \tilde{U}_i \subset \mathbb{C}$ is the local coordinate in U_i of the atlas on B.

Definition 4.17. Two bundles (F, π) and (F', π') are *isomorphic* if there is a fiber-preserving isomorphism $\varphi : F \to F'$

$$\begin{array}{ccc} F & \xrightarrow[\sim]{\varphi} & F' \\ & \searrow^{\pi} \quad \swarrow_{\pi'} & \\ & B & \end{array}$$

which induces a $\mathbb{C}$-linear isomorphism between the fibers $\pi^{-1}(\{x\})$ and $\pi'^{-1}(\{x\})$ for each $x \in B$.

There is a natural notion of *direct sum* of two vector bundles on B, also called their *Whitney sum*. Let $(U_i)_{i\in I}$ be a common trivializing covering for the vector bundles (F_1, π_1) and (F_2, π_2) on B, with trivializing maps

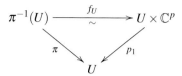

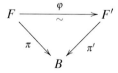

$$f_{1,i} : \pi_1^{-1}(U_i) \xrightarrow{\sim} U_i \times \mathbb{C}^m, \quad f_{2,i} : \pi_2^{-1}(U_i) \xrightarrow{\sim} U_i \times \mathbb{C}^n$$

respectively. Then $(F_1 \times F_2, \pi_1 \times \pi_2)$ is a vector bundle over $B \times B$, with $(U_i \times U_j)_{i,j}$ as a trivializing covering and with trivializing maps φ_{ij} on

$$(\pi_1 \times \pi_2)^{-1}(U_i \times U_j) = \pi_1^{-1}(U_i) \times \pi_2^{-1}(U_j)$$

given by

$$\pi_1^{-1}(U_i) \times \pi_2^{-1}(U_j) \xrightarrow{f_{1,i} \times f_{2,j}} (U_i \times \mathbb{C}^m) \times (U_j \times \mathbb{C}^n) \xrightarrow{(2,3)} (U_i \times U_j) \times \mathbb{C}^{m+n}$$

where $(2,3)$ denotes the map permuting the second and third factors. By definition, the *direct sum* of the bundles (F_1, π_1) and (F_2, π_2) is the so-called *pull-back* (G, π) of $(F_1 \times F_2, \pi_1 \times \pi_2)$ on B via the diagonal map $\delta : B \to B \times B$, defined so as to make the following diagram commutative

$$\begin{array}{ccc} G & \longrightarrow & F_1 \times F_2 \\ {\scriptstyle \pi}\downarrow & & \downarrow{\scriptstyle \pi_1 \times \pi_2} \\ B & \xrightarrow{\delta} & B \times B. \end{array}$$

More precisely, G is the subset (called a *fiber-product*)

$$G = \{(x, y_1, y_2) \in B \times F_1 \times F_2 \mid \pi_1(y_1) = \pi_2(y_2) = x\}$$

of $B \times F_1 \times F_2$. Let π denote the restriction to G of the projection of $B \times F_1 \times F_2$ on the first factor B. Let us show that (G, π) has indeed the structure of a vector bundle. On each open set $U_i \subset B$, the resulting map

$$\pi^{-1}(U_i) \subset U_i \times \pi_1^{-1}(U_i) \times \pi_2^{-1}(U_i) \xrightarrow{id \times f_{1,i} \times f_{2,i}} U_i \times (U_i \times \mathbb{C}^m) \times (U_i \times \mathbb{C}^n)$$

maps an element $y = (x, y_1, y_2) \in G$ to an element (x, x, t_1, x, t_2) where $t_1 \in \mathbb{C}^m$ and $t_2 \in \mathbb{C}^n$. We leave it as an exercise to prove that π together with the trivializing maps

$$\phi_i : \pi^{-1}(U_i) \to U_i \times \mathbb{C}^{m+n}$$

given by

$$\phi_i(y) = (x, t_1, t_2) \in U_i \times \mathbb{C}^{m+n}$$

defines a vector bundle structure on G over B, which is denoted $F_1 \oplus F_2$. Iterating this construction, one similarly defines the direct product of any finite number of vector bundles.

For any covering $\mathscr{U} = (U_i)_{i \in I}$ of B, let us write

$$U_{ij} = U_i \cap U_j, \quad U_{ijk} = U_i \cap U_J \cap U_k$$

whenever these intersections are nonempty.

Definition 4.18. An *analytic Lie group* is an analytic variety G, with a group structure such that the maps

(a) $G \times G \to G,\ (g_1, g_2) \mapsto g_1 g_2$
(b) $G \to G,\ g \mapsto g^{-1}$

are morphisms of analytic varieties.

Definition 4.19. If $\mathscr{U} = (U_i)_{i \in I}$ is a covering of B and G an analytic Lie group, a *$\mathscr{U}$-cocycle* for G is a family of morphisms $g_{ij} : U_{ij} \to G$ which satisfy for all i, j, k the *cocycle conditions* (CC) :

(a) $g_{ij}(x)^{-1} = g_{ji}(x)$ for all $x \in U_{ij}$,
(b) $g_{ij}(x)\, g_{jk}(x)\, g_{ki}(x) = x$ for all $x \in U_{ijk}$.

Let for instance F be a holomorphic vector bundle over B and $\mathscr{U} = (U_i, f_i)_{i \in I}$ a trivializing covering on B for F. It is easy to see that

$$f_i \circ f_j^{-1} : U_{ij} \times \mathbb{C}^p \to U_{ij} \times \mathbb{C}^p,$$

defined by composing f_j^{-1} by f_i

$$U_{ij} \times \mathbb{C}^p \xrightarrow{f_j^{-1}} \pi^{-1}(U_{ij}) \xrightarrow{f_i} U_{ij} \times \mathbb{C}^p$$

is fiber-preserving for all U_{ij}, that is, $f_i \circ f_j^{-1}$ maps any $(x, v) \in U_{ij} \times \mathbb{C}^p$ to some $(x, w) \in U_{ij} \times \mathbb{C}^p$. This defines a map

$$g_{ij} : U_{ij} \to \mathrm{GL}(p, \mathbb{C})$$

by

$$g_{ij}(x)(v) = w.$$

Exercise 4.20. Show that the family $g = (g_{ij})$ is a $\mathscr{U}$-cocycle for $\mathrm{GL}(p, \mathbb{C})$.

Conversely, we have the following result.

Proposition 4.21. *Let $\mathscr{U} = (U_i)_{i \in I}$ be a covering of B corresponding to an atlas and let $g = (g_{ij})$ be a $\mathscr{U}$-cocycle for* $\mathrm{GL}(p, \mathbb{C})$*. Then there is a holomorphic vector bundle F on B with a trivializing atlas $(U_i, f_i)_{i \in I}$ on B for F, which yields the cocycle g.*

From now on we will only consider cocycles for the group $\mathrm{GL}(p, \mathbb{C})$.

Definition 4.22. Let $\mathscr{U} = (U_i)_{i \in I}$ be a covering of B. Two $\mathscr{U}$-cocycles $g = (g_{ij})$ and $g' = (g'_{ij})$ are *equivalent* if there are morphisms $h_i : U_i \to \mathrm{GL}(p, \mathbb{C})$, $i \in I$, such that for all U_{ij} and $x \in U_{ij}$, the following holds

$$h_j(x)\, g'_{ji}(x) = g_{ji}(x)\, h_i(x).$$

This gives a criterion for holomorphic vector bundles to be isomorphic. Recall that a vector bundle is said to be trivial if it is isomorphic to the vector bundle $B \times \mathbb{C}^p$.

Proposition 4.23. *Two holomorphic vector bundles are isomorphic if and only if they are defined by equivalent $\mathscr{U}$-cocycles for some atlas on B with covering $\mathscr{U}$. A holomorphic vector bundle F defined by a $\mathscr{U}$-cocycle $g = (g_{ij})$, for some atlas on B with $\mathscr{U} = (U_i)_{i \in I}$, is trivial if and only if there are morphisms*

$$h_i : U_i \to \mathrm{GL}(p, \mathbb{C}), \quad i \in I$$

such that

$$h_j(x) = g_{ji}(x)\, h_i(x)$$

for all $x \in B$ and all $i, j \in I$ such that U_{ij} is nonempty.

Remark 4.24. For a holomorphic vector bundle to be trivial, the condition on the defining cocycle in Proposition 4.23 is that this cocycle is a so-called 'coboundary'.

Definition 4.25. Let F be a holomorphic vector bundle over B and $\pi : F \to B$ its structural map. For any open subset $U \subset B$ let $\Gamma(U, F)$ denote the set of *sections of F over U*, that is, the set of morphisms $s : U \to F$ such that $\pi \circ s = \mathrm{id}_U$.

Note that $\Gamma(U, F)$ is an $\mathscr{O}(U)$-module, where $\mathscr{O}(U)$ is the ring of holomorphic functions on U. The elements of $\Gamma(F) = \Gamma(B, F)$ are called the *global sections*.

Theorem 4.26. *A rank p holomorphic vector bundle is a trivial bundle if and only if it has p, $\mathbb{C}$-linearly independent, global sections.*

Example 4.27. The *tangent bundle* T_B over B is a line bundle, $p = 1$, defined as follows. Let $(U_i, z_i)_{i \in I}$ be an atlas on B, where $z_i : U_i \xrightarrow{\sim} \tilde{U}_i \subset \mathbb{C}$ is a local coordinate, and consider on each U_{ij} the map

$$g_{ij} = \frac{dz_i}{dz_j}$$

where $g_{ij}(a)$ is for each $a \in U_{ij}$ the usual derivative at $z_j(a)$ of the complex function $z_i \circ z_j{}^{-1}$ on $z_j(U_{ij}) \subset \mathbb{C}$. Locally on a chart (U_i, z_i) the sections of F over U_i are given by the derivations on $\tilde{U}_i$, that is

$$\Gamma(U_i, T_B) = \mathscr{O}(\tilde{U}_i)\frac{d}{dz_i} = \left\{ \alpha(z_i)\frac{d}{dz_i},\ \alpha \in \mathscr{O}(\tilde{U}_i) \right\}.$$

The *cotangent bundle* T_B^* is defined by the cocycle $g^* = (g_{ij}^*)$ where $g_{ij}^* = g_{ji}$ for all $i, j \in I$. The sections here are given by the differential 1-forms on $\tilde{U}_i$, that is

$$\Gamma(U_i, T_B^*) = \left\{ \alpha(z_i)\, dz_i,\ \alpha \in \mathscr{O}(\tilde{U}_i) \right\}.$$

Example 4.28. The determinant bundle of a bundle is defined as follows. Let F be a holomorphic vector bundle over B, defined by a $\mathscr{U}$-cocycle $g = (g_{ij})$ on some trivializing covering $\mathscr{U} = (U_i)_{i \in I}$. It is easy to see that the maps

$$\gamma_{ij} = U_{ij} \to \mathbb{C}^*$$

defined by

$$\gamma_{ij}(x) = \det\left(g_{ij}(x)\right)$$

for all $x \in U_{ij}$, satisfy the cocycle conditions (CC) of Definition 4.19 p. 97, and thus define a line bundle $\det F$ over B called the determinant bundle of F.

4.2.2 Holomorphic vector bundles on $\overline{\mathbb{C}}$

We will refer to the following result as FTB (fundamental theorem for vector bundles on $\overline{\mathbb{C}}$).

Theorem 4.29 (FTB). *If $\Omega \subset \overline{\mathbb{C}}$ is a proper domain ($\Omega \neq \overline{\mathbb{C}}$) in $\overline{\mathbb{C}}$, then any holomorphic vector bundle over Ω is trivial.*

Proof. The result holds more generally over any non-compact Riemann surface (cf.[For81, Theorem 30.4]). Here is a sketch of proof for the particular case of a proper domain $\Omega \subset \overline{\mathbb{C}}$ (cf. [AB94, p. 55]). The subset Ω is a *Stein space* (cf. [Fstn11, Section 2.2]) since in particular it is an open Riemann surface. One uses the key fact that any holomorphic vector bundle on a Stein space is holomorphically trivial if it is topologically trivial, following famous theorems by Cartan and Grauert (cf. [Gr57], [Gr58], [Car58, Theorem A p. 102]). These illustrate the so-called *Oka-Grauert principle* ([FoLa10, Corollary 3.2], [Fstn11, Theorem 7.2.1]) following which, on a Stein space, a problem can roughly speaking "be solved by holomorphic functions if it can be solved by continuous functions". Note that any holomorphic vector bundle F over Ω is orientable as a real vector bundle (this is due to the fact that the $\mathbb{C}$-linear transition maps, considered as $\mathbb{R}$-linear maps, have a positive determinant so that any given orientation on a fiber extends continuously over Ω). Assume for example that $\mathbb{C} \setminus \Omega$ is finite. Since Ω deformation retracts to a bouquet of circles, F is topologically hence holomorphically trivial. (We leave it as an exercise to show that the restriction of F to each circle is trivial, see for instance [Pan05, Example 3.5]). □

An immediate consequence of this theorem is the following.

Corollary 4.30. *A holomorphic vector bundle over $\overline{\mathbb{C}}$ is determined by a cocycle*

$$g_{0\infty} : U_{0\infty} = \mathbb{C}^* \to \mathrm{GL}(p, \mathbb{C}).$$

for the trivializing atlas $U_0 = \mathbb{C}$, $U_\infty = \overline{\mathbb{C}} \setminus \{0\}$.

Example 4.31. Let $\mathscr{O}(k)$ denote the line bundle over $\overline{\mathbb{C}}$ defined by the cocycle

$$g_{0\infty} = z^{-k} = t^k$$

where $t = 1/z$. The trivializing covering is

$$U_0 = \mathbb{C} \xrightarrow{z} \mathbb{C}, \quad U_\infty = \overline{\mathbb{C}} \setminus \{0\} \xrightarrow{t=1/z} \mathbb{C}.$$

The vector bundle $\mathcal{O}(-2)$ is the tangent bundle over $\overline{\mathbb{C}}$ and $\mathcal{O}(2)$ the cotangent bundle. The vector bundle $\mathcal{O}(k)$ is trivial if and only if $k = 0$.

The following result describes all holomorphic vector bundles on $\overline{\mathbb{C}}$ in terms of line bundles.

Theorem 4.32 (Birkhoff–Grothendieck). *Any rank p holomorphic vector bundle F over $\overline{\mathbb{C}}$ is a direct sum of line bundles*

$$F \simeq \mathcal{O}(k_1) \oplus \ldots \oplus \mathcal{O}(k_p),$$

that is, F is defined by a cocycle $g_{0\infty} = \mathrm{diag}(z^{k_1}, \ldots, z^{k_p})$, $k_i \in \mathbb{Z}$.

A proof of this famous theorem can be found in [AB94, Section 3.3].

Definition 4.33. The *degree* of a holomorphic vector bundle F over $\overline{\mathbb{C}}$, denoted $\deg(F)$, is by definition

$$\deg(F) = \sum_{i=1}^{p} k_i.$$

In particular, if F is a bundle over $\overline{\mathbb{C}}$ defined by the cocycle

$$g_{0\infty} = \mathrm{diag}(z^{k_1}, \ldots, z^{k_p})$$

the associated determinant bundle $\det F$ is defined by the cocycle

$$\gamma_{0\infty} = z^{k_1 + \ldots + k_p}$$

and hence

$$\det F = \mathcal{O}(\deg(F)), \quad \deg(\det F) = \deg(F).$$

4.2.3 Connections

In this section we shall relate holomorphic vector bundles to differential equations via connections.

Definition 4.34. Let F be a holomorphic vector bundle of rank p over B and $\mathcal{U} = (U_i, f_i)_{i \in I}$ a trivializing atlas with local coordinate z_i on U_i for each $i \in I$. A *meromorphic connection* ∇ on F is a family of meromorphic differential systems

$$\frac{dy}{dz_i} = A_i(z_i)\, y \qquad (S_i)$$

of order p such that on each U_{ij} the systems (S_i) and (S_j) are gauge transforms of each other, that is,

$$A_i = \frac{dg_{ij}}{dz_i} g_{ij}^{-1} + g_{ij} A_j g_{ij}^{-1}, \tag{4.6}$$

via the defining cocycle $g = (g_{ij})$ corresponding to the trivializing covering $\mathscr{U}$. (For general results about analytic connections see [At57]). The gauge transformation is $y = g_{ij}\tilde{y}$, where y is the 'dependent variable', or unknown solution, of (S_i) on U_i, and $\tilde{y}$ the corresponding dependent variable of (S_j) on U_j. By a *connection* we will always mean a *meromorphic connection* if not otherwise specified.

One may equivalently define a connection as a map

$$\Gamma(F) \xrightarrow{\nabla} \Gamma(T_B^* \otimes F)$$

such that

$$\nabla(fs) = df \otimes s + f\nabla s$$

for all $f \in \mathscr{O}(B)$ and $s \in \Gamma(F)$, since locally on a trivializing open subset U_i such a map can be written

$$\nabla(s) = ds - A_i(z)s$$

for any section $s = (s_1, \ldots, s_p)^T \in \mathscr{O}(U_i)^p$ of F over U_i.

Definition 4.35. A *horizontal section* of ∇ is a section $s \in \Gamma(F)$ such that

$$\nabla(s) = 0.$$

A *local* horizontal section is a section $s \in \Gamma(U, F)$ for some open subset $U \subset B$, such that $\nabla_U(s) = 0$ for the restriction ∇_U of the connection ∇ to U. Over $U_i \in \mathscr{U}$, a horizontal section of the connection is the same as a solution of the differential system (S_i).

Over $\mathbb{C}$ and $\overline{\mathbb{C}}$, more can be said about connections on holomorphic vector bundles.

Remark 4.36. Any holomorphic vector bundle F on $\mathbb{C}$ is trivial by the FTB. A connection on F is therefore defined by a single meromorphic differential system on $\mathbb{C}$.

Theorem 4.37. *It is equivalent to define a meromorphic connection on a holomorphic vector bundle over $\overline{\mathbb{C}}$, or a differential system on $\mathbb{C}$ with coefficients that are rational functions.*

Proof. Let F be a holomorphic vector bundle on $\overline{\mathbb{C}}$. By Theorem 4.32 p. 100, it is of the form

$$F \simeq \mathscr{O}(k_1) \oplus \ldots \oplus \mathscr{O}(k_p).$$

Let

$$\frac{dy}{dx} = A(x)y \tag{S_0}$$

be a differential system with rational coefficients and (S_∞) the gauge transform of (S_0) by

$$g_{0\infty} = \operatorname{diag}(x^{k_1}, \ldots, x^{k_p}).$$

Formula (4.6) shows that (S_∞) is a meromorphic differential system, which implies that (S_0) and (S_∞) together define a meromorphic connection on F. Conversely, let ∇ be a meromorphic connection on F. Locally on $U_0 = \mathbb{C}$ it is given by

$$\frac{dy}{dx} = A_0(x)y \qquad (S_0)$$

where A_0 is meromorphic, and on $U_\infty = \overline{\mathbb{C}} \setminus \{0\}$ by

$$\frac{dy}{dz} = A_\infty(z)y \qquad (S_\infty)$$

where A_∞ is meromorphic in $\mathbb{C}$ as a function of the local coordinate $z = \frac{1}{x}$ of U_∞. Since (S_∞) is the gauge transform of (S_0) by $g_{0\infty}$ we have

$$A_0(x) = \frac{dg_{0\infty}}{dx} g_{0\infty}^{-1} + g_{0\infty} A_\infty(\frac{1}{x}).$$

Since $A_\infty(\frac{1}{x})$ is meromorphic at ∞, this shows that $A_0(x)$ is also meromorphic at ∞, hence rational since it is meromorphic on $\mathbb{C}$. $\square$

Definition 4.38. Let ∇ be a meromorphic connection on a holomorphic vector bundle F over B. An element $b \in B$ is a *singular point* for ∇ if for a trivializing covering $\mathscr{U} = (U_i)_{i \in I}$ and $b \in U_i$, b is a singular point of the differential system (S_i) defined by ∇ on U_i. The *type* of the singular point is the type (regular, or irregular) of the corresponding singular point of (S_i) on U_i. It is called a *logarithmic pole* if it is a Fuchsian singular point of (S_i).

Example 4.39 (Trace of a connection). A connection ∇ on a vector bundle F induces a connection on the determinant bundle $\det F$, in the following way. Locally on U_i, the connection is described by a system

$$(S_i): \ \frac{dy}{dz_i} = A_i(z_i)y$$

where z_i is the local coordinate. We know that whenever some $p \times p$ matrix function Y_i satisfies (S_i), its determinant $\det Y_i$ satisfies

$$\frac{d}{dz_i}(\det Y_i) = \operatorname{tr}(A_i(z_i)) \det Y_i.$$

The family of differential equations

$$(E_i): \quad w'(z_i) = \operatorname{tr}(A_i(z_i))\, w, \ \ i \in I$$

where for convenience here $(\ )'$ stands for d/dz_i, defines a connection on $\det F$ called the *trace of the connection* ∇. To see this, let $g = (g_{ij})$ be the cocycle defining F. We have seen that $\gamma = (\gamma_{ij})$, $\gamma_{ij} = \det g_{ij}$, is the defining cocycle of $\det F$. We need to verify that (E_j) is the gauge transform of (E_i) by γ_{ij}. We have

$$A_i = (g'_{ij})\,g_{ij}^{-1} + g_{ij}A_j\,g_{ij}^{-1},$$

hence

$$\mathrm{tr}(A_i) = \mathrm{tr}(g'_{ij}\,g_{ij}^{-1}) + \mathrm{tr}(A_j) = \gamma'_{ij}\,\gamma_{ij}^{-1} + \gamma_{ij}\,\mathrm{tr}(A_j)\,\gamma_{ij}^{-1}$$

since the γ_{ij} are scalars. We will write $\mathrm{tr}\nabla$ for the trace of the connection ∇, which is a connection on $\det F$.

We now consider Fuchsian connections over $\overline{\mathbb{C}}$, that is, connections with logarithmic poles only.

Exercise 4.40. Consider a system

$$\frac{dy}{dx} = A(x)y \qquad (S)$$

where $A(x)$ is holomorphic in $\mathbb{C}\setminus\{a_1,\dots,a_n\}$ and meromorphic at each a_i.

(a) Show that (S) is Fuchsian on $\overline{\mathbb{C}}$ if and only if there are constant matrices A_i such that

$$A(x) = \sum_{i=1}^{n} \frac{A_i}{x-a_i}$$

(b) Show that ∞ is non-singular if and only if

$$\sum_{i=1}^{n} A_i = 0.$$

Theorem 4.41. *Let F be a holomorphic rank p vector bundle over $\overline{\mathbb{C}}$. If F can be endowed with a Fuchsian connection, then $\deg(F) = 0$, that is, the determinant bundle $\det F$ is trivial.*

Proof. By Theorem 4.37 the Fuchsian connection ∇ is defined by a system

$$y' = A(x)y \qquad (S)$$

with rational coefficients in x and we may assume, modulo a change of local coordinate, that ∞ is a non-singular point. Following Exercise 4.40 we can write the coefficient matrix as

$$A(x) = \sum_{i=1}^{n} \frac{A_i}{x-a_i}, \quad \text{with} \quad \sum_{i=1}^{n} A_i = 0.$$

Let $F = \oplus_{i=1}^{p}\mathscr{O}(k_i)$ be the decomposition of F as a direct sum of line bundles, following Theorem 4.32 p. 100. The determinant bundle is

$$\det F = \mathscr{O}(k)$$

where $k = \sum_i k_i$ is the degree of F. The connection $\mathrm{tr}\nabla$ on $\det F$ induced by ∇ is defined by the scalar differential equation $y' = \mathrm{tr}\,(A(x))\,y$ on the chart $U_0 = \mathbb{C}$, with

$$\operatorname{tr}(A(x)) = \sum_{i=1}^{n} \frac{\operatorname{tr}(A_i)}{x-a_i} = \sum_{i=1}^{n} \frac{s_i}{x-a_i}$$

where

$$s_i = \sum_{j=1}^{p} \beta_i^j.$$

The β_i^j are the Levelt exponents, which in the Fuchsian case are the eigenvalues of the residue matrix A_i. The description of $\det F$ on the chart U_∞ is given by the change of variable $z = 1/x$,

$$\frac{dy}{dz} = -\left(\sum_{i=1}^{n} \frac{s_i}{(1-a_i z)\, z}\right) y$$

followed by the change of dependent variable $\tilde{y} = g_{\infty 0}\, y = x^{-k}\, y = z^k\, y$,

$$\frac{d\tilde{y}}{dz} = \left(\frac{k}{z} - \sum_{i=1}^{n} \frac{s_i}{(1-a_i z)\, z}\right) \tilde{y}.$$

Since in this chart the image 0 of ∞ (by the local coordinate) is non-singular, the residue at 0 equals 0, that is,

$$k - \sum_{i=1}^{n} s_i = 0.$$

It follows from the Fuchs relation $\sum_{i=1}^{n} s_i = 0$ for Fuchsian systems (cf. Corollary 4.13 p. 93) that $k = 0$, which ends the proof. □

4.3 The Riemann-Hilbert problem

In Section 3.1 p. 75 we gave an outline of history of the Riemann-Hilbert problem. In the present section we will sketch the ideas of Plemelj's proof (cf. [Ple64]), corrected by Treibich Kohn in [Tr83] and revisited by Bolibrukh (cf. [Bo92], [Bo92], [Bo95a], [Bo95(3)]). The tools introduced by Bolibrukh enabled him to produce his famous counterexamples and a number of sufficient conditions to solve the RH problem. These may hopefully lead one day to the full characterization of realizable monodromy representations. We keep notation from Section 4.2.

4.3.1 The monodromy of a connection

Let ∇ be a meromorphic connection on a holomorphic vector bundle F over B and Σ the set of its singular points. By the local existence of solutions, there is a trivializing chart $(U_i, f_i)_{i\in I}$ of F over $B \setminus \Sigma$ such that over each U_i there are p linearly independent holomorphic sections $s_1^i,\dots,s_p^i$ with $\nabla(s_k^i)=0$ for $k=1,\dots,p$ (that is, horizontal sections[1]). The p-tuple

$$s^i=(s_1^i,\dots,s_p^i)$$

corresponds to a fundamental solution of the corresponding differential system on U_i.

This gives a new trivialization of the vector bundle F by means of the horizontal sections s^i, with the constant cocycles $g_{ij}=s^j(s^i)^{-1}$.

Let us fix $x_0\in B_\Sigma=B\setminus\Sigma$ and a loop γ from x_0 in B_Σ, and cover the image of γ by trivializing open sets $U_0,U_1,\dots,U_r$ of the trivializing chart above, with $x_0\in U_0$ and $U_i\cap U_{i+1}\neq\emptyset$ for $i=0,\dots,r-1$ and $U_r\cap U_0\neq\emptyset$.

Since $s^0=s^1g_{1,0}$ on $U_{1,2}$ the fundamental section s^0 extends to U_1, renamed as $s^1g_{1,0}$. Again, since $s^1=s^2g_{2,1}$ on $U_{1,2}$, the fundamental section s^0 (renamed on U_1) extends to U_2, renamed as $s^2g_{2,1}g_{1,0}$. Iterating this procedure we see that s_0 extends along γ all the way to U_r where it is renamed $s^r g_{r,r-1}\cdots g_{2,1}g_{1,0}$ and back to U_0 as s^0C_γ where

$$C_\gamma=g_{0,r}g_{r,r-1}\cdots g_{2,1}g_{1,0}.$$

One can show that C_γ is independent of the procedure and only depends on the homotopy class of γ in B_Σ. It is called the *monodromy matrix* of ∇ with respect to the germ at x_0 of the horizontal sections s^0.

4.3.2 Connections with a given monodromy

Fix $p,n\in\mathbb{N}^*$ and a representation

$$\chi:\pi_1(U_\Sigma;x_0)\longrightarrow \mathrm{GL}(p,\mathbb{C}) \tag{4.7}$$

of the fundamental group of $U_\Sigma=\overline{\mathbb{C}}\setminus\{a_1,\dots,a_n\}$ and let G_i denote the image by χ of the elementary loop-class $[\gamma_i]$ around a_i, for $i=1,\dots,n$. In order to tentatively solve the Riemann-Hilbert problem for this representation, one has to define a connection on a *trivial* holomorphic bundle on $\overline{\mathbb{C}}$ with the given monodromy representation. Let us first construct a holomorphic bundle and a connection with this monodromy representation, then look for conditions under which this bundle be holomorphically trivial. In the next sections we will use vector bundles as well as principal fiber bundles.

Definition 4.42. Let G be an analytic Lie group and B a connected one-dimensional analytic variety. A *principal fiber bundle* over B with structural group G, or *principal G-bundle*, is an analytic variety P with a free, transitive right action of G and a surjective morphism $\pi:P\to B$ such that

[1] We use upper indices, following Bolibrukh's notation.

(a) for each $x \in B$ there is an open neighborhood $U \subset B$ of x and an isomorphism f_U such that the following diagram commutes

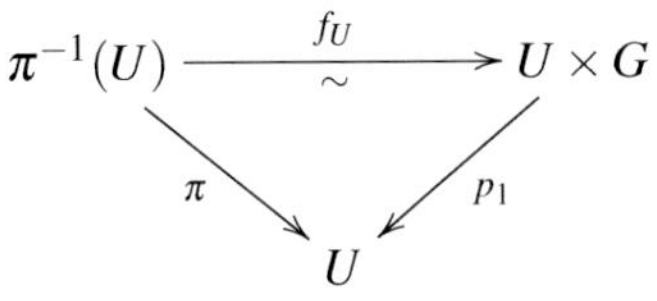

(b) the morphism f_U is G-equivariant, that is,

$$f_U(\tilde{x}g) = f_U(\tilde{x})\,g$$

for all $g \in G$ and $\tilde{x} \in \pi^{-1}(U)$, where the action of G on $U \times G$ is by right translation on the right factor.

Definition 4.43. The *universal covering* $\widetilde{U}_\Sigma$ of U_Σ with respect to a fixed base-point $x_0 \in U_\Sigma$ is the set of homotopy classes of all paths from x_0 in U_Σ.

Remark 4.44. In the special case $\Sigma = \{0, \infty\}$ the universal covering $\widetilde{U}_\Sigma$ of $U_\Sigma = \mathbb{C}^*$ is the *Riemann surface* $\widetilde{\mathbb{C}}$ *of the logarithm*, see Section 6.7 p. 195.

The universal covering $\widetilde{U}_\Sigma$ is in particular a principal $\pi_1(U_\Sigma; x_0)$-bundle over U_Σ. The covering map

$$\pi : \widetilde{U}_\Sigma \to U_\Sigma$$

maps the class of a path α (from x_0) to the endpoint $x \in U_\Sigma$ of this path, and the fundamental group $\pi_1(U_\Sigma; x_0)$ acts to the right on $\widetilde{U}_\Sigma$ by

$$[\alpha] \mapsto [\alpha\gamma]$$

for $[\gamma] \in \pi_1(U_\Sigma; x_0)$ and $[\alpha] \in \widetilde{U}_\Sigma$. As in Definition (1.30) p. 17, the composition $\alpha\gamma$ is defined by

$$\alpha\gamma = \begin{cases} \gamma(2t) & \text{if } 0 \le t \le 1/2 \\ \alpha(2t-1) & \text{if } 1/2 \le t \le 1. \end{cases}$$

Locally over an open disk D in U_Σ we have an isomorphism

$$\pi^{-1}(D) \simeq D \times \pi_1(U_\Sigma; x_0)$$

which can be described as follows. Let a be the center of D and let $\tilde{a}$ be any path from x_0 to a. For any $x \in D$ let $\tilde{r}$ denote the radial path from a to x in D and let $\tilde{x} = \tilde{r}\tilde{a}$. The isomorphism is given by

$$\alpha \mapsto (x, \tilde{x}^{-1}\alpha)$$

for any element α of $\pi^{-1}(D)$, that is, (the homotopy class of) a path α from x_0 to $x \in D$.

We are going to construct a vector bundle F over U_Σ, together with its *frame bundle* P, which by definition is the principal $\mathrm{GL}(p,\mathbb{C})$-bundle whose fibers are the $\mathbb{C}$-linear isomorphims of the fibers of F. Then we will define a connection ∇ on F with the prescribed monodromy and see how to extend F and ∇ to $\overline{\mathbb{C}}$ in order to solve the Riemann-Hilbert problem under certain conditions. Consider the quotients

$$F = (\widetilde{U_\Sigma} \times \mathbb{C}^p)/\sim \quad \text{and} \quad P = \widetilde{U_\Sigma} \times \mathrm{GL}(p,\mathbb{C})/\sim$$

of $\widetilde{U_\Sigma} \times \mathbb{C}^p$ and $\widetilde{U_\Sigma} \times \mathrm{GL}(p,\mathbb{C})$ respectively, where for any $(\alpha, v), (\beta, w) \in \widetilde{U_\Sigma} \times \mathbb{C}^p$ (respectively $\widetilde{U_\Sigma} \times \mathrm{GL}(p,\mathbb{C})$) the equivalence relation is given by

$$(\alpha, v) \sim (\beta, w) \ \text{ if}$$

(a) $\pi(\alpha) = \pi(\beta)$, that is, $\alpha = \beta\gamma$ for some loop γ from x_0 in U_Σ and
(b) $w = \chi([\gamma])\, v$ for this γ.

In other words, for any $(\alpha, v) \in \widetilde{U_\Sigma} \times \mathbb{C}^p$ (respectively $\widetilde{U_\Sigma} \times \mathrm{GL}(p,\mathbb{C})$) and any loop γ from x_0 in U_Σ, we identify

$$(\alpha\gamma,\, v) \quad \text{and} \quad (\alpha,\, \chi([\gamma])\, v).$$

One can show that F is a vector bundle, and P a principal $\mathrm{GL}(p,\mathbb{C})$-bundle over U_Σ, namely the frame-bundle of F whose fibers over U_{ij} are the $\mathbb{C}$-linear isomorphisms between the fibers of F over U_i and U_j respectively, for a trivializing covering $(U_i)_{i\in I}$. More precisely, the vector bundle structure of F over U_Σ is defined as follows. Locally over an open disk D of U_Σ, an element of

$$F = (\widetilde{U_\Sigma} \times \mathbb{C}^p)/\sim$$

can be identified with the class of a triple

$$(x, [\gamma], u) \in D \times \pi_1(U_\Sigma; x_0) \times \mathbb{C}^p$$

via the isomorphism

$$\pi^{-1}(D) \simeq D \times \pi_1(U_\Sigma; x_0).$$

Let $\tilde{\pi} : F \to U_\Sigma$ denote the projection map on U_Σ induced by

$$\tilde{\pi}((x, [\gamma], u)) = x.$$

Locally over D the map

$$f_D : \tilde{\pi}^{-1}(D) \longrightarrow D \times \mathbb{C}^p$$

defined by

$$f_D(\langle x, [\gamma], u\rangle) = \langle x,\, \chi([\gamma])u\rangle, \tag{4.8}$$

where $\langle x, [\gamma], u\rangle$, for each $x \in D$, denotes the class of $(x, [\gamma], u)$, is the fiber-preserving isomorphism as in Definition 4.15 p. 94. The (common) defining cocycles of the bundles F and P are constant. To see this on F for instance, let $(U_i)_{i\in I}$ be a trivializing covering of U_Σ for F by open disks, with the trivializing maps $f_i = f_{U_i}$. The cocycle g_{ij} on $U_i \cap U_j$ is given by

$$U_{ij} \times \mathbb{C}^p \xleftarrow{f_j} \tilde{\pi}^{-1}(U_{ij}) \xrightarrow{f_i} U_{ij} \times \mathbb{C}^p$$

$$(x,v) \longmapsto (x, g_{ij}(v))$$

where $g_{ij} \in \mathrm{Im}\chi$ by (4.8). Since the g_{ij} are continuous and $\mathrm{Im}\chi$ is a discrete subgroup of $\mathrm{GL}(p,\mathbb{C})$, the g_{ij} must be constant.

Throughout this section we assume that there is at least one prescribed singular point for the Riemann-Hilbert problem, that is, $n \geq 1$. Since its transition morphisms g_{ij} are constant, we can define a connection ∇ on F in a natural way by

$$\frac{dy}{dz_i} = 0$$

on each U_i and the gauge transformation (4.6) by g_{ij} on each intersection U_{ij}. This is a holomorphic connection and it is not difficult to see that it has, by construction, the given monodromy. Since F is a bundle over a proper open domain U_Σ of $\overline{\mathbb{C}}$ (we assumed $n \geq 1$) this bundle is trivial and its connection represented by some actual, differential system over U_Σ. The problem remains to extend the bundle F together with the connection ∇ holomorphically to $\overline{\mathbb{C}}$ and to see whether this extension is (holomorphically) trivial.

Note that horizontal sections of F with respect to a connection on F correspond to (vector) solutions of the corresponding differential equation, whereas fundamental solutions of this differential equation yield sections of P ('horizontal' with respect to the same connection, meaning that each column of the fundamental solution represents a horizontal section of F). Since a differential system is completely determined by a fundamental solution we will work simultaneously with the vector bundle F and its frame-bundle P.

Notation 4.45 *For any function f on $\widetilde{U}_\Sigma$ the action of $\pi_1(U_\Sigma; x_0)$ will be written*

$$\gamma^* f(\tilde{x}) = f(\tilde{x}\gamma)$$

for $[\gamma] \in \pi_1(U_\Sigma; x_0)$ and $\tilde{x} \in \widetilde{U}_\Sigma$. As before $\langle \tilde{x}, g \rangle$ denotes the class in P of an element $(\tilde{x}, g) \in \widetilde{U}_\Sigma \times \mathrm{GL}(p,\mathbb{C})$ and the map $\tilde{\pi} : P \to U_\Sigma$ is given by $\tilde{\pi}(\langle \tilde{x}, g \rangle) = x$.

The map $T : \widetilde{U}_\Sigma \to P$ defined by $T(\tilde{x}) = \langle \tilde{x}, I \rangle$ commutes with π and $\tilde{\pi}$

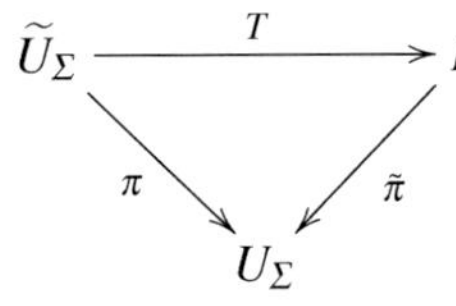

and the actions, on each side, of the structural groups $\pi_1(U_\Sigma; x_0)$ and $\mathrm{GL}(p,\mathbb{C})$ respectively, that is,

$$T(\tilde{x}\gamma) = T(\tilde{x})\,\chi([\gamma]) = \langle \tilde{x}, \chi([\gamma]) \rangle = \langle \tilde{x}\gamma, I \rangle$$

for all $[\gamma] \in \pi_1(U_\Sigma; x_0)$ and $\tilde{x} \in \widetilde{U}_\Sigma$.

The following simple observation leads to a useful formula, which we will label $(TV\widetilde{Y})$ in analogy to Bolibrukh's original, mnemonic notation. For any holomorphic section V of $\tilde{\pi}$ and $\tilde{x} \in \widetilde{U}_\Sigma$ there is namely a unique $\widetilde{Y}(\tilde{x}) \in \mathrm{GL}(p, \mathbb{C})$ such that

$$T(\tilde{x}) = V(\pi(\tilde{x}))\widetilde{Y}(\tilde{x}) \qquad (TV\widetilde{Y})$$

(cf. [Bo92] p. 141).

We can use this formula to define a one-to-one correspondence between the set of (global) holomorphic sections V of $\tilde{\pi}$ and the set of fundamental solutions $\widetilde{Y}$ on $\widetilde{U}_\Sigma$ of meromorphic systems (S) over $\overline{\mathbb{C}}$ with the given monodromy, as follows.

Lemma 4.46. *With notation as above, for any holomorphic section V of $\tilde{\pi}$ the matrix function $\widetilde{Y}$ given by $(TV\widetilde{Y})$ is a holomorphic function from $\widetilde{U}_\Sigma$ to $\mathrm{GL}(p, \mathbb{C})$ with the given monodromy χ.*

Proof. The function $\widetilde{Y}(\tilde{x})$ is holomorphic since V is. For each $[\gamma] \in \pi_1(U_\Sigma; x_0)$ we have

$$\begin{aligned}
\gamma^* T(\tilde{x}) &= T(\tilde{x}\gamma) = \langle \tilde{x}\gamma, I \rangle = \langle \tilde{x}, \chi([\gamma])\, I \rangle \\
&= \langle \tilde{x}, I \rangle\, \chi([\gamma]) = T(\tilde{x})\, \chi([\gamma]) \\
&= V\,(\pi(\tilde{x}))\, \widetilde{Y}(\tilde{x}) \chi([\gamma]),
\end{aligned}$$

and applying $(TV\widetilde{Y})$ we get

$$\gamma^* T(\tilde{x}) = \gamma^* (V\,(\pi(\tilde{x}))\, Y(\tilde{x})) = V\,(\pi(\tilde{x}))\, \widetilde{Y}(\tilde{x}\gamma),$$

hence

$$\widetilde{Y}(\tilde{x}\gamma) = \widetilde{Y}(\tilde{x})\, \chi([\gamma])$$

for all $\tilde{x} \in \widetilde{U}_\Sigma$, which proves that $\widetilde{Y}$ has the given monodromy. □

Conversely, for any holomorphic function $\widetilde{Y} : \widetilde{U}_\Sigma \mapsto \mathrm{GL}(p, \mathbb{C})$ with the given monodromy χ, the $(TV\widetilde{Y})$ formula gives a holomorphic section V of $\tilde{\pi}$, by the following fact.

Lemma 4.47. *The fonction*

$$\tilde{x} \mapsto T(\tilde{x})\, \widetilde{Y}(\tilde{x})^{-1}$$

is constant on $\pi^{-1}(\pi(\tilde{x}))$.

Proof. For any $\tilde{x}\gamma \in \pi^{-1}(\pi(\tilde{x}))$ we have

$$\begin{aligned}
T(\tilde{x}\gamma)\, \widetilde{Y}(\tilde{x}\gamma)^{-1} &= T(\tilde{x})\chi([\gamma])\,(\widetilde{Y}(\tilde{x})\chi([\gamma]))^{-1} \\
&= T(\tilde{x})\chi([\gamma])\chi([\gamma])^{-1}\widetilde{Y}(\tilde{x})^{-1} \\
&= T(\tilde{x})\, \widetilde{Y}(\tilde{x})^{-1}
\end{aligned}$$

which proves the result. □

The section $V : U_\Sigma \to P$ satisfying $(TV\widetilde{Y})$ is thus well-defined by

$$V(x) = T(\tilde{x})\widetilde{Y}(\tilde{x})^{-1}$$

for all $x \in U_\Sigma$, where $\tilde{x} \in \widetilde{U}_\Sigma$ is such that $\pi(\tilde{x}) = x$. Since the bundles F and P are holomorphically trivial by the FTB p. 99, there exist such sections and the corresponding differential systems have the given monodromy. How can we know whether their prescribed singular points are Fuchsian? The problem is now to extend F and ∇ on $\overline{\mathbb{C}}$ (and the global horizontal sections) on $\overline{\mathbb{C}}$ so as to get a differential system with regular singular points only, as many of them Fuchsian as possible.

4.3.3 Plemelj's result

Assume (S) is a Fuchsian differential system on $\overline{\mathbb{C}}$ with the given monodromy. Then (S) defines a holomorphic section V of $\tilde{\pi}$ and this section corresponds by $(TV\widetilde{Y})$ to a fundamental solution $\widetilde{Y}$ of (S) on $\widetilde{U}_\Sigma$, which locally induces fundamental solutions of (S) on U_Σ. We know that on a neighborhood $\mathscr{O}_i$ of $a_i \in \Sigma$ (with a local coordinate x) the system (S) has a fundamental solution of the Levelt form

$$U_i(x)x^{\Lambda_i}x^{E_i}$$

where the matrix $E_i = (1/2\pi\mathrm{i})\log G_i$ is upper block triangular (each block corresponds to a rootspace, in the solution space, of the action of the monodromy) and $\Lambda_i = \mathrm{diag}(\lambda_i^1, \dots, \lambda_i^p)$ where $\lambda_i^k \in \mathbb{Z}$ for all i,k and $\lambda_i^k \geq \lambda_i^{k+1}$ whenever λ_i^k and λ_i^{k+1} belong to the same block[2] (the blocks of Λ_i are shaped according to those of E_i) and U_i is holomorphically invertible at a_i, that is, $\det(U_i(0)) \neq 0$, by Levelt's criterion (Thm. 4.7 p. 90) for Fuchsian singular points. Thus there is an invertible matrix $X_i \in \mathrm{GL}(p, \mathbb{C})$ such that

$$Y(x) = X_i\left(U_i(x)x^{\Lambda_i}x^{E_i}\right)X_i^{-1} \tag{4.9}$$

locally at a_i and such that for all $i = 1, \dots, n$, the matrix $X_i^{-1}G_iX_i$ is block upper triangular.

Conversely, let us follow Bolibrukh's terminology (see [Bo95(3)]) and call any data $\mathscr{A}$ of families $\{X_i\}$ and $\{\Lambda_i\}$ of matrices with the above properties an admissible family. More precisely :

Definition 4.48. For a given monodromy representation (4.7) the family

$$\mathscr{A} = \{X_i, \Lambda_i\}_{1\leq i\leq n}$$

of matrices is an *admissible family* if for each i

[2] Here too we use upper indices in λ_i^j, following Bolibrukh's original notation.

(a) $X_i^{-1} G_i X_i$ is upper block triangular

(b) $\Lambda_i = \mathrm{diag}(\lambda_i^1, \ldots, \lambda_i^p)$, where $\lambda_i^k \in \mathbb{Z}$ for all i,k and $\lambda_i^k \geq \lambda_i^{k+1}$ whenever λ_i^k and λ_i^{k+1} belong to the same block (with respect to the diagonal block-structure of $X_i^{-1} G_i X_i$).

With any admissible family $\mathscr{A}$ one can actually construct an extension $\widetilde{F}$ (and $\widetilde{P}$ simultaneously) over $\overline{\mathbb{C}}$ of the bundle F (resp. P) and endow $\widetilde{F}$ with a meromorphic connection $\widetilde{\nabla}$ whose set Σ of prescribed singular points consists of Fuchsian singular points only (or logarithmic poles, in the usual terminology for connections). Note that, as shown before, all such logarithmic extensions are obtained this way.

The idea of the construction (of an extension $\widetilde{F}$ with a connection) is the following. Let $\mathscr{A} = \{X_i, \Lambda_i\}_{1 \leq i \leq n}$ be a given admissible family with respect to the monodromy representation (4.7). Let

$$E_i = \frac{1}{2\pi \mathrm{i}} \log(X_i^{-1} G_i X_i) \tag{4.10}$$

for each prescribed monodromy matrix G_i. If x denotes the local coordinate centered at a_i, let us for each value of x fix an arbitrary element $\tilde{y}_x$ in the fiber $\pi^{-1}(\{x\})$, then define $\widetilde{Y}$ on $\widetilde{U}_\Sigma$ by

$$\widetilde{Y}(\tilde{x}) = X_i \left(x^{\Lambda_i} \tilde{y}_x^{E_i} \right) X_i^{-1} \chi(\sigma),$$

where $x = \pi(\tilde{x})$ and $\sigma \in \pi_1(U_\Sigma; x_0)$ is given by $\tilde{x} = \tilde{y}_x \sigma$. It is easy to prove that this definition does not depend on the choice of $\tilde{y}_x$ in the fiber of x. Locally at each a_i we can thus apply the $(TV\widetilde{Y})$ formula to simultaneously extend F and P at $a_i \in \Sigma$ as holomorphic bundles over this chart, using the global section V. Then we define a connection on (the extension of) F, locally on each chart at a_i, by the differential system satisfied by $\widetilde{Y}$. This connection has regular singular points only, all Fuchsian but one. At $a_i \in \Sigma$ the $(TV\widetilde{Y})$ formula indeed provides a holomorphic section, let us call it V^{Λ_i} (it plays the role of V), of P over the punctured neighborhood $\mathscr{O}_i \setminus \{a_i\}$. This section is then used to glue together the bundle P and the trivial bundle $\mathscr{O}_i \times \mathrm{GL}(p, \mathbb{C})$ over $\mathscr{O}_i$ by identifying

$$V^{\Lambda_i}(x) \quad \text{with} \quad (x, I)$$

for all x, then

$$V^{\Lambda_i}(x) g \in \pi^{-1}(\{x\}) \quad \text{with} \quad (x, g)$$

for all $g \in \mathrm{GL}(p, \mathbb{C})$, thus extending the section V^{Λ_i} as a section V over the full neighborhood $\mathscr{O}_i$.

Let us consider a description of the principal fiber bundle $\widetilde{U}_\Sigma$ by means of an open covering $\mathscr{U} = (U_j)_j$, trivializing maps

$$h_j : \pi^{-1}(U_j) \to U_j \times \pi_1(U_\Sigma; x_0)$$

and constant cocycles

$$h_{ij} : U_{ij} \to \pi_1(U_\Sigma; x_0)$$

(since the image of these continuous functions is discrete). It follows from the construction of the bundles F and P that they are both defined by the cocycles $g_{ij} = \chi \circ h_{ij}$. Since these cocycles are constant (this was already shown in Section 4.3.2) we can define a connection ∇ on F, namely the connection which on each U_i is represented by the differential system $dy/dz_i = 0$. Then, on the extension $\widetilde{F}$ on $\overline{\mathbb{C}}$ of the vector bundle F one defines an extension $\widetilde{\nabla}$ of the connection ∇ that is compatible with the local glueing. Locally over $\mathcal{O}_i$, this extension is given by the differential system satisfied by

$$Y_i(z) = X_i z^{\Lambda_i} z^{E_i} X_i^{-1} \tag{4.11}$$

where z is the local coordinate at a_i. This connection clearly has a logarithmic singular point at a_i. (One has to carefully check that starting with this system on some open $U_k \cap \mathcal{O}_i \neq \emptyset$, its gauge transforms by the constant cocycles fit together to define a connection $\widetilde{\nabla}$ on $\widetilde{F}$ extending the connection ∇ on F, with logarithmic singular points and the prescribed monodromy). Since $\widetilde{P}$ is holomorphically trivial over, say, $\overline{\mathbb{C}} \setminus \{a_1\}$ by the FTB p. 99, any holomorphic section of this restriction of $\widetilde{P}$ yields by $(TV\widetilde{Y})$ a fundamental solution Y of a meromorphic differential system (S) with Fuchsian singular points at $\{a_2, \dots, a_n\}$ and the prescribed monodromy χ. Plemelj's theorem, revisited and completed by Treibich Kohn [Tr83] is the following. By (S) is meant a differential system with coefficients in $\mathbb{C}(x)$.

Theorem 4.49 (Plemelj, Treibich Kohn). *With notation of this section:*

(a) *There exists a differential system (S) with prescribed singular points in Σ and the given monodromy representation χ, the singular points of which are all Fuchsian but possibly one regular singular point.*
(b) *If one of the given monodromy matrices $G_i = \chi([\gamma_i])$ is diagonalizable, then χ is realizable as the monodromy representation of a Fuchsian system.*

We will present the proof by Bolibrukh of this theorem. It relies on the following result, referred to as *Kimura's lemma* by Bolibrukh, who actually reworked a lemma [Sib90, p. 83] attributed by Sibuya to Kimura [Ki71, p. 225].

Lemma 4.50. *Let $K = \mathrm{diag}(k_1, \dots, k_p)$ with integers k_i, and let $V(x)$ be some $p \times p$ matrix function which is holomorphically invertible at 0. Then there exists a matrix function $\Gamma(1/x)$ which is holomorphically invertible on $\overline{\mathbb{C}} \setminus \{0\}$, meromorphic at 0 and such that* $\det \Gamma = 1$ *and*

$$\Gamma(\frac{1}{x}) x^K V(x) = \hat{V}(x) x^{K'}$$

in the neighbourhood of 0, where $\hat{V}$ is holomorphically invertible at 0 and K' is diagonal, obtained by some permutation of the diagonal elements of K.

Note that the entries of Γ are rational functions by Theorem 1.2 p. 5, actually polynomials in $1/x$ since they are holomorphic in U_∞. We will admit the proof of this lemma for which we refer to ([Ilya07, Lemma 16.36 p. 282] and only sketch the proof of Theorem 4.49.

Proof. (of Theorem 4.49). With notation of this section, consider any admissible family $\mathscr{A}$ with $S_1 = I$ and $\Lambda_1 = 0$, where we assume for convenience that $a_1 = 0$. Let $\widetilde{F}$ and $\widetilde{P}$ denote the vector bundle and principal fiber bundle respectively, constructed over $\overline{\mathbb{C}}$ by means of $\mathscr{A}$. By the Birkhoff–Grothendieck theorem we have

$$\widetilde{F} \simeq \mathscr{O}(k_1) \oplus \ldots \oplus \mathscr{O}(k_p)$$

for some integers $k_1 \leq \ldots \leq k_p$. We recall that $\widetilde{F}$ is described by the atlas $U_0 = \mathbb{C}$, $U_\infty = \overline{\mathbb{C}} \setminus \{0\}$ and the glueing cocycle $g_{0\infty} = x^K$ where $K = \mathrm{diag}(k_1, \ldots, k_p)$. The extended principal fiber bundle $\widetilde{P}$ has the corresponding description

$$\widetilde{P}|_{U_0} \simeq U_0 \times \mathrm{GL}(p, \mathbb{C}) \quad \text{and} \quad \widetilde{P}|_{U_\infty} \simeq U_\infty \times \mathrm{GL}(p, \mathbb{C})$$

with the same glueing cocycle $g_{0\infty}$. Consider the global meromorphic section of $\widetilde{P}$ defined locally by

$$\widetilde{V}_0 = f_0^{-1} \circ V_0, \quad \widetilde{V}_\infty = f_\infty \circ V_\infty,$$

where f_0 and f_∞ are the trivializing homeomorphisms of $\widetilde{P}$

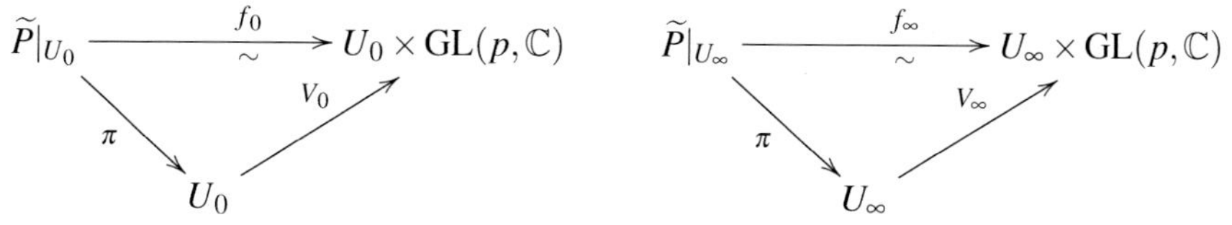

and

$$V_0(z) = (z, z^K), \quad V_\infty(z) = (z, I),$$

where z in each case denotes the local coordinate ($z = x$ in U_0, $z = 1/x$ in U_∞). Since $V_0 = g_{0\infty} V_\infty$, the local sections $\widetilde{V}_0$ and $\widetilde{V}_\infty$ glue together into a global section $\widetilde{V}$ of $\widetilde{P}$ which is holomorphic in U_∞ and meromorphic at 0. By the $(TV\widetilde{Y})$ formula the section $\widetilde{V}$ of $\widetilde{P}$, holomorphic in U_∞, defines a fundamental matrix $\widetilde{Y}$,

$$T(\tilde{x}) = \widetilde{V}(\pi(\tilde{x}))\widetilde{Y}(\tilde{x}) \tag{4.12}$$

such that $\widetilde{Y}$ is a fundamental solution of a differential system over U_∞ with Fuchsian singular points at $a_2, \ldots, a_p$ (since these are logarithmic for the connection we defined on $\widetilde{P}$, in the above construction). Let us show that $a_1 = 0$ is a regular singular point of this system. At $a_1 = 0$, we have $Y_1 = x^{E_1}$ by (4.11) p. 112 since $X_1 = I$ and $\Lambda_1 = 0$ in $\mathscr{A}$. The glueing section used to extend P at a_1 is therefore V_1 defined by

$$T(\tilde{x}) = V_1(\pi(\tilde{x}))\, Y_1(\tilde{x}) = V_1(\pi(\tilde{x}))\, x^{E_1}$$

with $E_1 = (1/2\pi \mathrm{i}) \log G_1$. By construction (glueing P locally with the trivial bundle via V_1) the section V_1 is holomorphic at $a_1 = 0$. Since the section $\widetilde{V}$ is meromorphic at $a_1 = 0$, there exists a matrix W, meromorphic at $a_1 = 0$, such that $V_1 = \widetilde{V} W$. Thus, locally at $a_1 = 0$ (with the local coordinate x) we get by $(TV\widetilde{Y})$ that

$$\widetilde{Y}(\tilde{x}) = W(x)x^{E_1}$$

with $x = \pi(\tilde{x})$. This proves that the differential system satisfied by $\widetilde{Y}$ is regular singular at $a_1 = 0$ by Corollary 1.37 p. 20.

To prove the second part of the theorem, assume that G_1, at $a_1 = 0$, is the diagonalizable monodromy matrix and, via a gauge transformation by a constant invertible matrix (which does not affect the type of the singular points) that G_1 is diagonal. We choose as before an admissible family $\mathscr{A}$ with $X_1 = I$ and $\Lambda_1 = 0$ to construct the fiber bundle $\widetilde{P}$. The resulting system (S), defined as before, has a fundamental matrix Y which locally at $a_1 = 0$ can be written as

$$Y(x) = W(x)x^{E_1}$$

where $E_1 = (1/2\pi \mathrm{i})\log G_1$ is diagonal and W is meromorphic at $a_1 = 0$. Let us show that the cocycle $g_{\infty 0} = x^{-K}$ is in fact equivalent to W as a cocycle defining $\widetilde{P}$. With notation as before, we have $V_0 = g_{0\infty}V_\infty$, that is,

$$V_1 W^{-1} = g_{0\infty}V_\infty.$$

Since V_1 is a holomorphic section of $\widetilde{P}$ over U_0 and V_∞ is a holomorphic section over U_∞, this implies that W^{-1} is a cocycle $\tilde{g}_{0\infty}$ which is equivalent to $g_{0\infty}$ and that W is its inverse cocycle $\tilde{g}_{\infty 0}$. It also implies that there exist matrix functions Γ and W_1, holomorphically invertible in U_∞ and U_0 respectively, such that

$$\Gamma(x)W(x) = x^{-K}W_1(x),$$

(cf. [IlYa07], Lemma 17.39 p. 305). If we transform (S) into a system $(\widetilde{S})$ via $\tilde{Y} = \Gamma Y$, this will not affect the Fuchsian singular points $a_2,\ldots,a_n$ but only possibly the singular point $a_1 = 0$ in the neighborhood of which $\widetilde{Y}$ has the expression

$$\widetilde{Y}(x) = x^{-K}W_1(x)x^{E_1}. \tag{4.13}$$

By Lemma 4.50 there is a matrix Γ_1, polynomial in $1/x$, such that

$$\Gamma_1(\frac{1}{x})x^{-K}W_1(x) = W_2(x)x^L$$

where W_2 is holomorphically invertible at 0 and L is a constant diagonal matrix. If we transform the system $(\widetilde{S})$ satisfied by $\widetilde{Y}$ via $\widetilde{Y}_1 = \Gamma_1\widetilde{Y}$, this will only affect the singular point $a_1 = 0$ since Γ_1 is holomorphically invertible in U_∞. Thus, we get a new system $(\widetilde{S}_1)$ with the coefficient matrix

$$\begin{aligned}\frac{d\widetilde{Y}_1}{dx}\widetilde{Y}_1^{-1} &= \frac{dW_2}{dx}(W_2)^{-1} + \frac{1}{x}W_2(L + x^L E_1 x^{-L})(W_2)^{-1}\\ &= \frac{dW_2}{dx}(W_2)^{-1} + \frac{1}{x}W_2(L + E_1)(W_2)^{-1}\end{aligned}$$

since the diagonal matrices x^L and E_1 commute. This shows that $a_1 = 0$ is a Fuchsian singular point of $(\widetilde{S})$ and that χ can be realized as announced by a Fuchsian system with singular points in Σ. □

To conclude this chapter let us sketch the proof of a theorem due to Bolibrukh and Kostov independenlty (cf. [Bo92], [Kos92]) which gives another sufficient condition to solve the RH problem.

Theorem 4.51 (Bolibrukh, Kostov). *Any irreducible representation χ can be realized by a Fuchsian system.*

Proof. Consider a system

$$\frac{dY}{dx} = BY \qquad (S)$$

with regular singular points $\Sigma = \{a_1, \ldots, a_n\}$, all Fuchsian but a_1, and assume that this system was obtained by means of an admissible family $\mathscr{A}$ such that

$$\Lambda_2 = \ldots = \Lambda_n = 0, \quad \Lambda_1 = \mathrm{diag}(\lambda_1, \ldots, \lambda_p)$$

with integers $\lambda_1 \geq \ldots \geq \lambda_p$. We moreover assume that

$$\lambda_i - \lambda_{i+1} > p(n-2)$$

for $i = 1, \ldots, p-1$, and that $a_1 = 0$ and $\infty \notin \Sigma$. If we follow the construction used in the proof of Theorem 4.49 p. 112, we get a fundamental solution of (S) of the form (4.13) p. 114 which in the neighborhood of $a_1 = 0$ is of the form

$$Y(x) = x^K V(x) x^{\Lambda_1} x^{E_1}$$

where V is holomorphically invertible at 0 and $K = \mathrm{diag}(k_1, \ldots, k_p)$ with $k_1 \geq \ldots \geq k_p$. (We have renamed $-K$ as K). Let us show that

$$k_i - k_{i+1} \leq n-2 \quad \text{for} \quad i = 1, \ldots, p-1.$$

Assume that $k_\ell - k_{\ell+1} > n-2$ for some ℓ. Locally at $a_1 = 0$, the coefficient matrix $B = (b_{ij})$ of (S) is

$$\begin{aligned} B(x) &= \frac{dY}{dx} Y^{-1} = \frac{K}{x} + x^K \left(\frac{dV}{dx} V^{-1} + \frac{V}{x} (\Lambda_1 + x^{\Lambda_1} E_1 x^{-\Lambda_1}) V^{-1} \right) x^{-K} \\ &= \frac{K}{x} + x^K \widetilde{B}(x) x^{-K} \end{aligned}$$

where $\widetilde{B} = (\tilde{b}_{ij})$ is defined by this formula. Then either $\widetilde{B} = 0$ or $\widetilde{B}$ has a pole of order 1 at 0. For $i \neq j$ we have $b_{ij} = \tilde{b}_{ij} x^{k_i - k_j}$. If $b_{ij} \neq 0$ for some i, j such that $i \leq \ell < j$ then $k_i - k_j \geq k_\ell - k_{\ell+1} > n-2$. This shows that b_{ij} has a zero of order $> n-3$ at $a_1 = 0$. Moreover, the system has coefficients that are meromorphic at all a_i and hence are rational functions. Since the system is Fuchsian at $a_2, \ldots, a_n$, the order of these poles of b_{ij} is ≤ 1 and since ∞ is not a singular point of the system, it must be a zero of order ≥ 2 of b_{ij} (this is easily seen by the change of variable $x = 1/t$ in (S)). We know from basic facts in complex analysis that since b_{ij} is a rational function, the sum of orders of the zeros of b_{ij} is equal to the sum of orders of its poles in $\overline{\mathbb{C}}$. From the inequalities above we get that the sum of orders of the zeros of b_{ij} is $> (n-3)+2$, whereas the sum of orders of the poles is $\leq (n-1)$. This is impossible, hence $b_{ij} = 0$ for $i \leq \ell < j$. The matrix B has therefore a block form

$$B = \begin{pmatrix} * & 0 \\ * & B_1 \end{pmatrix}$$

and this shows that the system (S), as well as its monodromy representation χ, is reducible. This contradicts the assumption and proves that $k_i - k_{i+1} \leq n-2$ for all $i = 1, \ldots, p-1$. By Lemma 4.50 applied to $x^K V(x)$ in

$$Y(x) = x^K V(x) x^{\Lambda_1} x^{E_1}$$

there exist a polynomial matrix Γ, a holomorphically invertible matrix W at 0 and a constant diagonal matrix $\widehat{K}$ obtained from K by some permutation of its entries, such that

$$\Gamma Y = W x^{\widehat{K}} x^{\Lambda_1} x^{E_1}.$$

If $\widehat{K} = \mathrm{diag}(\hat{k}_1, \ldots, \hat{k}_p)$, we have

$$|\hat{k}_i - \hat{k}_{i+1}| < (n-2)(p-1)$$

since $\widehat{K}$ is a permutation of K. Assume that the entries of the diagonal matrix $\widehat{K} + \Lambda_1$ satisfy

$$\hat{k}_i + \lambda_i > \hat{k}_{i+1} + \lambda_{i+1} \tag{4.14}$$

for all $i = 1, \ldots, p-1$. Then $\widehat{Y} = \Gamma Y$ is a Levelt fundamental solution of the corresponding transformed system $(\widehat{S})$ and since W is holomorphically invertible at $a_1 = 0$, this proves that a_1 is a Fuchsian singular point of the new system $(\widehat{S})$ (whose singular points at all other points remain Fuchsian as well). Actually, (4.14) follows easily from

$$\lambda_i - \lambda_{i+1} > (n-2)p \quad \text{and} \quad |\hat{k}_i - \hat{k}_{i+1}| \leq (n-2)(p-1)$$

and this ends the proof. □

4.4 Solutions to exercises of Chapter 4

Exercise 4.2 p. 88 Let $w : K \to \mathbb{Z} \cup \{\infty\}$ be defined by $w(0) = \infty$ and

$$w(f) = \sup\{k \in \mathbb{Z} \mid \lim_{x \to 0} f(x) x^{-\lambda} = 0 \text{ for all } \lambda < k\}$$

for $f \in K^*$. Let

$$f(x) = \sum_{i \geq d} \alpha_i x^i \in \mathbb{C}(\{x\}), \quad \alpha_d \neq 0.$$

Then $w(f) = d$ and it is easy to see that for all $\lambda \in \mathbb{Z}$, $m \in \mathbb{Z}$ and $k = 1, \ldots, s$,

$$\lim_{x \to 0} f(x) x^{\rho^k} (\log x)^m x^{-\lambda} = 0$$

whenever

$$\lim_{x \to 0} f(x) = 0,$$

which proves that $v = w$ on K, and the extensions of v and w also coincide on $\mathscr{X}$.

Exercise 4.20 p. 97 For $(x, v), (x, w) \in U_{ij} \times \mathbb{C}^p$, we have

$$f_i \circ f_j^{-1}(x, v) = (x, w) \iff f_j \circ f_i^{-1}(x, w) = (x, v),$$

that is,

$$g_{ij}(x)(v) = w \iff g_{ji}(x)(w) = v$$

which proves that

$$g_{ij}(x)^{-1} = g_{ji}(x)$$

for all $x \in U_{ij}$.

Exercise 4.40 p. 103 To prove (a) assume that the system

$$\frac{dy}{dx} = A(x)y \qquad (S)$$

is Fuchsian on $\overline{\mathbb{C}}$. Let A_i, for each $i = 1, \ldots, n$, be the residue matrix of $A(x)$ at a_i (which is a simple pole of A) and let

$$T(x) = A(x) - \sum_{i=1}^{n} \frac{A_i}{x - a_i}.$$

By definition of the matrices A_i the matrix function $T(x)$ is holomorphic at each a_i, hence in $\mathbb{C}$. The change of variable $x = 1/z$ permutes 0 and ∞ and replaces (S) by

$$\frac{du}{dz} = -\frac{1}{z^2} A\left(\frac{1}{z}\right) u \qquad (S_u)$$

with $u(z) = y(1/z)$. Since ∞ is either non-singular or a Fuchsian singular point, we have an expansion of $A(1/z)$ of the form

$$A\left(\frac{1}{z}\right) = \sum_{k \geq 1} M_k z^k$$

as z tends to 0, where M_k are constant matrices, and $M_1 = 0$ if and only if 0 is a non-singular point of (S_u), that is, if ∞ is a non-singular point of (S). The holomorphic matrix

$$T(x) = \sum_{k \geq 1} \frac{M_k}{x^k} - \sum_{i=1}^{n} \frac{A_i}{x - a_i}$$

tends to 0 as $x \to \infty$, which implies that $T = 0$ identically in $\overline{\mathbb{C}}$ by Liouville's theorem and A has the expected form.

To prove (b) note that the coefficient of x^{-1} in the expansion of $T(x)$ at infinity must be zero:

$$M_1 - \sum_{i=1}^{n} A_i = 0,$$

hence $\sum_{i=1}^{n} A_i = M_1$. This implies that $\sum_i A_i = 0$ if and only if ∞ is a non-singular point of (S).

References

ACMS53. Ahlfors, L., Calabi, E., Morse, M., Sarlo, L., Spencer, D. (eds), Contributions to the Theory of Riemann Surfaces. Annals of Math. Studies, Princeton (1953)

AB94. Anosov, D. V., Bolibrukh, A. A.: The Rieman-Hilbert problem. Aspects of mathematics, vol. 22. A Publication from the Steklov Institute of Mathematics, Vieweg und Sohn (1994)

At57. Atiyah, M. F.: Complex analytic connections in fiber bundles. Trans. Amer. Math. Soc. **85**, 181–207 (1957)

Bo90. Bolibrukh, A. A.: The Rieman-Hilbert problem. Russian Math. Surveys **45-2**, 1–47 (1990)

Bo92. Bolibrukh, A. A.: On sufficient conditions for the positive solvability of the Riemann-Hilbert problem. Math. Notes. Acad. Sci. USSR **51-1**, 110–117 (1992)

Bo92. Bolibrukh, A. A.: Fuchsian systems with reducible monodromy and the Riemann-Hilbert Problem. In: Borisovich, Yu. G., Gliklikh, Yu. E. (eds): Global analysis–Studies and Applications V. Lect. Notes Math. vol. 1520, 139–155, Springer (1992)

Bo95a. Bolibrukh, A. A.: The Riemann-Hilbert problem and Fuchsian differential equations on the Riemann sphere. Proceedings of the ICM (Zürich 1994), 1159–1168, Birkhäuser, (1995)

Bo95b. Bolibrukh, A. A.: Hilbert's twenty first problem for linear Fuchsian equations. Proc. Steklov Inst. Math. **203**, 29–35 (1995)

Bo95(3). Bolibrukh, A. A.: The 21st Hilbert Problem for Linear Fuchsian Systems. Proc. Steklov Inst. Math. **206**, 1–145 (1995)

Bo02. Bolibrukh, A. A.: The Riemann-Hilbert problem on a compact Riemann surface. Proc. Steklov Inst. Math. **228-3**, 47–60 (2002)

BIK04. Bolibrukh, A. A., Its, A. R., Kapaev, A. A.: On the Riemann-Hilbert inverse monodromy problem and the Painlevé equations. Algebra i Analiz **16-1**, 121–162 (2004)

Car58. Cartan, H.: Espaces fibrés analytiques. Symposium Internacional de Topologia Algebraica, Mexico, 97–121 (1958)

Cor04. Corel, E.: On Fuchs's relation for linear differential systems. Compositio Math. **140**, 1367–1398 (2004)

For81. Forster, O.: Lectures on Riemann Surfaces. Graduate Texts in Mathematics **81**, Springer (1981)

Fstn11. Forstnerič, F.: Stein manifolds and holomorphic mappings: The Homotopy Principle in Complex Analysis. Ergebnisse der Mathematik und ihrer Grenzgebiete, Springer (2011)

FoLa10. Forstnerič, F., Lárusson, F.: Survey of Oka Theory. New York J. Math. **17a**, 1-28 (2011)

Go04. Gontsov, R. R.: Refined Fuchs inequalities for systems of linear differential equations. Izvestiya: Mathematics **68-2**, 259–272 (2004)

Gr57. Grauert, H.: Holomorphe Funktionen mit Werten in komplexen Lieschen Gruppen. Math. Ann. **133**, 450–472 (1957)

Gr58. Grauert, H.: Analytische Faserungen über holomorph-vollständigen Räumen. Math. Ann. **135**, 263–273 (1958)

He06. Heu Berlinger, V.: Zum Riemann-Hilbertschen Problem. Diplom Arbeit, nach der Vorlesung 'Problème de Riemann-Hilbert' gelesen von Frank Loray an der Université de Rennes 1 (2006)

IlYa07. Ilyashenko, Yu., Yakovenko, S.: Lectures on Analytic Differential Equations. A.M.S. Graduate Studies in Mathematics, vol. 86 , American Mathematical Society (2007)

Ki71. Kimura, T.: On the Riemann problem on Riemann surfaces. In: Urabe, M. (ed.): Japan-US Seminar on Ordinary Differential and Functional Equations, Kyoto, Japan. Lect. Notes Math. vol. 243, 219–228, Springer (1971)

Kos92. Kostov, V. P.: Fuchsian linear systems on $\mathbb{CP}^1$ and the Riemann-Hilbert problem. C. R. Acad. Sci. Paris Sr. I Math. **315-2**, 143–148 (1992)

Le61. Levelt, A. H. M.: Hypergeometric functions. Nederl. Akad. Wet.,Proc. Ser. A, **64** 361–372, 373–385, 386–396, 397–403 (1961)

Na85. Narashiman, R.: Analysis on Real and Complex Manifolds. Third Edition. North-Holland Mathematical Library (1985)

Pan05. Pansu, P.: Fibrés vectoriels, chapitre 6, cours de DEA, Université de Paris Sud (2005) http://www.math.u-psud.fr/~pansu/web_dea/chapitre6.pdf

Ple64. Plemelj, J.: Problems in the sense of Riemann and Klein. Tracts in Mathematics vol. 16. Interscience Publishers (1964)

RS89. Ramis, J.-P., Sibuya, Y.: Hukuhara domains and fundamental existence and uniqueness theorems for asymptotic solutions of Gevrey type. Asymptotic Analysis **2**, 39–94 (1989)

Sauz15. Sauzin, D.: Divergent series, Summability and Resurgence I. Monodromy and Resurgence. Introduction to 1- summability and resurgence. Lect. Notes Math., vol. 2153, 123–294, Springer (2016)

Sib90. Sibuya, Y.: Linear Differential Equations in the Complex Domain: Problems of Analytic Continuation. Translations of Mathematical Monographs vol. 82, American mathematical Society (1990)

St99. Steenrod, N.: The Topology of Fiber Bundles. Princeton Mathematical Series vol. 14. Princeton Landmarks in Mathematics, Princeton University Press (1999)

Tr83. Treibich-Kohn, A.: Un résultat de Plemelj. In: Boutet de Montvel, L., Douady, A., Verdier, J.-L. (eds): Séminaire ENS, Mathématiques et Physique. Progress in Math. **37**, 307–312, Birkaüser (1983)

Wh72. Whitney, H.: Complex Analytic Varieties. Series in Mathematics, Addison-Wesley (1972)

Part II
Introduction to 1-Summability and Resurgence

David Sauzin

The occurrence of divergent power series in relation with irregular singular points of systems of linear differential equations has been mentioned in the first part of this volume. This second part is an introduction to the systematic study of two classes of possibly divergent series, 1-summable series and resurgent series, which appear in the context of irregular singular points of a variety of equations (differential, difference, differential-difference equations, linear or not), but also in other areas of mathematical analysis.

The opening chapter is an introduction to 1-summability. The presentation we adopt here relies on the formal Borel transform and the Laplace transform along an arbitrary direction of the complex plane. A power series to which one can apply the Borel transform and then the Laplace transform in an arc of directions is said to be 1-summable in that arc; then, one can attach to it a Borel-Laplace sum, i.e. a holomorphic function defined in a sufficiently wide sector (the opening of which is π plus the length of the arc) and asymptotic to that power series in the Gevrey sense.

The next chapter is an introduction to Écalle's resurgence theory. It is the core of this part of the volume. A power series is said to be resurgent when its Borel transform is convergent and has good analytic continuation properties: there may be singularities but these must be isolated. The analysis of these singularities, by means of the so-called alien calculus, allows one to compare the various Borel-Laplace sums attached to the same resurgent 1-summable series. In the context of analytic difference or differential equations, this sheds light on the Stokes phenomenon.

A few elementary or classical examples are given a thorough treatment (among which the Euler series, the Stirling series, and a less known example by Poincaré). Examples of linear differential equations which give rise to resurgent solutions (notably the Airy equation) are also discussed, in line with the first part of the volume; examples of resurgence in non-linear differential equations are also shown. Special attention is devoted to the operations needed in non-linear problems: 1-summable series as well as resurgent series are shown to form algebras which are stable by composition. As an application, the last chapter describes the resurgent approach to the classification of tangent-to-identity germs of holomorphic diffeomorphisms in the simplest case.

With the chapter on 1-summability, the reader will be well prepared for the second volume of the book *Divergent Series, Summability and Resurgence* [Lod16], which is devoted to the more general theories of k-summability and multisummability; there, the reader will also find a different approach to the classification of tangent-to-identity germs. The chapter on resurgence will prepare the reader for the third volume [Del16], in which resurgent methods are applied to the first Painlevé equation.

Throughout this part, we use the notations

$$\mathbb{N} = \{0,1,2,\dots\}, \qquad \mathbb{N}^* = \{1,2,3,\dots\}$$

for the set of non-negative integers and the set of positive integers, and

$$\mathbb{R}^+ = \{x \in \mathbb{R} \mid x \geq 0\}.$$

Chapter 5
Borel-Laplace Summation

5.1 Prologue

5.1.1 At the beginning of the second volume of his *New methods of celestial mechanics* [Poi87], H. Poincaré dedicates two pages to elucidating *"a kind of misunderstanding between geometers and astronomers about the meaning of the word* convergence*"*. He proposes a simple example, namely the two series

$$\sum \frac{1000^n}{n!} \quad \text{and} \quad \sum \frac{n!}{1000^n}. \tag{5.1}$$

He says that, for geometers (i.e. mathematicians), the first one is convergent because the term for $n = 1.000.000$ is much smaller than the term for $n = 999.999$, whereas the second one is divergent because the general term is unbounded (indeed, the $(n+1)$-th term is obtained from the nth one by multiplying either by $1000/n$ or by $n/1000$). On the contrary, according to Poincaré, astronomers will consider the first series as divergent because the general term is an increasing function of n for $n \leq 1000$, and they will consider the second one as convergent because the first 1000 terms decrease rapidly.

Poincaré then proposes to reconcile both points of view by clarifying the role that divergent series (in the sense of geometers) can play in the approximation of certain functions. He mentions the example of the classical Stirling series, for which the absolute value of the general term is first a decreasing function of n and then an increasing function; this is a divergent series and still, Poincaré says, *"when stopping at the least term one gets a representation of Euler's gamma function, with greater accuracy if the argument is larger"*. This is the origin of the modern theory of asymptotic expansions.[1]

---

[1] In fact, Poincaré's observations go even beyond this, in direction of least term summation for Gevrey series, but we shall not discuss the details of all this in the present volume; the interested reader may consult [Ram93], [Ram12b], [Ram12a].

C. Mitschi, D. Sauzin, *Divergent Series, Summability and Resurgence I*,
Lecture Notes in Mathematics 2153, DOI 10.1007/978-3-319-28736-2_5

5.1.2 In this volume (and in the other two volumes of this book as well), we focus on formal series given as power series expansions, like the Stirling series for instance, rather than on numerical series. Thus, we would rather present Poincaré's simple example (5.1) in the form of two formal series

$$\sum_{n\geq 0} \frac{1000^n}{n!} t^n \quad \text{and} \quad \sum_{n\geq 0} \frac{n!}{1000^n} t^n, \tag{5.2}$$

the first of which has infinite radius of convergence, while the second has zero radius of convergence. For us, *divergent series* will usually mean a formal power series with zero radius of convergence.

Our aim in this chapter is to discuss the Borel-Laplace summation process as a way of obtaining a function from a (possibly divergent) formal series, the relation between the original formal series and this function being a particular case of asymptotic expansion of Gevrey type. For instance, this will be illustrated on Euler's gamma function and the Stirling series (see Section 5.11). In the next chapter, we shall describe in this example and others the phenomenon for which J. Écalle coined the name *resurgence* at the beginning of the 1980s and give a brief introduction to this beautiful theory.

5.2 An example by Poincaré

Before stating the basic definitions and introducing the tools with which we shall work throughout this chapter, we want to give an example of a divergent formal series $\widetilde{\phi}(t)$ arising in connection with a holomorphic function $\phi(t)$ (later on, we shall come back to this example and see how the general theory helps to understand it). Up to changes in the notation this example is taken from Poincaré's discussion of divergent series, still at the beginning of [Poi87].

Fix $w \in \mathbb{C}$ with $0 < |w| < 1$ and consider the series of functions of the complex variable t

$$\phi(t) = \sum_{k\geq 0} \phi_k(t), \qquad \phi_k(t) = \frac{w^k}{1+kt}. \tag{5.3}$$

This series is uniformly convergent in any compact subset of

$$U := \mathbb{C}^* \setminus \big\{ -1, -\tfrac{1}{2}, -\tfrac{1}{3}, \dots \big\},$$

as is easily checked, thus its sum ϕ is holomorphic in U.

We can even check that ϕ is meromorphic in $\mathbb{C}^*$ with a simple pole at every point of the form $-\frac{1}{k}$ with $k \in \mathbb{N}^*$: Indeed, $\mathbb{C}^*$ can be written as the union of the open sets

$$\Omega_N = \{t \in \mathbb{C} \mid |t| > 1/N\}$$

for all $N \geq 1$; for each N, the finite sum $\phi_0 + \phi_1 + \cdots + \phi_N$ is meromorphic in Ω_N with simple poles at $-1, -\frac{1}{2}, \ldots, -\frac{1}{N-1}$, on the other hand the functions ϕ_k are holomorphic in Ω_N for all $k \geq N+1$, with $|\phi_k(t)| \leq \frac{|w|^k}{k|t+1/k|} \leq \left(\frac{1}{N} - \frac{1}{N+1}\right)^{-1} \frac{|w|^k}{k}$, whence the uniform convergence and the holomorphy in Ω_N of $\phi_{N+1} + \phi_{N+2} + \cdots$ follow, and consequently the meromorphy of ϕ.

We now show how this function ϕ gives rise to a divergent formal series when t approaches 0. For each $k \in \mathbb{N}$, we have a convergent Taylor expansion at the origin

$$\phi_k(t) = \sum_{n\geq 0} (-1)^n w^k k^n t^n \quad \text{for } |t| \text{ small enough.}$$

Since for each $n \in \mathbb{N}$ the numerical series

$$b_n = \sum_{k\geq 0} k^n w^k \tag{5.4}$$

is convergent, one might be tempted to recombine the (convergent) Taylor expansions of the ϕ_k's as $\phi(t)$"="$\sum\limits_k \Big(\sum\limits_n (-1)^n w^k k^n t^n\Big)$"="$\sum\limits_n (-1)^n \Big(\sum\limits_k k^n w^k\Big) t^n$, which amounts to considering the well-defined formal series

$$\widetilde{\phi}(t) = \sum_{n\geq 0} (-1)^n b_n t^n \tag{5.5}$$

as a Taylor expansion at 0 for $\phi(t)$. But it turns out that *this formal series is divergent!*

Indeed, the coefficients b_n can be considered as functions of the complex variable $w = \mathrm{e}^s$, for w in the unit disc or, equivalently, for $\Re e\, s < 0$; we have $b_0 = (1-w)^{-1} = (1-\mathrm{e}^s)^{-1}$ and $b_n = \left(w\frac{d}{dw}\right)^n b_0 = \left(\frac{d}{ds}\right)^n b_0$. Now, if $\widetilde{\phi}(t)$ had nonzero radius of convergence, there would exist $A, B > 0$ such that $|b_n| \leq AB^n$ and the formal series

$$F(\zeta) = \sum (-1)^n b_n \frac{\zeta^n}{n!} \tag{5.6}$$

would have infinite radius of convergence, whereas, recognizing the Taylor formula of b_0 with respect to the variable s, we see that

$$F(\zeta) = \sum (-1)^n \frac{\zeta^n}{n!} \left(\frac{d}{ds}\right)^n b_0 = (1-\mathrm{e}^{s-\zeta})^{-1}$$

has a finite radius of convergence ($F(\zeta)$ is in fact the Taylor expansion at 0 of a meromorphic function with poles on $s + 2\pi\mathrm{i}\mathbb{Z}$, thus this radius of convergence is $\operatorname{dist}(s, 2\pi\mathrm{i}\mathbb{Z})$).

Now the question is to understand the relation between the divergent formal series $\widetilde{\phi}(t)$ and the function $\phi(t)$ we started from. We shall see in this course that the Borel-Laplace summation is a way of going from $\widetilde{\phi}(t)$ to $\phi(t)$, that $\widetilde{\phi}(t)$ is the asymptotic expansion of $\phi(t)$ as $|t| \to 0$ in a very precise sense and we shall explain what resurgence means in this example.

Remark 5.1. We can already observe that the moduli of the coefficients b_n satisfy

$$|b_n| \le AB^n n!, \qquad n \in \mathbb{N}, \tag{5.7}$$

for appropriate constants A and B (independent of n). Such inequalities are called 1-*Gevrey estimates* for the formal series $\tilde{\phi}(t) = \sum b_n t^n$ (see Definition 5.8). For the specific example of the coefficients (5.4), inequalities (5.7) can be obtained by reverting the last piece of reasoning: since the meromorphic function $F(\zeta)$ is holomorphic for $|\zeta| < d = \text{dist}(s, 2\pi\text{i}\mathbb{Z})$ and $b_n = (-1)^n F^{(n)}(0)$, the Cauchy inequalities yield (5.7) with any $B > 1/d$.

Remark 5.2. The function ϕ we started with is not holomorphic (nor meromorphic) in any neighbourhood of 0, because of the accumulation at the origin of the sequence of simple poles $-1/k$; it would thus have been quite surprising to find a positive radius of convergence for $\tilde{\phi}$.

5.3 The differential algebra $\left(\mathbb{C}[[z^{-1}]], \partial\right)$

5.3.1 It will be convenient for us to set $z = 1/t$ in order to "work at ∞" rather than at the origin. At the level of formal expansions, this simply means that we shall deal with expansions involving non-positive integer powers of the indeterminate. We denote by

$$\mathbb{C}[[z^{-1}]] = \Big\{ \varphi(z) = \sum_{n\ge 0} a_n z^{-n}, \text{ with any } a_0, a_1, \ldots \in \mathbb{C} \Big\}$$

the set of all these formal series. This is a complex vector space, and also an algebra when we take into account the Cauchy product:

$$\Big(\sum_{n\ge 0} a_n z^{-n}\Big)\Big(\sum_{n\ge 0} b_n z^{-n}\Big) = \sum_{n\ge 0} c_n z^{-n}, \qquad c_n = \sum_{p+q=n} a_p b_q.$$

The natural derivation

$$\partial = \frac{\mathrm{d}}{\mathrm{d}z} \tag{5.8}$$

makes it a *differential algebra*; this simply means that we have singled out a $\mathbb{C}$-linear map which satisfies the Leibniz rule

$$\partial(\varphi\psi) = (\partial\varphi)\psi + \varphi(\partial\psi), \qquad \varphi, \psi \in \mathbb{C}[[z^{-1}]]. \tag{5.9}$$

If we return to the variable t and define $D = -t^2 \frac{\mathrm{d}}{\mathrm{d}t}$, we obviously get an isomorphism of differential algebras between $(\mathbb{C}[[z^{-1}]], \partial)$ and $(\mathbb{C}[[t]], D)$ by mapping $\sum a_n z^{-n}$ to $\sum a_n t^n$.

5.3.2 The standard valuation (or "order") on $\mathbb{C}[[z^{-1}]]$ is the map

$$\mathrm{val}\colon\ \mathbb{C}[[z^{-1}]] \to \mathbb{N} \cup \{\infty\} \tag{5.10}$$

defined by $\mathrm{val}(0) = \infty$ and $\mathrm{val}(\varphi) = \min\{n \in \mathbb{N} \mid a_n \neq 0\}$ for $\varphi = \sum a_n z^{-n} \neq 0$.

For $\nu \in \mathbb{N}$, we shall use the notation

$$z^{-\nu}\mathbb{C}[[z^{-1}]] = \Big\{ \sum_{n \geq \nu} a_n z^{-n}, \text{ with any } a_\nu, a_{\nu+1}, \ldots \in \mathbb{C} \Big\}. \tag{5.11}$$

This is precisely the set of all $\varphi \in \mathbb{C}[[z^{-1}]]$ such that $\mathrm{val}(\varphi) \geq \nu$. In particular, from the viewpoint of the ring structure, $\mathfrak{I} = z^{-1}\mathbb{C}[[z^{-1}]]$ is the maximal ideal of $\mathbb{C}[[z^{-1}]]$; its elements will often be referred to as "formal series without constant term".

Observe that

$$\mathrm{val}(\partial\varphi) \geq \mathrm{val}(\varphi) + 1, \qquad \varphi \in \mathbb{C}[[z^{-1}]], \tag{5.12}$$

with equality if and only if $\varphi \in z^{-1}\mathbb{C}[[z^{-1}]]$.

5.3.3 It is an exercise to check that the formula

$$d(\varphi, \psi) = 2^{-\mathrm{val}(\psi - \varphi)}, \qquad \varphi, \psi \in \mathbb{C}[[z^{-1}]], \tag{5.13}$$

defines a distance and that $\mathbb{C}[[z^{-1}]]$ then becomes a complete metric space. The topology induced by this distance is called the Krull topology or the topology of the formal convergence (or the $\mathfrak{I}$-adic topology). It provides a simple way of using the language of topology to describe certain algebraic properties.

We leave it to the reader to check that a sequence $(\varphi_p)_{p\in\mathbb{N}}$ of formal series is a Cauchy sequence if and only if, for each $n \in \mathbb{N}$, the sequence of the nth coefficients is stationary: there exists an integer μ_n such that $\mathrm{coeff}_n(\varphi_p)$ is the same complex number α_n for all $p \geq \mu_n$. The limit $\lim\limits_{p\to\infty} \varphi_p$ is then simply $\sum\limits_{n\geq 0} \alpha_n z^{-n}$. (This property of formal convergence of (φ_p) has no relation with any topology on the field of coefficients, except with the discrete one).

In practice, we shall use the fact that a series of formal series $\sum \varphi_p$ is formally convergent if and only if there is a sequence of integers $\nu_p \xrightarrow[p\to\infty]{} \infty$ such that $\varphi_p \in z^{-\nu_p}\mathbb{C}[[z^{-1}]]$ for all p. Each coefficient of the sum $\varphi = \sum \varphi_p$ is then given by a finite sum: the coefficient of z^{-n} in φ is $\mathrm{coeff}_n(\varphi) = \sum\limits_{p\in M_n} \mathrm{coeff}_n(\varphi_p)$, where M_n denotes the finite set $\{p \mid \nu_p \leq n\}$.

Exercise 5.3. Check that, as claimed above, (5.13) defines a distance which makes $\mathbb{C}[[z^{-1}]]$ a complete metric space; check that the subspace $\mathbb{C}[z^{-1}]$ of polynomial formal series is dense. Show that, for the Krull topology, $\mathbb{C}[[z^{-1}]]$ is a topological ring (i.e. addition and multiplication are continuous) but not a topological $\mathbb{C}$-algebra (the scalar multiplication is not). Show that ∂ is a contraction for the distance (5.13).

5.3.4 As an illustration of the use of the Krull topology, let us define the *composition operators* by means of formally convergent series.

Given $\varphi, \chi \in \mathbb{C}[[z^{-1}]]$, we observe that $\operatorname{val}(\partial^p \varphi) \geq \operatorname{val}(\varphi) + p$ (by repeated use of (5.12)), hence $\operatorname{val}(\chi^p \partial^p \varphi) \geq \operatorname{val}(\varphi) + p$ and the series

$$\varphi \circ (\mathrm{id} + \chi) := \sum_{p \geq 0} \frac{1}{p!} \chi^p \partial^p \varphi \tag{5.14}$$

is formally convergent. Moreover

$$\operatorname{val}\big(\varphi \circ (\mathrm{id} + \chi)\big) = \operatorname{val}(\varphi). \tag{5.15}$$

We leave it as an exercise for the reader to check that, for fixed χ, the operator $\Theta\colon \varphi \mapsto \varphi \circ (\mathrm{id} + \chi)$ is a continuous automorphism of algebra (i.e. a $\mathbb{C}$-linear invertible map, continuous for the Krull topology, such that $\Theta(\varphi\psi) = (\Theta\varphi)(\Theta\psi)$).

A particular case is the shift operator

$$T_c\colon \mathbb{C}[[z^{-1}]] \to \mathbb{C}[[z^{-1}]], \quad \varphi(z) \mapsto \varphi(z+c) \tag{5.16}$$

with any $c \in \mathbb{C}$ (the operator T_c is even a differential algebra automorphism, i.e. an automorphism of algebra which commutes with the differential ∂).

The counterpart of these operators in $\mathbb{C}[[t]]$ via the change of indeterminate $t = z^{-1}$ is $\phi(t) \mapsto \phi(\frac{t}{1+ct})$ for the shift operator and, more generally for the composition with $\mathrm{id} + \chi$, $\phi \mapsto \phi \circ F$ with $F(t) = \frac{t}{1+tG(t)}$, $G(t) = \chi(t^{-1})$. See Sections 5.14–5.16 for more on the composition of formal series at ∞ (in particular for associativity).

Exercise 5.4 (Substitution into a power series). Check that, for $\varphi(z) \in z^{-1}\mathbb{C}[[z^{-1}]]$, the formula

$$H(t) = \sum_{p \geq 0} h_p t^p \mapsto H \circ \varphi(z) := \sum_{p \geq 0} h_p \big(\varphi(z)\big)^p$$

defines a homomorphism of algebras from $\mathbb{C}[[t]]$ to $\mathbb{C}[[z^{-1}]]$, i.e. a linear map Θ such that $\Theta 1 = 1$ and $\Theta(H_1 H_2) = (\Theta H_1)(\Theta H_2)$ for all H_1, H_2.

Exercise 5.5. Put the Krull topology on $\mathbb{C}[[t]]$ and use it to define the composition operator $C_F\colon \phi \mapsto \phi \circ F$ for any $F \in t\mathbb{C}[[t]]$; check that C_F is an algebra endomorphism of $\mathbb{C}[[t]]$. Prove that any algebra endomorphim Θ of $\mathbb{C}[[t]]$ is of this form. (Hint: justify that $\phi \in t\mathbb{C}[[t]] \iff \forall \alpha \in \mathbb{C}^*,\ \alpha + \phi$ invertible $\implies \forall \alpha \in \mathbb{C}^*,\ \alpha + \Theta\phi$ invertible; deduce that $F := \Theta t \in t\mathbb{C}[[t]]$; then, for any $\phi \in \mathbb{C}[[t]]$ and $k \in \mathbb{N}$, show that $\operatorname{val}(\Theta\phi - C_F\phi) \geq k$ by writing $\phi = P + t^k \psi$ with a polynomial P and conclude.)

5.4 The formal Borel transform and the space of 1-Gevrey formal series $\mathbb{C}[[z^{-1}]]_1$

5.4.1 We now define a map on the space $z^{-1}\mathbb{C}[[z^{-1}]]$ of formal series without constant term (recall the notation (5.11)):

Definition 5.6. The *formal Borel transform* is the linear map $\mathscr{B}\colon z^{-1}\mathbb{C}[[z^{-1}]] \to \mathbb{C}[[\zeta]]$ defined by

$$\mathscr{B}\colon \tilde{\varphi} = \sum_{n=0}^{\infty} a_n z^{-n-1} \mapsto \widehat{\varphi} = \sum_{n=0}^{\infty} a_n \frac{\zeta^n}{n!}.$$

In other words, we simply shift the powers by one unit and divide the nth coefficient by $n!$. Changing the name of the indeterminate from z (or z^{-1}) into ζ is only a matter of convention, however we strongly advise against keeping the same symbol.

The motivation for introducing $\mathscr{B}$ will appear in Sections 5.6 and 5.7 with the use of the Laplace transform.

The map $\mathscr{B}$ is obviously a linear isomorphism between the spaces $z^{-1}\mathbb{C}[[z^{-1}]]$ and $\mathbb{C}[[\zeta]]$. Let us see what happens with the *convergent* formal series of the first of these spaces. We say that $\tilde{\varphi}(z) \in \mathbb{C}[[z^{-1}]]$ is "convergent at ∞" (or simply "convergent") if the associated formal series $\tilde{\phi}(t) = \tilde{\varphi}(1/z) \in \mathbb{C}[[t]]$ has positive radius of convergence. The set of convergent formal series at ∞ is denoted $\mathbb{C}\{z^{-1}\}$; the ones without constant term form a subspace denoted by

$$z^{-1}\mathbb{C}\{z^{-1}\} \subset \mathbb{C}\{z^{-1}\}.$$

Lemma 5.7. *Let $\tilde{\varphi} \in z^{-1}\mathbb{C}[[z^{-1}]]$. Then $\tilde{\varphi} \in z^{-1}\mathbb{C}\{z^{-1}\}$ if and only if its formal Borel transform $\widehat{\varphi} = \mathscr{B}\tilde{\varphi}$ has infinite radius of convergence and defines an entire function of bounded exponential type, i.e. there exist $A, c > 0$ such that $|\widehat{\varphi}(\zeta)| \le A\,\mathrm{e}^{c|\zeta|}$ for all $\zeta \in \mathbb{C}$.*

Proof. Let $\tilde{\varphi}(z) = \sum_{n\ge 0} a_n z^{-n-1}$. This formal series is convergent if and only if there exist $A, c > 0$ such that, for all $n \in \mathbb{N}$, $|a_n| \le Ac^n$.

If it is so, then $|a_n \zeta^n/n!| \le A|c\zeta|^n\, n!$, whence the conclusion follows.

Conversely, suppose $\widehat{\varphi} = \mathscr{B}\tilde{\varphi}$ sums to an entire function satisfying $|\widehat{\varphi}(\zeta)| \le A\,\mathrm{e}^{c|\zeta|}$ for all $\zeta \in \mathbb{C}$ and fix $n \in \mathbb{N}$. We have $a_n = \widehat{\varphi}^{(n)}(0)$ and, applying the Cauchy inequality with a circle $C(0,R) = \{\, \zeta \in \mathbb{C} \mid |\zeta| = R \,\}$, we get

$$|a_n| \le \frac{n!}{R^n} \max_{C(0,R)} |\widehat{\varphi}| \le \frac{n!}{R^n} A\,\mathrm{e}^{cR}.$$

Choosing $R = n$ and using $n! = 1 \times 2 \times \cdots \times n \le n^n$, we obtain $|a_n| \le A(\mathrm{e}^c)^n$, which concludes the proof. □

The most basic example is the geometric series

$$\widetilde{\chi}_c(z) = z^{-1}(1-cz^{-1})^{-1} = T_{-c}(z^{-1}) \tag{5.17}$$

convergent for $|z| > |c|$, where $c \in \mathbb{C}$ is fixed. Its formal Borel transform is the exponential series

$$\widehat{\chi}_c(\zeta) = \mathrm{e}^{c\zeta}. \tag{5.18}$$

5.4.2 In fact, we shall be more interested in formal series of $\mathbb{C}[[\zeta]]$ having positive but not necessarily infinite radius of convergence. They will correspond to power expansions in z^{-1} satisfying Gevrey estimates similar to the ones encountered in Remark 5.1:

Definition 5.8. We call 1-*Gevrey formal series* any formal series $\widetilde{\varphi}(z) = \sum_{n\geq 0} a_n z^{-n} \in \mathbb{C}[[z^{-1}]]$ for which there exist $A, B > 0$ such that $|a_n| \leq AB^n n!$ for all $n \geq 0$. We denote by $\mathbb{C}[[z^{-1}]]_1$ the vector space formed of all 1-Gevrey formal series.

Lemma 5.9. *Let $\widetilde{\varphi} \in z^{-1}\mathbb{C}[[z^{-1}]]$ and $\widehat{\varphi} = \mathscr{B}\widetilde{\varphi} \in \mathbb{C}[[\zeta]]$. Then $\widehat{\varphi} \in \mathbb{C}\{\zeta\}$ (i.e. the formal series $\widehat{\varphi}(\zeta)$ has positive radius of convergence) if and only if $\widetilde{\varphi} \in \mathbb{C}[[z^{-1}]]_1$.*

Proof. Obvious. □

In other words, a formal series without constant term is 1-Gevrey if and only if its formal Borel transform is convergent. The space of 1-Gevrey formal series without constant term will be denoted $z^{-1}\mathbb{C}[[z^{-1}]]_1 = \mathscr{B}^{-1}\big(\mathbb{C}\{\zeta\}\big)$, thus

$$\mathbb{C}[[z^{-1}]]_1 = \mathbb{C} \oplus z^{-1}\mathbb{C}[[z^{-1}]]_1. \tag{5.19}$$

5.4.3 We leave it to the reader to check the following elementary properties:

Lemma 5.10. *If $\widetilde{\varphi} \in z^{-1}\mathbb{C}[[z^{-1}]]$ and $\widehat{\varphi} = \mathscr{B}\widetilde{\varphi} \in \mathbb{C}[[\zeta]]$, then*

- *$\partial\widetilde{\varphi} \in z^{-2}\mathbb{C}[[z^{-1}]]$ and $\mathscr{B}(\partial\widetilde{\varphi}) = -\zeta\widehat{\varphi}(\zeta)$,*
- *$T_c\widetilde{\varphi} \in z^{-1}\mathbb{C}[[z^{-1}]]$ and $\mathscr{B}(T_c\widetilde{\varphi}) = \mathrm{e}^{-c\zeta}\widehat{\varphi}(\zeta)$ for any $c \in \mathbb{C}$,*
- *$\mathscr{B}(z^{-1}\widetilde{\varphi}) = \int_0^\zeta \widehat{\varphi}(\zeta_1)\,\mathrm{d}\zeta_1$,*
- *if $\widetilde{\varphi} \in z^{-2}\mathbb{C}[[z^{-1}]]$ then $\mathscr{B}(z\widetilde{\varphi}) = \dfrac{\mathrm{d}\widehat{\varphi}}{\mathrm{d}\zeta}$.*

In the third property, the integration in the right-hand side is to be interpreted termwise. The second property can be used to deduce (5.18) from the fact that, according to (5.17), $\widetilde{\chi}_c = T_{-c}(\widetilde{\chi}_0)$ and $\widetilde{\chi}_0 = z^{-1}$ has Borel tranform $= 1$.

Since $\frac{\mathrm{e}^{-\zeta}-1}{\zeta}$ is invertible in $\mathbb{C}[[\zeta]]$ and in $\mathbb{C}\{\zeta\}$, the second property implies

Corollary 5.11. *Given $\widetilde{\psi} \in z^{-2}\mathbb{C}[[z^{-1}]]$, with Borel transform $\widehat{\psi}(\zeta) \in \zeta\mathbb{C}[[\zeta]]$, the equation*

$$\widetilde{\varphi}(z+1) - \widetilde{\varphi}(z) = \widetilde{\psi}(z)$$

admits a unique solution $\widetilde{\varphi}$ in $z^{-1}\mathbb{C}[[z^{-1}]]$, whose Borel transform is given by

$$\widehat{\varphi}(\zeta) = \frac{1}{\mathrm{e}^{-\zeta}-1}\widehat{\psi}(\zeta).$$

If $\widetilde{\psi}(z)$ is 1-Gevrey, then so is the solution $\widetilde{\varphi}(z)$.

5.5 The convolution in $\mathbb{C}[[\zeta]]$ and in $\mathbb{C}\{\zeta\}$

5.5.1 The convolution product, denoted by the symbol $*$, is defined as the push-forward by $\mathscr{B}$ of the Cauchy product:

Definition 5.12. Given two formal series $\widehat{\varphi}, \widehat{\psi} \in \mathbb{C}[[\zeta]]$, their *convolution product* is $\widehat{\varphi} * \widehat{\psi} := \mathscr{B}(\widetilde{\varphi}\widetilde{\psi})$, where $\widetilde{\varphi} = \mathscr{B}^{-1}\widehat{\varphi}$, $\widetilde{\psi} = \mathscr{B}^{-1}\widehat{\psi}$.

At the level of coefficients, we thus have

$$\widehat{\varphi} = \sum_{n\geq 0} a_n \frac{\zeta^n}{n!}, \quad \widehat{\psi} = \sum_{n\geq 0} b_n \frac{\zeta^n}{n!} \quad \Longrightarrow \quad \widehat{\varphi} * \widehat{\psi} = \sum_{n\geq 0} c_n \frac{\zeta^{n+1}}{(n+1)!}$$

$$\text{with } c_n = \sum_{p+q=n} a_p b_q. \tag{5.20}$$

The convolution product is bilinear, commutative and associative in $\mathbb{C}[[\zeta]]$ (because the Cauchy product is bilinear, commutative and associative in $z^{-1}\mathbb{C}[[z^{-1}]]$). It has no unit in $\mathbb{C}[[\zeta]]$ (since the Cauchy product, when restricted to $z^{-1}\mathbb{C}[[z^{-1}]]$, has no unit). One remedy consists in *adjoining* a unit: consider the vector space $\mathbb{C} \times \mathbb{C}[[\zeta]]$, in which we denote the element $(1,0)$ by δ; we can write this space as $\mathbb{C}\delta \oplus \mathbb{C}[[\zeta]]$ if we identify the subspace $\{0\} \times \mathbb{C}[[\zeta]]$ with $\mathbb{C}[[\zeta]]$. Defining the product by

$$(a\delta + \widehat{\varphi}) * (b\delta + \widehat{\psi}) := ab\delta + a\widehat{\psi} + b\widehat{\varphi} + \widehat{\varphi} * \widehat{\psi},$$

we extend the convolution law of $\mathbb{C}[[\zeta]]$ and get a unital algebra $\mathbb{C}\delta \oplus \mathbb{C}[[\zeta]]$ in which $\mathbb{C}[[\zeta]]$ is embedded; by setting

$$\mathscr{B}1 := \delta,$$

we extend $\mathscr{B}$ as an algebra isomorphism between $\mathbb{C}[[z^{-1}]]$ and $\mathbb{C}\delta \oplus \mathbb{C}[[\zeta]]$. The formula

$$\widehat{\partial} \colon a\delta + \widehat{\varphi}(\zeta) \mapsto -\zeta\widehat{\varphi}(\zeta) \tag{5.21}$$

defines a derivation of $\mathbb{C}\delta \oplus \mathbb{C}[[\zeta]]$ and the extended $\mathscr{B}$ appears as an isomorphism of differential algebras

$$\mathscr{B} \colon \big(\mathbb{C}[[z^{-1}]], \partial\big) \xrightarrow{\sim} \big(\mathbb{C}\delta \oplus \mathbb{C}[[\zeta]], \widehat{\partial}\big)$$

(simple consequence of the first property in Lemma 5.10). It induces a linear isomorphism

$$\mathscr{B} \colon \mathbb{C}[[z^{-1}]]_1 \xrightarrow{\sim} \mathbb{C}\delta \oplus \mathbb{C}\{\zeta\} \tag{5.22}$$

(in view of (5.19) and Lemma 5.9), which is in fact an algebra isomorphism: we shall see in Lemma 5.14 that $\mathbb{C}\delta \oplus \mathbb{C}\{\zeta\}$ is a subalgebra of $\mathbb{C}\delta \oplus \mathbb{C}[[\zeta]]$, and hence $\mathbb{C}[[z^{-1}]]_1$ is a subalgebra of $\mathbb{C}[[z^{-1}]]$.

Remark 5.13. For $c \in \mathbb{C}$, the formula

$$\widehat{T}_c \colon a\delta + \widehat{\varphi}(\zeta) \mapsto a\delta + \mathrm{e}^{-c\zeta}\widehat{\varphi}(\zeta) \tag{5.23}$$

defines a differential algebra automorphism of $\big(\mathbb{C}\delta \oplus \mathbb{C}[[\zeta]], \widehat{\partial}\big)$, which is the counterpart of the operator T_c via the extended Borel transform.

5.5.2 When particularized to convergent formal series of the indeterminate ζ, the convolution can be given a more analytic description:

Lemma 5.14. *Consider two convergent formal series $\widehat{\varphi}, \widehat{\psi} \in \mathbb{C}\{\zeta\}$. Let $R > 0$ be smaller than the radius of convergence of each of them and denote by Φ and Ψ the holomorphic functions defined by $\widehat{\varphi}$ and $\widehat{\psi}$ in the disc $D(0,R) = \{\, \zeta \in \mathbb{C} \mid |\zeta| < R \,\}$. Then the formula*

$$\Phi * \Psi(\zeta) = \int_0^\zeta \Phi(\zeta_1)\Psi(\zeta - \zeta_1)\,\mathrm{d}\zeta_1 \tag{5.24}$$

defines a function $\Phi * \Psi$ holomorphic in $D(0,R)$ which is the sum of the formal series $\widehat{\varphi} * \widehat{\psi}$ (the radius of convergence of which is thus at least R).

Proof. By assumption, the power series

$$\widehat{\varphi}(\zeta) = \sum_{n\geq 0} a_n \frac{\zeta^n}{n!} \quad \text{and} \quad \widehat{\psi}(\zeta) = \sum_{n\geq 0} b_n \frac{\zeta^n}{n!}$$

sum to $\Phi(\zeta)$ and $\Psi(\zeta)$ for any ζ in $D(0,R)$.

Formula (5.24) defines a function holomorphic in $D(0,R)$, since $\Phi * \Psi(\zeta) = \int_0^1 F(s,\zeta)\,\mathrm{d}s$ with

$$(s,\zeta) \mapsto F(s,\zeta) = \zeta\Phi(s\zeta)\Psi\big((1-s)\zeta\big) \tag{5.25}$$

continuous in s, holomorphic in ζ and bounded in $[0,1] \times D(0,R')$ for any $R' < R$.

Now, manipulating $F(s,\zeta)$ as a product of absolutely convergent series, we write

$$F(s,\zeta) = \sum_{p,q\geq 0} a_p b_q \frac{(s\zeta)^p}{p!} \frac{\big((1-s)\zeta\big)^q}{q!} \zeta = \sum_{n\geq 0} F_n(s)\zeta^{n+1}$$

with $F_n(s) = \sum_{p+q=n} a_p b_q \frac{s^p}{p!}\frac{(1-s)^q}{q!}$; the elementary identity $\int_0^1 \frac{s^p}{p!}\frac{(1-s)^q}{q!}\,\mathrm{d}s = \frac{1}{(p+q+1)!}$ yields $\int_0^1 F_n(s)\,\mathrm{d}s = \frac{c_n}{(n+1)!}$ with $c_n = \sum_{p+q=n} a_p b_q$, hence

$$\Phi * \Psi(\zeta) = \sum_{n\geq 0} c_n \frac{\zeta^{n+1}}{(n+1)!}$$

for any $\zeta \in D(0,R)$; recognizing in the right-hand side the sum of the formal series $\widehat{\varphi} * \widehat{\psi}$ defined by (5.20), we conclude that this formal series has radius of convergence $\geq R$ and sums to $\Phi * \Psi$. □

For instance, since $\mathscr{B}z^{-1} = 1$, we can rewrite the left-hand side in the third property of Lemma 5.10 as $(1 * \widehat{\varphi})(\zeta)$ and, if $\tilde{\varphi}(z) \in z^{-1}\mathbb{C}[[z^{-1}]]_1$, the integral $\int_0^{\zeta} \widehat{\varphi}(\zeta_1)\,d\zeta_1$ in the right-hand side can now be given its usual analytical meaning: it is the antiderivative of $\widehat{\varphi}$ which vanishes at 0.

We usually make no difference between a convergent formal series $\widehat{\varphi}$ and the holomorphic function Φ that it defines in a neighbourhood of the origin; for instance we usually denote them by the same symbol and consider that the convolution law defined by the integral (5.24) coincides with the restriction to $\mathbb{C}\{\zeta\}$ of the convolution law of $\mathbb{C}[[\zeta]]$. However, as we shall see from Section 6.1 onward, things get more complicated when we consider the analytic continuation in the large of such holomorphic functions. Think for instance of a convergent $\widehat{\varphi}(\zeta)$ which is the Taylor expansion at 0 of a function holomorphic in $\mathbb{C} \setminus \Omega$, where Ω is a discrete subset of $\mathbb{C}^*$ (e.g. a function which is meromorphic in $\mathbb{C}$ and regular at 0): in this case $\widehat{\varphi}$ has an analytic continuation in $\mathbb{C} \setminus \Omega$ whereas, as a rule, its antiderivative $1 * \widehat{\varphi}$ has only a multiple-valued continuation there...

5.5.3 We end this section with an example which is simple (because it deals with explicit entire functions of ζ) but useful:

Lemma 5.15. *Let $p, q \in \mathbb{N}$ and $c \in \mathbb{C}$. Then*

$$\left(\frac{\zeta^p}{p!}e^{c\zeta}\right) * \left(\frac{\zeta^q}{q!}e^{c\zeta}\right) = \frac{\zeta^{p+q+1}}{(p+q+1)!}e^{c\zeta}. \tag{5.26}$$

Proof. One could compute the convolution integral e.g. by induction on q, but one can also notice that $\frac{\zeta^p}{p!}e^{c\zeta}$ is the formal Borel transform of $T_{-c}z^{-p-1}$ (by virtue of the second property in Lemma 5.10), therefore the left-hand side of (5.26) is the Borel transform of $(T_{-c}z^{-p-1})(T_{-c}z^{-q-1}) = T_{-c}z^{-p-q-2}$. □

5.6 The Laplace transform along $\mathbb{R}^+$

The Laplace transform of a function $\widehat{\varphi}\colon \mathbb{R}^+ \to \mathbb{C}$ is the function $\mathscr{L}^0\widehat{\varphi}$ defined by the formula

$$(\mathscr{L}^0\widehat{\varphi})(z) = \int_0^{+\infty} e^{-z\zeta}\,\widehat{\varphi}(\zeta)\,d\zeta. \tag{5.27}$$

Here we assume $\widehat{\varphi}$ continuous (or at least locally integrable on $\mathbb{R}^{*+}$ and integrable on $[0,1]$) and

$$|\widehat{\varphi}(\zeta)| \leq A\,e^{c_0\zeta}, \qquad \zeta \geq 1, \tag{5.28}$$

for some constants $A > 0$ and $c_0 \in \mathbb{R}$, so that the above integral makes sense for any complex number z in the half-plane

$$\Pi_{c_0} := \{ z \in \mathbb{C} \mid \Re e\, z > c_0 \}.$$

Standard theorems ensure that $\mathscr{L}^0\widehat{\varphi}$ is holomorphic in Π_{c_0} (because $|e^{-z\zeta}| = e^{-\zeta \Re e\, z} \le e^{-c_1\zeta}$ for any $z \in \Pi_{c_1}$, hence, for any $c_1 > c_0$, we can find $\Phi\colon \mathbb{R}^+ \to \mathbb{R}^+$ integrable and independent of z such that $|e^{-z\zeta}\widehat{\varphi}(\zeta)| \le \Phi(\zeta)$ and deduce that $\mathscr{L}^0\widehat{\varphi}$ is holomorphic on Π_{c_1}).

Lemma 5.16. *For any $n \in \mathbb{N}$, $\mathscr{L}^0\big(\frac{\zeta^n}{n!}\big)(z) = z^{-n-1}$ on Π_0.*

Proof. The function $\mathscr{L}^0\big(\frac{\zeta^n}{n!}\big)$ is holomorphic in Π_{c_0} for any $c_0 > 0$, thus in Π_0. The reader can check by induction on n that $\int_0^{+\infty} e^{-s}s^n\,ds = n!$ and deduce the result for $z > 0$ by the change of variable $\zeta = s/z$, and then for $z \in \Pi_0$ by analytic continuation. □

In fact, for any complex number ν such that $\Re e\,\nu > 0$, $\mathscr{L}^0\big(\frac{\zeta^{\nu-1}}{\Gamma(\nu)}\big) = z^{-\nu}$ for $z \in \Pi_0$, where Γ is Euler's gamma function (see Section 5.11).

We leave it to the reader to check

Lemma 5.17. *Let $\widehat{\varphi}$ as above, $\varphi := \mathscr{L}^0\widehat{\varphi}$ and $c \in \mathbb{C}$. Then each of the functions $-\zeta\widehat{\varphi}(\zeta)$, $e^{-c\zeta}\widehat{\varphi}(\zeta)$ or $1 * \widehat{\varphi}(\zeta) = \int_0^\zeta \widehat{\varphi}(\zeta_1)\,d\zeta_1$ satisfies estimates of the form* (5.28) *and*

- $\mathscr{L}^0(-\zeta\widehat{\varphi}) = \dfrac{d\varphi}{dz}$,
- $\mathscr{L}^0(e^{-c\zeta}\widehat{\varphi}) = \varphi(z+c)$,
- $\mathscr{L}^0(1 * \widehat{\varphi}) = z^{-1}\varphi(z)$,
- *if moreover $\widehat{\varphi}$ is continuously derivable on $\mathbb{R}^+$ with $\frac{d\widehat{\varphi}}{d\zeta}$ satisfying estimates of the form* (5.28)*, then* $\mathscr{L}^0\left(\dfrac{d\widehat{\varphi}}{d\zeta}\right) = z\varphi(z) - \widehat{\varphi}(0)$.

Remark 5.18. Assume that $\widehat{\varphi}\colon \mathbb{R}^+ \to \mathbb{C}$ is bounded and locally integrable. Then $\mathscr{L}^0\widehat{\varphi}$ is holomorphic in $\{\Re e\, z > 0\}$. If one assumes moreover that $\mathscr{L}^0\widehat{\varphi}$ extends holomorphically to a neighbourhood of $\{\Re e\, z \ge 0\}$, then the limit of $\int_0^T \widehat{\varphi}(\zeta)\,d\zeta$ as $T \to \infty$ exists and equals $(\mathscr{L}^0\widehat{\varphi})(0)$; see [Zag97] for a proof of this statement and its application to a remarkably short proof of the Prime Number Theorem (less than three pages!).

5.7 The fine Borel-Laplace summation

5.7.1 We shall be particularly interested in the Laplace transforms of functions that are analytic in a neighbourhood of $\mathbb{R}^+$ and that we view as analytic continuations of holomorphic germs at 0.

Definition 5.19. We call *half-strip* any set of the form $S_\delta = \{\, \zeta \in \mathbb{C} \mid \operatorname{dist}(\zeta, \mathbb{R}^+) < \delta \,\}$ with a $\delta > 0$. For $c_0 \in \mathbb{R}$, we denote by $\mathscr{N}_{c_0}(\mathbb{R}^+)$ the set consisting of all convergent formal series $\widehat{\varphi}(\zeta)$ defining a holomorphic function near 0 which extends analytically to a half-strip S_δ with

$$|\widehat{\varphi}(\zeta)| \le A\,\mathrm{e}^{c_0|\zeta|}, \qquad \zeta \in S_\delta,$$

where A is a positive constant (we use the same symbol $\widehat{\varphi}$ to denote the function in S_δ and the power series which is its Taylor expansion at 0). We also set

$$\mathscr{N}(\mathbb{R}^+) = \bigcup_{c_0 \in \mathbb{R}} \mathscr{N}_{c_0}(\mathbb{R}^+)$$

(increasing union).

Theorem 5.20. *Let $\widehat{\varphi} \in \mathscr{N}_{c_0}(\mathbb{R}^+)$, $c_0 \ge 0$. Set $a_n := \widehat{\varphi}^{(n)}(0)$ for every $n \in \mathbb{N}$ and $\varphi = \mathscr{L}^0 \widehat{\varphi}$. Then for any $c_1 > c_0$ there exist $L, M > 0$ such that*

$$|\varphi(z) - a_0 z^{-1} - a_1 z^{-2} - \cdots - a_{N-1} z^{-N}| \le L M^N N! |z|^{-N-1}$$
$$\text{for all } z \in \Pi_{c_1} \text{ and } N \in \mathbb{N}. \qquad (5.29)$$

Proof. Without loss of generality we can assume $c_0 > 0$. Let $\delta, A > 0$ be as in Definition 5.19. We first apply the Cauchy inequalities in the discs $D(\zeta, \delta)$ of radius δ centred on the points $\zeta \in \mathbb{R}^+$:

$$|\widehat{\varphi}^{(n)}(\zeta)| \le \frac{n!}{\delta^n} \sup_{D(\zeta,\delta)} |\widehat{\varphi}| \le n! \delta^{-n} A' \mathrm{e}^{c_0 \zeta}, \qquad \zeta \in \mathbb{R}^+,\ n \in \mathbb{N}, \qquad (5.30)$$

where $A' = A\,\mathrm{e}^{c_0 \delta}$. In particular, the coefficient $a_N = \widehat{\varphi}^{(N)}(0)$ satisfies

$$|a_N| \le N! \delta^{-N} A' \qquad (5.31)$$

for any $N \in \mathbb{N}$. Let us introduce the function

$$R(\zeta) := \widehat{\varphi}(\zeta) - a_0 - a_1 \zeta - \cdots - a_N \frac{\zeta^N}{N!},$$

which belongs to $\mathscr{N}_{c_0}(\mathbb{R}^+)$ (because $c_0 > 0$) and has Laplace transform

$$\mathscr{L}^0 R(z) = \varphi(z) - a_0 z^{-1} - a_1 z^{-2} - \cdots - a_N z^{-N-1}.$$

Since $0 = R(0) = R'(0) = \cdots = R^{(N)}(0)$, the last property in Lemma 5.17 implies $\mathscr{L}^0 R(z) = z^{-1} \mathscr{L}^0 R'(z) = z^{-2} \mathscr{L}^0 R''(z) = \cdots = z^{-N-1} \mathscr{L}^0 R^{(N+1)}(z)$ and, taking into account $R^{(N+1)} = \widehat{\varphi}^{(N+1)}$, we end up with

$$\varphi(z) - a_0 z^{-1} - \cdots - a_{N-1} z^{-N} = a_N z^{-N-1} + z^{-N-1} \mathscr{L}^0 \widehat{\varphi}^{(N+1)}(z).$$

For $z \in \Pi_{c_1}$, $|\mathscr{L}^0(\mathrm{e}^{c_0\zeta})(z)| \le \frac{1}{\Re e\, z - c_0} \le \frac{1}{c_1 - c_0}$, thus inequality (5.30) implies that $|\mathscr{L}^0\widehat{\varphi}^{(N+1)}(z)| \le (N+1)!\delta^{-N-1}\frac{A'}{c_1-c_0} \le N!(2/\delta)^N\frac{A'}{\delta(c_1-c_0)}$. Together with (5.31), this yields the conclusion with $M = 2/\delta$ and $L = A'\big(1 + \frac{1}{\delta(c_1-c_0)}\big)$. □

5.7.2 Here we see the link between the Laplace transform of analytic functions and the formal Borel transform: the Taylor series at 0 of $\widehat{\varphi}(\zeta)$ is $\sum a_n \frac{\zeta^n}{n!}$, thus the finite sum which is subtracted from $\varphi(z)$ in the left-hand side of (5.29) is nothing but a partial sum of the formal series $\widetilde{\varphi}(z) := \mathscr{B}^{-1}\widehat{\varphi} = \sum a_n z^{-n-1} \in z^{-1}\mathbb{C}[[z^{-1}]]_1$.

The connection between the formal series $\widetilde{\varphi}$ and the function φ which is expressed by (5.29) is a particular case of a kind of asymptotic expansion, called 1-Gevrey asymptotic expansion. Let us make this more precise:

Definition 5.21. Given $\mathscr{D} \subset \mathbb{C}^*$ unbounded, a function $\phi\colon \mathscr{D} \to \mathbb{C}$ and a formal series $\widetilde{\phi}(z) = \sum_{n\ge0} c_n z^{-n} \in \mathbb{C}[[z^{-1}]]$, we say that ϕ *admits* $\widetilde{\phi}$ *as uniform asymptotic expansion in* $\mathscr{D}$ if there exists a sequence of positive numbers $(K_N)_{N\in\mathbb{N}}$ such that

$$\left|\phi(z) - c_0 - c_1 z^{-1} - \cdots - c_{N-1} z^{-(N-1)}\right| \le K_N |z|^{-N}, \qquad z \in \mathscr{D}, \quad N \in \mathbb{N}. \quad (5.32)$$

We then use the notation

$$\phi(z) \sim \widetilde{\phi}(z) \quad \text{uniformly for } z \in \mathscr{D}.$$

If there exist $L, M > 0$ such that (5.32) holds with the sequence $K_N = LM^N N!$, then we say that ϕ *admits* $\widetilde{\phi}$ *as uniform* 1*-Gevrey asymptotic expansion in* $\mathscr{D}$ and we use the notation

$$\phi(z) \sim_1 \widetilde{\phi}(z) \quad \text{uniformly for } z \in \mathscr{D}.$$

The reader is referred to [Lod16] for more on asymptotic expansions. As for now, we content ourselves with observing that, given ϕ and $\mathscr{D}$,

- there can be at most one formal series $\widetilde{\phi}$ such that $\phi(z) \sim \widetilde{\phi}(z)$ uniformly for $z \in \mathscr{D}$;
- if $\phi(z) \sim_1 \widetilde{\phi}(z)$ uniformly for $z \in \mathscr{D}$, then $\widetilde{\phi} \in \mathbb{C}[[z^{-1}]]_1$.

(Indeed, if (5.32) holds, then the coefficients of $\widetilde{\phi}$ are inductively determined by

$$c_N = \lim_{\substack{|z|\to\infty \\ z\in\mathscr{D}}} z^N \rho_N(z), \qquad \rho_N(z) := \phi(z) - \sum_{n=0}^{N-1} c_n z^{-n}$$

because $|\rho_N(z) - c_N z^{-N}| \le K_{N+1} z^{-N-1}$, and it follows that $|c_N| \le K_N$.)

Theorem 5.20 can be rephrased as:

If $\widehat{\varphi} \in \mathscr{N}_{c_0}(\mathbb{R}^+)$ *with* $c_0 \ge 0$*, then the function* $\varphi := \mathscr{L}^0\widehat{\varphi}$ *(which is holomorphic in* Π_{c_0}*) and the formal series* $\widetilde{\varphi} := \mathscr{B}^{-1}\widehat{\varphi}$ *(which belong to* $z^{-1}\mathbb{C}[[z^{-1}]]_1$*) satisfy*

$$\varphi(z) \sim_1 \widetilde{\varphi}(z) \quad \textit{uniformly for } z \in \Pi_{c_1} \qquad (5.33)$$

for any $c_1 > c_0$.

5.7.3 Theorem 5.20 can be exploited as a tool for "resummation": if it is the formal series $\tilde{\varphi}(z) \in z^{-1}\mathbb{C}[[z^{-1}]]_1$ which is given in the first place, we may apply the formal Borel transform to get $\hat{\varphi}(\zeta) \in \mathbb{C}\{\zeta\}$; if it turns out that $\hat{\varphi}$ belongs to the subspace $\mathscr{N}(\mathbb{R}^+)$ of $\mathbb{C}\{\zeta\}$, then we can apply the Laplace transform and get a holomorphic function $\varphi(z)$ which admits $\tilde{\varphi}(z)$ as 1-Gevrey asymptotic expansion. This process, which allows us to go from the formal series $\tilde{\varphi}(z)$ to the function $\varphi = \mathscr{L}^0\mathscr{B}\tilde{\varphi}$, is called *fine Borel-Laplace summation* (in the direction of $\mathbb{R}^+$).

The above proof of Theorem 5.20 is taken from [Mal95], in which the reader will also find a converse statement (see also Nevanlinna's theorem in the second volume of this book [Lod16, Th. 5.3.9]): given $\tilde{\varphi} \in z^{-1}\mathbb{C}[[z^{-1}]]$, the mere existence of a holomorphic function φ which admits $\tilde{\varphi}$ as uniform 1-Gevrey asymptotic expansion in a half-plane of the form Π_{c_1} entails that $\mathscr{B}\tilde{\varphi} \in \mathscr{N}(\mathbb{R}^+)$; moreover, such a holomorphic function φ is then unique[2] (we skip the proof of these facts, to be found in the second volume [Lod16]). In this situation, the holomorphic function $\varphi(z)$ can be viewed as a kind of sum of $\tilde{\varphi}(z)$, although this formal series may be divergent, and the formal series $\tilde{\varphi}$ itself is said to be *fine-summable in the direction of* $\mathbb{R}^+$.

If we start with a convergent formal series, say $\tilde{\varphi}(z) \in z^{-1}\mathbb{C}\{z^{-1}\}$ supposed to be convergent for $|z| > c_0$, then the reader can check that $\mathscr{B}\tilde{\varphi} \in \mathscr{N}_{c_1}(\mathbb{R}^+)$ for any $c_1 > c_0$, thus $\tilde{\varphi}(z)$ is fine-summable and $\mathscr{L}^0\mathscr{B}\tilde{\varphi}$ is holomorphic in the half-plane Π_{c_0}. We shall see in Section 5.9 that $\mathscr{L}^0\mathscr{B}\tilde{\varphi}$ is nothing but the restriction to Π_{c_0} of the ordinary sum of $\tilde{\varphi}(z)$.

5.7.4 The formal series without constant term that are fine-summable in the direction of $\mathbb{R}^+$ clearly form a linear subspace of $z^{-1}\mathbb{C}[[z^{-1}]]_1$. To cover the case where there is a non-zero constant term, we make use of the convolution unit $\delta = \mathscr{B}1$ introduced in Section 5.5. We extend the Laplace transform by setting $\mathscr{L}^0\delta := 1$ and, more generally,

$$\mathscr{L}^0(a\,\delta + \hat{\varphi}) := a + \mathscr{L}^0\hat{\varphi}$$

for a complex number a and a function $\hat{\varphi}$.

Definition 5.22. A formal series of $\mathbb{C}[[z^{-1}]]$ is said to be *fine-summable in the direction of* $\mathbb{R}^+$ if it can be written in the form $\tilde{\varphi}_0(z) = a + \tilde{\varphi}(z)$ with $a \in \mathbb{C}$ and $\tilde{\varphi} \in \mathscr{B}^{-1}(\mathscr{N}(\mathbb{R}^+))$, i.e. if its formal Borel transform $\mathscr{B}\tilde{\varphi}_0 = a\,\delta + \hat{\varphi}(\zeta)$ belongs to the subspace $\mathbb{C}\,\delta \oplus \mathscr{N}(\mathbb{R}^+)$ of $\mathbb{C}\,\delta \oplus \mathbb{C}[[\zeta]]$. Its Borel sum is then defined as the function $\mathscr{L}^0(a\,\delta + \hat{\varphi})$, which is holomorphic in a half-plane Π_c (choosing $c \in \mathbb{R}$ large enough).

The operator of *Borel-Laplace summation in the direction of* $\mathbb{R}^+$ is defined as the composition $\mathscr{S}^0 := \mathscr{L}^0 \circ \mathscr{B}$ acting on all such formal series $\tilde{\varphi}_0(z)$.

Clearly, as a consequence of Theorem 5.20 and Definition 5.22, we have

---

[2] However, if the 1-Gevrey asymptotic expansion property holds in a sector of opening less than π, the uniqueness statement is false; in fact, in this case, there are holomorphic functions which are "1-Gevrey flat", i.e. with 1-Gevrey asymptotic expansion equal to 0, but the remarkable fact is that they are exponentially small—once more the reader is referred to the second volume [Lod16].

Corollary 5.23. *If $\widetilde{\varphi}_0 \in \mathbb{C}[[z^{-1}]]$ is fine-summable in the direction of $\mathbb{R}^+$, then there exists $c > 0$ such that the function $\mathscr{S}^0\widetilde{\varphi}_0$ is holomorphic in Π_c and satisfies*

$$\mathscr{S}^0\widetilde{\varphi}_0(z) \sim_1 \widetilde{\varphi}_0(z) \quad \textit{uniformly for } z \in \Pi_c.$$

Remark 5.24. Beware that Π_c is usually not the maximal domain of holomorphy of the Borel sum $\mathscr{S}^0\widetilde{\varphi}_0$: it often happens that this function admits analytic continuation in a much larger domain and, in that case, Π_c may or may not be the maximal domain of validity of the uniform 1-Gevrey asymptotic expansion property.

5.7.5 We now indicate a simple result of stability under convolution:

Theorem 5.25. *The space $\mathscr{N}(\mathbb{R}^+)$ is a subspace of $\mathbb{C}\{\zeta\}$ stable by convolution. Moreover, if $c_0 \in \mathbb{R}$ and $\widehat{\varphi}, \widehat{\psi} \in \mathscr{N}_{c_0}(\mathbb{R}^+)$, then $\widehat{\varphi} * \widehat{\psi} \in \mathscr{N}_{c_1}(\mathbb{R}^+)$ for every $c_1 > c_0$ and*

$$\mathscr{L}^0(\widehat{\varphi} * \widehat{\psi}) = (\mathscr{L}^0\widehat{\varphi})(\mathscr{L}^0\widehat{\psi}) \tag{5.34}$$

in the half-plane Π_{c_0}.

Corollary 5.26. *The space $\mathbb{C} \oplus \mathscr{B}^{-1}\big(\mathscr{N}(\mathbb{R}^+)\big)$ of all fine-summable formal series in the direction of $\mathbb{R}^+$ is a subalgebra of $\mathbb{C}[[z^{-1}]]$ which contains the convergent formal series. The operator of Borel-Laplace summation $\mathscr{S}^0$ satisfies*

$$\mathscr{S}^0\left(\frac{\mathrm{d}\widetilde{\varphi}_0}{\mathrm{d}z}\right) = \frac{\mathrm{d}}{\mathrm{d}z}(\mathscr{S}^0\widetilde{\varphi}_0), \qquad \mathscr{S}^0\big(\widetilde{\varphi}_0(z+c)\big) = (\mathscr{S}^0\widetilde{\varphi}_0)(z+c) \tag{5.35}$$

$$\mathscr{S}^0(\widetilde{\varphi}_0\widetilde{\psi}_0) = (\mathscr{S}^0\widetilde{\varphi}_0)(\mathscr{S}^0\widetilde{\psi}_0) \tag{5.36}$$

for any $c \in \mathbb{C}$ and fine-summable formal series $\widetilde{\varphi}_0$, $\widetilde{\psi}_0$.

Later, we shall see that Borel-Laplace summation is also compatible with the non-linear operation of composition of formal series.

Proof of Theorem 5.25. Suppose $\widehat{\varphi}, \widehat{\psi} \in \mathscr{N}(\mathbb{R}^+)$, with $\widehat{\varphi}$ holomorphic in a half-strip $S_{\delta'}$ in which $|\widehat{\varphi}(\zeta)| \le A' \mathrm{e}^{c_0'|\zeta|}$, and $\widehat{\psi}$ holomorphic in a half-strip $S_{\delta''}$ in which $|\widehat{\psi}(\zeta)| \le A'' \mathrm{e}^{c_0''|\zeta|}$. Let $\delta = \min\{\delta', \delta''\}$ and $c_0 = \max\{c_0', c_0''\}$.

We write $\widehat{\chi}(\zeta) = \int_0^1 F(s,\zeta)\,\mathrm{d}s$ with $F(s,\zeta) = \zeta\widehat{\varphi}(s\zeta)\widehat{\psi}\big((1-s)\zeta\big)$ and argue as in the proof of Lemma 5.14: F is continuous in s and holomorphic in ζ for $(s,\zeta) \in [0,1] \times S_\delta$, with

$$|F(s,\zeta)| \le |\zeta| A'A'' \mathrm{e}^{c_0's|\zeta| + c_0''(1-s)|\zeta|} \le A'A''|\zeta|\mathrm{e}^{c_0|\zeta|}. \tag{5.37}$$

In particular F is bounded in $[0,1] \times C$ for any compact subset C of S_δ, thus $\widehat{\chi}$ is holomorphic in S_δ. Inequality (5.37) implies $|\widehat{\chi}(\zeta)| \le A'A''|\zeta|\mathrm{e}^{c_0|\zeta|} = O\big(\mathrm{e}^{c_1|\zeta|}\big)$ for any $c_1 > c_0$, hence $\widehat{\chi} \in \mathscr{N}_{c_1}(\mathbb{R}^+)$. The identity (5.34) follows from Fubini's theorem. □

Proof of Corollary 5.26. Let $\tilde{\varphi}_0 = a\delta + \tilde{\varphi}$ and $\tilde{\psi}_0 = b + \tilde{\psi}$ with $a,b \in \mathbb{C}$ and $\tilde{\varphi}, \tilde{\psi} \in z^{-1}\mathbb{C}[[z^{-1}]]$. We already mentioned the fact that if $\tilde{\varphi} \in z^{-1}\mathbb{C}\{z^{-1}\}$ then $\tilde{\varphi}$ is fine-summable, thus $\tilde{\varphi}_0$ is fine-summable in that case.

Suppose $\tilde{\varphi}, \tilde{\psi} \in \mathscr{B}^{-1}\big(\mathscr{N}(\mathbb{R}^+)\big)$. Property (5.35) follows from Lemmas 5.10 and 5.17, since the constant a is killed by $\frac{\mathrm{d}}{\mathrm{d}z}$ and left invariant by T_c. Since $\tilde{\varphi}_0\tilde{\psi}_0 = ab + a\tilde{\psi} + b\tilde{\varphi} + \tilde{\varphi}\tilde{\psi}$ has formal Borel transform $ab\,\delta + a\hat{\psi} + b\hat{\varphi} + \hat{\varphi} * \hat{\psi}$, Theorem 5.25 implies that $\tilde{\varphi}_0\tilde{\psi}_0 \in \mathbb{C} \oplus \mathscr{B}^{-1}\big(\mathscr{N}(\mathbb{R}^+)\big)$ and, since $\mathscr{S}^0(ab) = ab$, property (5.36) follows by linearity from Lemma 5.14 and Theorem 5.25 applied to $\mathscr{B}\tilde{\varphi} * \mathscr{B}\tilde{\psi}$. □

5.8 The Euler series

The Euler series $\tilde{\Phi}^{\mathrm{E}}(t) = \sum_{n\geq 0}(-1)^n n!\, t^{n+1}$ is a classical example of divergent formal series. We write it "at ∞" as

$$\tilde{\varphi}^{\mathrm{E}}(z) = \sum_{n\geq 0}(-1)^n n!\, z^{-n-1}. \tag{5.38}$$

Clearly, its Borel transform is the geometric series

$$\hat{\varphi}^{\mathrm{E}}(\zeta) = \sum_{n\geq 0}(-1)^n \zeta^n = \frac{1}{1+\zeta}, \tag{5.39}$$

which is convergent in the unit disc and sums to a meromorphic function. The divergence of $\tilde{\varphi}^{\mathrm{E}}(z)$ is reflected in the non-entireness of $\hat{\varphi}^{\mathrm{E}}$, which has a pole at -1 (cf. Lemma 5.7).

Observe that $\tilde{\Phi}^{\mathrm{E}}(t)$ can be obtained as the unique formal solution to a differential equation, the so-called Euler equation:

$$t^2\frac{\mathrm{d}\tilde{\Phi}}{\mathrm{d}t} + \tilde{\Phi} = t.$$

With our change of variable $z = 1/t$, the Euler equation becomes $-\partial\tilde{\varphi} + \tilde{\varphi} = z^{-1}$; applying the formal Borel transform to the equation itself is an efficient way of checking the formula for $\hat{\varphi}^{\mathrm{E}}(\zeta)$: a formal series without constant term $\tilde{\varphi}$ is solution if and only if its Borel transform $\hat{\varphi}$ satisfies $(\zeta+1)\hat{\varphi}(\zeta) = 1$ (by Lemma 5.10) and, since $1+\zeta$ is invertible in the ring $\mathbb{C}[[\zeta]]$, the only possibility is $\hat{\varphi}^{\mathrm{E}}(\zeta) = (1+\zeta)^{-1}$.

Formula (5.39) shows that $\hat{\varphi}^{\mathrm{E}}(\zeta)$ is holomorphic and bounded in a neighbourhood of $\mathbb{R}^+$ in $\mathbb{C}$, hence $\hat{\varphi}^{\mathrm{E}} \in \mathscr{N}_0(\mathbb{R}^+)$. The Euler series is thus fine-summable in the direction of $\mathbb{R}^+$ and has a Borel sum

$$\varphi^{\mathrm{E}} = \mathscr{L}^0\mathscr{B}\tilde{\varphi}^{\mathrm{E}}$$

holomorphic in the half-plane $\Pi_0 = \{\Re e\, z > 0\}$. The first part of (5.35) shows that this function φ^{E} is a solution of the Euler equation in the variable z.

Remark 5.27. The series $\widetilde{\Phi}^{\mathrm{E}}(t)$ appears in Euler's famous 1760 article *De seriebus divergentibus*, in which Euler introduces it as a tool in one of his methods to study the divergent numerical series

$$1 - 1! + 2! - 3! + \cdots,$$

which he calls Wallis' series—see [Bar79] and [Ram12b]. Following Euler, we may adopt $\varphi^{\mathrm{E}}(1) \simeq 0.59634736\ldots$ as the numerical value to be assigned this divergent series.

The discussion of this example continues in Section 5.10; in particular, we shall see how Borel sums can be defined in other half-planes than the ones bisected by $\mathbb{R}^+$ and that φ^{E} admits an analytic continuation outside Π_0 (cf. Remark 5.24).

5.9 Varying the direction of summation

5.9.1 Let $\theta \in \mathbb{R}$. By $\mathrm{e}^{\mathrm{i}\theta}\mathbb{R}^+$ we mean the oriented half-line which can be parametrised as $\{\xi\,\mathrm{e}^{\mathrm{i}\theta},\ \xi \in \mathbb{R}^+\}$. Correspondingly, we define the Laplace transform of a function $\widehat{\varphi}\colon \mathrm{e}^{\mathrm{i}\theta}\mathbb{R}^+ \to \mathbb{C}$ by the formula

$$(\mathscr{L}^\theta \widehat{\varphi})(z) = \int_0^{+\infty} \mathrm{e}^{-z\xi\,\mathrm{e}^{\mathrm{i}\theta}}\,\widehat{\varphi}(\xi\,\mathrm{e}^{\mathrm{i}\theta})\mathrm{e}^{\mathrm{i}\theta}\,\mathrm{d}\xi, \tag{5.40}$$

with obvious adaptations of the assumptions we had at the beginning of Section 5.6, in particular $|\widehat{\varphi}(\zeta)| \le A\,\mathrm{e}^{c_0|\zeta|}$ for $\zeta \in \mathrm{e}^{\mathrm{i}\theta}[1,+\infty)$, so that $\mathscr{L}^\theta\widehat{\varphi}$ is a well-defined function holomorphic in a half-plane

$$\Pi^\theta_{c_0} := \{z \in \mathbb{C} \mid \Re e(z\,\mathrm{e}^{\mathrm{i}\theta}) > c_0\}.$$

Since $\langle z, w\rangle := \Re e(z\bar{w})$ defines the standard real scalar product on $\mathbb{C} \simeq \mathbb{R} \oplus \mathrm{i}\mathbb{R}$, we see that $\Pi^\theta_{c_0}$ is the half-plane bisected by the half-line $\mathrm{e}^{-\mathrm{i}\theta}\mathbb{R}^+$ obtained from $\Pi_{c_0} = \Pi^0_{c_0}$ by the rotation of angle $-\theta$.

The operator $\mathscr{L}^\theta$ is the Laplace transform in the direction θ; the reader can check that it satisfies properties analogous to those explained in Sections 5.6 and 5.7 for $\mathscr{L}^0$.

Definition 5.28. A formal series $\widetilde{\varphi}_0(z) \in \mathbb{C}[[z^{-1}]]$ is said to be *fine-summable in the direction* θ if it can be written $\widetilde{\varphi}_0 = a + \widetilde{\varphi}$ with $a \in \mathbb{C}$ and $\widetilde{\varphi} \in \mathscr{B}^{-1}\big(\mathscr{N}(\mathrm{e}^{\mathrm{i}\theta}\mathbb{R}^+)\big)$, where the space $\mathscr{N}(\mathrm{e}^{\mathrm{i}\theta}\mathbb{R}^+)$ is defined by replacing S_δ with $S^\theta_\delta := \{\zeta \in \mathbb{C} \mid \operatorname{dist}(\zeta, \mathrm{e}^{\mathrm{i}\theta}\,\mathbb{R}^+) < \delta\}$ in Definition 5.19 (see Figure 6.7 on p. 206).

The Laplace transform $\mathscr{L}^\theta$ is well-defined in $\mathscr{N}(e^{i\theta}\mathbb{R}^+)$; we extend it as a linear map on $\mathbb{C}\,\delta \oplus \mathscr{N}(e^{i\theta}\mathbb{R}^+)$ by setting $\mathscr{L}^\theta\delta := 1$ and define the *Borel-Laplace summation operator* as the composition

$$\mathscr{S}^\theta := \mathscr{L}^\theta \circ \mathscr{B} \tag{5.41}$$

acting on all fine-summable formal series in the direction θ. There is an analogue of Corollary 5.23:

If $\tilde{\varphi}_0 \in \mathbb{C}[[z^{-1}]]$ is fine-summable in the direction θ, then there exists $c > 0$ such that the function $\mathscr{S}^\theta\tilde{\varphi}_0$ is holomorphic in Π_c^θ and satisfies

$$\mathscr{S}^\theta\tilde{\varphi}_0(z) \sim_1 \tilde{\varphi}_0(z) \quad \textit{uniformly for } z \in \Pi_c^\theta.$$

There is also an analogue of Corollary 5.26: the product of two fine-summable formal series is fine-summable and $\mathscr{S}^\theta$ satisfies properties analogous to (5.35) and (5.36).

5.9.2 The case of a function $\hat{\varphi}$ holomorphic in a sector is of particular interest, we thus give a new definition in the spirit of Definitions 5.19 and 5.28, replacing half-strips by sectors:

Definition 5.29. Let I be an open interval of $\mathbb{R}$ and $\gamma\colon I \to \mathbb{R}$ a locally bounded function.[3] For any locally bounded function $\alpha\colon I \to \mathbb{R}^+$, we denote by $\mathscr{N}(I,\gamma,\alpha)$ the set consisting of all convergent formal series $\hat{\varphi}(\zeta)$ defining a holomorphic function near 0 which extends analytically to the open sector $\{\xi\, e^{i\theta} \mid \xi > 0,\ \theta \in I\}$ and satisfies

$$|\hat{\varphi}(\xi\, e^{i\theta})| \le \alpha(\theta)\, e^{\gamma(\theta)\xi}, \qquad \xi > 0,\ \theta \in I.$$

We denote by $\mathscr{N}(I,\gamma)$ the set of all $\hat{\varphi}(\zeta)$ for which there exists a locally bounded function α such that $\hat{\varphi} \in \mathscr{N}(I,\gamma,\alpha)$. We denote by $\mathscr{N}(I)$ the set of all $\hat{\varphi}(\zeta)$ for which there exists a locally bounded function γ such that $\hat{\varphi} \in \mathscr{N}(I,\gamma)$.

For example, in view of (5.39), the Borel transform $\hat{\varphi}^{\mathrm{E}}(\zeta)$ of the Euler series belongs to $\mathscr{N}(I,0,\alpha)$ with $I = (-\pi,\pi)$ and

$$\alpha(\theta) = \begin{cases} 1 & \text{if } |\theta| \le \pi/2, \\ 1/|\sin\theta| & \text{else.} \end{cases}$$

Clearly, if $\hat{\varphi} \in \mathscr{N}(I,\gamma)$ and $\theta \in I$, then $z \mapsto (\mathscr{L}^\theta\hat{\varphi})(z)$ is defined and holomorphic in $\Pi_{\gamma(\theta)}^\theta$.

Lemma 5.30. *Let γ and I be as in Definition 5.29. Then, for every $\theta \in I$, there exists a number $c = c(\theta)$ such that $\mathscr{N}(I,\gamma) \subset \mathscr{N}_c(e^{i\theta}\mathbb{R}^+)$; one can choose c to be the supremum of γ on an arbitrary neighbourhood of θ.*

---

[3] A function $\gamma\colon I \to \mathbb{R}$ is said to be locally bounded if any point θ of I admits a neighbourhood on which γ is bounded. Equivalently, the function is bounded on any compact subinterval of I.

The proof is left as an exercise.

Lemma 5.30 shows that a $\widehat{\varphi}$ belonging to $\mathscr{N}(I,\gamma)$ is the Borel transform of a formal series $\widetilde{\varphi}(z)$ which is fine-summable in any direction $\theta \in I$; for each $\theta \in I$, we get a function $\mathscr{L}^{\theta}\widehat{\varphi}$ holomorphic in the half-plane $\Pi^{\theta}_{\gamma(\theta)}$, with the property of uniform 1-Gevrey asymptotic expansion

$$\mathscr{L}^{\theta}\widehat{\varphi}(z) \sim_1 \widetilde{\varphi}(z) \quad \text{uniformly for } z \in \Pi^{\theta}_{\gamma'(\theta)},$$

where $\gamma'(\theta) > 0$ is large enough to be larger than a local bound of γ. We now show that these various functions match, at least if the length of I is less than π, so that we can glue them and define a Borel sum of $\widetilde{\varphi}(z)$ holomorphic in the union of all the half-planes $\Pi^{\theta}_{\gamma(\theta)}$.

Lemma 5.31. *Suppose $\widehat{\varphi} \in \mathscr{N}(I,\gamma)$ with γ and I as in Definition 5.29 and suppose*

$$\theta_1, \theta_2 \in I, \qquad 0 < \theta_2 - \theta_1 < \pi.$$

Then $\Pi^{\theta_1}_{\gamma(\theta_1)} \cap \Pi^{\theta_2}_{\gamma(\theta_2)}$ is a non-empty sector in restriction to which the functions $\mathscr{L}^{\theta_1}\widehat{\varphi}$ and $\mathscr{L}^{\theta_2}\widehat{\varphi}$ coincide.

Proof. The non-emptiness of the intersection of the half-planes $\Pi^{\theta_1}_{\gamma(\theta_1)}$ and $\Pi^{\theta_2}_{\gamma(\theta_2)}$ is an elementary geometric fact which follows from the assumption $0 < \theta_2 - \theta_1 < \pi$: this set is the sector $\mathscr{D} = \{ z_* + r\mathrm{e}^{\mathrm{i}\theta} \mid r > 0,\ \theta \in (-\theta_1 - \frac{\pi}{2}, -\theta_2 + \frac{\pi}{2}) \}$, where $\{z_*\}$ is the intersection of the lines $\mathrm{e}^{-\mathrm{i}\theta_1}\big(\gamma(\theta_1) + \mathrm{i}\mathbb{R}\big)$ and $\mathrm{e}^{-\mathrm{i}\theta_2}\big(\gamma(\theta_2) + \mathrm{i}\mathbb{R}\big)$.

Let $\alpha\colon I \to \mathbb{R}^+$ be a locally bounded function such that $\widehat{\varphi} \in \mathscr{N}(I,\gamma,\alpha)$. Let $c = \sup_{[\theta_1,\theta_2]} \gamma$ and $A = \sup_{[\theta_1,\theta_2]} \alpha$ (both c and A are finite by the local boundedness assumption). By the identity theorem for holomorphic functions, it is sufficient to check that $\mathscr{L}^{\theta_1}\widehat{\varphi}$ and $\mathscr{L}^{\theta_2}\widehat{\varphi}$ coincide on the set $\mathscr{D}_1 = \Pi^{\theta_1}_{c+1} \cap \Pi^{\theta_2}_{c+1}$, since $\mathscr{D}_1$ is a non-empty sector contained in $\mathscr{D}$.

Let $z \in \mathscr{D}_1$. We have $\Re e(z\mathrm{e}^{\mathrm{i}\theta}) > c+1$ for all $\theta \in [\theta_1,\theta_2]$ (simple geometric property, or property of the superlevel sets of the cosine function) thus, for any $\zeta \in \mathbb{C}^*$,

$$\arg\zeta \in [\theta_1,\theta_2] \implies |\mathrm{e}^{-z\zeta}\widehat{\varphi}(\zeta)| \le A\,\mathrm{e}^{-|\zeta|}. \tag{5.42}$$

The two Laplace transforms can be written

$$\mathscr{L}^{\theta_j}\widehat{\varphi}(z) = \int_0^{\mathrm{e}^{\mathrm{i}\theta_j}\infty} \mathrm{e}^{-z\zeta}\widehat{\varphi}(\zeta)\,\mathrm{d}\zeta = \lim_{R\to\infty}\int_0^{R\mathrm{e}^{\mathrm{i}\theta_j}} \mathrm{e}^{-z\zeta}\widehat{\varphi}(\zeta)\,\mathrm{d}\zeta, \qquad j = 1,2,$$

but, for each $R > 0$, the Cauchy theorem implies

$$\left(\int_0^{R\mathrm{e}^{\mathrm{i}\theta_2}} - \int_0^{R\mathrm{e}^{\mathrm{i}\theta_1}}\right)\mathrm{e}^{-z\zeta}\widehat{\varphi}(\zeta)\,\mathrm{d}\zeta = \int_C \mathrm{e}^{-z\zeta}\widehat{\varphi}(\zeta)\,\mathrm{d}\zeta, \qquad C = \{R\mathrm{e}^{\mathrm{i}\theta} \mid \theta \in [\theta_1,\theta_2]\}$$

and, by (5.42), this difference has a modulus $\le AR(\theta_2-\theta_1)\mathrm{e}^{-R}$, hence it tends to 0 as $R \to \infty$. □

5.9.3 Lemma 5.31 allows us to glue toghether the various Laplace transforms:

Definition 5.32. For I open interval of $\mathbb{R}$ of length $|I| \le \pi$ and $\gamma\colon I \to \mathbb{R}$ locally bounded, we define

$$\mathscr{D}(I,\gamma) = \bigcup_{\theta \in I} \Pi^{\theta}_{\gamma(\theta)},$$

which is an open subset of $\mathbb{C}$ (see Figure 5.1), and, for any $\widehat{\varphi} \in \mathscr{N}(I,\gamma)$, we define a function $\mathscr{L}^I \widehat{\varphi}$ holomorphic in $\mathscr{D}(I,\gamma)$ by

$$\mathscr{L}^I \widehat{\varphi}(z) = \mathscr{L}^\theta \widehat{\varphi}(z) \quad \text{with } \theta \in I \text{ such that } z \in \Pi^{\theta}_{\gamma(\theta)}$$

for any $z \in \mathscr{D}(I,\gamma)$.

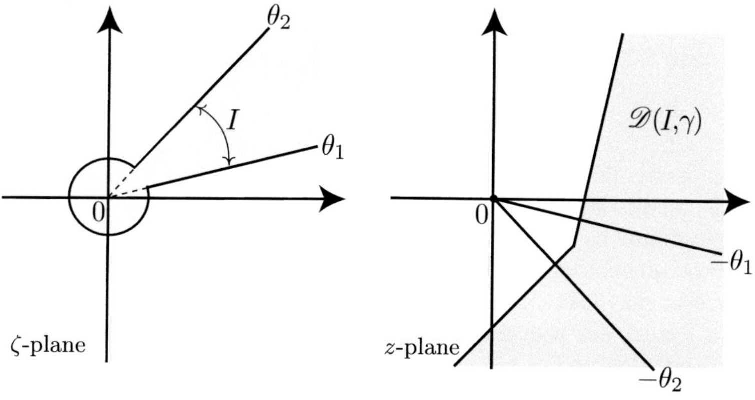

Fig. 5.1: 1-*summability in an arc of directions.* Left: $\widehat{\varphi}(\zeta) \in \mathscr{N}(I,\gamma)$ is holomorphic in the union of a disc and a sector. Right: the domain $\mathscr{D}(I,\gamma)$ where $\mathscr{L}^I \widehat{\varphi}(z)$ is holomorphic.

Observe that, for a given $z \in \mathscr{D}(I,\gamma)$, there are infinitely many possible choices for θ, which all give the same result by virtue of Lemma 5.31; $\mathscr{D}(I,\gamma)$ is a "sectorial neighbourhood of ∞" centred on the ray $\arg z = -\theta^*$ with aperture $\pi + |I|$, where θ^* denotes the midpoint of I, in the sense that, for every $\varepsilon > 0$, it contains a sector bisected by the half-line of direction $-\theta^*$ with opening $\pi + |I| - \varepsilon$ (see [CNP93]).

We extend the definition of the linear map $\mathscr{L}^I$ to $\mathbb{C}\delta \oplus \mathscr{N}(I,\gamma)$ by setting $\mathscr{L}^I \delta := 1$.

Definition 5.33. Given an open interval I, we say that a formal series $\widetilde{\varphi}_0(z) \in \mathbb{C}[[z^{-1}]]$ is 1-*summable in the directions of I* if $\mathscr{B}\widetilde{\varphi}_0 \in \mathbb{C}\delta \oplus \mathscr{N}(I)$. The *Borel-Laplace summation operator* is defined as the composition

$$\mathscr{S}^I := \mathscr{L}^I \circ \mathscr{B} \tag{5.43}$$

acting on all such formal series, which produces functions holomorphic in sectorial neighbourhoods of ∞ of the form $\mathscr{D}(I,\gamma)$, with locally bounded functions $\gamma\colon I \to \mathbb{R}$.

There is an analogue of Corollary 5.26: the product of two formal series which are 1-summable in the directions of I is itself 1-summable in these directions, as a consequence of Lemma 5.30 and of the stability under multiplication of fine-summable series, and the properties (5.35) and (5.36) hold for the summation operator $\mathscr{S}^I$ too.

As for the property of asymptotic expansion, it takes the following form: if $\widetilde{\varphi}_0(z)$ is 1-summable in the directions of I, then there exists $\gamma\colon I \to \mathbb{R}$ locally bounded such that

$$J \text{ relatively compact subinterval of } I \quad \Longrightarrow \quad \mathscr{S}^I\widetilde{\varphi}_0(z) \sim_1 \widetilde{\varphi}_0(z) \quad \text{uniformly for } z \in \mathscr{D}(J,\gamma_{|J})$$

(use Theorem 5.20 and Lemma 5.30). We introduce the notation

$$\mathscr{S}^I\widetilde{\varphi}_0(z) \sim_1 \widetilde{\varphi}_0(z) \quad \text{for } z \in \mathscr{D}(I,\gamma) \tag{5.44}$$

for this property, thus dropping the adverb "uniformly". Indeed we cannot claim that $\mathscr{S}^I\widetilde{\varphi}_0$ admits $\widetilde{\varphi}_0$ as uniform 1-Gevrey asymptotic expansion for $z \in \mathscr{D}(I,\gamma)$ (this might simply be wrong for any locally bounded function $\gamma\colon I \to \mathbb{R}$): uniform estimates are guaranteed only when restricting to relatively compact subintervals.

The reader may check that the above definition of 1-summability in an arc of directions I coincides with the definition of k-summability in the directions of I given in [Lod16] when $k = 1$.

Remark 5.34. Suppose that $\widetilde{\varphi}_0(z) \in \mathscr{B}^{-1}\big(\mathbb{C}\,\delta \oplus \mathscr{N}(I,\gamma)\big)$, so that the Borel sum $\varphi_0(z) = \mathscr{S}^I\widetilde{\varphi}_0(z)$ is holomorphic in $\mathscr{D}(I,\gamma)$ with the asymptotic property (5.44). Of course it may happen that $\widetilde{\varphi}_0$ is 1-summable in the directions of an interval which is larger than I, in which case there will be an analytic continuation for φ_0 with 1-Gevrey asymptotic expansion in a sectorial neighbourhood of ∞ of aperture larger than $\pi + |I|$. But even if it is not so it may happen that φ_0 admits analytic continuation outside $\mathscr{D}(I,\gamma)$.

An interesting phenomenon which may occur in that case is the so-called Stokes phenomenon: *the asymptotic behaviour at ∞ of the analytic continuation of φ_0 may be totally different of what it was in the directions of $\mathscr{D}(I,\gamma)$*, typically one may encounter oscillatory behaviour along the limiting directions $-\theta^* \pm \frac{1}{2}\big(\pi + |I|\big)$ (where θ^* is the midpoint of I) and exponential growth beyond these directions. Examples can be found in Section 5.10 (Euler series: Remark 5.36 and Exercise 5.37) and § 5.13.3 (exponential of the Stirling series). In the case of the solutions of linear differential equations, this is related to the Stokes matrices already encountered in the first part of this volume; see § 6.14.3 for such an example (Airy equation).

5.9.4 What if $|I| > \pi$? First observe that, if $|I| \geq 2\pi$, then $\mathscr{N}(I)$ coincides with the set of entire functions of bounded exponential type and the corresponding formal series in z are precisely the convergent ones by Lemma 5.7:

$$|I| \geq 2\pi \quad \Longrightarrow \quad \mathscr{B}^{-1}\big(\mathbb{C}\delta \oplus \mathscr{N}(I)\big) = \mathbb{C}\{z^{-1}\}.$$

This case will be dealt with in § 5.9.5. We thus suppose $\pi < |I| < 2\pi$.

For $\widehat{\varphi} \in \mathscr{N}(I,\gamma)$, we can still define a family of holomorphic functions $\varphi_\theta := \mathscr{L}^\theta \widehat{\varphi}$ holomorphic on $\pi_\theta := \Pi^\theta_{\gamma(\theta)}$ $(\theta \in I)$, with the property that

$$0 < \theta_2 - \theta_1 < \pi \quad \Longrightarrow \quad \pi_{\theta_1} \cap \pi_{\theta_2} \neq \emptyset \text{ and } \varphi_{\theta_1} \equiv \varphi_{\theta_2} \text{ on } \pi_{\theta_1} \cap \pi_{\theta_2},$$

but the trouble is that also for $\pi < \theta_2 - \theta_1 < 2\pi$ is the intersection of half-planes $\pi_{\theta_1} \cap \pi_{\theta_2}$ non-empty and then nothing guarantees that φ_{θ_1} and φ_{θ_2} match on this intersection.

The remedy consists in lifting the half-planes π_θ and their union $\mathscr{D}(I,\gamma)$ to the Riemann surface of the logarithm $\widetilde{\mathbb{C}} = \{r\underline{\mathrm{e}}^{\mathrm{i}t} \mid r > 0,\ t \in \mathbb{R}\}$; the reader is referred to Section 6.7 for the definition of $\widetilde{\mathbb{C}}$ and the notation $\underline{\mathrm{e}}^{\mathrm{i}t}$ which represents a point "above" the complex number $\mathrm{e}^{\mathrm{i}t}$.

We thus suppose $\gamma(\theta) > 0$, so that π_θ is the set of all complex numbers $z = r\mathrm{e}^{\mathrm{i}t}$ with $r > \gamma(\theta)$ and $t \in \big(-\theta - \arccos\frac{\gamma(\theta)}{r}, -\theta + \arccos\frac{\gamma(\theta)}{r}\big)$ (and adding any integer multiple of 2π to t yields the same z), we set

$$\widetilde{\pi}_\theta := \{r\underline{\mathrm{e}}^{\mathrm{i}t} \in \widetilde{\mathbb{C}} \mid r > \gamma(\theta),\ t \in \big(-\theta - \arccos\tfrac{\gamma(\theta)}{r}, -\theta + \arccos\tfrac{\gamma(\theta)}{r}\big)\}$$

and $\widetilde{\mathscr{D}}(I,\gamma) := \bigcup_{\theta\in I} \widetilde{\pi}_\theta$ (this time $r\underline{\mathrm{e}}^{\mathrm{i}t}$ and $r\underline{\mathrm{e}}^{\mathrm{i}(t+2\pi m)}$ are regarded as different points of $\widetilde{\mathbb{C}}$), and we now define φ_θ as the holomorphic function of $\widetilde{\pi}_\theta$ obtained by composing the canonical projection $r\underline{\mathrm{e}}^{\mathrm{i}t} \in \widetilde{\pi}_\theta \mapsto r\mathrm{e}^{\mathrm{i}t} \in \pi_\theta$ and the Laplace transform $\mathscr{L}^\theta \widehat{\varphi}$.

By gluing the various φ_θ's we now get a function, which we denote by $\mathscr{L}^I \widehat{\varphi}$, that is holomorphic in $\widetilde{\mathscr{D}}(I,\gamma) \subset \widetilde{\mathbb{C}}$. The overlap between the half-planes π_{θ_1} and π_{θ_2} for $\theta_2 - \theta_1 > \pi$ is no longer a problem since their lifts $\widetilde{\pi}_{\theta_1}$ and $\widetilde{\pi}_{\theta_2}$ do not intersect (they do not lie in the same sheet of $\widetilde{\mathbb{C}}$) and $\mathscr{L}^I \widehat{\varphi}$ may behave differently on them.[4]

Therefore, one can extend Definition 5.33 to the case of an interval I of length $> \pi$ and define 1*-summability in the directions of I and the summation operator $\mathscr{S}^I$ the same way, except that the Borel sum $\mathscr{S}^I \widetilde{\varphi}_0$ of a* 1*-summable formal series $\widetilde{\varphi}_0$ is now a function holomorphic in an open subset of the Riemann surface of the logarithm $\widetilde{\mathbb{C}}$.*

5.9.5 As already announced, the Borel sum of a convergent formal series coincides with its ordinary sum:

[4] Notice that $\mathscr{N}(I,\gamma) = \mathscr{N}(2\pi + I,\gamma)$, but the functions $\mathscr{L}^\theta \widehat{\varphi}$ and $\mathscr{L}^{\theta+2\pi} \widehat{\varphi}$ must now be considered as different: they are a priori defined in domains $\widetilde{\pi}_\theta$ and $\widetilde{\pi}_{\theta+2\pi}$ which do not intersect in $\widetilde{\mathbb{C}}$. Besides, it may happen that $\mathscr{L}^\theta \widehat{\varphi}$ admit an analytic continuation in a part of $\widetilde{\pi}_{\theta+2\pi}$ which does not coincide with $\mathscr{L}^{\theta+2\pi} \widehat{\varphi}$.

Lemma 5.35. *Suppose $\tilde{\varphi}_0 \in \mathbb{C}\{z^{-1}\}$ and call $\varphi_0(z)$ the holomorphic function it defines for $|z|$ large enough. Then $\tilde{\varphi}_0$ is* 1*-summable in the directions of any interval I and $\mathscr{S}^I\tilde{\varphi}_0$ coincides with φ_0.*

Proof. Let $\tilde{\varphi}_0 = a + \tilde{\varphi}$ with $a \in \mathbb{C}$ and $\tilde{\varphi}(z) = \sum a_n z^{-n-1}$, so $\varphi(z) = a + \sum a_n z^{-n-1}$ for $|z|$ large enough. By Lemma 5.7, $\hat{\varphi} = \mathscr{B}\tilde{\varphi}$ is a convergent formal series summing to an entire function and there exists $c > 0$ such that $\hat{\varphi} \in \mathscr{N}_c(\mathrm{e}^{\mathrm{i}\theta}\mathbb{R}^+)$ for all $\theta \in \mathbb{R}$. Lemma 5.31 allows us to glue together the Laplace transforms $\mathscr{L}^\theta\hat{\varphi}$: we get one function φ_* holomorphic in $\bigcup_{\theta\in\mathbb{R}} \Pi_c^\theta = \{\,|z| > c\,\}$, with the asymptotic expansion property $\varphi_*(z) \sim_1 \tilde{\varphi}(z)$ uniformly for $\{\,|z| > c_1\,\}$ for any $c_1 > c$.

The function $\Phi_* \colon t \mapsto \varphi_*(1/t)$ is thus holomorphic in the punctured disc $\{\,0 < |t| < 1/c\,\}$. Inequality (5.32) with $N = 0$ shows that Φ_* is bounded, thus the origin is a removable singularity and Φ_* is holomorphic at $t = 0$. Now inequality (5.32) with $N = 1, 2, \dots$ shows that $\sum a_n t^{n+1}$ is the Taylor expansion at 0 of $\Phi_*(t)$, hence $a + \varphi_*(1/t) \equiv \varphi_0(1/t)$. □

5.10 Return to the Euler series

As already mentioned (right after Definition 5.29), $\hat{\varphi}^{\mathrm{E}} \in \mathscr{N}(I, 0)$ with $I = (-\pi, \pi)$. We can thus extend the domain of analyticity of $\varphi^{\mathrm{E}} = \mathscr{L}^0\hat{\varphi}^{\mathrm{E}}$, a priori holomorphic in $\pi_0 = \{\,\Re e\, z > 0\,\}$, by gluing the Laplace transforms $\mathscr{L}^\theta\hat{\varphi}^{\mathrm{E}}$, $-\pi < \theta < \pi$, each of which is holomorphic in the open half-plane π_θ bisected by the ray of direction $-\theta$ and having the origin on its boundary. But if we take no precaution this yields a multiple-valued function: there are two possible values for $\Re e\, z < 0$, according as one uses θ close to π or to $-\pi$.

A first way of presenting the situation consists in considering the subinterval $J^+ = (0, \pi)$, the Borel sum $\varphi^+ = \mathscr{S}^{J^+}\tilde{\varphi}^{\mathrm{E}}$ holomorphic in $\mathscr{D}(J^+, 0) = \mathbb{C} \setminus \mathrm{i}\mathbb{R}^+$ which extends analytically φ^{E} there, and $J^- = (-\pi, 0)$, $\varphi^- = \mathscr{S}^{J^-}\tilde{\varphi}^{\mathrm{E}}$ analytic continuation of φ^{E} in $\mathscr{D}(J^-, 0) = \mathbb{C} \setminus \mathrm{i}\mathbb{R}^-$. See the first two parts of Figure 5.2.

The intersection of the domains $\mathbb{C} \setminus \mathrm{i}\mathbb{R}^+$ and $\mathbb{C} \setminus \mathrm{i}\mathbb{R}^-$ has two connected components, the half-planes $\{\,\Re e\, z > 0\,\}$ and $\{\,\Re e\, z < 0\,\}$; both functions φ^+ and φ^- coincide with φ^{E} on the former, whereas a simple adaptation of the proof of Lemma 5.31 involving Cauchy's residue theorem yields

$$\Re e\, z < 0 \implies \varphi^+(z) - \varphi^-(z) = 2\pi\mathrm{i}\,\mathrm{e}^z. \tag{5.45}$$

(This corresponds to the cohomological viewpoint presented in the second volume of this book [Lod16]: (φ^+, φ^-) defines a 0-cochain.)

Another way of putting it is to declare that $\varphi^{\mathrm{E}} = \mathscr{S}^I\tilde{\varphi}^{\mathrm{E}}$ is a holomorphic function on

$$\tilde{\mathscr{D}}(I, 0) = \{\, z \in \tilde{\mathbb{C}} \mid -\tfrac{3\pi}{2} < \arg z < \tfrac{3\pi}{2} \,\}$$

(see Section 5.9.4) and to rewrite (5.45) as

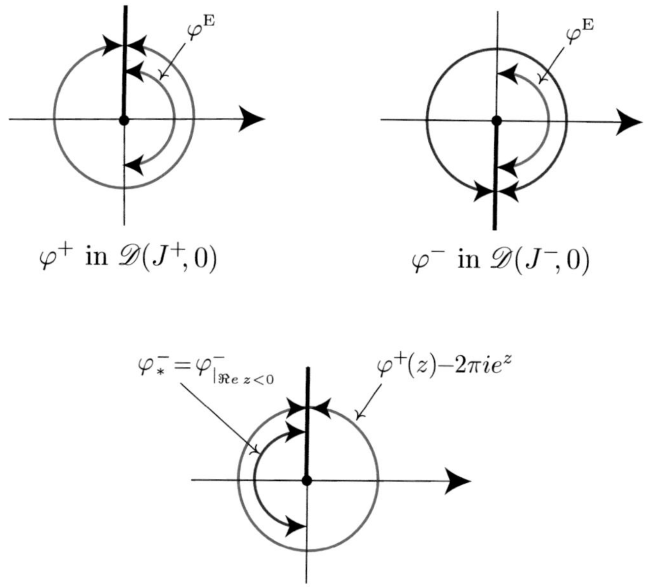

Fig. 5.2: *Borel sums of the Euler series.* Top: $\varphi^\pm$ extends φ^{E} in the cut plane $\mathscr{D}(J^\pm,0)$. Bottom: $\varphi^+(z) - 2\pi \mathrm{i}\, \mathrm{e}^z$ extends $\varphi_*^- = \varphi^-_{|\{\Re e\, z<0\}}$ in the cut plane $\mathscr{D}(J^+,0)$.

$$\frac{\pi}{2} < \arg z < \frac{3\pi}{2} \implies \varphi^{\mathrm{E}}(z\underline{\mathrm{e}}^{-2\pi \mathrm{i}}) - \varphi^{\mathrm{E}}(z) = 2\pi \mathrm{i}\, \mathrm{e}^z. \tag{5.46}$$

Remark 5.36. **[Stokes phenomenon for φ^{E}.]** Let us consider the restriction φ_*^- of the above function φ^- to the left half-plane $\{\Re e\, z < 0\}$. Using (5.45) we can write it as $\varphi^+(z) - 2\pi \mathrm{i}\, \mathrm{e}^z$, where φ^+ is holomorphic in an open sector bisected by $\mathrm{i}\mathbb{R}^-$, namely the cut plane $\mathscr{D}^+ = \mathbb{C} \setminus \mathrm{i}\mathbb{R}^+$, and the other term is an entire function: this provides the analytic continuation of φ_*^- through the cut $\mathrm{i}\mathbb{R}^-$ to the whole of $\mathscr{D}^+$. See the third part of Figure 5.2.

Observe that $\varphi^+ \sim_1 \widetilde{\varphi}^{\mathrm{E}}$ in $\mathscr{D}^+$, in particular it tends to 0 at ∞ along the directions contained in $\mathscr{D}^+$, while the exponential e^z oscillates along $\mathrm{i}\mathbb{R}^-$ and is exponentially growing in the right half-plane: we see that, for φ^-, the asymptotic behaviour encoded by $\widetilde{\varphi}^{\mathrm{E}}$ in the left half-plane breaks when we cross the limiting direction $\mathrm{i}\mathbb{R}^-$; the asymptotic behaviour of the analytic continuation is oscillatory on $\mathrm{i}\mathbb{R}^-$ (up to a correction which tends to 0) and after the crossing we find exponential growth.

A similar analysis can be performed with $\varphi_*^+ = \varphi^+_{|\{\Re e\, z<0\}}$ when one crosses $\mathrm{i}\mathbb{R}^+$, writing it as $\varphi^-(z) + 2\pi \mathrm{i}\, \mathrm{e}^z$. This is a manifestation of the Stokes phenomenon evoked in Remark 5.34.

Exercise 5.37. Use (5.46) to prove that φ^{E} is the restriction to $\widetilde{\mathscr{D}}(I,0)$ of a function which is holomorphic in the whole of $\widetilde{\mathbb{C}}$. (Hint: Show that the formula $z \in \widetilde{\mathbb{C}} \mapsto$

$\varphi(z) := \varphi^{\mathrm{E}}(z\underline{\mathrm{e}}^{-2\pi \mathrm{i} m}) - 2\pi \mathrm{i} m \mathrm{e}^{z}$ if $m \in \mathbb{Z}$ and $\arg z \in (2\pi m - \frac{3\pi}{2}, 2\pi m + \frac{3\pi}{2})$ makes sense.) In which sectors of $\widetilde{\mathbb{C}}$ is the Euler series asymptotic to this function?

Exercise 5.38. What kind of singularity has $\varphi^{\mathrm{E}}(z)$ when $|z| \to 0$? (Hint: Find an elementary function $L(z)$ such that $L(z\underline{\mathrm{e}}^{-2\pi\mathrm{i}}) - L(z) = -2\pi\mathrm{i}\,\mathrm{e}^{z}$ and consider $\varphi^{\mathrm{E}} + L$.)

Observe that the Euler equation $-\frac{\mathrm{d}\varphi}{\mathrm{d}z} + \varphi = z^{-1}$ is a non-homogeneous linear differential equation; the solutions of the associated homogeneous equation are the functions $\lambda\,\mathrm{e}^{z}$, $\lambda \in \mathbb{C}$. By virtue of the general properties of the summation operator $\mathscr{S}^{\theta}$, any Borel sum of $\widetilde{\varphi}^{\mathrm{E}}$ is an analytic solution of the Euler equation. In particular, the Borel sums φ^{+} and φ^{-} are solutions each in its own domain of definition; on formula (5.45) we can check that their restrictions to $\{\Re e\, z < 0\}$ differ by a solution of the homogeneous equation, as should be. In fact, any two branches of the analytic continuation of φ^{E} differ by an integer multiple of $2\pi\mathrm{i}\,\mathrm{e}^{z}$. Among all the solutions of the Euler equation, φ^{E} can be characterised as the only one which tends to 0 when $z \to \infty$ along a ray of direction $\in (-\frac{\pi}{2}, \frac{\pi}{2})$ (whereas, in the directions of $(\frac{\pi}{2}, \frac{3\pi}{2})$, this is no longer a distinctive property of φ^{E}: all the solutions tend to 0 in those directions!).

Exercise 5.39. How can one use the so-called method of "variation of constants" to find directly an integral formula for the solution φ^{E} of the Euler equation?

5.11 The Stirling series

The Stirling series is a classical example of divergent formal series, which is connected to Euler's gamma function. The latter is the holomorphic function defined by the formula

$$\Gamma(z) = \int_0^{+\infty} t^{z-1} \mathrm{e}^{-t}\,\mathrm{d}t \tag{5.47}$$

for any $z \in \mathbb{C}$ with $\Re e\, z > 0$ (so as to ensure the convergence of the integral). Integrating by parts, one gets the functional equation

$$\Gamma(z+1) = z\Gamma(z). \tag{5.48}$$

This equation provides the analytic continuation of Γ for $z \in \mathbb{C} \setminus (-\mathbb{N})$ in the form

$$\Gamma(z) = \frac{\Gamma(z+n)}{z(z+1)\cdots(z+n-1)} \qquad \text{for } \Re e\, z > -n, \tag{5.49}$$

for any non-negative integer n; thus Γ is meromorphic in $\mathbb{C}$ with simple poles at the non-positive integers. Since $\Gamma(1) = 1$, the functional equation also shows that

$$\Gamma(n+1) = n!, \qquad n \in \mathbb{N}. \tag{5.50}$$

Our starting point will be Stirling's formula for the restriction of Γ to the positive real axis:

Lemma 5.40.

$$\left(\frac{x}{2\pi}\right)^{\frac{1}{2}} x^{-x} \mathrm{e}^{x} \Gamma(x) \xrightarrow[x\to+\infty]{} 1. \tag{5.51}$$

Proof. This is an exercise in real analysis (and, as such, the following proof has nothing to do with the rest of the text!). In view of the functional equation, it is sufficient to prove that the function

$$f(x) := \frac{\Gamma(x+1)}{x^{x+\frac{1}{2}}\mathrm{e}^{-x}} = \int_0^{+\infty} \frac{t^x \mathrm{e}^{-t}}{x^x \mathrm{e}^{-x}} \frac{\mathrm{d}t}{x^{1/2}}$$

tends to $\sqrt{2\pi}$ as $x \to +\infty$. The idea is that the main contribution in this integral arises for t close to x and that, for $t = x+s$ with $s \to 0$, $\frac{t^x\mathrm{e}^{-t}}{x^x\mathrm{e}^{-x}} \sim \exp(-\frac{s^2}{2x})$ and $\int_{-x}^{+\infty} \exp(-\frac{s^2}{2x})\frac{\mathrm{d}s}{x^{1/2}} = \int_{-\sqrt{x}}^{+\infty} \exp(-\frac{\xi^2}{2})\,\mathrm{d}\xi$, which converges to

$$\int_{-\infty}^{+\infty} \mathrm{e}^{-\xi^2/2}\,\mathrm{d}\xi = \sqrt{2\pi} \tag{5.52}$$

as $x \to +\infty$. We now provide estimates to convert this into rigorous arguments.

We shall always assume $x \geq 1$. The change of variable $t = x + \xi\sqrt{x}$ yields

$$f(x) = \int_{-\infty}^{+\infty} \mathrm{e}^{g(x,\xi)}\,\mathrm{d}\xi, \quad \text{with } g(x,\xi) := \left(x \log\left(1+\frac{\xi}{\sqrt{x}}\right) - \xi\sqrt{x}\right)\mathbb{1}_{\{\xi>-\sqrt{x}\}}. \tag{5.53}$$

Integrating $\frac{1}{1+\sigma} = 1 - \frac{\sigma}{1+\sigma} = 1 - \sigma + \frac{\sigma^2}{1+\sigma}$, we get $\log(1+\tau) = \tau - \int_0^\tau \frac{\sigma\,\mathrm{d}\sigma}{1+\sigma} = \tau - \tau^2/2 + \int_0^\tau \frac{\sigma^2\,\mathrm{d}\sigma}{1+\sigma}$ for any $\tau > -1$, whence

$$g(x,\xi) = -x\int_0^{\xi/\sqrt{x}} \frac{\sigma\,\mathrm{d}\sigma}{1+\sigma} = -\frac{\xi^2}{2} + x\int_0^{\xi/\sqrt{x}} \frac{\sigma^2\,\mathrm{d}\sigma}{1+\sigma} \tag{5.54}$$

for any $\xi > -\sqrt{x}$. Since $\int_0^\tau \frac{\sigma^2\,\mathrm{d}\sigma}{1+\sigma} = O(\tau^3)$ as $\tau \to 0$, the last part of (5.54) shows that

$$g(x,\xi) \xrightarrow[x\to+\infty]{} -\xi^2/2 \quad \text{for each } \xi \in \mathbb{R}.$$

We shall use the first part of (5.54) to show that

(i) for $-\sqrt{x} < \xi \leq 0$, $g(x,\xi) \leq -\xi^2/2$, whence $\mathrm{e}^{g(x,\xi)} \leq G_1(\xi) := \mathrm{e}^{-\xi^2/2}$;
(ii) for $0 \leq \xi \leq \sqrt{x}$, $g(x,\xi) \leq -\xi^2/4$, whence $\mathrm{e}^{g(x,\xi)} \leq G_2(\xi) := \mathrm{e}^{-\xi^2/4}$;
(iii) for $\xi \geq \sqrt{x}$, $g(x,\xi) \leq -\xi/4$, whence $\mathrm{e}^{g(x,\xi)} \leq G_3(\xi) := \mathrm{e}^{-|\xi|/4}$.

This is sufficient to conclude by means of Lebesgue's dominated convergence theorem, since this will yield $\mathrm{e}^{g(x,\xi)} \leq G_1(\xi) + G_2(\xi) + G_3(\xi)$ for all $x \geq 1$ and $\xi \in \mathbb{R}$ and the function $G_1 + G_2 + G_3$ is independent of x and integrable on $\mathbb{R}$, thus (5.53) implies $f(x) \xrightarrow[x\to+\infty]{} \int_{-\infty}^{+\infty} \lim_{x\to+\infty} \mathrm{e}^{g(x,\xi)}\,\mathrm{d}\xi$ and (5.52) yields the final result.

– Proof of (i): Assume $-\sqrt{x} < \xi \leq 0$. Changing σ into $-\sigma$ and integrating the inequality $\frac{\sigma}{1-\sigma} \geq \sigma$ over $\sigma \in \left[0, |\xi|/\sqrt{x}\right]$, we get $g(x,\xi) = -x\int_0^{|\xi|/\sqrt{x}} \frac{\sigma\,\mathrm{d}\sigma}{1-\sigma} \leq -|\xi|^2/2$.

– Proof of (ii): Assume $0 \le \xi \le \sqrt{x}$, observe that $\frac{\sigma}{1+\sigma} \ge \frac{\sigma}{2}$ for $0 \le \sigma \le \xi/\sqrt{x}$ and integrate.

– Proof of (iii): Assume $\xi \ge \sqrt{x} \ge 1$. We compute $\int_0^{\xi/\sqrt{x}} \frac{\sigma\,\mathrm{d}\sigma}{1+\sigma} = \frac{\xi}{\sqrt{x}} - \log\left(1+\frac{\xi}{\sqrt{x}}\right) \ge \frac{\xi}{4\sqrt{x}}$, hence $g(x,\xi) \le -\frac{1}{4}\xi\sqrt{x} \le -\frac{\xi}{4}$. □

Observe that the left-hand side of (5.51) extends to a holomorphic function in a cut plane:

$$\lambda(z) := \frac{1}{\sqrt{2\pi}} z^{\frac{1}{2}-z} \mathrm{e}^z \Gamma(z), \qquad z \in \mathbb{C} \setminus \mathbb{R}^- \tag{5.55}$$

(using the principal branch of the logarithm (6.25) to define $z^{\frac{1}{2}-z} := \mathrm{e}^{(\frac{1}{2}-z)\mathrm{Log}\, z}$; in fact, λ has a meromorphic continuation to the Riemann surface of the logarithm $\widetilde{\mathbb{C}}$ defined in Section 6.7).

Theorem 5.41. *Let $I = (-\frac{\pi}{2}, \frac{\pi}{2})$. The above function λ can be written $\mathrm{e}^{\mathscr{S}^I \widetilde{\mu}}$, where $\widetilde{\mu}(z) \in z^{-1}\mathbb{C}[[z^{-1}]]$ is a divergent odd formal series which is 1-summable in the directions of I, whose formal Borel transform belongs to $\mathscr{N}(I,0)$ and is explicitly given by*

$$\widehat{\mu}(\zeta) = \zeta^{-2}\left(\frac{\zeta}{2}\coth\frac{\zeta}{2} - 1\right), \qquad \zeta \in \mathbb{C} \setminus (\Delta^+ \cup \Delta^-) \tag{5.56}$$

where $\Delta^{\pm}$ is the half-line $\pm 2\pi\mathrm{i}[1,+\infty)$, and whose Borel sum $\mathscr{S}^I\widetilde{\mu}$ is holomorphic in the cut plane $\mathscr{D}(I,0) = \mathbb{C} \setminus \mathbb{R}^-$.

It is the formal series $\widetilde{\mu}(z)$, the asymptotic expansion of $\log \lambda(z)$, that is usually called the Stirling series.

Exercise 5.42. Compute the Taylor expansion of the right-hand side of (5.56) in terms of the Bernoulli numbers B_{2k} defined by $\frac{\zeta}{\mathrm{e}^\zeta - 1} = 1 - \frac{1}{2}\zeta + \sum_{k\ge 1} \frac{B_{2k}}{(2k)!}\zeta^{2k}$ (so $B_2 = 1/6$, $B_4 = -1/30$, $B_6 = 1/42$, etc.). Deduce that

$$\widetilde{\mu}(z) = \sum_{k\ge 1} \frac{B_{2k}}{2k(2k-1)} z^{-2k+1} = \frac{1}{12}z^{-1} - \frac{1}{360}z^{-3} + \frac{1}{1260}z^{-5} + \cdots. \tag{5.57}$$

We shall see in § 5.13.3 that one can pass from $\widetilde{\mu}$ to its exponential and get an improvement of (5.51) in the form of

Corollary 5.43 (Refined Stirling formula). *The formal series $\widetilde{\lambda}(z) := \mathrm{e}^{\widetilde{\mu}(z)}$ is 1-summable in the directions of $(-\frac{\pi}{2}, \frac{\pi}{2})$ and its Borel sum is the function λ, with*

$$\lambda(z) = \frac{1}{\sqrt{2\pi}} z^{\frac{1}{2}-z}\mathrm{e}^z\Gamma(z) \sim_1 \widetilde{\lambda}(z) = 1 + \sum_{n\ge 0} g_n z^{-n-1}$$

$$\textit{uniformly for } |z| > c \textit{ and } \arg z \in (-\beta, \beta) \tag{5.58}$$

for any $c > 0$ and $\beta \in (0,\pi)$, with rationals $g_0, g_1, g_2, \ldots$ computable in terms of the Bernoulli numbers:

$$\begin{aligned}
g_0 &= \tfrac{1}{2}B_2\\
g_1 &= \tfrac{1}{8}B_2^2\\
g_2 &= \tfrac{1}{48}B_2^3 + \tfrac{1}{12}B_4\\
g_3 &= \tfrac{1}{384}B_2^4 + \tfrac{1}{24}B_2B_4\\
g_4 &= \tfrac{1}{3840}B_2^5 + \tfrac{1}{96}B_2^2B_4 + \tfrac{1}{30}B_6\\
&\vdots
\end{aligned}$$

Inserting the numerical values of the Bernoulli numbers,[5] we get

$$\Gamma(z) \sim_1 \mathrm{e}^{-z} z^{z-\frac{1}{2}} \sqrt{2\pi}\Big(1 + \tfrac{1}{12}z^{-1} + \tfrac{1}{288}z^{-2} - \tfrac{139}{51840}z^{-3} - \tfrac{571}{2488320}z^{-4} + \tfrac{163879}{209018880}z^{-5} + \cdots\Big) \quad (5.59)$$

uniformly in the domain specified in (5.58).

Proof of Theorem 5.41. **a)** We first consider $\lambda(x) = \frac{1}{\sqrt{2\pi}}x^{\frac{1}{2}-x}\mathrm{e}^x\Gamma(x)$ for $x>0$. The functional equation (5.48) yields

$$\lambda(x+1) = (1+x^{-1})^{-\frac{1}{2}-x}\mathrm{e}\,\lambda(x).$$

Formula (5.47) shows that, for $x>0$, $\Gamma(x)>0$ thus also $\lambda(x)>0$ and we can define

$$\mu(x) := \log\lambda(x), \qquad x>0. \quad (5.60)$$

This function is a particular solution of the linear difference equation

$$\mu(x+1) - \mu(x) = \psi(x), \quad (5.61)$$

where $\psi(x) := \log\big((1+x^{-1})^{-\frac{1}{2}-x}\mathrm{e}\big) = 1 - (\frac{1}{2}+x)\log(1+x^{-1})$.

b) Using the principal branch of the logarithm (6.25), holomorphic in $\mathbb{C}\setminus\mathbb{R}^-$, we see that ψ is the restriction to $(0,+\infty)$ of a function which is holomorphic in $\mathbb{C}\setminus[-1,0]$:

$$\psi(z) = -\frac{1}{2}\mathrm{Log}\,(1+z^{-1}) + z\big(z^{-1} - \mathrm{Log}\,(1+z^{-1})\big).$$

We observe that ψ is holomorphic at ∞ (i.e. $t\mapsto\psi(1/t)$ is holomorphic at the origin); moreover $\psi(z) = O(z^{-2})$ and its Taylor series at ∞ is

$$\widetilde{\psi}(z) = \frac{1}{2}\widetilde{L}(z) + z\big(z^{-1} + \widetilde{L}(z)\big) \in z^{-2}\mathbb{C}\{z^{-1}\}, \qquad \widetilde{L}(z) := -\sum_{n\geq 1}\frac{(-1)^{n-1}}{n}z^{-n}.$$

[5] and extending the notation "$\sim_1$" used in (5.33) or (5.44) by writing $F(z)\sim_1 G(z)\widetilde{\varphi}_0(z)$ whenever $F(z)/G(z)\sim_1\widetilde{\varphi}_0(z)$

With a view to applying Corollary 5.11, we compute the Borel transform $\widehat{\psi} = \mathscr{B}\widetilde{\psi}$: using $\widehat{L}(\zeta) = -\sum_{n\geq 1}(-\zeta)^{n-1}/n! = \zeta^{-1}(\mathrm{e}^{-\zeta}-1)$ and the last property in Lemma 5.10, we get

$$\widehat{\psi}(\zeta) = \frac{1}{2}\widehat{L}(\zeta) + \frac{\mathrm{d}}{\mathrm{d}\zeta}(1+\widehat{L}) = \frac{1}{2}\zeta^{-1}(\mathrm{e}^{-\zeta}-1) - \zeta^{-2}(\mathrm{e}^{-\zeta}-1) - \zeta^{-1}\mathrm{e}^{-\zeta}.$$

c) Corollary 5.11 shows that the difference equation $\widetilde{\varphi}(z+1) - \widetilde{\varphi}(z) = \widetilde{\psi}(z)$ has a unique solution in $z^{-1}\mathbb{C}[[z^{-1}]]$, whose Borel transform is

$$-\zeta^{-2} + \frac{1}{2}\zeta^{-1} - \zeta^{-1}\frac{\mathrm{e}^{-\zeta}}{\mathrm{e}^{-\zeta}-1} = \zeta^{-2}\left(-1+\zeta\left(\frac{1}{2}+\frac{1}{\mathrm{e}^{\zeta}-1}\right)\right) = \widehat{\mu}(\zeta),$$

where $\widehat{\mu}(\zeta)$ is defined by (5.56). The formal series $\widehat{\mu}(\zeta)$ is convergent and defines an even holomorphic function which extends analytically to $\mathbb{C}\setminus(\Delta^+\cup\Delta^-)$ (in fact, it even extends meromorphically to $\mathbb{C}$, with simple poles on $2\pi\mathrm{i}\mathbb{Z}^*$).

d) Let us check that $\widehat{\mu} \in \mathscr{N}(I,0)$ with $I = (-\frac{\pi}{2}, \frac{\pi}{2})$. For $\theta_0 \in (0, \frac{\pi}{2})$, we shall bound $|\widehat{\mu}|$ in the sector $\Sigma = \{\,\xi\,\mathrm{e}^{\mathrm{i}\theta} \mid \xi \geq 0,\ \theta \in [-\theta_0, \theta_0]\,\}$. Let $\varepsilon := \min\{\pi, 2\pi\cos\theta_0\}$, so that Σ does not intersect the discs $D(\pm 2\pi\mathrm{i}, \varepsilon)$. Since $\varepsilon > 0$, the number

$$A(\varepsilon) := \sup\left\{\left|\coth\frac{\zeta}{2}\right|,\ \zeta \in \mathbb{C}\setminus\bigcup_{m\in\mathbb{Z}} D(2\pi\mathrm{i}m, \varepsilon)\right\}$$

is finite, because $\zeta \mapsto \coth\frac{\zeta}{2}$ is $2\pi\mathrm{i}$-periodic, continuous in the closed set $\{\,|\Im\mathrm{m}\,\zeta| \leq \pi\,\}\setminus D(0,\varepsilon)$ and tends to ± 1 as $\Re\mathrm{e}\,\zeta \to \pm\infty$; A is in fact a decreasing function of ε. For $\zeta \in \Sigma\setminus D(0,1)$, we have $|\widehat{\mu}(\zeta)| \leq \frac{1}{2}|\zeta|^{-1}A(\varepsilon) + |\zeta|^{-2} \leq A(\varepsilon)+1$. Since $\widehat{\mu}$ is holomorphic in the disc $D(0,2\pi)$, the number $B := \sup\{|\widehat{\mu}(\zeta)|,\ \zeta \in D(0,1)\}$ is finite too, and we end up with

$$|\widehat{\mu}(\zeta)| \leq \max\{A(\varepsilon)+1, B\}, \qquad \zeta \in \Sigma,$$

whence we can conclude $\widehat{\mu} \in \mathscr{N}(I,0,\alpha)$ with $\alpha(\theta) = \max\{A(\varepsilon(\theta))+1, B\}$, $\varepsilon(\theta) = \min\{\pi, 2\pi|\cos\theta|\}$.

e) On the one hand, we have a solution $x \mapsto \mu(x)$ of equation (5.61): $\mu(x+1) - \mu(x) = \psi(x)$; this solution is defined for $x > 0$ and Stirling's formula (5.51) implies that $\mu(x)$ tends to 0 as $x \to +\infty$.

On the other hand, we have a formal solution $\widetilde{\mu}(z)$ to the equation $\widetilde{\mu}(z+1) - \widetilde{\mu}(z) = \widetilde{\psi}(z)$, which is 1-summable, with a Borel sum $\mu^+(z) := \mathscr{S}^I\widetilde{\mu}(z)$ holomorphic in $\mathscr{D}(I,0) = \mathbb{C}\setminus\mathbb{R}^-$. The property (5.35) for the summation operator $\mathscr{S}^I$ implies that

$$\mu^+(z+1) - \mu^+(z) = \mathscr{S}^I\widetilde{\psi}(z), \qquad z \in \mathbb{C}\setminus\mathbb{R}^-.$$

But $\widetilde{\psi}$ is the convergent Taylor expansion of ψ at ∞, $\mathscr{S}^I\widetilde{\psi}$ is nothing but the analytic continuation of $\psi_{|(0,+\infty)}$. The restriction of μ^+ to $(0,+\infty)$ is thus a solution to the same difference equation (5.61). Moreover, the 1-Gevrey asymptotic property implies that $\mu^+(x)$ tends to 0 as $x \to +\infty$.

The difference $x \mapsto \Delta(x) := \mu^+(x) - \mu(x)$ thus satisfies $\Delta(x+1) - \Delta(x) = 0$ and it tends to 0 as $x \to +\infty$, hence $\Delta \equiv 0$. □

Remark 5.44. Our chain of reasoning consisted in considering $\log \lambda_{|(0,+\infty)}$ and obtaining its analytic continuation to $\mathbb{C} \setminus \mathbb{R}^-$ in the form $\mathscr{S}^I \widetilde{\mu}$. As a by-product, we deduce that the holomorphic function λ does not vanish on $\mathbb{C} \setminus \mathbb{R}^-$ (being the exponential of a holomorphic function), hence the function Γ itself does not vanish on $\mathbb{C} \setminus \mathbb{R}^-$, nor does its meromorphic continuation anywhere in the complex plane in view of (5.49).

Exercise 5.45. Show that

$$\Gamma(z) = \lim_{n \to +\infty} \frac{n^z \mathrm{e}^{-z}}{z(z+1)\cdots(z+n)}$$

for every $z \in \mathbb{C} \setminus (-\mathbb{N})$. (Hint: $\lambda(z+n) \xrightarrow[n \to +\infty]{} 1$ by Theorem 5.41.)

The formal series $\widetilde{\mu}(z)$ is odd because $\widehat{\mu}(\zeta)$ is even and the Borel transform $\mathscr{B}$ shifts the powers by one unit. This does not imply that $\mathscr{S}^I \widetilde{\mu}$ is odd! The direct consequence of the oddness of $\widetilde{\mu}$ is rather the following: $\widetilde{\mu}$ is 1-summable in the directions of $J = (\frac{\pi}{2}, \frac{3\pi}{2})$ and the Borel sums $\mu^+ = \mathscr{S}^I \widetilde{\mu}$ and $\mu^- = \mathscr{S}^J \widetilde{\mu}$ are related by

$$\mu^-(z) = -\mu^+(-z), \qquad z \in \mathbb{C} \setminus \mathbb{R}^+,$$

because a change of variable in the Laplace integral yields $\mathscr{L}^\theta \widehat{\mu}(z) = -\mathscr{L}^{\theta+\pi} \widehat{\mu}(-z)$. The function μ^- is in fact another solution of the difference equation (5.61).

Exercise 5.46. – With the notations of Remark 5.44, prove that

$$\mu^+(z) - \mu^-(z) = \sum_{m \geq 1} \frac{1}{m} \mathrm{e}^{-2\pi \mathrm{i} m z}, \qquad \Im\mathrm{m}\, z < 0$$

by means of a residue computation (taking advantage of the existence of a meromorphic continuation to $\mathbb{C}$ for $\widehat{\mu}(\zeta)$, with simple poles on $2\pi\mathrm{i}\mathbb{Z}^*$, according to (5.56)).

– Deduce that, when we increase $\arg z$ above π or diminish it below $-\pi$, the function $\mu^+(z)$ has a multiple-valued analytic continuation with logarithmic singularities at negative integers.

– Deduce that $\lambda(z) = \frac{1}{(1-\mathrm{e}^{-2\pi \mathrm{i} z})\lambda(-z)}$ for $\Im\mathrm{m}\, z < 0$, thus the restriction $\lambda_{|\{\Im\mathrm{m}\, z<0\}}$ extends meromorphically to $\mathbb{C} \setminus \mathbb{R}^+$ with simple poles at the negative integers.

– Compute the residue of this meromorphic continuation at a negative integer $-k$ and check that the result is consistent with formula (5.55) and the fact that the residue of the simple pole of Γ at $-k$ is $(-1)^k/k!$. (Answer: $-\frac{\mathrm{i} k^{k+\frac{1}{2}} \mathrm{e}^{-k}}{k! \sqrt{2\pi}}$.)

– Repeat the previous computations with $\Im m\, z > 0$. Does one obtain the same meromorphic continuation to $\mathbb{C}\setminus\mathbb{R}^+$ for $\lambda_{|\{\Im m\, z>0\}}$? (Answer: no! But why?)
– Prove the reflection formula

$$\Gamma(z)\Gamma(1-z) = \frac{\pi}{\sin(\pi z)}. \tag{5.62}$$

Exercise 5.47. Using (5.48), write a functional equation for the logarithmic derivative $\psi(z) := \Gamma'(z)/\Gamma(z)$. Is there any solution of this equation in $\mathbb{C}[[z^{-1}]]$? Using the principal branch of the logarithm (6.25) and taking for granted that $\chi(z) := \psi(z) - \operatorname{Log} z$ tends to 0 as z tends to $+\infty$ along the real axis, show that $\chi(z)$ is the Borel sum of a 1-summable formal series (to be computed explicitly).

5.12 Return to Poincaré's example

In Section 5.2, we saw Poincaré's example of a meromorphic function $\phi(t)$ of $\mathbb{C}^*$ giving rise to a divergent formal series $\widetilde{\phi}(t)$ (formulas (5.3) and (5.5)). There, $w = \mathrm{e}^s$ was a parameter, with $|w| < 1$, i.e. $\Re e\, s < 0$, and we had

$$\phi(t) = \sum_{k\ge 0} \frac{w^k}{1+kt}, \qquad \widetilde{\phi}(t) = \sum_{n\ge 0} a_n t^n$$

with well-defined coefficients $a_n = (-1)^n b_n$ depending on s.

To investigate the relationship between $\phi(t)$ and $\widetilde{\phi}(t)$, we now set

$$\varphi^{\mathrm{P}}(z) = z^{-1}\phi(z^{-1}) = \sum_{k\ge 0} \frac{w^k}{z+k}, \qquad \widetilde{\varphi}^{\mathrm{P}}(z) = z^{-1}\widetilde{\phi}(z^{-1}) = \sum_{n\ge 0} a_n z^{-n-1} \tag{5.63}$$

(to place ourselves at ∞ and get rid of the constant term) so that φ^{P} is a meromorphic function of $\mathbb{C}$ with simple poles at non-positive integers and $\widetilde{\varphi}^{\mathrm{P}}(z) \in z^{-1}\mathbb{C}[[z^{-1}]]$. The formal Borel transform $\widehat{\varphi}^{\mathrm{P}}(\zeta)$ of $\widetilde{\varphi}^{\mathrm{P}}(z)$ was already computed under the name $F(\zeta)$ (see formula (5.6) and the paragraph which contains it):

$$\widehat{\varphi}^{\mathrm{P}}(\zeta) = \frac{1}{1-\mathrm{e}^{s-\zeta}}. \tag{5.64}$$

The natural questions are now: Is $\widetilde{\varphi}^{\mathrm{P}}$ 1-summable in any arc of directions and is φ^{P} its Borel sum? We shall see that the answers are affirmative, with the help of a difference equation:

Lemma 5.48. *The function φ^{P} of* (5.63) *satisfies the functional equation*

$$\varphi(z) - w\varphi(z+1) = z^{-1}. \tag{5.65}$$

For any $z_0 \in \mathbb{C} \setminus \mathbb{R}^-$, the restriction of φ^{P} to the half-line $z_0 + \mathbb{R}^+$ is the only bounded solution of (5.65) *on this half-line.*

Proof. We easily see that $w\varphi^{\mathrm{P}}(z+1) = \sum \frac{w^{k+1}}{z+1+k} = \varphi^{\mathrm{P}}(z) - z^{-1}$ for any $z \in \mathbb{C} \setminus (-\mathbb{N})$. The boundedness of φ^{P} on the half-lines stems from the fact that, for $z \in z_0 + \mathbb{R}^+$ and $k \in \mathbb{N}$, $|z+k| \geq |\Im\mathrm{m}(z+k)| = |\Im\mathrm{m}\, z_0|$ and, if $\Im\mathrm{m}\, z_0 = 0$, $|z+k| \geq z_0 > 0$, hence, in all cases, $\left|\frac{w^k}{z+k}\right| \leq A(z_0)|w|^k$ with $A(z_0) > 0$ independent of z.

As for the uniqueness: suppose φ_1 and φ_2 are bounded functions on $z_0 + \mathbb{R}^+$ which solve (5.65), then $\psi := \varphi_2 - \varphi_1$ is a bounded solution of the equation $\psi(z) - w\psi(z+1) = 0$, which implies $\psi(z) = w^n \psi(z+n)$ for any $z \in z_0 + \mathbb{R}^+$ and $n \in \mathbb{N}$; we get $\psi(z) = 0$ by taking the limit as $n \to \infty$. □

But equation (5.65), written in the form $\varphi - wT_1\varphi = z^{-1}$, can also be considered in $\mathbb{C}[[z^{-1}]]$.

Lemma 5.49. *The formal series $\tilde{\varphi}^{\mathrm{P}}$ of* (5.63) *is the unique solution of* (5.65) *in* $\mathbb{C}[[z^{-1}]]$.

Proof. It is clear that the constant term of any formal solution of (5.65) must vanish. We thus consider a formal series $\tilde{\varphi}(z) \in z^{-1}\mathbb{C}[[z^{-1}]]$. Let us denote its formal Borel transform by $\hat{\varphi}(\zeta) \in \mathbb{C}[[\zeta]]$; in view of the second property of Lemma 5.10, $\tilde{\varphi}$ is solution of (5.65) if and only if $(1 - w\mathrm{e}^{-\zeta})\hat{\varphi}(\zeta) = 1$. There is a unique solution because $1 - w\mathrm{e}^{-\zeta}$ is invertible in $\mathbb{C}[[\zeta]]$ (recall that $w \neq 1$ by assumption) and its Borel transform is $(1 - w\mathrm{e}^{-\zeta})^{-1}$, which according to (5.64) coincides with $\hat{\varphi}^P(\zeta)$ (recall that $w = \mathrm{e}^s$). □

Theorem 5.50. *The formal series $\tilde{\varphi}^{\mathrm{P}}$ is 1-summable in the directions of $I = (-\frac{\pi}{2}, \frac{\pi}{2})$ and fine-summable in the directions $\pm\frac{\pi}{2}$, with $\hat{\varphi}^{\mathrm{P}} \in \mathscr{N}(I,0) \cap \mathscr{N}_0(\mathrm{i}\mathbb{R}^+) \cap \mathscr{N}_0(-\mathrm{i}\mathbb{R}^-)$. Its Borel sum $\mathscr{S}^I\tilde{\varphi}^P$ coincides with the function φ^{P} in $\mathscr{D}(I,0) = \mathbb{C} \setminus \mathbb{R}^-$.*

Let $\omega_k = s - 2\pi\mathrm{i}k$ for $k \in \mathbb{Z}$. Then, for each $k \in \mathbb{Z}$, the formal series $\tilde{\varphi}^{\mathrm{P}}$ is 1-summable in the directions of $J_k = (\arg\omega_k, \arg\omega_{k+1}) \subset (\frac{\pi}{2}, \frac{3\pi}{2})$, with $\hat{\varphi}^{\mathrm{P}} \in \mathscr{N}(J_k,\gamma)$, $\gamma(\theta) := \cos\theta$, thus $\mathscr{D}(J_k,\gamma)$ is a sectorial neighbourhood of ∞ containing the real half-line $(-\infty, 1)$ (see Figure 5.3). The Borel sum of $\tilde{\varphi}^{\mathrm{P}}$ in the directions of J_k is a solution of equation (5.65) *which differs from φ^{P} by*

$$\varphi^{\mathrm{P}}(z) - \mathscr{S}^{J_k}\tilde{\varphi}^{\mathrm{P}}(z) = 2\pi\mathrm{i}\frac{\mathrm{e}^{-\omega_k z}}{1 - \mathrm{e}^{-2\pi\mathrm{i}z}} = -2\pi\mathrm{i}\frac{\mathrm{e}^{-\omega_{k+1} z}}{1 - \mathrm{e}^{2\pi\mathrm{i}z}}. \tag{5.66}$$

Remark 5.51. As a consequence of (5.66), we rediscover the fact that φ^{P} not only is holomorphic in $\mathbb{C} \setminus \mathbb{R}^-$ but also extends to a meromorphic function of $\mathbb{C}$, with simple poles at non-positive integers (because we can express it as the sum of $2\pi\mathrm{i}\frac{\mathrm{e}^{-sz}}{1-\mathrm{e}^{-2\pi\mathrm{i}z}}$, meromorphic on $\mathbb{C}$, and $\mathscr{S}^{J_0}\tilde{\varphi}^{\mathrm{P}}$, holomorphic in a sectorial neighbourhood of ∞ which contains $\mathbb{R}^-$). Similarly, each function $\mathscr{S}^{J_k}\tilde{\varphi}^{\mathrm{P}}$ is meromorphic in $\mathbb{C}$, with simple poles at the positive integers.

In the course of the proof of formula (5.66), it will be clear that its right-hand side is exponentially flat at ∞ in the appropriate directions, as one might expect since it has 1-Gevrey asymptotic expansion reduced to 0 (cf. footnote 2 on p. 137). This right-hand side is of the form $\psi(z) = \mathrm{e}^{-sz}\chi(z)$ with a 1-periodic function χ; it is easy to check that this is the general form of the solution of the homogeneous difference equation $\psi(z) - w\psi(z+1) = 0$.

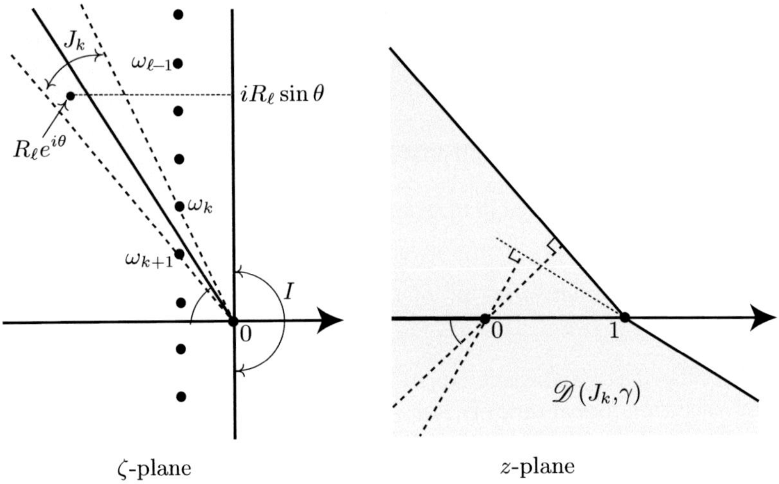

Fig. 5.3: *Borel-Laplace summation for Poincaré's example.*

The proof of Theorem 5.50 makes use of

Lemma 5.52. *Let $\sigma \in (0, -\Re e\, s)$ and $\delta > 0$. Then there exist $A = A(\sigma) > 0$ and $B = B(\delta) > 0$ such that, for any $\zeta \in \mathbb{C}$,*

$$\Re e\, \zeta \geq -\sigma \quad \Longrightarrow \quad |\widehat{\varphi}^{\mathrm{P}}(\zeta)| \leq A, \tag{5.67}$$

$$\operatorname{dist}(\zeta, s + 2\pi\mathrm{i}\mathbb{Z}) \geq \delta \quad \Longrightarrow \quad |\widehat{\varphi}^{\mathrm{P}}(\zeta)| \leq B\mathrm{e}^{\Re e\, \zeta}. \tag{5.68}$$

Lemma 5.52 implies Theorem 5.50. Inequality (5.67) implies that

$$\widehat{\varphi}^{\mathrm{P}} \in \mathscr{N}(I,0) \cap \mathscr{N}_0(\mathrm{i}\mathbb{R}^+) \cap \mathscr{N}_0(-\mathrm{i}\mathbb{R}^-),$$

whence the first summability statements follow. Lemma 5.49 and the property (5.35) for the summation operator $\mathscr{S}^I$ imply that $\mathscr{S}^I \widetilde{\varphi}^{\mathrm{P}}$ is a solution of (5.65); this solution is bounded on the half-line $[1, +\infty)$, because of the property (5.44) (in fact it tends to 0 on any half-line of the form $z_0 + \mathbb{R}^+$), thus it coincides with φ^{P} by virtue of Lemma 5.48.

Since $\Re e\, \zeta = \gamma(\arg\zeta)|\zeta|$, inequality (5.68) implies that

$$\widehat{\varphi}^{\mathrm{P}} \in \mathscr{N}(J_k, \gamma, \alpha_k)$$

with $\alpha_k\colon\ \theta \in J_k \mapsto B\big(\delta_k(\theta)\big)$, $\delta_k(\theta) = \min\big\{\operatorname{dist}(\omega_k, e^{i\theta}\mathbb{R}^+), \operatorname{dist}(\omega_{k+1}, e^{i\theta}\mathbb{R}^+)\big\}$, whence the 1-summability in the directions of J_k follows. Again, the Borel sum is a solution of the difference equation (5.65), a priori defined and holomorphic in $\mathscr{D}(J_k,\gamma)$, which is the union of the half-planes $\Pi^\theta_{\gamma(\theta)}$ for $\theta \in J_k$; one can check that each of these half-planes has the point 1 on its boundary and that the intersection $\mathscr{D}$ of $\mathscr{D}(J_k,\gamma)$ with $\mathbb{C}\setminus\mathbb{R}^-$ is connected. Thus, to conclude, it is sufficient to prove (5.66) for z belonging to one of the open subdomains $\mathscr{D}_1^+ := \Pi^\theta_{\gamma(\theta)+1} \cap \Pi_1^{\pi/2}$ or $\mathscr{D}_1^- := \Pi^\theta_{\gamma(\theta)+1} \cap \Pi_1^{-\pi/2}$, with an arbitrary $\theta \in J_k$ (none of them is empty).

Without loss of generality we can suppose $\theta \neq \pi$. If $\theta \in (\frac{\pi}{2}, \pi)$, we proceed as follows: for any integer $\ell \leq k$, the horizontal line through the midpoint of $(\omega_\ell, \omega_{\ell-1})$ cuts the half-lines $e^{i\theta}\mathbb{R}^+$ and $i\mathbb{R}^+$ in the points $R_\ell e^{i\theta}$ and $iR_\ell \sin\theta$, where R_ℓ is a positive real number which tends to $+\infty$ as $\ell \to \infty$ (see Figure 5.3). Thus, for $z \in \mathscr{D}_1^+$, we have

$$\varphi^{\mathrm{P}}(z) = \mathscr{L}^{\pi/2}\widehat{\varphi}^{\mathrm{P}}(z) = \lim_{\ell\to\infty} \int_0^{iR_\ell \sin\theta} e^{-z\zeta}\,\widehat{\varphi}^{\mathrm{P}}(\zeta)\,d\zeta,$$

$$\mathscr{S}^{J_k}\widetilde{\varphi}^{\mathrm{P}}(z) = \mathscr{L}^{\theta}\widehat{\varphi}^{\mathrm{P}}(z) = \lim_{\ell\to\infty} \int_0^{R_\ell e^{i\theta}} e^{-z\zeta}\,\widehat{\varphi}^{\mathrm{P}}(\zeta)\,d\zeta.$$

Formula (5.64) shows that $\widehat{\varphi}^{\mathrm{P}}$ is meromorphic, with simple poles at the points ω_m, $m \in \mathbb{Z}$, and residue $= 1$ at each of these poles. Cauchy's Residue Theorem implies that, for each $\ell \leq k$,

$$\left(\int_0^{iR_\ell\sin\theta} - \int_0^{R_\ell e^{i\theta}}\right) e^{-z\zeta}\,\widehat{\varphi}^{\mathrm{P}}(\zeta)\,d\zeta = 2\pi i \sum_{m=\ell}^{k} e^{-\omega_m z} + \int_{L_\ell} e^{-z\zeta}\,\widehat{\varphi}^{\mathrm{P}}(\zeta)\,d\zeta, \quad (5.69)$$

where L_ℓ is the line-segment $[R_\ell e^{i\theta}, iR_\ell\sin\theta]$. As in the proof of Lemma 5.31, we have

$$\arg\zeta \in \big[\tfrac{\pi}{2}, \theta\big] \quad\Longrightarrow\quad |e^{-z\zeta}| \leq e^{-|\zeta|(\gamma(\theta)+1)} = e^{-\Re e\,\zeta - |\zeta|}$$

(we have used $1 \geq \gamma(\theta)+1$), thus

$$\zeta \in L_\ell \quad\Longrightarrow\quad |e^{-z\zeta}\,\widehat{\varphi}^{\mathrm{P}}(\zeta)| \leq B(\pi)\,e^{-|\zeta|} \leq B(\pi)\,e^{-R_\ell\sin\theta}.$$

Hence the integral in the right-hand side of (5.69) tends to 0 and we are left with the geometric series $e^{-\omega_k z} + e^{-\omega_{k-1} z} + \cdots = e^{-\omega_k z}\sum_{n\geq 0} e^{-2\pi i n z}$ (since $-\omega_m z = -\omega_k z - 2\pi i(k-m)z$), which yields (5.66).

If $\theta \in (\pi, \frac{3\pi}{2})$, we rather take $\ell \geq k+1$ and $z \in \mathscr{D}_1^-$ and end up with

$$\varphi^{\mathrm{P}}(z) - \mathscr{S}^{J_k}\widetilde{\varphi}^{\mathrm{P}}(z) = \left(\int_0^{-i\infty} - \int_0^{e^{i\theta}\infty}\right) e^{-z\zeta}\,\widehat{\varphi}^{\mathrm{P}}(\zeta)\,d\zeta =$$

$$-2\pi i \sum_{m=k+1}^{\infty} e^{-\omega_m z} = -2\pi i e^{-\omega_k z} \sum_{n\geq 1} e^{2\pi i n z},$$

which yields the same formula. □

Proof of Lemma 5.52. In view of (5.64), for $\Re e\,\zeta \geq -\sigma$ we have $|e^{s-\zeta}| \leq e^{\sigma+\Re e\, s} < 1$ and inequality (5.67) thus holds with $A = (1-e^{\sigma+\Re e\, s})^{-1}$.

Formula (5.64) can be rewritten as $\widehat{\varphi}^{\mathrm{P}}(\zeta) = \frac{e^{\zeta}}{e^{\zeta}-e^{s}}$. Let $C_\delta := \{\,\zeta \in \mathbb{C} \mid \operatorname{dist}(\zeta, s+2\pi i\mathbb{Z}) \geq \delta\,\}$ and $F(\zeta) := |e^{\zeta} - e^{s}|$. The function F is $2\pi i$-periodic and does not vanish on C_δ; since $F(\zeta)$ tends to $+\infty$ as $\Re e\,\zeta \to +\infty$ and to $|w|$ as $\Re e\,\zeta \to -\infty$, we can find $R>0$ such that $F(\zeta) \geq |w|/2$ for $|\Re e\,\zeta| \geq R$, while $M := \min\{\,F(\zeta) \mid \zeta \in C_\delta, |\Re e\,\zeta| \leq R, |\Im m\,\zeta| \leq \pi\,\}$ is a well-defined positive number by compactness; (5.68) follows with $B = \max\{2/|w|, 1/M\}$. □

5.13 Non-linear operations with 1-summable formal series

5.13.1 The stability under multiplication of the space of 1-summable formal series associated with an interval I was already mentioned (right after Definition 5.33), but it is often useful to have more quantitative information on what happens in the variable ζ, which amounts to controlling better the convolution products.

Lemma 5.53. *Suppose that $\theta \in \mathbb{R}$ and we are given locally integrable functions $\widehat{\varphi}_1, \widehat{\varphi}_2 \colon e^{i\theta}\mathbb{R}^+ \to \mathbb{C}$ and $\Phi_1, \Phi_2 \colon \mathbb{R}^+ \to \mathbb{R}^+$ such that*

$$|\widehat{\varphi}_j(\zeta)| \leq \Phi_j(|\zeta|), \qquad \zeta \in e^{i\theta}\mathbb{R}^+$$

for $j=1,2$ and Φ_1, Φ_2 are integrable on $[0,1]$. Then the convolution products $\widehat{\varphi}_3 = \widehat{\varphi}_1 * \widehat{\varphi}_2$ and $\Phi_3 = \Phi_1 * \Phi_2$ defined by formula (5.24) satisfy

$$|\widehat{\varphi}_3(\zeta)| \leq \Phi_3(|\zeta|), \qquad \zeta \in e^{i\theta}\mathbb{R}^+.$$

Proof. Write $\widehat{\varphi}_3(\zeta)$ as $\int_0^1 \widehat{\varphi}_1(s\zeta)\widehat{\varphi}_2((1-s)\zeta)\,\zeta\,ds$ and $\Phi_3(\xi)$ as $\int_0^1 \Phi_1(s\xi)\Phi_2((1-s)\xi)\,\xi\,ds$. □

Lemma 5.54. *Suppose Δ is an open subset of $\mathbb{C}$ which is star-shaped with respect to 0 (i.e. it is non-empty and, for every $\zeta \in \Delta$, the line-segment $[0,\zeta]$ is included in Δ). Suppose $\widehat{\varphi}_1$ and $\widehat{\varphi}_2$ are holomorphic in Δ. Then their convolution product (which is well defined since $0 \in \Delta$) is also holomorphic in Δ.*

Proof. The function $(s,\zeta) \mapsto \widehat{\varphi}_1(s\zeta)\widehat{\varphi}_2((1-s)\zeta)$ is continuous in s, holomorphic in ζ and bounded in $[0,1] \times K$ for any compact subset K of Δ. □

5.13.2 As an application, we show that 1-summability is compatible with the composition operator associated with a 1-summable formal series and with substitution into a convergent power expansion:

Theorem 5.55. *Suppose I is an open interval of $\mathbb{R}$, $\widetilde{\varphi}_0(z) = a + \widetilde{\varphi}(z)$ and $\widetilde{\psi}_0(z)$ are 1-summable formal series in the directions of I, with $a \in \mathbb{C}$ and $\widetilde{\varphi}(z) \in z^{-1}\mathbb{C}[[z^{-1}]]$, and $H(t) \in \mathbb{C}\{t\}$. Then the formal series $\widetilde{\psi}_0 \circ (\mathrm{id} + \widetilde{\varphi}_0)$ and $H \circ \widetilde{\varphi}$ are 1-summable in the directions of I and*

$$\mathscr{S}^I\big(\widetilde{\psi}_0 \circ (\mathrm{id} + \widetilde{\varphi}_0)\big) = (\mathscr{S}^I \widetilde{\psi}_0) \circ (\mathrm{id} + \mathscr{S}^I \widetilde{\varphi}_0), \qquad \mathscr{S}^I(H \circ \widetilde{\varphi}) = H \circ \mathscr{S}^I \widetilde{\varphi}. \quad (5.70)$$

More precisely, if $\mathscr{B}\widetilde{\varphi} \in \mathscr{N}(I, \gamma, \alpha)$ and $\mathscr{B}\widetilde{\psi}_0 \in \mathbb{C}\delta \oplus \mathscr{N}(I, \gamma)$ with $\alpha, \gamma \colon I \to \mathbb{R}$ locally bounded, $\alpha \geq 0$, and ρ is a positive number smaller than the radius of convergence of H, then

$$\mathscr{B}\big(\widetilde{\psi}_0 \circ (\mathrm{id} + \widetilde{\varphi}_0)\big) \in \mathbb{C}\delta \oplus \mathscr{N}(I, \gamma_1), \qquad \gamma_1 := \gamma + |a| + \sqrt{\alpha}, \quad (5.71)$$

$$\mathscr{B}(H \circ \widetilde{\varphi}) \in \mathbb{C}\delta \oplus \mathscr{N}(I, \gamma_2), \qquad \gamma_2 := \gamma + \rho^{-1}\alpha, \quad (5.72)$$

$$z \in \mathscr{D}(I, \gamma_1) \implies z + \mathscr{S}^I \widetilde{\varphi}_0(z) \in \mathscr{D}(I, \gamma), \qquad z \in \mathscr{D}(I, \gamma_2) \implies |\mathscr{S}^I \widetilde{\varphi}(z)| \leq \rho \quad (5.73)$$

and the identities in (5.70) *hold in $\mathscr{D}(\gamma_1, I)$ and $\mathscr{D}(\gamma_2, I)$ respectively.*

Proof. By assumption, $\widehat{\varphi} = \mathscr{B}\widetilde{\varphi} \in \mathscr{N}(I, \gamma, \alpha)$. The properties (5.73) are easily obtained as a consequence of

$$z \in \Pi^{\theta}_{\gamma_j(\theta)} \implies \mathscr{S}^I \widetilde{\varphi}(z) = \mathscr{L}^{\theta} \widehat{\varphi}(z) \text{ and } |\mathscr{L}^{\theta} \widehat{\varphi}(z)| \leq \frac{\alpha(\theta)}{\gamma_j(\theta) - \gamma(\theta)} \quad (5.74)$$

for any $\theta \in I$ and $j = 1, 2$.

Let $\widetilde{\psi}_0(z) = b + \widetilde{\psi}(z)$ and $H = c + h(t)$ with $b, c \in \mathbb{C}$ and $\widetilde{\psi}(z) \in z^{-1}\mathbb{C}[[z^{-1}]]$, $h(t) \in t\mathbb{C}\{t\}$, so that

$$\widetilde{\psi}_0 \circ (\mathrm{id} + \widetilde{\varphi}_0) = b + \widetilde{\lambda}, \qquad \widetilde{\lambda} := \widetilde{\psi} \circ (\mathrm{id} + \widetilde{\varphi}_0), \quad (5.75)$$

$$H \circ \widetilde{\varphi} = c + \widetilde{\mu}, \qquad \widetilde{\mu} := h \circ \widetilde{\varphi}. \quad (5.76)$$

We recall that $\widetilde{\lambda}$ and $\widetilde{\mu}$ are defined by the formally convergent series of formal series

$$\widetilde{\lambda} = \sum_{k \geq 0} \frac{1}{k!} (\partial^k \widetilde{\psi})(\widetilde{\varphi}_0)^k, \qquad \widetilde{\mu} = \sum_{k \geq 1} h_k \widetilde{\varphi}^k, \quad (5.77)$$

where we use the notation $h(t) = \sum_{k \geq 1} h_k t^k$.

Correspondingly, in we have formally convergent series of formal series in $\mathbb{C}[[\zeta]]$: for instance, the Borel transform of $\widetilde{\mu}$ is

$$\widehat{\mu} = \sum_{k \geq 1} h_k \widehat{\varphi}^{*k}, \quad \text{where } \widehat{\varphi}^{*k} = \underbrace{\widehat{\varphi} * \cdots * \widehat{\varphi}}_{k \text{ factors}} \in \zeta^{k-1}\mathbb{C}[[\zeta]]. \quad (5.78)$$

But the series in the right-hand side of (5.78) can be viewed as a series of holomorphic functions, since $\widehat{\varphi}$ is holomorphic in the union of a disc $D(0, R)$ and of the sector $\Sigma = \{\, \xi\, \mathrm{e}^{\mathrm{i}\theta} \mid \xi > 0,\ \theta \in I \,\}$: the open set $D(0, R) \cup \Sigma$ is star-shaped with respect to 0, thus Lemma 5.54 applies and each $\widehat{\varphi}^{*k}$ is holomorphic in $D(0,R)\cup\Sigma$. We shall prove the normal convergence of this series of functions in each compact subset of $D(0,R)\cup\Sigma$ and provide appropriate bounds.

Choosing $R>0$ smaller than the radius of convergence of $\widehat{\varphi}$, we have

$$|\widehat{\varphi}(\zeta)|\le A, \qquad \zeta\in D(0,R),$$

$$|\widehat{\varphi}(\zeta)|\le \Phi_\theta(\xi):=\alpha(\theta)\,\mathrm{e}^{\gamma(\theta)\xi}, \qquad \zeta\in\Sigma,$$

with a positive number A, using the notations $\xi=|\zeta|$ and $\theta=\arg\zeta$ in the second case. The computation of $\Phi_\theta^{*k}(\xi)$ is easy, since Φ_θ can be viewed as the restriction to $\mathbb{R}^+$ of the Borel transform of $\alpha(\theta)T_{-\gamma(\theta)}(z^{-1})$; Lemma 5.53 thus yields

$$|\widehat{\varphi}^{*k}(\zeta)|\le A^k\frac{\xi^{k-1}}{(k-1)!}, \qquad \zeta\in D(0,R), \tag{5.79}$$

$$|\widehat{\varphi}^{*k}(\zeta)|\le \Phi_\theta^{*k}(\xi)=\alpha(\theta)^k\frac{\xi^{k-1}}{(k-1)!}\mathrm{e}^{\gamma(\theta)\xi}, \qquad \zeta\in\Sigma. \tag{5.80}$$

These inequalities, together with the fact that there exists $B>0$ such that $|h_k|\le B\rho^{-k}$ for all $k\ge1$ (because ρ is smaller than the radius of convergence of H), imply that the series of functions $\sum h_k\widehat{\varphi}^{*k}$ is uniformly convergent in every compact subset of $D(0,R)\cup\Sigma$; the sum of this series is a holomorphic function whose Taylor coefficients at 0 coincide with those of $\widehat{\mu}$, hence $\widehat{\mu}(\zeta)\in\mathbb{C}\{\zeta\}$ and $\widehat{\mu}$ extends analytically to $D(0,R)\cup\Sigma$.

Inequalities (5.80) also show that, for $\zeta\in\Sigma$,

$$|h_k\widehat{\varphi}^{*k}(\zeta)|\le\alpha(\theta)B\rho^{-1}\frac{\big(\rho^{-1}\alpha(\theta)\xi\big)^{k-1}}{(k-1)!}\mathrm{e}^{\gamma(\theta)\xi},$$

hence $|\widehat{\mu}(\zeta)|\le\alpha(\theta)B\rho^{-1}\exp\big((\gamma(\theta)+\rho^{-1}\alpha(\theta))\xi\big)$, i.e. $\widehat{\mu}\in\mathscr{N}(I,\gamma+\rho^{-1}\alpha)$. The dominated convergence theorem shows that, for each $\theta\in I$ and $z\in\Pi^\theta_{\gamma_2(\theta)}$, $\mathscr{L}^\theta\widehat{\mu}(z)$ coincides with the convergent sum of the series

$$\sum h_k(\mathscr{L}^\theta\widehat{\varphi}^{*k})(z)=\sum h_k\big(\mathscr{L}^\theta\widehat{\varphi}(z)\big)^k,$$

which is $h\big(\mathscr{L}^\theta\widehat{\varphi}(z)\big)$, whence $\mathscr{S}^I\widetilde{\mu}(z)\equiv h\big(\mathscr{S}^I\widetilde{\varphi}(z)\big)$.

We now move on to the case of $\widetilde{\lambda}$. Without loss of generality we can suppose that $a=0$, i.e. that there is no translation term in $\widetilde{\varphi}_0$, since $\widetilde{\lambda}=(T_a\widetilde{\psi})\circ(\mathrm{id}+\widetilde{\varphi})$, thus it will be sufficient to apply the translationless case of (5.70) and (5.71) to $T_a\widetilde{\psi}\in\mathscr{B}^{-1}\big(\mathscr{N}(I,\gamma+|a|)\big)$: the identity (5.35) for $\mathscr{S}^I$ will yield $\mathscr{S}^I\Big((T_a\widetilde{\psi})\circ(\mathrm{id}+\widetilde{\varphi})\Big)=(\mathscr{S}^IT_a\widetilde{\psi})\circ(\mathrm{id}+\mathscr{S}^I\widetilde{\varphi})=(\mathscr{S}^I\widetilde{\psi})\circ(\mathrm{id}+a)\circ(\mathrm{id}+\mathscr{S}^I\widetilde{\varphi})=(\mathscr{S}^I\widetilde{\psi})\circ(\mathrm{id}+a+\mathscr{S}^I\widetilde{\varphi})$.

When $a=0$, in view of (5.77) and the first property in Lemma 5.10, the formal series $\widehat{\lambda}:=\mathscr{B}\widetilde{\lambda}\in\mathbb{C}[[\zeta]]$ is given by the formally convergent series of formal series

$$\widehat{\lambda} = \sum_{k\geq 0} \widehat{\chi}_k, \qquad \widehat{\chi}_k := \frac{1}{k!}\big((-\zeta)^k\widehat{\psi}\big) * \widehat{\varphi}^{*k}.$$

We now view the right-hand side as a series of holomorphic functions. Diminishing R if necessary so as to make it smaller than the radius of convergence of $\widehat{\psi}$ and taking $\alpha'\colon I \to \mathbb{R}^+$ locally bounded such that $\widehat{\psi} \in \mathscr{N}(I,\gamma,\alpha')$, we can find $A' > 0$ such that

$$|\widehat{\psi}(\zeta)| \leq A', \qquad \zeta \in D(0,R),$$

$$|\widehat{\psi}(\zeta)| \leq \Psi_\theta(\xi) := \alpha'(\theta)\,\mathrm{e}^{\gamma(\theta)\xi}, \qquad \zeta \in \Sigma.$$

Lemma 5.53 and 5.54 show that the $\widehat{\chi}_k$'s are holomorphic in $D(0,R)\cup\Sigma$ and satisfy

$$|\widehat{\chi}_k(\zeta)| \leq A'\frac{\xi^k}{k!} * A^k\frac{\xi^{k-1}}{(k-1)!} = A'A^k\frac{\xi^{2k}}{(2k)!}, \qquad \zeta \in D(0,R), \tag{5.81}$$

$$|\widehat{\chi}_k(\zeta)| \leq \Big(\frac{\xi^k}{k!}\Psi_\theta\Big) * \Phi_\theta^{*k}(\xi) = \alpha'(\theta)\alpha^k(\theta)\frac{\xi^{2k}}{(2k)!}\mathrm{e}^{\gamma(\theta)\xi}, \quad \zeta \in \Sigma \tag{5.82}$$

(we used (5.79), (5.80) and (5.26)). The series $\sum\widehat{\chi}_k$ is thus uniformly convergent in the compact subsets of $D(0,R)\cup\Sigma$ and sums to a holomorphic function, whose Taylor series at 0 is $\widehat{\lambda}$. Hence we can view $\widehat{\lambda}$ as a holomorphic function and the last inequalities imply that

$$|\widehat{\lambda}(\zeta)| \leq \alpha'(\theta)\cosh\big(\sqrt{\alpha(\theta)}\xi\big)\mathrm{e}^{\gamma(\theta)\xi} \leq \alpha'(\theta)\mathrm{e}^{(\sqrt{\alpha(\theta)}+\gamma(\theta))\xi}$$

for $\zeta\in\Sigma$. This yields $\widehat{\lambda}\in\mathscr{N}(I,\gamma+\sqrt{\alpha})$ and, since $\mathscr{L}^\theta\widehat{\chi}_k = \frac{1}{k!}\big(\big(\frac{\mathrm{d}}{\mathrm{d}z}\big)^k\mathscr{L}^\theta\widehat{\psi}\big)(\mathscr{L}^\theta\widehat{\varphi})^k$ (use the first property in Lemma 5.17 and the identity (5.34) for $\mathscr{L}^\theta$), the dominated convergence theorem yields $\mathscr{S}^I\widetilde{\lambda} = (\mathscr{S}^I\widetilde{\psi})\circ(\mathrm{id}+\mathscr{S}^I\widetilde{\varphi})$. □

Exercise 5.56. Prove the following multivariate version of the result on substitution in a convergent series: suppose that $r\geq 1$, $H(t_1,\dots,t_r)\in\mathbb{C}\{t_1,\dots,t_r\}$, I is an open interval of $\mathbb{R}$ and $\widetilde{\varphi}_1(z),\dots,\widetilde{\varphi}_r(z)\in z^{-1}\mathbb{C}[[z^{-1}]]$ are 1-summable in the directions of I; then the formal series

$$\widetilde{\chi}(z) := H\big(\widetilde{\varphi}_1(z),\dots,\widetilde{\varphi}_r(z)\big)$$

is 1-summable in the directions of I and $\mathscr{S}^I\widetilde{\chi} = H\circ(\mathscr{S}^I\widetilde{\varphi}_1,\dots,\mathscr{S}^I\widetilde{\varphi}_r)$.

5.13.3 *Proof of Corollary 5.43.* As a consequence of Theorem 5.55, using $H(t) = \mathrm{e}^t$, we obtain the 1-summability in the directions of $I = (-\frac{\pi}{2},\frac{\pi}{2})$ of the exponential $\widetilde{\lambda}$ of the Stirling series $\widetilde{\mu}$, whence the refined Stirling formula (5.58) for

$$\lambda = \mathrm{e}^{\mathscr{S}^I\widetilde{\mu}} = \mathscr{S}^I\widetilde{\lambda}.$$

□

Exercise 5.57. We just obtained that

$$\Gamma(z) \sim_1 \mathrm{e}^{-z} z^{z-\frac{1}{2}} \sqrt{2\pi}\Big(1+\sum_{k\geq 0} g_k z^{-k-1}\Big) \quad \text{uniformly for } |z|>c \text{ and } \arg z \in (-\beta,\beta),$$

for any $c>0$ and $\beta \in (0,\pi)$ (with the extended notation of footnote 5). Show that

$$\frac{1}{\Gamma(z)} \sim_1 \frac{1}{\sqrt{2\pi}} \mathrm{e}^{z} z^{-z+\frac{1}{2}} \Big(1+\sum_{k\geq 0} (-1)^{k+1} g_k z^{-k-1}\Big)$$

$$\text{uniformly for } |z|>c \text{ and } \arg z \in (-\beta,\beta)$$

for the same values of c and β.

Remark 5.58. Since $n! = n\Gamma(n)$ by (5.50) and (5.48), we get

$$n! \sim \frac{n^n\sqrt{2\pi n}}{\mathrm{e}^n}\Big(1+\frac{1}{12n}+\frac{1}{288n^2}-\frac{139}{51840n^3}-\frac{571}{2488320n^4}+\frac{163879}{209018880n^5}+\cdots\Big)$$

$$\frac{1}{n!} \sim \frac{\mathrm{e}^n}{n^n\sqrt{2\pi n}}\Big(1-\frac{1}{12n}+\frac{1}{288n^2}+\frac{139}{51840n^3}-\frac{571}{2488320n^4}-\frac{163879}{209018880n^5}+\cdots\Big)$$

uniformly for $n \in \mathbb{N}^*$. See [DA09] for a direct proof.

Remark 5.59. In accordance with Remark 5.34, we observe a kind of Stokes phenomenon for the function λ: it is a priori holomorphic in the cut plane $\mathbb{C}\setminus\mathbb{R}^-$, or equivalently in the sector $\{-\pi<\arg z<\pi\}$ of the Riemann surface of the logarithm $\widetilde{\mathbb{C}}$, but Exercise 5.46 gives the "reflection formula" $\lambda(z) = \frac{1}{(1-\mathrm{e}^{-2\pi \mathrm{i} z})\lambda(\mathrm{e}^{\mathrm{i}\pi}z)}$ for $-\pi<\arg z<0$, which yields a meromorphic continuation for λ in the larger sector $\{-2\pi<\arg z<\pi\}$ (with the points $k\mathrm{e}^{-\mathrm{i}\pi}$, $k\in\mathbb{N}^*$, as only poles); the asymptotic property $\lambda(z)\sim_1\widetilde{\lambda}(z)$ is valid in the directions of $(-\pi,\pi)$ but not in those of $(-2\pi,-\pi]$: the ray $\mathrm{e}^{-\mathrm{i}\pi}\mathbb{R}^+$ is singular and the reflection formula implies that, in the directions of $(-2\pi,-\pi)$, $\lambda(z)\sim -\mathrm{e}^{2\pi \mathrm{i} z}$, which is exponentially small (and $\mathrm{e}^{-2\pi \mathrm{i} z}\lambda(z)\sim_1 -\widetilde{\lambda}(z)$ there).

In fact, iterating the reflection formula we find a meromorphic continuation to the whole of $\widetilde{\mathbb{C}}$, with a "monodromy relation" $\lambda(z) = -\mathrm{e}^{2\pi \mathrm{i} z}\lambda(z\underline{\mathrm{e}}^{2\pi \mathrm{i}})$ (with the notations of Section 6.7). Outside the singular rays, the asymptotic behaviour is given by

$$\lambda(z) = (-1)^n \mathrm{e}^{-2\pi \mathrm{i} n z}\lambda(z\underline{\mathrm{e}}^{-2\pi \mathrm{i} n}) \sim_1 (-1)^n \mathrm{e}^{-2\pi \mathrm{i} n z}\widetilde{\lambda}(z)$$

uniformly for $|z|$ large enough and $2\pi n-\beta<\arg z<2\pi n+\beta$, with arbitrary $n\in\mathbb{Z}$ and $\beta\in(0,\pi)$. Except in the initial sector of definition ($n=0$), we thus find exponential decay and exponential growth alternating at each crossing of a singular ray $\underline{\mathrm{e}}^{(2n-1)\mathrm{i}\pi}\mathbb{R}^+$ or of a ray $\underline{\mathrm{e}}^{2n\mathrm{i}\pi}\mathbb{R}^+$ on which the behaviour is oscillatory, according to the sign of $n\,\Im\mathrm{m}\, z$ (since $|\mathrm{e}^{-2\pi \mathrm{i} n z}| = \mathrm{e}^{2\pi n\,\Im\mathrm{m}\, z}$).

The last properties can also be deduced from formula (5.55).

5.13.4 We leave it to the reader to adapt the results of this section to fine-summable formal series in a direction θ.

5.14 Germs of holomorphic diffeomorphisms

A *holomorphic local diffeomorphism around* 0 is a holomorphic map $F\colon U \to \mathbb{C}$, where U is an open neighbourhood of 0 in $\mathbb{C}$, such that $F(0)=0$ and $F'(0)\neq 0$. The local inversion theorem shows that there is an open neighbourhood V of 0 contained in U such that $F(V)$ is open and F induces a biholomorphism from V to $F(V)$. When we are not too much interested in the precise domains U or V but are ready to replace them by smaller neighrbouhoods of 0, we may consider the *germ of* F *at* 0. This means that we consider the equivalence class of F for the following equivalence relation: two holomorphic local diffeomorphisms are equivalent if there exists an open neighbourhood of 0 on which their restrictions coincide.

It is easy to see that a germ of holomorphic diffeomorphism at 0 can be identified with the Taylor series at 0 of any of its representatives. Moreover, our equivalence relation is compatible with the composition and the inversion of holomorphic local diffeomorphisms. Consequently, the germs of holomorphic diffeomorphisms at 0 make up a (nonabelian) group, isomorphic to

$$\{F(t)\in t\mathbb{C}\{t\} \mid F'(0)\neq 0\} = \Big\{F(t) = \sum_{n\geq 1} c_n t^n \in \mathbb{C}\{t\} \mid c_1\neq 0\Big\}.$$

The coefficient $c_1 = F'(0)$ is called the "multiplier" of F. Obviously, for two germs of holomorphic diffeomorphisms F and G, $(F\circ G)'(0) = F'(0)G'(0)$. Therefore, the germs F of holomorphic diffeomorphisms at 0 such that $F'(0)=1$ make up a subgroup; such germs are said to be "tangent-to-identity".

Germs of holomorphic diffeomorphisms can also be considered at ∞: via the inversion $t\mapsto z=1/t$, a germ $F(t)$ at 0 is conjugate to $f(z)=1/F(1/z)$. From now on, we focus on the tangent-to-identity case

$$F(t) = t-\sigma t^2-\tau t^3+\cdots = t(1-\sigma t-\tau t^2+\cdots)\in\mathbb{C}\{t\} \qquad (\sigma,\tau\in\mathbb{C}). \quad (5.83)$$

This amounts to considering germs of holomorphic diffeomorphisms at ∞ of the form

$$f(z) = z(1-\sigma z^{-1}-\tau z^{-2}+\cdots)^{-1} = z+\sigma+(\tau+\sigma^2)z^{-1}+\cdots \in \mathrm{id}+\mathbb{C}\{z^{-1}\}. \quad (5.84)$$

For such a germ f, there exists $c>0$ large enough and a representative which is an injective holomorphic function in $\{\,|z|>c\,\}$. We use the notations

$$\mathscr{G} := \mathrm{id}+\mathbb{C}\{z^{-1}\}$$

for the group of tangent-to-identity germs of holomorphic diffeomorphisms at ∞, and

$$\mathscr{G}_\sigma := \mathrm{id} + \sigma + z^{-1}\mathbb{C}\{z^{-1}\}$$

when we want to keep track of the coefficient σ in (5.84). Notice that, if $f_1 \in \mathscr{G}_{\sigma_1}$ and $f_2 \in \mathscr{G}_{\sigma_2}$, then $f_1 \circ f_2 \in \mathscr{G}_{\sigma_1+\sigma_2}$.

5.15 Formal diffeomorphisms

Even if we are interested in properties of the group $\mathscr{G}$, or even of a single element of $\mathscr{G}$, it is useful (as we shall see in Sections 7.1–7.6) to drop the convergence requirement and consider the larger set

$$\widetilde{\mathscr{G}} = \mathrm{id} + \mathbb{C}[[z^{-1}]].$$

This is the set of *formal tangent-to-identity diffeomorphisms at* ∞, which we view as a complete metric space by means of the distance

$$d(\tilde{f},\tilde{h}) := 2^{-\mathrm{val}(\tilde{\chi}-\tilde{\varphi})}, \qquad \tilde{f} = \mathrm{id} + \tilde{\varphi},\ \tilde{h} = \mathrm{id} + \tilde{\chi}, \qquad \tilde{\varphi},\tilde{\chi} \in \mathbb{C}[[z^{-1}]],$$

as we did for $\mathbb{C}[[z^{-1}]]$ in § 5.3.3. Notice that $\mathscr{G}$ appears as a dense subset of $\widetilde{\mathscr{G}}$. We also use the notation

$$\widetilde{\mathscr{G}}_\sigma = \mathrm{id} + \sigma + z^{-1}\mathbb{C}[[z^{-1}]] = \big\{ \tilde{f}(z) = z + \sigma + \tilde{\varphi}(z) \mid \tilde{\varphi} \in z^{-1}\mathbb{C}[[z^{-1}]] \big\} \subset \widetilde{\mathscr{G}}$$

for any $\sigma \in \mathbb{C}$. Via the inversion $z \mapsto 1/z$, the elements of $\widetilde{\mathscr{G}}$ are conjugate to formal tangent-to-identity diffeomorphisms at 0, i.e. formal series of the form (5.83) but without the convergence condition (the corresponding $F(t)$ is in $\mathbb{C}[[t]]$ but not necessarily in $\mathbb{C}\{t\}$); the elements of $\widetilde{\mathscr{G}}_\sigma$ are conjugate to formal series of the form $F(t) = t - \sigma t^2 + \cdots \in \mathbb{C}[[t]]$, by the formal analogue of (5.84).

Theorem 5.60. *The set* $\widetilde{\mathscr{G}}$ *is a nonabelian topological group for the composition law*

$$(\tilde{f},\tilde{h}) = (\mathrm{id}+\tilde{\varphi}, \mathrm{id}+\tilde{\chi}) \mapsto \tilde{f} \circ \tilde{h} := \mathrm{id} + \tilde{\chi} + \tilde{\varphi} \circ (\mathrm{id}+\tilde{\chi}), \tag{5.85}$$

for $\tilde{\varphi},\tilde{\chi} \in \mathbb{C}[[z^{-1}]]$, *with* $\tilde{\varphi} \circ (\mathrm{id}+\tilde{\chi})$ *defined by* (5.14). *The subset*

$$\widetilde{\mathscr{G}}_0 = \mathrm{id} + z^{-1}\mathbb{C}[[z^{-1}]]$$

is a subgroup of $\widetilde{\mathscr{G}}$.

Notice that the definition (5.85) of the composition law in $\widetilde{\mathscr{G}}$ can also be written

$$\tilde{f} \circ \tilde{h} = \sum_{k \geq 0} \frac{1}{k!} \tilde{\chi}^k \partial^k \tilde{f}, \qquad \tilde{h} = \mathrm{id} + \tilde{\chi}, \tag{5.86}$$

with the convention $\partial^0 \widetilde{f} = \widetilde{f} = \mathrm{id} + \widetilde{\varphi}$, $\partial \widetilde{f} = 1 + \partial \widetilde{\varphi}$ and $\partial^k \widetilde{f} = \partial^k \widetilde{\varphi}$ for $k \geq 2$.

Proof of Theorem 5.60. The composition (5.85) is a continuous map $\widetilde{\mathscr{G}} \times \widetilde{\mathscr{G}} \to \widetilde{\mathscr{G}}$ because, for $\widetilde{f}, \widetilde{f}^*, \widetilde{h}, \widetilde{h}^* \in \widetilde{\mathscr{G}}$, formula (5.86) implies

$$\widetilde{f} \circ \widetilde{h}^* - \widetilde{f} \circ \widetilde{h} = (\widetilde{h}^* - \widetilde{h}) \int_0^1 \partial \widetilde{f} \circ \big((1-t)\widetilde{h} + t\widetilde{h}^*\big)\, \mathrm{d}t \tag{5.87}$$

(where $\partial \widetilde{f} \circ \big((1-t)\widetilde{h} + t\widetilde{h}^*\big)$ is a formal series whose coefficients depend polynomially on t and integration is meant coefficient-wise); this is a formal series of valuation $\geq \operatorname{val}(\widetilde{h}^* - \widetilde{h})$, by virtue of (5.15), hence the difference

$$\widetilde{f}^* \circ \widetilde{h}^* - \widetilde{f} \circ \widetilde{h} = (\widetilde{f}^* - \widetilde{f}) \circ \widetilde{h}^* + \widetilde{f} \circ \widetilde{h}^* - \widetilde{f} \circ \widetilde{h}$$

is a formal series of valuation $\geq \min\big\{ \operatorname{val}(\widetilde{f}^* - \widetilde{f}), \operatorname{val}(\widetilde{h}^* - \widetilde{h}) \big\}$ (using again (5.15)), i.e.

$$d(\widetilde{f} \circ \widetilde{h}, \widetilde{f}^* \circ \widetilde{h}^*) \leq \max\big\{ d(\widetilde{f}, \widetilde{f}^*), d(\widetilde{h}, \widetilde{h}^*) \big\}.$$

The subset $\widetilde{\mathscr{G}}_0$ is clearly stable by composition.

The composition law of $\widetilde{\mathscr{G}}$, when restricted to $\mathscr{G}$, boils down to the composition of holomorphic germs which is associative ($\mathscr{G}$ is a group) and $\mathscr{G}$ is a dense subset of $\widetilde{\mathscr{G}}$, thus composition is associative in $\widetilde{\mathscr{G}}$ too. It is not commutative in $\widetilde{\mathscr{G}}$ since it is not commutative in $\mathscr{G}$. The element id is clearly a unit for composition in $\widetilde{\mathscr{G}}$ thus we only need to show that there is a well-defined continuous inverse map $\widetilde{h} \in \widetilde{\mathscr{G}} \mapsto \widetilde{h}^{\circ(-1)} \in \widetilde{\mathscr{G}}$ and that this map leaves $\widetilde{\mathscr{G}}_0$ invariant.

We first show that every element $\widetilde{h} \in \widetilde{\mathscr{G}}$ has a unique left inverse $\mathscr{L}(\widetilde{h})$. Given $\widetilde{h} = \mathrm{id} + \widetilde{\chi}$, the equation $\widetilde{f} \circ \widetilde{h} = \mathrm{id}$ is equivalent to the fixed-point equation

$$\widetilde{f} = \mathscr{C}(\widetilde{f}), \qquad \mathscr{C}(\widetilde{f}) := \mathrm{id} - (\widetilde{f} \circ \widetilde{h} - \widetilde{f}) = \mathrm{id} - \widetilde{\chi} \int_0^1 \partial \widetilde{f} \circ (\mathrm{id} + t\widetilde{\chi})\, \mathrm{d}t \tag{5.88}$$

(we have used (5.87) to get the last expression of $\mathscr{C}$). The map $\mathscr{C}\colon \widetilde{\mathscr{G}} \to \widetilde{\mathscr{G}}$ is a contraction of our complete metric space, because the difference

$$\mathscr{C}(\widetilde{f}^*) - \mathscr{C}(\widetilde{f}) = -\widetilde{\chi} \int_0^1 \partial(\widetilde{f}^* - \widetilde{f}) \circ (\mathrm{id} + t\widetilde{\chi})\, \mathrm{d}t \tag{5.89}$$

has valuation $\geq \operatorname{val}(\widetilde{f}^* - \widetilde{f}) + 1$ (because of (5.15): $\operatorname{val}\big(\partial(\widetilde{f}^* - \widetilde{f}) \circ (\mathrm{id} + t\widetilde{\chi})\big) = \operatorname{val}\big(\partial(\widetilde{f}^* - \widetilde{f})\big) \geq \operatorname{val}(\widetilde{f}^* - \widetilde{f}) + 1$ for each t), hence $d\big(\mathscr{C}(\widetilde{f}), \mathscr{C}(\widetilde{f}^*)\big) \leq \frac{1}{2} d(\widetilde{f}, \widetilde{f}^*)$. The Banach fixed-point theorem implies that there is a unique solution $\widetilde{f} = \mathscr{L}(\widetilde{h})$, obtained as the limit of the Cauchy sequence $\mathscr{L}_n(\widetilde{h}) := \underbrace{\mathscr{C} \circ \cdots \circ \mathscr{C}}_{n \text{ times}}(0)$ as $n \to \infty$.

We observe that, if $\widetilde{h} \in \widetilde{\mathscr{G}}_0$, then $\mathscr{C}(\widetilde{\mathscr{G}}) \subset \widetilde{\mathscr{G}}_0$, thus $\mathscr{L}_n(\widetilde{h}) \in \widetilde{\mathscr{G}}_0$ for each $n \geq 0$ and clearly $\mathscr{L}(\widetilde{h}) \in \widetilde{\mathscr{G}}_0$ in that case.

The fact that each element has a unique left inverse implies that each element is invertible: given $\widetilde{h} \in \widetilde{\mathscr{G}}$, its left inverse $\widetilde{f} := \mathscr{L}(\widetilde{h})$ is also a right inverse because $\widetilde{h}^* := \mathscr{L}(\widetilde{f})$ satisfies $\widetilde{h}^* = \widetilde{h}^* \circ (\widetilde{f} \circ \widetilde{h}) = (\widetilde{h}^* \circ \widetilde{f}) \circ \widetilde{h} = \widetilde{h}$, i.e. $\widetilde{h} \circ \widetilde{f} = \mathrm{id}$.

Finally, we check that $\mathscr{L}$ is continuous. For $\widetilde{h}, \widetilde{h}^* \in \widetilde{\mathscr{G}}$, we denote by $\mathscr{C}, \mathscr{C}^*$ the corresponding maps defined by (5.88). For any $\widetilde{f}, \widetilde{f}^*$,

$$\operatorname{val}\big(\mathscr{C}^*(\widetilde{f}) - \mathscr{C}(\widetilde{f})\big) = \operatorname{val}(\widetilde{f} \circ \widetilde{h} - \widetilde{f} \circ \widetilde{h}^*) \geq \operatorname{val}(\widetilde{h}^* - \widetilde{h})$$

(as already deduced from (5.87)), while

$$\operatorname{val}\big(\mathscr{C}^*(\widetilde{f}^*) - \mathscr{C}^*(\widetilde{f})\big) \geq \operatorname{val}(\widetilde{f}^* - \widetilde{f}) + 1$$

(as already deduced from (5.89)), hence

$$d\big(\mathscr{C}(\widetilde{f}), \mathscr{C}^*(\widetilde{f}^*)\big) \leq \max\big\{d(\widetilde{h}, \widetilde{h}^*), \frac{1}{2} d(\widetilde{f}, \widetilde{f}^*)\big\}.$$

It follows by induction that $d\big(\mathscr{L}_n(\widetilde{h}), \mathscr{L}_n(\widetilde{h}^*)\big) = d\big(\mathscr{C}(\mathscr{L}_{n-1}(\widetilde{h})), \mathscr{C}^*(\mathscr{L}_{n-1}(\widetilde{h}^*))\big) \leq d(\widetilde{h}, \widetilde{h}^*)$ for every $n \geq 1$, hence $d\big(\mathscr{L}(\widetilde{h}), \mathscr{L}(\widetilde{h}^*)\big) \leq d(\widetilde{h}, \widetilde{h}^*)$. □

Notice that $\widetilde{\mathscr{G}}_0 = \{\widetilde{f} \in \widetilde{\mathscr{G}} \mid d(\mathrm{id}, \widetilde{f}) \leq \frac{1}{2}\} = \{\widetilde{f} \in \widetilde{\mathscr{G}} \mid d(\mathrm{id}, \widetilde{f}) < 1\}$ is a closed ball as well as an open ball, thus it is both closed and open for the Krull topology of $\widetilde{\mathscr{G}}$.

5.16 Inversion in the group $\widetilde{\mathscr{G}}$

There is an explicit formula for the inverse of an element of $\widetilde{\mathscr{G}}$, which is a particular case of the Lagrange reversion formula (adapted to our framework):

Theorem 5.61. *For any $\widetilde{\chi} \in \mathbb{C}[[z^{-1}]]$, the inverse of $\widetilde{h} = \mathrm{id} + \widetilde{\chi}$ can be written as the formally convergent series of formal series*

$$(\mathrm{id} + \widetilde{\chi})^{\circ(-1)} = \mathrm{id} + \sum_{k \geq 1} \frac{(-1)^k}{k!} \partial^{k-1}(\widetilde{\chi}^k). \tag{5.90}$$

The proof of Theorem 5.61 will make use of

Lemma 5.62. *Let $\widetilde{\chi} \in \mathbb{C}[[z^{-1}]]$ and $n \geq 1$. Then, for any $\widetilde{\psi} \in \mathbb{C}[[z^{-1}]]$,*

$$\sum_{k=0}^{n} (-1)^k \binom{n}{k} \widetilde{\chi}^{n-k} \partial^{n-1}(\widetilde{\chi}^k \widetilde{\psi}) = 0. \tag{5.91}$$

Proof of Lemma 5.62. Let us call $H_n \widetilde{\psi}$ the left-hand side of (5.91). We have $H_1 \widetilde{\psi} = \widetilde{\chi} \partial^0 \widetilde{\psi} - \partial^0(\widetilde{\chi} \widetilde{\psi}) = 0$. It is thus sufficient to prove the recursive formula

$$H_{n+1}\widetilde{\psi} = -\partial H_n(\widetilde{\chi}\widetilde{\psi}) + \widetilde{\chi}\,\partial H_n\widetilde{\psi} - n(\partial\widetilde{\chi})H_n\widetilde{\psi}.$$

To this end, we use the convention $\binom{n}{-1} = \binom{n}{n+1} = 0$ and compute

$$\begin{aligned} -\partial H_n(\widetilde{\chi}\widetilde{\psi}) &= \sum_{k=-1}^{n}(-1)^{k+1}\binom{n}{k}\partial\left[\widetilde{\chi}^{n-k}\,\partial^{n-1}(\widetilde{\chi}^{k+1}\widetilde{\psi})\right] \\ &= \sum_{k=0}^{n+1}(-1)^k\binom{n}{k-1}\partial\left[\widetilde{\chi}^{n+1-k}\,\partial^{n-1}(\widetilde{\chi}^{k}\widetilde{\psi})\right] \end{aligned}$$

(shifting the summation index to get the last expression), while

$$\widetilde{\chi}\,\partial H_n\widetilde{\psi} = \sum_{k=0}^{n+1}(-1)^k\binom{n}{k}\widetilde{\chi}\,\partial\left[\widetilde{\chi}^{n-k}\,\partial^{n-1}(\widetilde{\chi}^{k}\widetilde{\psi})\right].$$

The Leibniz rule yields

$$\begin{aligned} -\partial H_n(\widetilde{\chi}\widetilde{\psi}) + \widetilde{\chi}\,\partial H_n\widetilde{\psi} &= \sum_{k=0}^{n+1}(-1)^k\left[\binom{n}{k-1} + \binom{n}{k}\right]\widetilde{\chi}^{n+1-k}\,\partial^{n}(\widetilde{\chi}^{k}\widetilde{\psi}) \\ &+ \sum_{k=0}^{n+1}(-1)^k\left[(n+1-k)\binom{n}{k-1} + (n-k)\binom{n}{k}\right]\widetilde{\chi}^{n-k}(\partial\widetilde{\chi})\partial^{n-1}(\widetilde{\chi}^{k}\widetilde{\psi}). \end{aligned}$$

The expression in the former bracket is $\binom{n+1}{k}$, hence the first sum is nothing but $H_{n+1}\widetilde{\psi}$; the expression in the latter bracket is n times $\binom{n-1}{k} + \binom{n-1}{k-1} = \binom{n}{k}$, hence the second sum is $n(\partial\widetilde{\chi})H_n\widetilde{\psi}$. □

Proof of Theorem 5.61. Let $\widetilde{h} = \mathrm{id} + \widetilde{\chi} \in \widetilde{\mathscr{G}}$. Lemma 5.62 shows that the right-hand side of (5.90) defines a left inverse for $\widetilde{h}$. Indeed, denoting by $\widetilde{f} = \mathrm{id} + \widetilde{\varphi}$ this right-hand side, we have

$$\widetilde{f}\circ\widetilde{h} - \mathrm{id} = \widetilde{\chi} + \widetilde{\varphi}\circ(\mathrm{id}+\widetilde{\chi}) = \widetilde{\chi} + \sum_{\ell\geq 0,k\geq 1}\frac{(-1)^k}{k!\ell!}\widetilde{\chi}^{\ell}\,\partial^{k+\ell-1}(\widetilde{\chi}^k) = \sum_{n\geq 1}\frac{1}{n!}\widetilde{H}_n$$

with $\widetilde{H}_n = \sum(-1)^k\binom{n}{k}\widetilde{\chi}^{\ell}\,\partial^{n-1}(\widetilde{\chi}^k)$, the last sum running over all pairs of non-negative integers (k,ℓ) such that $k+\ell = n$ (absorbing the first $\widetilde{\chi}$ in $\widetilde{H}_1$ and taking care of $k=0$ according as $n=1$ or $n\geq 2$; formal summability legitimates our Fubini-like manipulation), then Lemma 5.62 with $\widetilde{\psi} = 1$ says that $\widetilde{H}_n = 0$ for every $n \geq 1$. □

Exercise 5.63 (Lagrange reversion formula). Prove that, with the same convention as in (5.86),

$$\widetilde{f}\circ\widetilde{h}^{\circ(-1)} = \widetilde{f} + \sum_{k\geq 1}\frac{(-1)^k}{k!}\partial^{k-1}(\widetilde{\chi}^k\partial\widetilde{f}), \qquad \widetilde{h} = \mathrm{id}+\widetilde{\chi}.$$

(Hint: Use Lemma 5.62 with $\widetilde{\psi} = \partial(\widetilde{f} - \mathrm{id}) = -1 + \partial\widetilde{f}$.)

Exercise 5.64. Let $h = \mathrm{id} + \chi \in \mathscr{G}$, i.e. with $\chi \in \mathbb{C}\{z^{-1}\}$. We can thus choose $c_0, M > 0$ such that $|\chi(z)| \le M$ for $|z| \ge c_0$. Show that $h^{\circ(-1)}(z)$ is convergent for $|z| \ge c_0 + M$. (Hint: Given $\delta > M$, use the Cauchy inequalities to bound $|\partial^{k-1}(\chi^k)(z)|$ for $|z| > c_0 + \delta$.)

5.17 The group of 1-summable formal diffeomorphisms

Among all formal tangent-to-identity diffeomorphisms, we now distinguish those which are 1-summable in an arc of directions.

Definition 5.65. Let I be an open interval of $\mathbb{R}$. Let $\gamma, \alpha \colon I \to \mathbb{R}$ be locally bounded functions with $\alpha \ge 0$. For any $\sigma \in \mathbb{C}$ we define

$$\widetilde{\mathscr{G}}(I,\gamma,\alpha) := \{\, \widetilde{f} = \mathrm{id} + \widetilde{\varphi}_0 \mid \widetilde{\varphi}_0 \in \mathscr{B}^{-1}\big(\mathbb{C}\,\delta \oplus \mathscr{N}(I,\gamma,\alpha)\big) \,\}$$

$$\widetilde{\mathscr{G}}(I,\gamma) := \{\, \widetilde{f} = \mathrm{id} + \widetilde{\varphi}_0 \mid \widetilde{\varphi}_0 \in \mathscr{B}^{-1}\big(\mathbb{C}\,\delta \oplus \mathscr{N}(I,\gamma)\big) \,\},$$

$$\widetilde{\mathscr{G}}(I) := \{\, \widetilde{f} = \mathrm{id} + \widetilde{\varphi}_0 \mid \widetilde{\varphi}_0 \in \mathscr{B}^{-1}\big(\mathbb{C}\,\delta \oplus \mathscr{N}(I)\big) \,\}$$

and $\widetilde{\mathscr{G}}_\sigma(I,\gamma,\alpha) := \widetilde{\mathscr{G}}(I,\gamma,\alpha) \cap \widetilde{\mathscr{G}}_\sigma$, $\widetilde{\mathscr{G}}_\sigma(I,\gamma) := \widetilde{\mathscr{G}}(I,\gamma) \cap \widetilde{\mathscr{G}}_\sigma$, $\widetilde{\mathscr{G}}_\sigma(I) := \widetilde{\mathscr{G}}(I) \cap \widetilde{\mathscr{G}}_\sigma$. We extend the definition of the Borel summation operator $\mathscr{S}^I$ to $\widetilde{\mathscr{G}}(I)$ by setting

$$\widetilde{f} = \mathrm{id} + \widetilde{\varphi}_0 \in \widetilde{\mathscr{G}}(I,\gamma) \quad \Longrightarrow \quad \mathscr{S}^I \widetilde{f}(z) = z + \mathscr{S}^I \widetilde{\varphi}^0(z), \qquad z \in \mathscr{D}(I,\gamma).$$

For $|I| \ge 2\pi$, $\widetilde{\mathscr{G}}(I)$ coincides with the group $\mathscr{G}$ of holomorphic tangent-to-identity diffeomorphisms and $\mathscr{S}^I$ is the ordinary summation operator for Taylor series at ∞, but

$$|I| < 2\pi \quad \Longrightarrow \quad \mathscr{G} \subsetneq \widetilde{\mathscr{G}}(I) \subsetneq \widetilde{\mathscr{G}}.$$

For $\widetilde{f} \in \widetilde{\mathscr{G}}(I)$, the function $\mathscr{S}^I \widetilde{f}$ is holomorphic in a sectorial neighbourhood of ∞ (but not in a full neighbourhood of ∞ if $\widetilde{f} \notin \mathscr{G}$); we shall see that it defines an injective transformation in a domain of the form $\mathscr{D}(I,\gamma)$. We first study composition and inversion in $\widetilde{\mathscr{G}}(I)$.

Theorem 5.66. *Let I be an open interval of $\mathbb{R}$ and $\gamma, \alpha \colon I \to \mathbb{R}$ be locally bounded functions with $\alpha \ge 0$. Let $\sigma, \tau \in \mathbb{C}$ and $\widetilde{f} \in \widetilde{\mathscr{G}}_\sigma(I,\gamma,\alpha)$, $\widetilde{g} \in \widetilde{\mathscr{G}}_\tau(I,\gamma)$. Then*

$$\widetilde{g} \circ \widetilde{f} \in \widetilde{\mathscr{G}}_{\sigma+\tau}(I,\gamma_1) \quad \textit{with } \gamma_1 := \gamma + |\sigma| + \sqrt{\alpha},$$

the function $\mathscr{S}^I \widetilde{f}$ maps $\mathscr{D}(I,\gamma_1)$ in $\mathscr{D}(I,\gamma)$ and

$$\mathscr{S}^I(\widetilde{g} \circ \widetilde{f}) = (\mathscr{S}^I \widetilde{g}) \circ (\mathscr{S}^I \widetilde{f}) \quad \textit{on } \mathscr{D}(I,\gamma_1).$$

Proof. Apply Theorem 5.55 to $\widetilde{\varphi}_0 := \widetilde{f} - \mathrm{id}$ and $\widetilde{\psi}_0 := \widetilde{g} - \mathrm{id}$. □

Theorem 5.67. *Let $\widetilde{f} \in \widetilde{\mathscr{G}}_\sigma(I,\gamma,\alpha)$. Then*

$$\widetilde{h} := \widetilde{f}^{\circ(-1)} \in \widetilde{\mathscr{G}}_{-\sigma}(I,\gamma^*,\alpha) \quad \text{with } \gamma^* := \gamma + |\sigma| + 2\sqrt{\alpha}$$

and

$$\mathscr{S}^I\widetilde{f}\big(\mathscr{D}(I,\gamma_1)\big) \subset \mathscr{D}(I,\gamma^*), \qquad (\mathscr{S}^I\widetilde{h}) \circ \mathscr{S}^I\widetilde{f} = \mathrm{id} \ \text{ on } \mathscr{D}(I,\gamma_1), \tag{5.92}$$

$$\mathscr{S}^I\widetilde{h}\big(\mathscr{D}(I,\gamma_2)\big) \subset \mathscr{D}(I,\gamma), \qquad (\mathscr{S}^I\widetilde{f}) \circ \mathscr{S}^I\widetilde{h} = \mathrm{id} \ \text{ on } \mathscr{D}(I,\gamma_2), \tag{5.93}$$

with $\gamma_1 := \gamma + 2|\sigma| + (1+\sqrt{2})\sqrt{\alpha}$ and $\gamma_2 := \gamma + |\sigma| + (1+\sqrt{2})\sqrt{\alpha}$.

Moreover, $\mathscr{S}^I\widetilde{f}$ is injective on $\mathscr{D}\big(I,\gamma+(1+\sqrt{2})\sqrt{\alpha}\big)$.

Proof. We first assume $\widetilde{f} \in \widetilde{\mathscr{G}}_0(I,\gamma,\alpha)$. By (5.90), we have $\widetilde{h} = \mathrm{id} + \widetilde{\chi}$ with $\widetilde{\chi}$ given by a formally convergent series in $z^{-1}\mathbb{C}[[z^{-1}]]$:

$$\widetilde{\chi} = \sum_{k\ge1} \widetilde{\chi}_k, \qquad \widetilde{\chi}_k = \frac{(-1)^k}{k!}\partial^{k-1}(\widetilde{\varphi}^k).$$

Correspondingly, $\mathscr{B}\widetilde{\chi}$ is given by a formally convergent series in $\mathbb{C}[[\zeta]]$:

$$\widehat{\chi} = \sum_{k\ge1} \widehat{\chi}_k, \qquad \widehat{\chi}_k = -\frac{\zeta^{k-1}}{k!}\widehat{\varphi}^{*k}$$

(beware that the last expression involves multiplication by $-\frac{\zeta^{k-1}}{k!}$, not convolution!). We argue as in the proof of Theorem 5.55 and view $\widehat{\chi}$ as a series of holomorphic functions in the union of a disc $D(0,R)$ and a sector Σ in which $\widehat{\varphi}$ itself is holomorphic; inequalities (5.79) and (5.80) yield

$$|\widehat{\chi}_k(\zeta)| \le A^k \frac{\xi^{2(k-1)}}{k!(k-1)!}, \qquad \zeta \in D(0,R), \tag{5.94}$$

$$|\widehat{\chi}_k(\zeta)| \le \alpha(\theta)^k \frac{\xi^{2(k-1)}}{k!(k-1)!}\mathrm{e}^{\gamma(\theta)\xi}, \qquad \zeta \in \Sigma, \tag{5.95}$$

where $\xi = |\zeta|$ and $\theta = \arg\zeta$. The series of holomorphic functions $\sum\widehat{\chi}_k$ is thus uniformly convergent in every compact subset of $D(0,R)\cup\Sigma$ and its sum is a holomorphic function whose Taylor series at 0 is $\widehat{\chi}$. Therefore $\widehat{\chi} \in \mathbb{C}\{\zeta\}$ extends analytically to $D(0,R)\cup\Sigma$; moreover, since $\frac{1}{k!(k-1)!} \le \frac{1}{k}\frac{2^{2(k-1)}}{(2(k-1))!}$, (5.95) yields

$$|\widehat{\chi}(\zeta)| \le \sum_{k\ge1} \frac{\alpha(\theta)^k}{k}\frac{(2\xi)^{2(k-1)}}{(2(k-1))!}\mathrm{e}^{\gamma(\theta)\xi} \le \alpha(\theta)\,\mathrm{e}^{(\gamma(\theta)+2\sqrt{\alpha(\theta)})\xi}$$

for $\zeta \in \Sigma$. Hence $\widetilde{h} \in \widetilde{\mathscr{G}}_0(I,\gamma+2\sqrt{\alpha},\alpha)$ when $\sigma = 0$.

In the general case, we observe that $\widetilde{f} = (\mathrm{id}+\sigma)\circ\widetilde{g}$ with $\widetilde{g} := (\mathrm{id}-\sigma)\circ\widetilde{f} \in \widetilde{\mathscr{G}}_0(I,\gamma,\alpha)$, thus $\widetilde{g}^{\circ(-1)} = \mathrm{id}+\widetilde{\chi} \in \widetilde{\mathscr{G}}_0(I,\gamma+2\sqrt{\alpha},\alpha)$ and $\widetilde{h} = \widetilde{f}^{\circ(-1)} = \widetilde{g}^{\circ(-1)} \circ (\mathrm{id}-\sigma) = \mathrm{id}-\sigma+T_{-\sigma}\widetilde{\chi}$, which implies $\widetilde{h} \in \widetilde{\mathscr{G}}(I,\gamma+2\sqrt{\alpha}+|\sigma|,\alpha)$ by the second property in Lemma 5.10.

Since $\widetilde{h}\circ\widetilde{f} = \widetilde{f}\circ\widetilde{h} = \mathrm{id}$, we can apply Theorem 5.66 and get $(\mathscr{S}^I\widetilde{h})\circ\mathscr{S}^I\widetilde{f} = \mathrm{id}$ and $(\mathscr{S}^I\widetilde{f})\circ\mathscr{S}^I\widetilde{h} = \mathrm{id}$ in appropriate domains; in fact, by analytic continuation, these identities will hold in any domain $\mathscr{D}(I,\gamma+\delta_1)$, resp. $\mathscr{D}(I,\gamma+\delta_2)$, such that

$$\mathscr{S}^I\widetilde{f}\big(\mathscr{D}(I,\gamma+\delta_1)\big) \subset \mathscr{D}(I,\gamma^*), \qquad \mathscr{S}^I\widetilde{h}\big(\mathscr{D}(I,\gamma+\delta_2)\big) \subset \mathscr{D}(I,\gamma).$$

Writing $\widetilde{f} = \mathrm{id}+\sigma+\widetilde{\varphi}$ with $\mathscr{B}\widetilde{\varphi} \in \mathscr{N}(I,\gamma,\alpha)$, with the help of (5.74) one can easily show that $\delta_1 = \gamma_1-\gamma$ and $\delta_2 = \gamma_2-\gamma$ satisfy this.

For the injectivity statement, we write again $\widetilde{f} = (\mathrm{id}+\sigma)\circ\widetilde{g}$ and apply the previous result to $\widetilde{g} \in \widetilde{\mathscr{G}}_0(I,\gamma,\alpha)$. The function $\mathscr{S}^I\widetilde{g}$ maps

$$\mathscr{D} := \mathscr{D}\big(I,\gamma+(1+\sqrt{2})\sqrt{\alpha}\big)$$

in the domain $\mathscr{D}\big(I,\gamma+2\sqrt{\alpha}\big)$, on which $\mathscr{S}^I(\widetilde{g}^{\circ(-1)})$ is well-defined, and we have $\mathscr{S}^I(\widetilde{g}^{\circ(-1)})\circ\mathscr{S}^I\widetilde{g} = \mathrm{id}$ on $\mathscr{D}$, therefore $\mathscr{S}^I\widetilde{g}$ is injective on $\mathscr{D}$, and so is the function $\mathscr{S}^I\widetilde{f} = \sigma+\mathscr{S}^I\widetilde{g}$. □

Corollary 5.68. *For any open interval I, $\widetilde{\mathscr{G}}(I)$ and $\widetilde{\mathscr{G}}_0(I)$ are subgroups of $\widetilde{\mathscr{G}}$.*

Exercise 5.69. Consider the set $\mathrm{id}+\mathbb{C}[[z^{-1}]]_1$ of 1-Gevrey tangent-to-identity formal diffeomorphisms, so that

$$\widetilde{\mathscr{G}}(I) \subsetneq \mathrm{id}+\mathbb{C}[[z^{-1}]]_1 \subsetneq \widetilde{\mathscr{G}}.$$

Show that $\mathrm{id}+\mathbb{C}[[z^{-1}]]_1$ is a subgroup of $\widetilde{\mathscr{G}}$. (Hint: Recall that $\mathbb{C}[[z^{-1}]]_1$ coincides $\mathscr{B}^{-1}\big(\mathbb{C}\,\delta\oplus\mathbb{C}\{\zeta\}\big)$ and imitate the previous chain of reasoning.)

We shall see in Section 7.3 how 1-summable formal diffeomorphisms occur in the study of a holomorphic germ $f \in \mathscr{G}_1$.

References

Bar79. E. J. Barbeau. Euler subdues a very obstreperous series. *Amer. Math. Monthly*, 86(5):356–372, 1979.

CNP93. B. Candelpergher, J.-C. Nosmas, and F. Pham. *Approche de la résurgence*. Actualités Mathématiques. [Current Mathematical Topics]. Hermann, Paris, 1993.

DA09. V. De Angelis. Stirling's series revisited. *Amer. Math. Monthly*, 116(9):839–843, 2009.

Del16. E. Delabaere. *Divergent Series, summability and resurgence. Volume 3: Resurgent Methods and the First Painlevé Equation.*, volume 2155 of *Lecture Notes in Mathematics*. Springer, Heidelberg, 2016.

Lod16. M. Loday-Richaud. *Divergent Series, summability and resurgence. Volume 2: Simple and multiple summability.*, volume 2154 of *Lecture Notes in Mathematics*. Springer, Heidelberg, 2016.

Mal95. B. Malgrange. Sommation des séries divergentes. *Exposition. Math.*, 13(2-3):163–222, 1995.

Poi87. H. Poincaré. *Les méthodes nouvelles de la mécanique céleste. Tome II.* Les Grands Classiques Gauthier-Villars. [Gauthier-Villars Great Classics]. Librairie Scientifique et Technique Albert Blanchard, Paris, 1987. Méthodes de MM. Newcomb, Gyldén, Lindstedt et Bohlin. [The methods of Newcomb, Gyldén, Lindstedt and Bohlin], Reprint of the 1893 original, Bibliothèque Scientifique Albert Blanchard. [Albert Blanchard Scientific Library].

Ram93. J.-P. Ramis. Séries divergentes et théories asymptotiques. *Bull. Soc. Math. France*, 121(Panoramas et Syntheses, suppl.):74, 1993.

Ram12a. J.-P. Ramis. Les développements asymptotiques après Poincaré: continuité et... divergences. *Gaz. Math.*, 134:17–36, 2012.

Ram12b. J.-P. Ramis. Poincaré et les développements asymptotiques (premìere partie). *Gaz. Math.*, 133:33–72, 2012.

Zag97. D. Zagier. Newman's short proof of the prime number theorem. *Amer. Math. Monthly*, 104(8):705–708, 1997.

Chapter 6
Resurgent Functions and Alien Calculus

6.1 Resurgent functions, resurgent formal series

Among 1-Gevrey formal series, we have distinguished the subspace of those which are 1-summable in a given arc of directions and studied it in Sections 5.9–5.17. We shall now study another subspace of $\mathbb{C}[[z^{-1}]]_1$, which consists of "resurgent formal series". As in the case of 1-summability, we make use of the algebra isomorphism (5.22)

$$\mathscr{B}\colon\ \mathbb{C}[[z^{-1}]]_1 \xrightarrow{\sim} \mathbb{C}\delta \oplus \mathbb{C}\{\zeta\}$$

and give the definition not directly in terms of the formal series themselves, but rather in terms of their formal Borel transforms, for which, beyond convergence near the origin, we shall require a certain property of analytic continuation.

For any $R > 0$ and $\zeta_0 \in \mathbb{C}$ we use the notations

$$D(\zeta_0, R) := \{\, \zeta \in \mathbb{C} \mid |\zeta - \zeta_0| < R \,\}, \tag{6.1}$$

$$\mathbb{D}_R := D(0,R), \qquad \mathbb{D}_R^* := \mathbb{D}_R \setminus \{0\}. \tag{6.2}$$

Definition 6.1. Let Ω be a non-empty closed discrete subset of $\mathbb{C}$, let $\widehat{\varphi}(\zeta) \in \mathbb{C}\{\zeta\}$ be a holomorphic germ at the origin. We say that $\widehat{\varphi}$ is an *Ω-continuable germ* if there exists $R > 0$ not larger than the radius of convergence of $\widehat{\varphi}$ such that $\mathbb{D}_R^* \cap \Omega = \emptyset$ and $\widehat{\varphi}$ admits analytic continuation along any path of $\mathbb{C} \setminus \Omega$ originating from any point of $\mathbb{D}_R^*$. See Figure 6.1. We use the notation

$$\widehat{\mathscr{R}}_\Omega := \{\, \text{all } \Omega\text{-continuable germs} \,\} \subset \mathbb{C}\{\zeta\}. \tag{6.3}$$

We call *Ω-resurgent function* any element of $\mathbb{C}\,\delta \oplus \widehat{\mathscr{R}}_\Omega$, i.e. any element of $\mathbb{C}\,\delta \oplus \mathbb{C}\{\zeta\}$ of the form $c\,\delta + \widehat{\varphi}$ with $c =$ a complex number and $\widehat{\varphi} =$ an Ω-continuable germ.

We call *Ω-resurgent formal series* any $\widetilde{\varphi}_0(z) \in \mathbb{C}[[z^{-1}]]_1$ whose formal Borel transform is an Ω-resurgent function, i.e. any $\widetilde{\varphi}_0$ belonging to

C. Mitschi, D. Sauzin, *Divergent Series, Summability and Resurgence I*,
Lecture Notes in Mathematics 2153, DOI 10.1007/978-3-319-28736-2_6

$$\widetilde{\mathscr{R}}_\Omega := \mathscr{B}^{-1}\big(\mathbb{C}\,\delta \oplus \widehat{\mathscr{R}}_\Omega\big) \subset \mathbb{C}[[z^{-1}]]_1. \tag{6.4}$$

Remark 6.2. In the above definition, by "path" we mean a continuous function $\gamma\colon J \to \mathbb{C}\setminus\Omega$, where J is any compact interval $[a,b]$ of $\mathbb{R}$. The reader is referred to Section 1.1.2 of the first part of this volume (p. 5) for the notion of analytic continuation; there, the interval of definition of paths was always taken to be $[0,1]$, but this difference is innocuous[1] and, from now on, it will be convenient not to take the same J is in all cases. Moreover, from now on, all our paths will be assumed piecewise continuously differentiable; this is no loss of generality since the property of being a path of analytic continuation of a given holomorphic germ is open in the uniform norm topology.

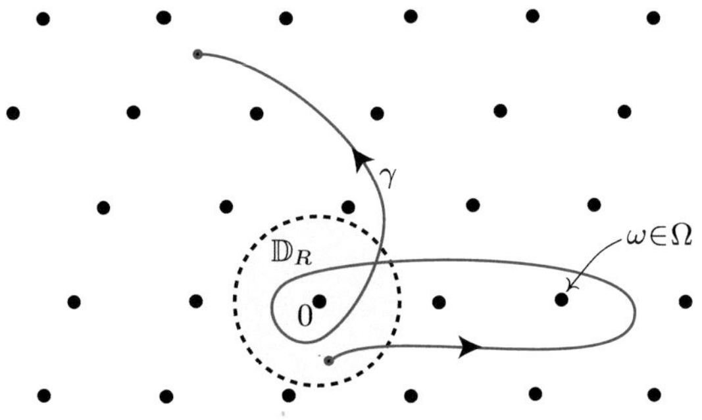

Fig. 6.1: *Ω-continuability.* Any path γ starting in $\mathbb{D}_R^*$ and contained in $\mathbb{C}\setminus\Omega$ must be a path of analytic continuation for $\widehat{\varphi} \in \widehat{\mathscr{R}}_\Omega$.

Our definitions of Ω-continuable functions and Ω-resurgent series are particular cases of Écalle's definition of "continuability without a cut" (or "endless continuability") for germs, and "resurgence" for formal series. In this text, we give ourselves a set Ω so as to prescribe in advance the possible location of the singularities of the analytic continuation of $\widehat{\varphi}$, whereas the theory is developed in [Éca85] without this restriction. Typical examples of set Ω with which we shall work are $\Omega = \mathbb{Z}$ or $2\pi\mathrm{i}\,\mathbb{Z}$.

Remark 6.3. Let $\rho(\Omega) := \min\big\{|\omega|,\ \omega \in \Omega\setminus\{0\}\big\}$. Any $\widehat{\varphi} \in \widehat{\mathscr{R}}_\Omega$ is a holomorphic germ at 0 with radius of convergence $\geq \rho(\Omega)$ and one can always take $R = \rho(\Omega)$ in Definition 6.1. In fact, given an arbitrary $\zeta_0 \in \mathbb{D}_{\rho(\Omega)}$, we have

[1] As is often the case with analytic continuation and Cauchy integrals, the precise parametrisation of γ will usually not matter, in the sense that we shall get the same result from two paths $\gamma\colon [a,b] \to \mathbb{C}\setminus\Omega$ and $\gamma'\colon [a',b'] \to \mathbb{C}\setminus\Omega$ which only differ by a change of parametrisation ($\gamma = \gamma'\circ\sigma$ with σ increasing homeomophism from $[a,b]$ to $[a',b']$).

$$\widehat{\varphi} \in \widehat{\mathscr{R}}_\Omega \iff \left| \begin{array}{l} \widehat{\varphi} \text{ germ of holomorphic function of } \mathbb{D}_{\rho(\Omega)} \text{ admitting} \\ \text{analytic continuation along any path } \gamma\colon [0,1] \to \mathbb{C} \\ \text{such that } \gamma(0) = \zeta_0 \text{ and } \gamma\big((0,1]\big) \subset \mathbb{C} \setminus \Omega \end{array} \right.$$

(even if $\zeta_0 = 0$ and $0 \in \Omega$: there is no need to avoid 0 at the beginning of the path, when we still are in the disc of convergence of $\widehat{\varphi}$).

Example 6.4. Trivially, any entire function of $\mathbb{C}$ defines an Ω-continuable germ; as a consequence, by Lemma 5.7,

$$\mathbb{C}\{z^{-1}\} \subset \widetilde{\mathscr{R}}_\Omega.$$

Other elementary examples of Ω-continuable germs are the functions which are holomorphic in $\mathbb{C} \setminus \Omega$ and regular at 0, like $\frac{1}{(\zeta-\omega)^m}$ with $m \in \mathbb{N}^*$ and $\omega \in \Omega \setminus \{0\}$.

Lemma 6.5. – *The Euler series* $\widetilde{\varphi}^{\mathrm{E}}(z)$ *defined by* (5.38) *belongs to* $\widetilde{\mathscr{R}}_{\{-1\}}$.
– *Given* $w = \mathrm{e}^s$ *with* $\Re\mathrm{e}\, s < 0$, *the series* $\widetilde{\varphi}^{\mathrm{P}}(z)$ *of Poincaré's example* (5.63) *belongs to* $\widetilde{\mathscr{R}}_\Omega$ *with* $\Omega := s + 2\pi\mathrm{i}\mathbb{Z}$.
– *The Stirling series* $\widetilde{\mu}(z)$ *of Theorem 5.41 (explicitly given by* (5.57)*) belongs to* $\widetilde{\mathscr{R}}_{2\pi\mathrm{i}\mathbb{Z}}$.

Proof. The Borel transforms of all these series have a meromorphic continuation:

- Euler: $\widehat{\varphi}^{\mathrm{E}}(\zeta) = (1+\zeta)^{-1}$ by (5.39).
- Poincaré: $\widehat{\varphi}^{\mathrm{P}}(\zeta) = \frac{1}{1-\mathrm{e}^{s-\zeta}}$ by (5.64).
- Stirling: $\widehat{\mu}(\zeta) = \zeta^{-2}\left(\frac{\zeta}{2}\coth\frac{\zeta}{2} - 1\right)$ by (5.56).

□

Exercise 6.6. Any $\{0\}$-continuable germ defines an entire function of $\mathbb{C}$. (Hint: view $\mathbb{C}$ as the union of a disc and two cut planes.)

Exercise 6.7. Give an example of a holomorphic germ at 0 which is not Ω-continuable for any non-empty closed discrete subset Ω of $\mathbb{C}$.

But in all the previous examples the Borel transform was single-valued, whereas the interest of Definition 6.1 is to authorize multiple-valuedness when following the analytic continuation. For instance, the exponential of the Stirling series $\widetilde{\lambda} = \mathrm{e}^{\widetilde{\mu}}$, which gives rise to the refined Stirling formula (5.58), has a Borel transform with a multiple-valued analytic continuation and belongs to $\widetilde{\mathscr{R}}_{2\pi\mathrm{i}\mathbb{Z}}$, although this is more difficult to check (see Sections 6.5 and 6.93). We now give elementary examples which illustrate multiple-valued analytic continuation.

Notation 6.8 *From now on, if* $\gamma\colon [a,b] \to \mathbb{C}$ *is a path and* $\widehat{\varphi}$ *is a holomorphic germ at* $\gamma(a)$ *which admits an analytic continuation along* γ, *we denote by*

$$\operatorname{cont}_\gamma \widehat{\varphi} \tag{6.5}$$

the resulting holomorphic germ at the endpoint $\gamma(b)$ (instead of $\widehat{\varphi}^{\gamma}$ as in Section 1.1.2 of the first part of this volume).

Example 6.9. Consider $\widehat{\varphi}(\zeta) = \sum\limits_{n\geq 1} \frac{\zeta^n}{n}$: this is a holomorphic germ belonging to $\widehat{\mathscr{R}}_{\{1\}}$ but its analytic continuation is not single-valued. Indeed, the disc of convergence of $\widehat{\varphi}$ is $\mathbb{D}_1$ and, for any $\zeta \in \mathbb{D}_1$, $\widehat{\varphi}(\zeta) = \int_0^\zeta \frac{d\xi}{1-\xi} = -\mathrm{Log}\,(1-\zeta)$ with the notation (6.25) for the principal branch of the logarithm, hence the analytic continuation of $\widehat{\varphi}$ along a path γ originating from 0, avoiding 1 and ending at a point ζ_1 is the holomorphic germ at ζ_1 explicitly given by

$$\mathrm{cont}_\gamma \widehat{\varphi}(\zeta) = \int_\gamma \frac{d\xi}{1-\xi} + \int_{\zeta_1}^{\zeta} \frac{d\xi}{1-\xi} \qquad (\zeta \text{ close enough to } \zeta_1),$$

which yields a multiple-valued function in $\mathbb{C} \setminus \{1\}$ (two paths from 0 to ζ_1 do not give rise to the same analytic continuation near ζ_1 unless they are homotopic in $\mathbb{C} \setminus \{1\}$). The germ $\widehat{\varphi}$ is Ω-continuable if and only if $1 \in \Omega$.

Example 6.10. A related example of $\{0,1\}$-continuable germ with mutivalued analytic continuation is given by $\sum\limits_{n\geq 0} \frac{\zeta^n}{n+1} = -\frac{1}{\zeta}\mathrm{Log}\,(1-\zeta)$, for which there is a branch holomorphic in the cut plane $\mathbb{C} \setminus [1,+\infty)$ and all the other branches[2] have a simple pole at 0. This germ is Ω-continuable if and only if $\{0,1\} \subset \Omega$.

Example 6.11. Let Ω be a non-empty closed discrete subset of $\mathbb{C}$. If $\omega \in \Omega \setminus \{0\}$ and $\widehat{\psi} \in \mathbb{C}\{\zeta\}$ extends analytically to $\mathbb{C} \setminus \Omega$, then one can take any branch of the logarithm $\mathscr{L}$ holomorphic near $-\omega$ and define by the formula

$$\widehat{\varphi}(\zeta) = \widehat{\psi}(\zeta)\mathscr{L}(\zeta - \omega)$$

a germ of $\widehat{\mathscr{R}}_\Omega$ with non-trivial monodromy around ω: the branches of the analytic continuation of $\widehat{\varphi}$ differ by integer multiples of $2\pi\mathrm{i}\,\widehat{\psi}$.

Example 6.12. If $\omega \in \mathbb{C}^*$ and $m \in \mathbb{N}^*$, then $\big(\mathscr{L}(\zeta-\omega)\big)^m \in \widehat{\mathscr{R}}_{\{\omega\}}$ for any branch of the logarithm $\mathscr{L}$; if moreover $\omega \neq -1$, then $\big(\mathscr{L}(\zeta-\omega)\big)^{-m} \in \widehat{\mathscr{R}}_{\{\omega,\omega+1\}}$.

Example 6.13. Given $\alpha \in \mathbb{C}$, the incomplete gamma function is defined for $z > 0$ by

$$\Gamma(\alpha, z) := \int_z^{+\infty} \mathrm{e}^{-t} t^{\alpha-1}\,\mathrm{d}t$$

---

[2] Given $\zeta_0 \in \mathbb{C}$ and a holomorphic germ f at ζ_0, what we call a *branch of the analytic continuation of f* or, simply, a *branch of f*, is a function g holomorphic in an open connected subset U of $\mathbb{C}$ for which there exists a path γ from ζ_0 to a point $\zeta_1 \in U$ such that $\mathrm{cont}_\gamma f$ exists and coincides with the germ of g at ζ_1. For instance, the logarithm $\sum_{n\geq 1} \frac{(-1)^{n-1}}{n}(\zeta-1)^n$ is a holomorphic germ at 1, a particular branch of which is the *principal branch of the logarithm* denoted by Log in (6.25); for any $\zeta_1 \in \mathbb{C}^*$ one can find a branch of the logarithm $\mathscr{L}$ holomorphic near ζ_1, and all the branches of the logarithm holomorphic near ζ_1 are of the form $\mathscr{L} + 2\pi\mathrm{i}m$, $m \in \mathbb{Z}$.

and it extends to a holomorphic function in $\mathbb{C}\setminus\mathbb{R}^-$ (notice that $\Gamma(\alpha,z) \xrightarrow[z\to 0]{} \Gamma(\alpha)$ if $\Re e\,\alpha>0$). The change of variable $t=z(\zeta+1)$ in the integral yields the formula

$$\Gamma(\alpha,z)=\mathrm{e}^{-z}z^{\alpha}(\mathscr{S}^I\widehat{\varphi}_\alpha)(z), \qquad \widehat{\varphi}_\alpha(\zeta):=(1+\zeta)^{\alpha-1}, \tag{6.6}$$

where $I=(-\frac{\pi}{2},\frac{\pi}{2})$ and we use the principal branch of the logarithm (6.25) to define the holomorphic function $(1+\zeta)^{\alpha-1}$ as $\mathrm{e}^{(\alpha-1)\mathrm{Log}\,(1+\zeta)}$. The germ $\widehat{\varphi}_\alpha$ is always $\{-1\}$-resurgent; it has multiple-valued analytic continuation if $\alpha\notin\mathbb{Z}$. Hence

$$z^{-\alpha}\mathrm{e}^{z}\Gamma(\alpha,z)\sim_1\widetilde{\varphi}_\alpha(z)=\sum_{n\geq 0}(\alpha-1)(\alpha-2)\cdots(\alpha-n)z^{-n-1}, \tag{6.7}$$

which is always a 1-summable and $\{-1\}$-resurgent formal series (a polynomial in z^{-1} if $\alpha\in\mathbb{N}^*$, a divergent formal series otherwise).

$\widehat{\mathscr{R}}_\Omega$ and $\widetilde{\mathscr{R}}_\Omega$ clearly are linear subspaces of $\mathbb{C}\{\zeta\}$ and $\mathbb{C}[[z^{-1}]]_1$. We end this section with elementary stability properties:

Lemma 6.14. *Let Ω be any non-empty closed discrete subset of $\mathbb{C}$. Let $\widehat{B}\in\widehat{\mathscr{R}}_\Omega$. Then multiplication by $\widehat{B}$ leaves $\widehat{\mathscr{R}}_\Omega$ invariant. In particular, for any $c\in\mathbb{C}$,*

$$\widehat{\varphi}(\zeta)\in\widehat{\mathscr{R}}_\Omega \implies -\zeta\widehat{\varphi}(\zeta)\in\widehat{\mathscr{R}}_\Omega \text{ and } \mathrm{e}^{-c\zeta}\widehat{\varphi}(\zeta)\in\widehat{\mathscr{R}}_\Omega.$$

The operator $\frac{\mathrm{d}}{\mathrm{d}\zeta}$ too leaves $\widehat{\mathscr{R}}_\Omega$ invariant.

As a consequence, $\widetilde{\mathscr{R}}_\Omega$ is stable by $\partial=\frac{\mathrm{d}}{\mathrm{d}z}$ and T_c. Moreover, if $\widetilde{\psi}\in\widetilde{\mathscr{R}}_\Omega\cap z^{-2}\mathbb{C}[[z^{-1}]]$, then $z\widetilde{\psi}\in\widetilde{\mathscr{R}}_\Omega$ and the solution in $z^{-1}\mathbb{C}[[z^{-1}]]$ of the difference equation

$$\widetilde{\varphi}(z+1)-\widetilde{\varphi}(z)=\widetilde{\psi}(z)$$

belongs to $\widetilde{\mathscr{R}}_{\Omega\cup 2\pi\mathrm{i}\mathbb{Z}^*}$.

Proof. Exercise (use the fact that multiplication by $\widehat{B}$ commutes with analytic continuation: the analytic continuation of $\widehat{B}\widehat{\varphi}$ along a path γ of $\mathbb{C}\setminus\Omega$ starting in $\mathbb{D}^*_{\rho(\Omega)}$ exists and equals $\widehat{B}(\zeta)\,\mathrm{cont}_\gamma\widehat{\varphi}(\zeta)$; then use Lemma 5.10, (5.21), (5.23) and Corollary 5.11). □

6.2 Analytic continuation of a convolution product: the easy case

Lemma 6.14 was dealing with the multiplication of two germs of $\mathbb{C}\{\zeta\}$, however we saw in Section 5.5 that the natural product in this space is convolution. The question of the stability of $\widehat{\mathscr{R}}_\Omega$ under convolution is much subtler. Let us begin with an easy case, which is already of interest:

Lemma 6.15. *Let Ω be any non-empty closed discrete subset of $\mathbb{C}$ and suppose $\widehat{B}$ is an entire function of $\mathbb{C}$. Then, for any $\widehat{\varphi} \in \widehat{\mathscr{R}}_\Omega$, the convolution product $\widehat{B} * \widehat{\varphi}$ belongs to $\widehat{\mathscr{R}}_\Omega$; its analytic continuation along a path γ of $\mathbb{C} \setminus \Omega$ starting from a point $\zeta_0 \in \mathbb{D}_{\rho(\Omega)}$ and ending at a point ζ_1 is the holomorphic germ at ζ_1 explicitly given by*

$$\operatorname{cont}_\gamma(\widehat{B} * \widehat{\varphi})(\zeta) = \int_0^{\zeta_0} \widehat{B}(\zeta - \xi)\widehat{\varphi}(\xi)\,\mathrm{d}\xi + \int_\gamma \widehat{B}(\zeta - \xi)\widehat{\varphi}(\xi)\,\mathrm{d}\xi \\ + \int_{\zeta_1}^{\zeta} \widehat{B}(\zeta - \xi)\widehat{\varphi}(\xi)\,\mathrm{d}\xi \quad (6.8)$$

for ζ close enough to ζ_1. As a consequence,

$$\widetilde{B}_0 \in \mathbb{C}\{z^{-1}\},\ \widetilde{\varphi}_0 \in \widetilde{\mathscr{R}}_\Omega \implies \widetilde{B}_0\widetilde{\varphi}_0 \in \widetilde{\mathscr{R}}_\Omega. \quad (6.9)$$

Remark 6.16. Formulas such as (6.8) require a word of caution: the value of $\widehat{B}(\zeta - \xi)$ is unambiguously defined whatever ζ and ξ are, but in the notation "$\widehat{\varphi}(\xi)$" it is understood that we are using the appropriate branch of the possibly multiple-valued function $\widehat{\varphi}$; in such a formula, what branch we are using is clear from the context:

- $\widehat{\varphi}$ is unambiguously defined in its disc of convergence D_0 (centred at 0) and the first integral thus makes sense for $\zeta_0 \in D_0$;
- in the second integral ξ is moving along γ which is a path of analytic continuation for $\widehat{\varphi}$, we thus consider the analytic continuation of $\widehat{\varphi}$ along the piece of γ between its origin and ξ;
- in the third integral, "$\widehat{\varphi}$" is to be understood as $\operatorname{cont}_\gamma \widehat{\varphi}$, the germ at ζ_1 resulting form the analytic continuation of $\widehat{\varphi}$ along γ, this integral then makes sense for any ζ at a distance from ζ_1 less than the radius of convergence of $\operatorname{cont}_\gamma \widehat{\varphi}$.

Using a parametrisation $\gamma\colon [0,1] \to \mathbb{C} \setminus \Omega$, with $\gamma(0) = \zeta_0$ and $\gamma(1) = \zeta_1$, and introducing the truncated paths $\gamma_s := \gamma_{|[0,s]}$ for any $s \in [0,1]$, the interpretation of the last two integrals in (6.8) is

$$\int_\gamma \widehat{B}(\zeta - \xi)\widehat{\varphi}(\xi)\,\mathrm{d}\xi := \int_0^1 \widehat{B}(\zeta - \gamma(s))(\operatorname{cont}_{\gamma_s} \widehat{\varphi})(\gamma(s))\tfrac{\mathrm{d}\gamma}{\mathrm{d}s}(s)\,\mathrm{d}s, \quad (6.10)$$

$$\int_{\zeta_1}^{\zeta} \widehat{B}(\zeta - \xi)\widehat{\varphi}(\xi)\,\mathrm{d}\xi := \int_{\zeta_1}^{\zeta} \widehat{B}(\zeta - \xi)(\operatorname{cont}_\gamma \widehat{\varphi})(\xi)\,\mathrm{d}\xi. \quad (6.11)$$

Proof of Lemma 6.15. The property (6.9) directly follows from the first statement: write $\widetilde{B}_0 = a + \widetilde{B}$ and $\widetilde{\varphi}_0 = b + \widetilde{\varphi}$ with $a, b \in \mathbb{C}$ and $\widetilde{A}, \widetilde{\varphi} \in z^{-1}\mathbb{C}[[z^{-1}]]$ and apply Lemma 5.7 to $\widetilde{B}$.

To prove the first statement, we use a parametrisation $\gamma\colon [0,1] \to \mathbb{C} \setminus \Omega$ and the truncated paths $\gamma_s := \gamma_{|[0,s]}$: we shall check that, for each $t \in [0,1]$, the formula

$$\widehat{\chi}_t(\zeta) := \int_0^{\zeta_0} \widehat{B}(\zeta-\xi)\widehat{\varphi}(\xi)\,\mathrm{d}\xi + \int_{\gamma_t} \widehat{B}(\zeta-\xi)\widehat{\varphi}(\xi)\,\mathrm{d}\xi + \int_{\gamma(t)}^{\zeta} \widehat{B}(\zeta-\xi)\widehat{\varphi}(\xi)\,\mathrm{d}\xi \tag{6.12}$$

(with the above conventions for the interpretation of "$\widehat{\varphi}(\xi)$" in the integrals) defines a holomorphic germ at $\gamma(t)$ which is the analytic continuation of $\widehat{B} * \widehat{\varphi}$ along γ_t.

The holomorphic dependence of the integrals upon the parameter ζ is such that $\zeta \mapsto \int_0^{\zeta_0} \widehat{B}(\zeta-\xi)\widehat{\varphi}(\xi)\,\mathrm{d}\xi + \int_{\gamma_t} \widehat{B}(\zeta-\xi)\widehat{\varphi}(\xi)\,\mathrm{d}\xi$ is an entire function of ζ, and $\zeta \mapsto \int_{\gamma(t)}^{\zeta} \widehat{B}(\zeta-\xi)\widehat{\varphi}(\xi)\,\mathrm{d}\xi$ is holomorphic for ζ in the disc of convergence D_t of $\mathrm{cont}_{\gamma_t}\widehat{\varphi}$. Therefore, (6.12) defines a family of analytic elements[3] $(D_t, \widehat{\chi}_t)$, $t \in [0,1]$, where each disc D_t is centred at $\gamma(t)$.

For t small enough, the truncated path γ_t is contained in D_0; then, for $\zeta \in D_0$, the Cauchy theorem implies that $\widehat{\chi}_t(\zeta)$ coincides with $\widehat{A} * \widehat{\varphi}(\zeta) = \int_0^{\zeta} \widehat{B}(\zeta-\xi)\widehat{\varphi}(\xi)\,\mathrm{d}\xi$ (since there is homotopy in D_0 between the path $t \in [0,1] \mapsto t\zeta$ and the path obtained by following the line-segment from 0 to ζ_0, then γ_t from ζ_0 to $\gamma(t)$, and then the line-segment from $\gamma(t)$ to ζ).

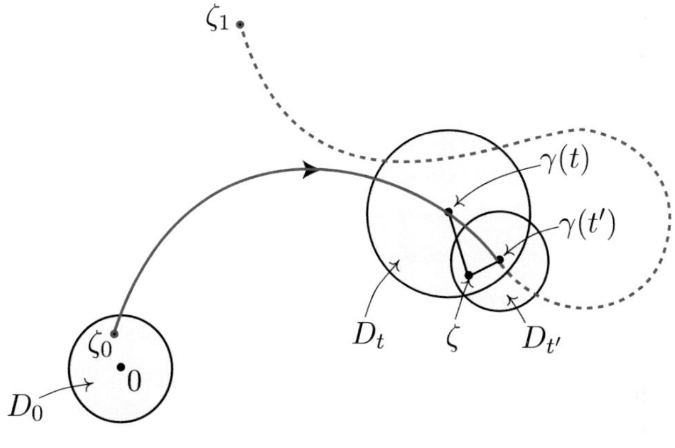

Fig. 6.2: Integration paths for the convolution in the easy case.

For every $t \in [0,1]$, there exists $\varepsilon > 0$ such that $\gamma\big((t-\varepsilon, t+\varepsilon) \cap [0,1]\big) \subset D_t$; by compactness, we can thus find $N \in \mathbb{N}^*$ and $0 = t_0 < t_1 < \cdots < t_N = 1$ so that $\gamma\big([t_j, t_{j+1}]\big) \subset D_{t_j}$ for every j. In view of Proposition 1.15 (p. 10), the proof will thus be complete if we check that, for any $t < t'$ in $[0,1]$,

$$\gamma\big([t,t']\big) \subset D_t \implies \widehat{\chi}_t \equiv \widehat{\chi}_{t'} \text{ in } D_t \cap D_{t'}.$$

[3] We say that (D, f) is an "analytic element" if D is an open non-empty disc and $f\colon D \to \mathbb{C}$ is a holomorphic function. The "pairs" of Section 1.1.2 of the first part of this volume (p. 5) correspond to the case where D is the disc of convergence of the germ of f at its centre.

This follows from the observation that, under the hypothesis $\gamma\big([t,t']\big) \subset D_t$,

$$s \in [t,t'] \text{ and } \xi \in D_t \cap D_s \implies \operatorname{cont}_{\gamma_s} \widehat{\varphi}(\xi) = \operatorname{cont}_{\gamma_t} \widehat{\varphi}(\xi),$$

thus, when computing $\widehat{\chi}_{t'}(\zeta)$ with $\zeta \in D_t \cap D_{t'}$, the third integral in (6.12) is

$$\int_{\gamma(t')}^{\zeta} \widehat{B}(\zeta-\xi) \operatorname{cont}_{\gamma_{t'}} \widehat{\varphi}(\xi)\, \mathrm{d}\xi = \int_{\gamma(t')}^{\zeta} \widehat{B}(\zeta-\xi) \operatorname{cont}_{\gamma_t} \widehat{\varphi}(\xi)\, \mathrm{d}\xi$$

and, interpreting the second integral of (6.12) as in (6.10), we can rewrite the difference $\widehat{\chi}_{t'}(\zeta) - \widehat{\chi}_t(\zeta)$ as

$$\int_t^{t'} \widehat{B}(\zeta-\gamma(s)) \big(\operatorname{cont}_{\gamma_s} \widehat{\varphi}\big)(\gamma(s)) \gamma'(s)\, \mathrm{d}s + \int_{\gamma(t')}^{\gamma(t)} \widehat{B}(\zeta-\xi) \big(\operatorname{cont}_{\gamma_t} \widehat{\varphi}\big)(\xi)\, \mathrm{d}\xi =$$

$$\int_t^{t'} \widehat{B}(\zeta-\gamma(s)) \big(\operatorname{cont}_{\gamma_t} \widehat{\varphi}\big)(\gamma(s)) \gamma'(s)\, \mathrm{d}s + \int_{\gamma(t')}^{\gamma(t)} \widehat{B}(\zeta-\xi) \big(\operatorname{cont}_{\gamma_t} \widehat{\varphi}\big)(\xi)\, \mathrm{d}\xi = 0$$

(see Figure 6.2). □

Remark 6.17. Lemma 6.15 can be used to prove the Ω-resurgence of certain formal series solutions of linear or non-linear functional equations: see Section 7.3 in this volume, the proof of Theorem 5.3.21 in the second volume of this book [Lod16], and Section 8 of [Sau09].

6.3 Analytic continuation of a convolution product: an example

We now wish to consider the convolution of two Ω-continuable holomorphic germs at 0 without assuming that any of them extends to an entire function. A first example will convince us that there is no hope to get stability under convolution if we do not impose that Ω be stable under addition.

Let $\omega_1, \omega_2 \in \mathbb{C}^*$ and

$$\widehat{\varphi}(\zeta) := \frac{1}{\zeta - \omega_1}, \qquad \widehat{\psi}(\zeta) := \frac{1}{\zeta - \omega_2}.$$

Their convolution product is

$$\widehat{\chi}(\zeta) := \widehat{\varphi} * \widehat{\psi}(\zeta) = \int_0^{\zeta} \frac{1}{(\xi-\omega_1)(\zeta-\xi-\omega_2)}\, \mathrm{d}\xi, \qquad |\zeta| < \min\big\{|\omega_1|, |\omega_2|\big\}.$$

The formula

$$\frac{1}{(\xi-\omega_1)(\zeta-\xi-\omega_2)} = \frac{1}{\zeta-\omega_1-\omega_2}\left(\frac{1}{\xi-\omega_1} + \frac{1}{\zeta-\xi-\omega_2}\right)$$

shows that, for any $\zeta \neq \omega_1 + \omega_2$ of modulus $< \min\{|\omega_1|, |\omega_2|\}$, one can write

$$\widehat{\chi}(\zeta) = \frac{1}{\zeta - \omega_1 - \omega_2}\big(L_1(\zeta) + L_2(\zeta)\big), \qquad L_j(\zeta) := \int_0^{\zeta} \frac{\mathrm{d}\xi}{\xi - \omega_j} \tag{6.13}$$

(with the help of the change of variable $\xi \mapsto \zeta - \xi$ in the case of L_2).

Removing the half-lines $\omega_j[1, +\infty)$ from $\mathbb{C}$, we obtain a cut plane Δ in which $\widehat{\chi}$ has a meromorphic continuation, because both L_1 and L_2 define holomorphic functions on

$$\Delta = \{\, \zeta \in \mathbb{C} \mid \zeta_j \notin [0, \zeta] \text{ for } j = 1, 2 \,\}.$$

In fact, we are not limited to Δ: we can follow the *meromorphic* continuation of $\widehat{\chi}$ along any path of $\mathbb{C}$ which avoids ω_1 and ω_2, because

$$L_j(\zeta) = -\int_0^{\zeta/\omega_j} \frac{\mathrm{d}\xi}{1 - \xi} = \operatorname{Log}\left(1 - \frac{\zeta}{\omega_j}\right) \in \widehat{\mathscr{R}}_{\{\omega_j\}}$$

(cf. example 6.9). We used the words "meromorphic continuation" and not "analytic continuation" because of the factor $\frac{1}{\zeta - \omega_1 - \omega_2}$ in (6.13). As far as analytic continuation is concerned, the obstructions are located at ω_1 because of L_1, ω_2 because of L_2 and $\omega_1 + \omega_2$ because of the factor $\frac{1}{\zeta - \omega_1 - \omega_2}$; the conclusion is thus only $\widehat{\chi} \in \widehat{\mathscr{R}}_{\Omega}$, with $\Omega := \{\omega_1, \omega_2, \omega_1 + \omega_2\}$.

– If $\omega := \omega_1 + \omega_2 \in \Delta$, i.e. if we cannot write $\omega_2 = s\omega_1$ with $s > 0$, then the principal branch of $\widehat{\chi}$ (i.e. its meromorphic continuation to Δ) has a removable singularity at ω, because $(L_1 + L_2)(\omega) = \int_0^{\omega} \frac{\mathrm{d}\xi}{\xi - \omega_1} + \int_0^{\omega} \frac{\mathrm{d}\xi}{\xi - \omega_2} = 0$ in that case (by the change of variable $\xi \mapsto \omega - \xi$ in one of the integrals). This is consistent with Lemma 5.54 (the set Δ is clearly star-shaped with respect to 0). But it is easy to see that this does not happen for all the branches of $\widehat{\chi}$: when considering all the paths γ going from 0 to ω and avoiding ω_1 and ω_2, we have

$$\operatorname{cont}_{\gamma} L_j(\omega) = \int_{\gamma} \frac{\mathrm{d}\xi}{\xi - \omega_j}, \qquad j = 1, 2,$$

hence $\frac{1}{2\pi\mathrm{i}}\big(\operatorname{cont}_{\gamma} L_1(\omega) + \operatorname{cont}_{\gamma} L_2(\omega)\big)$ is the sum of the winding numbers around ω_1 and ω_2 of the loop obtained by following γ from 0 to ω and then the line-segment from ω to 0; elementary geometry shows that this sum of winding numbers can take any integer value, but whenever this value is non-zero the corresponding branch of $\widehat{\chi}$ does have a pole at ω.

– The case $\omega \notin \Delta$ is slightly different. Then we can write $\omega_j = r_j \mathrm{e}^{\mathrm{i}\theta}$ with $r_1, r_2 > 0$ and consider the path γ_0 which follows the line-segment $[0, \omega]$ except that it circumvents ω_1 and ω_2 by small half-circles travelled anti-clockwise (notice that ω_1 and ω_2 may coincide)—see the left part of Figure 6.3; an easy computation yields

$$\operatorname{cont}_{\gamma_0} L_1(\omega) = \int_{-r_1}^{-1} \frac{\mathrm{d}\xi}{\xi} + \int_{\Gamma_0} \frac{\mathrm{d}\xi}{\xi} + \int_1^{r_2} \frac{\mathrm{d}\xi}{\xi},$$

where Γ_0 is the half-circle from -1 to 1 with radius 1 travelled anti-clockwise (see the right part of Figure 6.3), hence $\operatorname{cont}_{\gamma_0} L_1(\omega) = \ln\frac{r_2}{r_1} + \mathrm{i}\pi$, similarly $\operatorname{cont}_{\gamma_0} L_2(\omega) = \ln\frac{r_1}{r_2} + \mathrm{i}\pi$, therefore $\operatorname{cont}_{\gamma_0} L_1(\omega) + \operatorname{cont}_{\gamma_0} L_2(\omega) = 2\pi\mathrm{i}$ is non-zero and this again yields a branch of $\widehat{\chi}$ with a pole at ω (and infinitely many others by using other paths than γ_0).

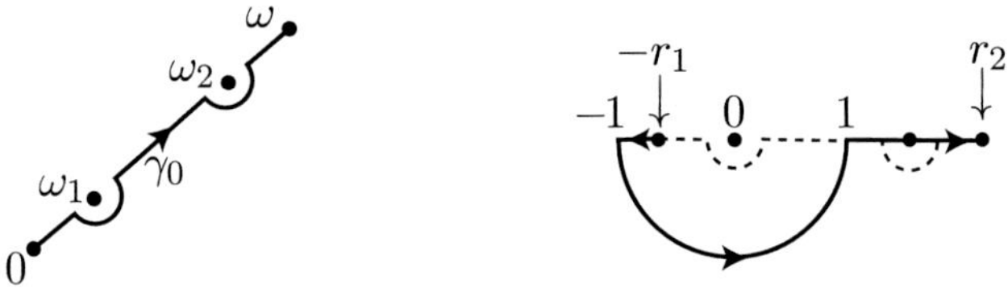

Fig. 6.3: Convolution of aligned poles.

In all cases, there are paths from 0 to $\omega_1 + \omega_2$ which avoid ω_1 and ω_2 and which are not paths of analytic continuation for $\widehat{\chi}$. This example thus shows that $\widehat{\mathscr{R}}_{\{\omega_1,\omega_2\}}$ is *not* stable under convolution: it contains $\widehat{\varphi}$ and $\widehat{\psi}$ but not $\widehat{\varphi} * \widehat{\psi}$.

Now, whenever Ω is not stable under addition, one can find $\omega_1, \omega_2 \in \Omega$ such that $\omega_1 + \omega_2 \notin \Omega$ and the previous example then yields $\widehat{\varphi}, \widehat{\psi} \in \widehat{\mathscr{R}}_\Omega$ with $\widehat{\varphi} * \widehat{\psi} \notin \widehat{\mathscr{R}}_\Omega$.

6.4 Analytic continuation of a convolution product: the general case

6.4.1 The main result of this section is

Theorem 6.18. *Let Ω be a non-empty closed discrete subset of $\mathbb{C}$. Then the space $\widehat{\mathscr{R}}_\Omega$ is stable under convolution if and only if Ω is stable under addition.*

The necessary and sufficient condition on Ω is satisfied by the typical examples $\mathbb{Z}$ or $2\pi\mathrm{i}\mathbb{Z}$, but also by $\mathbb{N}^*$, $\mathbb{Z}+\mathrm{i}\mathbb{Z}$, $\mathbb{N}^*+\mathrm{i}\mathbb{N}$ or $\{m+n\sqrt{2} \mid m,n \in \mathbb{N}^*\}$ for instance. An immediate consequence of Theorem 6.18 is

Corollary 6.19. *Let Ω be a non-empty closed discrete subset of $\mathbb{C}$. Then the space $\widetilde{\mathscr{R}}_\Omega$ of Ω-resurgent formal series is a subalgebra of $\mathbb{C}[[z^{-1}]]$ if and only if Ω is stable under addition.*

The necessity of the condition on Ω was proved in Section 6.3. In the rest of this section we shall prove that the condition is sufficient. However we shall restrict ourselves to the case where $0 \in \Omega$, because this will allow us to give a simpler proof. The reader is referred to [Sau13] for the proof in the general case.

6.4.2 We thus fix Ω closed, discrete, containing 0 and stable under addition. We begin with a new definition (see Figure 6.4):

Definition 6.20. A continuous map $H\colon I \times J \to \mathbb{C}$, where $I = [0,1]$ and J is a compact interval of $\mathbb{R}$, is called a *symmetric Ω-homotopy* if, for each $t \in J$,

$$s \in I \mapsto H_t(s) := H(s,t)$$

defines a path which satisfies

1. $H_t(0) = 0$,
2. $H_t\big((0,1]\big) \subset \mathbb{C} \setminus \Omega$,
3. $H_t(1) - H_t(s) = H_t(1-s)$ for every $s \in I$.

We then call *endpoint path* of H the path

$$\Gamma_H\colon\ t \in J \mapsto H_t(1).$$

Writing $J = [a,b]$, we call H_a (resp. H_b) the *initial path* of H (resp. its *final path*).

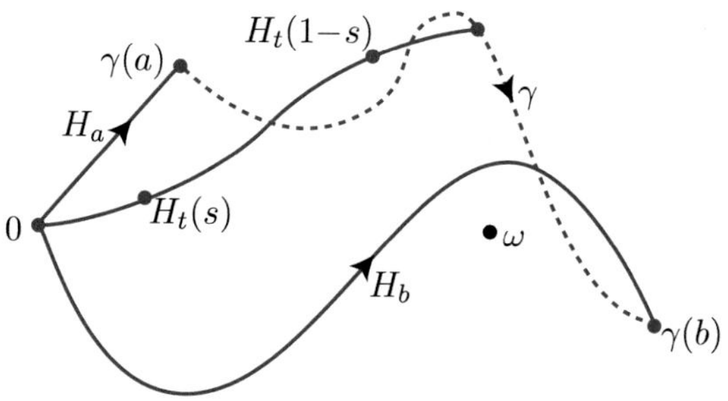

Fig. 6.4: *Symmetric Ω-homotopy* (H_a = initial path, H_b = final path, γ = endpoint path Γ_H).

The first two conditions imply that each path H_t is a path of analytic continuation for any $\widehat{\varphi} \in \widehat{\mathscr{R}}_\Omega$, in view of Remark 6.3.

We shall use the notation $H_{t|s}$ for the truncated paths $(H_t)_{|[0,s]}$, $s \in I$, $t \in J$ (analogously to what we did when commenting Lemma 6.15). Here is a technical statement we shall use:

Lemma 6.21. *For a symmetric Ω-homotopy H defined on $I \times J$, there exists $\delta > 0$ such that, for any $\widehat{\varphi} \in \widehat{\mathscr{R}}_\Omega$ and $(s,t) \in I \times J$, the radius of convergence of the holomorphic germ $\operatorname{cont}_{H_{t|s}} \widehat{\varphi}$ at $H_t(s)$ is at least δ.*

Proof. Let $\rho = \rho(\Omega)$ as in Remark 6.3. Consider

$$U := \{ (s,t) \in I \times J \mid H([0,s] \times \{t\}) \subset \mathbb{D}_{\rho/2} \}, \qquad K := I \times J \setminus U.$$

The continuity of H implies that K is a compact subset of $I \times J$, which is contained in $(0,1] \times J$ by property (i) of Definition 6.20. Thus $H(K)$ is a compact set, contained in $\mathbb{C} \setminus \Omega$ by property (ii), and $\delta := \min\{ \operatorname{dist}(H(K), \Omega), \rho/2 \} > 0$. Now, for any s and t,

- either $(s,t) \in U$, then the truncated path $H_{t|s}$ lies in $\mathbb{D}_{\rho/2}$, hence $\operatorname{cont}_{H_{t|s}} \widehat{\varphi}$ is a holomorphic germ at $H_t(s)$ with radius of convergence $\geq \frac{\rho}{2} \geq \delta$;
- or $(s,t) \in K$, and then $\operatorname{dist}(H_t(s), \Omega) \geq \delta$, which yields the same conclusion for the germ $\operatorname{cont}_{H_{t|s}} \widehat{\varphi}$.

□

The third condition in Definition 6.20 means that each path H_t is symmetric with respect to its midpoint $\frac{1}{2} H_t(1)$. Here is the motivation behind this requirement:

Lemma 6.22. *Let* $\gamma\colon [0,1] \to \mathbb{C} \setminus \Omega$ *be a path such that* $\gamma(0) \in \mathbb{D}_{\rho(\Omega)}$ *(as in Remark 6.3). If there exists a symmetric* Ω*-homotopy whose endpoint path coincides with* γ *and whose initial path is contained in* $\mathbb{D}_{\rho(\Omega)}$*, then any convolution product* $\widehat{\varphi} * \widehat{\psi}$ *with* $\widehat{\varphi}, \widehat{\psi} \in \widehat{\mathscr{R}}_\Omega$ *can be analytically continued along* γ.

Before proving it, we introduce the classical notion of "path concatenation" (also called "path composition"):

Definition 6.23. Given two paths $\gamma\colon [a,b] \to \mathbb{C}$ and $\gamma'\colon [a',b'] \to \mathbb{C}$ such that $\gamma(b) = \gamma'(a')$, their *concatenation* is the path

$$\gamma \centerdot \gamma'\colon [a, b + b' - a'] \to \mathbb{C} \tag{6.14}$$

defined by $\gamma \centerdot \gamma'(t) := \gamma(t)$ for $t \in [a,b]$ and $\gamma \centerdot \gamma'(t) := \gamma'(a' + t - b)$ for $t \in [b, b + b' - a']$.

In other words, $\gamma \centerdot \gamma'$ consists in following γ from $\gamma(a)$ to its terminal point $\gamma(b)$, which is supposed to coincide with $\gamma'(a')$, and then γ' from $\gamma'(a')$ to $\gamma'(b')$.

Remark 6.24. Recall that the precise parametrisation of our paths is of little interest (footnote 1), this is why a variant of the above definition is sometimes used: restricting to paths γ and γ' parametrised on the interval $[0,1]$ (or performing an affine change of parametrisation so that the parametrisation interval becomes $[0,1]$), one may prefer to parametrise $\gamma \centerdot \gamma'$ on $[0,1]$ as well and set $\gamma \centerdot \gamma'(t) := \gamma(2t)$ for $t \in [0, \frac{1}{2}]$ and $\gamma \centerdot \gamma'(t) := \gamma'(2(t - \frac{1}{2}))$ for $t \in [\frac{1}{2}, 1]$. For instance, Definition 1.30 of the first part of this volume ("loop product" p. 17) is a particular case of this variant.

Proof of Lemma 6.22. We assume that $\gamma = \Gamma_H$ with a symmetric Ω-homotopy H defined on $[0,1] \times J$. Let $\widehat{\varphi}, \widehat{\psi} \in \widehat{\mathscr{R}}_\Omega$ and, for $t \in J$, consider the formula

$$\widehat{\chi}_t(\zeta) = \int_{H_t} \widehat{\varphi}(\xi)\widehat{\psi}(\zeta-\xi)\,\mathrm{d}\xi + \int_{\gamma(t)}^{\zeta} \widehat{\varphi}(\xi)\widehat{\psi}(\zeta-\xi)\,\mathrm{d}\xi. \tag{6.15}$$

Recall that H_t is parametrised on $I=[0,1]$ and has terminal point $H_t(1)=\gamma(t)$, so we can make use of Definition 6.23 and write as well[4] $\widehat{\chi}_t(\zeta) = \int_{H_t \bullet [\gamma(t,\zeta)]} \widehat{\varphi}(\xi)\widehat{\psi}(\zeta-\xi)\,\mathrm{d}\xi$. We shall check that $\widehat{\chi}_t$ is a well-defined holomorphic germ at $\gamma(t)$ and that it provides the analytic continuation of $\widehat{\varphi} * \widehat{\psi}$ along γ.

a) The idea is that when ξ moves along H_t, $\xi = H_t(s)$ with $s \in I$, we can use for "$\widehat{\varphi}(\xi)$" the analytic continuation of $\widehat{\varphi}$ along the truncated path $H_{t|s}$; correspondingly, if ζ is close to $\gamma(t)$, then $\zeta - \xi$ is close to $\gamma(t) - \xi = H_t(1) - H_t(s) = H_t(1-s)$, thus for "$\widehat{\psi}(\zeta-\xi)$" we can use the analytic continuation of $\widehat{\psi}$ along $H_{t|1-s}$. In other words, setting $\zeta = \gamma(t)+\sigma$, we wish to interpret (6.15) as

$$\widehat{\chi}_t(\gamma(t)+\sigma) := \int_0^1 (\mathrm{cont}_{H_{t|s}}\widehat{\varphi})(H_t(s))(\mathrm{cont}_{H_{t|1-s}}\widehat{\psi})(H_t(1-s)+\sigma)H_t'(s)\,\mathrm{d}s$$
$$+ \int_0^1 (\mathrm{cont}_{H_t}\widehat{\varphi})(\gamma(t)+u\sigma)\widehat{\psi}((1-u)\sigma)\sigma\,\mathrm{d}u \tag{6.16}$$

(in the last integral, we have performed the change variable $\xi = \gamma(t)+u\sigma$; it is the germ of $\widehat{\psi}$ at the origin that we use there).

Lemma 6.21 provides $\delta > 0$ such that, by regular dependence of the integrals upon the parameter σ, the right-hand side of (6.16) is holomorphic for $|\sigma| < \delta$. We thus have a family of analytic elements $(D_t, \widehat{\chi}_t)$, $t \in J$, with $D_t := \{\zeta \in \mathbb{C} \mid |\zeta - \gamma(t)| < \delta\}$ (see footnote 3).

b) For t small enough, the path H_t is contained in $\mathbb{D}_{\rho(\Omega)}$ which is open and simply connected; then, for $|\zeta|$ small enough, the line-segment $[0,\zeta]$ and the concatenation $H_t \bullet [\gamma(t),\zeta]$ are homotopic in $\mathbb{D}_{\rho(\Omega)}$, hence the Cauchy theorem implies $\widehat{\chi}_t(\zeta) = \widehat{\varphi} * \widehat{\psi}(\zeta)$.

c) By uniform continuity, there exists $\varepsilon > 0$ such that, for any $t_0, t \in J$,

$$|t-t_0| \le \varepsilon \quad \Longrightarrow \quad |H_t(s) - H_{t_0}(s)| < \delta/2 \quad \text{for all } s \in I. \tag{6.17}$$

We now check that, for any t_0, t in J such that $t_0 \le t \le t_0 + \varepsilon$, we have $\widehat{\chi}_{t_0} \equiv \widehat{\chi}_t$ in $D(\gamma(t_0), \delta/2)$ (which is contained in $D_{t_0} \cap D_t$); in view of Proposition 1.15 of the first part of this volume (p. 10), this will be sufficient to complete the proof.

Let $t_0, t \in J$ be such that $t_0 \le t \le t_0 + \varepsilon$ and let $\zeta \in D(\gamma(t_0), \delta/2)$. By Lemma 6.21 and (6.17), we have for every $s \in I$

$$\mathrm{cont}_{H_{t|s}}\widehat{\varphi}\big(H_t(s)\big) = \mathrm{cont}_{H_{t_0|s}}\widehat{\varphi}\big(H_t(s)\big),$$
$$\mathrm{cont}_{H_{t|1-s}}\widehat{\psi}\big(\zeta - H_t(s)\big) = \mathrm{cont}_{H_{t_0|1-s}}\widehat{\psi}\big(\zeta - H_t(s)\big)$$

[4] For the sake of definiteness, we identify any line-segment $[\zeta_0, \zeta_1]$ in $\mathbb{C}$ with a path $t \in [0,1] \mapsto \zeta_0 + t(\zeta_1 - \zeta_0)$.

(for the latter identity, write $\zeta - H_t(s) = H_t(1-s) + \zeta - \gamma(t) = H_{t_0}(1-s) + \zeta - \gamma(t_0) + H_{t_0}(s) - H_t(s)$, thus this point belongs to $D\big(H_t(1-s),\delta\big) \cap D\big(H_{t_0}(1-s),\delta\big)$). Moreover, $[\gamma(t),\zeta] \subset D\big(\gamma(t_0),\delta/2\big)$ by convexity, hence $\operatorname{cont}_{H_t}\widehat{\varphi} \equiv \operatorname{cont}_{H_{t_0}}\widehat{\varphi}$ on this line-segment, and we can write

$$\widehat{\chi}_t(\zeta) = \int_0^1 (\operatorname{cont}_{H_{t_0|s}}\widehat{\varphi})(H_t(s))(\operatorname{cont}_{H_{t_0|1-s}}\widehat{\psi})(\zeta - H_t(s))H_t'(s)\,\mathrm{d}s$$
$$+ \int_{\gamma(t)}^{\zeta} (\operatorname{cont}_{H_{t_0}}\widehat{\varphi})(\xi)\widehat{\psi}(\zeta-\xi)\,\mathrm{d}\xi.$$

We then get $\widehat{\chi}_{t_0}(\zeta) = \widehat{\chi}_t(\zeta)$ from the Cauchy theorem, because H induces a homotopy between the concatenation $H_{t_0} \cdot [\gamma(t_0),\zeta]$ and the concatenation $H_t \cdot [\gamma(t),\zeta]$. □

Remark 6.25. With the notation of Definition 6.20, when the initial path H_a is a line-segment contained in $\mathbb{D}_{\rho(\Omega)}$, the final path H_b is what Écalle calls a "symmetrically contractible path" in [Éca81]. The proof of Lemma 6.22 shows that the analytic continuation of $\widehat{\varphi} * \widehat{\psi}$ until the endpoint $H_b(1) = \Gamma_H(b)$ can be computed by the usual integral taken over H_b:

$$\operatorname{cont}_\gamma(\widehat{\varphi} * \widehat{\psi})(\zeta) = \int_{H_b} \widehat{\varphi}(\xi)\widehat{\psi}(\zeta-\xi)\,\mathrm{d}\xi, \qquad \gamma = \Gamma_H,\ \zeta = \gamma(b) \tag{6.18}$$

(with appropriate interpretation, as in (6.16)). However, it usually cannot be computed as the same integral over $\gamma = \Gamma_H$ itself, even when the latter integral is well-defined).

6.4.3 In view of Lemma 6.22, the proof of Theorem 6.18 will be complete if we prove the following purely geometric result:

Lemma 6.26. *For any path $\gamma\colon I = [0,1] \to \mathbb{C} \setminus \Omega$ such that $\gamma(0) \in \mathbb{D}^*_{\rho(\Omega)}$, there exists a symmetric Ω-homotopy H on $I \times I$ whose endpoint path is γ and whose initial path is a line-segment, i.e. $\Gamma_H = \gamma$ and $H_0(s) \equiv s\gamma(0)$.*

Proof. Assume that γ is given as in the hypothesis of Lemma 6.26. We are looking for a symmetric Ω-homotopy whose initial path is imposed: it must be

$$s \in I \mapsto H_0(s) := s\gamma(0),$$

which satisfies the three requirements of Definition 6.20 at $t = 0$:

(i) $H_0(0) = 0$,
(ii) $H_0\big((0,1]\big) \subset \mathbb{C} \setminus \Omega$,
(iii) $H_0(1) - H_0(s) = H_0(1-s)$ for every $s \in I$.

The idea is to define a family of maps $(\Psi_t)_{t\in[0,1]}$ so that

$$H_t(s) := \Psi_t\big(H_0(s)\big), \qquad s \in I, \tag{6.19}$$

yield the desired homotopy. For that, it is sufficient that $(t,\zeta) \in [0,1] \times \mathbb{C} \mapsto \Psi_t(\zeta)$ be continuously differentiable (for the structure of real two-dimensional vector space of $\mathbb{C}$), $\Psi_0 = \mathrm{id}$ and, for each $t \in [0,1]$,

(i') $\Psi_t(0) = 0$,
(ii') $\Psi_t(\mathbb{C} \setminus \Omega) \subset \mathbb{C} \setminus \Omega$,
(iii') $\Psi_t\big(\gamma(0) - \zeta\big) = \Psi_t\big(\gamma(0)\big) - \Psi_t(\zeta)$ for all $\zeta \in \mathbb{C}$,
(iv') $\Psi_t\big(\gamma(0)\big) = \gamma(t)$.

In fact, the properties (i')–(iv') ensure that any initial path H_0 satisfying (i)–(iii) and ending at $\gamma(0)$ produces through (6.19) a symmetric Ω-homotopy whose endpoint path is γ. Consequently, we may assume without loss of generality that γ is C^1 on $[0,1]$ (then, if γ is only piecewise C^1, we just need to concatenate the symmetric Ω-homotopies associated with the various pieces).

The maps Ψ_t will be generated by the flow of a non-autonomous vector field $X(\zeta,t)$ associated with γ that we now define. We view $(\mathbb{C}, |\cdot|)$ as a real 2-dimensional Banach space and pick[5] a C^1 function $\eta\colon \mathbb{C} \to [0,1]$ such that

$$\{\,\zeta \in \mathbb{C} \mid \eta(\zeta) = 0\,\} = \Omega.$$

Observe that $D(\zeta,t) := \eta(\zeta) + \eta\big(\gamma(t) - \zeta\big)$ defines a C^1 function of (ζ,t) which satisfies

$$D(\zeta,t) > 0 \quad \text{for all } \zeta \in \mathbb{C} \text{ and } t \in [0,1]$$

because Ω is stable under addition; indeed, $D(\zeta,t) = 0$ would imply $\zeta \in \Omega$ and $\gamma(t) - \zeta \in \Omega$, hence $\gamma(t) \in \Omega$, which would contradict our assumptions. Therefore, the formula

$$X(\zeta,t) := \frac{\eta(\zeta)}{\eta(\zeta) + \eta\big(\gamma(t) - \zeta\big)}\gamma'(t) \tag{6.20}$$

defines a non-autonomous vector field, which is continuous in (ζ,t) on $\mathbb{C} \times [0,1]$, C^1 in ζ and has its partial derivatives continuous in (ζ,t). The Cauchy-Lipschitz theorem on the existence and uniqueness of solutions to differential equations applies to $\frac{\mathrm{d}\zeta}{\mathrm{d}t} = X(\zeta,t)$: for every $\zeta \in \mathbb{C}$ and $t_0 \in [0,1]$ there is a unique solution $t \mapsto \Phi^{t_0,t}(\zeta)$ such that $\Phi^{t_0,t_0}(\zeta) = \zeta$. The fact that the vector field X is bounded implies that $\Phi^{t_0,t}(\zeta)$ is defined for all $t \in [0,1]$ and the classical theory guarantees that $(t_0,t,\zeta) \mapsto \Phi^{t_0,t}(\zeta)$ is C^1 on $[0,1] \times [0,1] \times \mathbb{C}$.

Let us set $\Psi_t := \Phi^{0,t}$ for $t \in [0,1]$ and check that this family of maps satisfies (i')–(iv'). We have

---

[5] For instance pick a C^1 function $\varphi_0\colon \mathbb{R} \to [0,1]$ such that $\{\,x \in \mathbb{R} \mid \varphi_0(x) = 1\,\} = \{0\}$ and $\varphi_0(x) = 0$ for $|x| \geq 1$, and a bijection $\omega\colon \mathbb{N} \to \Omega$; then set $\delta_k := \operatorname{dist}\big(\omega(k), \Omega \setminus \{\omega(k)\}\big) > 0$ and $\sigma(\zeta) := \sum_{k=0}^{\infty} \varphi_0\big(\frac{4|\zeta - \omega(k)|^2}{\delta_k^2}\big)$: for each $\zeta \in \mathbb{C}$ there is at most one non-zero term in this series (because $k \neq \ell$, $|\zeta - \omega(k)| < \delta_k/2$ and $|\zeta - \omega(\ell)| < \delta_\ell/2$ would imply $|\omega(k) - \omega(\ell)| < (\delta_k + \delta_\ell)/2$, which would contradict $|\omega(k) - \omega(\ell)| \geq \delta_k$ and δ_ℓ), thus σ is C^1, takes its values in $[0,1]$ and satisfies $\{\,\zeta \in \mathbb{C} \mid \sigma(\zeta) = 1\,\} = \Omega$, therefore $\eta := 1 - \sigma$ will do.

$$X(\omega,t) = 0 \quad \text{for all } \omega \in \Omega, \tag{6.21}$$

$$X\big(\gamma(t) - \zeta, t\big) = \gamma'(t) - X(\zeta,t) \quad \text{for all } \zeta \in \mathbb{C} \tag{6.22}$$

for all $t \in [0,1]$ (by the very definition of X). Therefore

- (i') and (ii') follow from (6.21) which yields $\Phi^{t_0,t}(\omega) = \omega$ for every t_0 and t, whence $\Psi_t(0) = 0$ since $0 \in \Omega$, and from the non-autonomous flow property $\Phi^{t,0} \circ \Phi^{0,t} = \mathrm{id}$ (hence $\Psi_t(\zeta) = \omega$ implies $\zeta = \Phi^{t,0}(\omega) = \omega$);
- (iv') follows from the fact that $X\big(\gamma(t),t\big) = \gamma'(t)$, by (6.21) and (6.22) with $\zeta = 0$, using again that $0 \in \Omega$, hence $t \mapsto \gamma(t)$ is a solution of X;
- (iii') follows from (6.22): for any solution $t \mapsto \zeta(t)$, the curve $t \mapsto \xi(t) := \gamma(t) - \zeta(t)$ satisfies $\xi(0) = \gamma(0) - \zeta(0)$ and $\xi'(t) = \gamma'(t) - X\big(\zeta(t),t\big) = X\big(\xi(t),t\big)$, hence it is a solution: $\xi(t) = \Psi_t\big(\gamma(0) - \zeta(0)\big)$.

As explained above, formula (6.19) thus produces the desired symmetric Ω-homotopy. □

6.4.4 *Note on this section:* The presentation we adopted is influenced by [CNP93] (the example of Section 6.3 is taken from this book). Lemma 6.26, which is the key to the proof of Theorem 6.18 and which essentially relies on the use of the flow of the non-autonomous vector field (6.20), arose as an attempt to understand a related but more complicated (somewhat obscure!) construction which can be found in an appendix of [CNP93]. See [Éca81], [Éca85] and [Ou10] for other approaches to the stability under convolution of the space of resurgent functions.

For the proof of Lemma 6.26, according to [Éca81] and [CNP93], one can visualize the realization of a given path γ as the enpoint path Γ_H of a symmetric Ω-homotopy as follows: Let a point $\zeta = \gamma(t)$ move along γ (as t varies from 0 to 1) and remain connected to 0 by an extensible thread, with moving nails pointing downwards at each point of $\zeta - \Omega$, while fixed nails point upwards at each point of Ω (imagine for instance that the first nails are fastened to a moving rule and the last ones to a fixed rule). As t varies, the thread is progressively stretched but it has to meander between the nails. The path H_1 used as integration path for $\mathrm{cont}_\gamma(\widehat{\varphi} * \widehat{\psi})(\gamma(1))$ in formula (6.18) is given by the thread in its final form, when ζ has reached the extremity of γ; the paths H_t correspond to the thread at intermediary stages. See Figure 6.5 (or Figure 5 of [Sau12]). The point is that none of the moving nails $\zeta - \omega' \in \zeta - \Omega$ will ever collide with a fixed nail $\omega'' \in \Omega$ because we assumed that γ avoids $\{\omega' + \omega''\} \subset \Omega$.

6.4.5 *Asymmetric version of the result.* Theorem 6.18 admits a useful generalization, concerning the convolution product of two resurgent germs which do not belong to the same space of Ω-continuable germs:

Theorem 6.27. *Let Ω_1 and Ω_2 be non-empty closed discrete subsets of $\mathbb{C}$. Let*

$$\Omega := \Omega_1 \cup \Omega_2 \cup (\Omega_1 + \Omega_2),$$

where $\Omega_1 + \Omega_2 := \{\, \omega_1 + \omega_2 \mid \omega_1 \in \Omega_1,\ \omega_2 \in \Omega_2 \,\}$. If Ω is closed and discrete, then

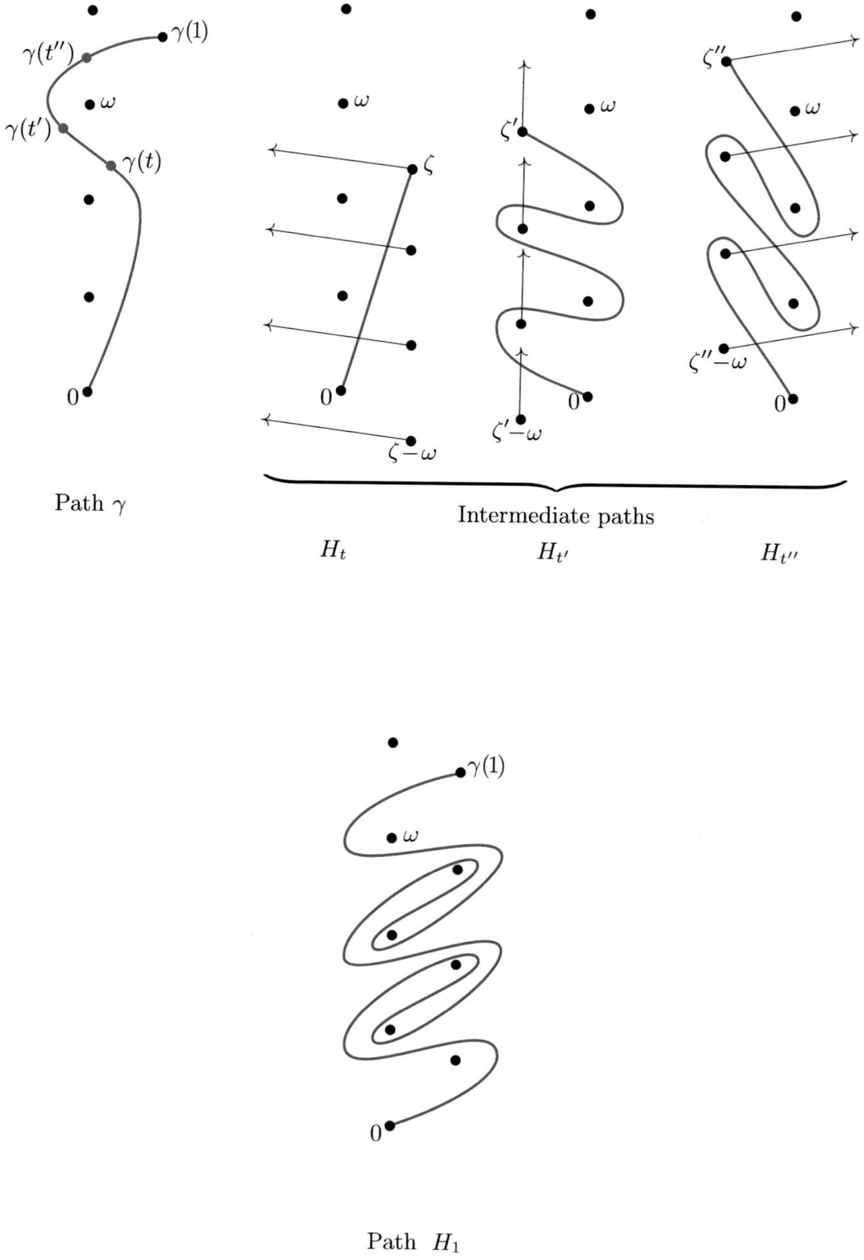

Fig. 6.5: From γ to the integration path H_1 used for $\mathrm{cont}_\gamma(\widehat{\varphi} * \widehat{\psi})(\gamma(1))$.

$$\widehat{\varphi} \in \widehat{\mathscr{R}}_{\Omega_1} \text{ and } \widehat{\psi} \in \widehat{\mathscr{R}}_{\Omega_2} \quad \Longrightarrow \quad \widehat{\varphi} * \widehat{\psi} \in \widehat{\mathscr{R}}_{\Omega}.$$

We shall content ourselves with giving hints about the proof when both Ω_1 and Ω_2 are assumed to contain 0, in which case

$$\Omega = \Omega_1 + \Omega_2$$

since both Ω_1 and Ω_2 are contained in $\Omega_1 + \Omega_2$ (the general case is obtained by adapting the arguments of [Sau13]). Assuming this, we generalize Definition 6.20 and Lemma 6.22:

Definition 6.28. A continuous map $H\colon I \times J \to \mathbb{C}$, where $I = [0,1]$ and J is a compact interval of $\mathbb{R}$, is called an (Ω_1, Ω_2)*-homotopy* if, for each $t \in J$, the paths $s \in I \mapsto H_t(s) := H(s,t)$ and $s \in I \mapsto H_t^*(s) := H_t(1) - H_t(1-s)$ satisfy

1. $H_t(0) = 0$,
2. $H_t\big((0,1]\big) \subset \mathbb{C} \setminus \Omega_1$,
3. $H_t^*\big((0,1]\big) \subset \mathbb{C} \setminus \Omega_2$.

We then call $t \in J \mapsto H_t(1)$ the *endpoint path* of H.

Lemma 6.29. *Let* $\gamma\colon [0,1] \to \mathbb{C} \setminus \Omega$ *be a path such that* $\gamma(0) \in \mathbb{D}_{\rho(\Omega_1)} \cap \mathbb{D}_{\rho(\Omega_2)}$. *Suppose that there exists an* (Ω_1, Ω_2)*-homotopy whose endpoint path coincides with* γ *and such that* $H_0(I) \subset \mathbb{D}_{\rho(\Omega_1)}$ *and* $H_0^*(I) \subset \mathbb{D}_{\rho(\Omega_2)}$. *Then any convolution product* $\widehat{\varphi} * \widehat{\psi}$ *with* $\widehat{\varphi} \in \widehat{\mathscr{R}}_{\Omega_1}$ *and* $\widehat{\psi} \in \widehat{\mathscr{R}}_{\Omega_2}$ *can be analytically continued along* γ.

Idea of the proof of Lemma 6.29. Mimick the proof of Lemma 6.22, replacing the right-hand side of (6.16) with

$$\int_0^1 (\operatorname{cont}_{H_{t|s}} \widehat{\varphi})(H_t(s))(\operatorname{cont}_{H^*_{t|1-s}} \widehat{\psi})(H_t^*(1-s) + \sigma) H_t'(s)\,\mathrm{d}s$$
$$+ \int_0^1 (\operatorname{cont}_{H_t} \widehat{\varphi})(\gamma(t) + u\sigma)\widehat{\psi}((1-u)\sigma)\sigma\,\mathrm{d}u$$

and showing that this expression is the value at $\gamma(t) + \sigma$ of a holomorphic germ, which is $\operatorname{cont}_{\gamma|t}(\widehat{\varphi} * \widehat{\psi})$. □

To conclude the proof of Theorem 6.27, it is thus sufficient to show

Lemma 6.30. *For any path* $\gamma\colon [0,1] \to \mathbb{C}$ *such that* $\gamma(0) \in \mathbb{D}^*_{\rho(\Omega_1)} \cap \mathbb{D}^*_{\rho(\Omega_2)}$ *and* $\gamma\big((0,1]\big) \subset \mathbb{C} \setminus \Omega$, *there exists an* (Ω_1, Ω_2)*-homotopy* H *on* $I \times [0,1]$ *whose endpoint path is* γ *and such that* $H_0(s) \equiv s\gamma(0)$.

Indeed, if this lemma holds true, then all such paths γ will be, by virtue of Lemma 6.29, paths of analytic continuation for our convolution products $\widehat{\varphi} * \widehat{\psi}$, which is the content of Theorem 6.27.

Idea of the proof of Lemma 6.30. It is sufficient to construct a family of maps $(\Psi_t)_{t \in [0,1]}$ such that $(t, \zeta) \in [0,1] \times \mathbb{C} \mapsto \Psi_t(\zeta) \in \mathbb{C}$ be continuously differentiable (for the structure of real two-dimensional vector space of $\mathbb{C}$), $\Psi_0 = \mathrm{id}$ and, for each $t \in [0,1]$,

(i') $\Psi_t(0) = 0$,
(ii') $\Psi_t(\mathbb{C} \setminus \Omega_1) \subset \mathbb{C} \setminus \Omega_1$,
(iii') the map $\zeta \in \mathbb{C} \mapsto \Psi_t^*(\zeta) := \gamma(t) - \Psi_t(\zeta)$ satisfies $\Psi_t^*(\mathbb{C} \setminus \Omega_2) \subset \mathbb{C} \setminus \Omega_2$,
(iv') $\Psi_t\big(\gamma(0)\big) = \gamma(t)$.

Indeed, the formula $H_t(s) := \Psi_t\big(s\gamma(0)\big)$ then yields the desired homotopy, with $H_t^*(s) = \Psi_t^*\big((1-s)\gamma(0)\big)$.

As in the proof of Lemma 6.26, the maps Ψ_t will be generated by the flow of a non-autonomous vector field associated with γ. We view $(\mathbb{C}, |\cdot|)$ as a real 2-dimensional Banach space and pick C^1 functions $\eta_1, \eta_2 \colon \mathbb{C} \to [0,1]$ such that

$$\{\zeta \in \mathbb{C} \mid \eta_j(\zeta) = 0\} = \Omega_j, \qquad j = 1,2.$$

Observe that $D(\zeta,t) := \eta_1(\zeta) + \eta_2\big(\gamma(t) - \zeta\big)$ defines a C^1 function of (ζ,t) which satisfies

$$D(\zeta,t) > 0 \quad \text{for all } \zeta \in \mathbb{C} \text{ and } t \in [0,1],$$

since $D(\zeta,t) = 0$ would imply $\zeta \in \Omega_1$ and $\gamma(t) - \zeta \in \Omega_2$, hence $\gamma(t) \in \Omega_1 + \Omega_2$, which would contradict our assumptions. Therefore, the formula

$$X(\zeta,t) := \frac{\eta_1(\zeta)}{\eta_1(\zeta) + \eta_2\big(\gamma(t) - \zeta\big)} \gamma'(t) \tag{6.23}$$

defines a non-autonomous vector field and the Cauchy-Lipschitz theorem applies to $\frac{\mathrm{d}\zeta}{\mathrm{d}t} = X(\zeta,t)$: for every $\zeta \in \mathbb{C}$ and $t_0 \in [0,1]$ there is a unique solution $t \in [0,1] \mapsto \Phi_X^{t_0,t}(\zeta)$ such that $\Phi_X^{t_0,t_0}(\zeta) = \zeta$; the flow map $(t_0,t,\zeta) \mapsto \Phi_X^{t_0,t}(\zeta)$ is C^1 on $[0,1] \times [0,1] \times \mathbb{C}$.

Setting $\Psi_t := \Phi_X^{0,t}$ for $t \in [0,1]$, one can check that this family of maps satisfies (i')–(iv') by mimicking the arguments in the proof of Lemma 6.26 and using the fact that the corresponding family of maps (Ψ_t^*) in (iii') can be obtained from the identity

$$\gamma(t) - \Phi_X^{0,t}(\zeta) = \Phi_{X^*}^{0,t}\big(\gamma(0) - \zeta\big),$$

where we denote by $(t_0,t,\zeta) \mapsto \Phi_{X^*}^{t_0,t}(\zeta)$ the flow map of the non-autonomous vector field

$$X^*(\zeta,t) := \gamma'(t) - X\big(\gamma(t) - \zeta, t\big) = \frac{\eta_2(\zeta)}{\eta_1\big(\gamma(t) - \zeta\big) + \eta_2(\zeta)} \gamma'(t).$$

□

6.5 Non-linear operations with resurgent formal series

From now on, we give ourselves a non-empty closed discrete subset Ω of $\mathbb{C}$ which is stable under addition.

We already mentioned the stability of $\widetilde{\mathscr{R}}_\Omega$ under certain linear difference or differential operators in Lemma 6.14. Now, with our assumption that Ω is stable under addition, we can obtain the stability of Ω-resurgent formal series under the non-linear operations which were studied in Sections 5.13 and 5.17. However this requires quantitative estimates for iterated convolutions whose proof is beyond the scope of the present text, we thus quote without proof the following

Lemma 6.31. *Let γ be a path of $\mathbb{C}\setminus\Omega$ starting from a point $\zeta_0 \in \mathbb{D}_{\rho(\Omega)}$ and ending at a point ζ_1. Let $R>0$ be such that $\overline{D(\zeta_1,R)} \subset \mathbb{C}\setminus\Omega$. Then there exist a positive number L and a set $\mathscr{C}$ of paths parametrized by $[0,1]$ and contained in $\mathbb{D}_L \setminus \Omega$ such that, for every $\widehat{\varphi} \in \widehat{\mathscr{R}}_\Omega$, the number*

$$\|\widehat{\varphi}\|_{\mathscr{C}} := \sup_{\widetilde{\gamma}\in\mathscr{C}} \left|\operatorname{cont}_{\widetilde{\gamma}} \widehat{\varphi}\big(\widetilde{\gamma}(1)\big)\right|$$

is finite, and there exist $A,B>0$ such that, for every $k \geq 1$ and $\widehat{\varphi},\widehat{\psi} \in \widehat{\mathscr{R}}_\Omega$, the iterated convolution products

$$\widehat{\varphi}^{*k} := \underbrace{\widehat{\varphi} * \cdots * \widehat{\varphi}}_{k \text{ factors}}$$

and $\widehat{\psi} * \widehat{\varphi}^{*k}$ (which admit analytic continuation along γ, according to Theorem 6.18) satisfy

$$|\operatorname{cont}_\gamma \widehat{\varphi}^{*k}(\zeta)| \leq A\frac{B^k}{k!}\big(\|\widehat{\varphi}\|_{\mathscr{C}}\big)^k,$$

$$|\operatorname{cont}_\gamma(\widehat{\psi} * \widehat{\varphi}^{*k})(\zeta)| \leq A\frac{B^k}{k!}\|\widehat{\psi}\|_{\mathscr{C}}\big(\|\widehat{\varphi}\|_{\mathscr{C}}\big)^k,$$

for every $\zeta \in \overline{D(\zeta_1,R)}$.

The proof can be found in [Sau15]. Taking this result for granted, we can show

Theorem 6.32. *Suppose that $\widetilde{\varphi}(z),\widetilde{\psi}(z),\widetilde{\chi}(z) \in \widetilde{\mathscr{R}}_\Omega$ and that $\widetilde{\chi}(z)$ has no constant term. Let $H(t) \in \mathbb{C}\{t\}$. Then*

$$\widetilde{\psi}\circ(\mathrm{id}+\widetilde{\varphi}) \in \widetilde{\mathscr{R}}_\Omega, \qquad H\circ\widetilde{\chi} \in \widetilde{\mathscr{R}}_\Omega. \tag{6.24}$$

Proof. We can write $\widetilde{\varphi} = a + \widetilde{\varphi}_1$, $\widetilde{\psi} = b + \widetilde{\psi}_1$, where $a,b \in \mathbb{C}$ and $\widetilde{\varphi}_1$ and $\widetilde{\psi}_1$ have no constant term. With notations similar to those of the proof of Theorem 5.55, we write the first formal series in (6.24) as $b+\widetilde{\lambda}(z)$ and the second one as $c+\widetilde{\mu}(z)$, where $c = H(0)$. Since $\widetilde{\lambda} = (T_a\widetilde{\psi}_1)\circ(\mathrm{id}+\widetilde{\varphi}_1)$, where $T_a\widetilde{\psi}_1$ is Ω-resurgent (by Lemma 6.14) and has no constant term, we see that it is sufficient to deal with the case $a = b = 0$; from now on we thus suppose $\widetilde{\varphi} = \widetilde{\varphi}_1$ and $\widetilde{\psi} = \widetilde{\psi}_1$. Then

$$\widetilde{\lambda} = \widetilde{\psi} \circ (\mathrm{id} + \widetilde{\varphi}) = \sum_{k \geq 0} \frac{1}{k!} (\partial^k \widetilde{\psi}) \widetilde{\varphi}^k, \qquad \widetilde{\mu} = \sum_{k \geq 1} h_k \widetilde{\chi}^k$$

where $H(t) = c + \sum_{k \geq 1} h_k t^k$ with $|h_k| \leq CD^k$ for some $C, D > 0$ independent of k, and the corresponding formal Borel transforms are

$$\widehat{\lambda} = \sum_{k \geq 0} \frac{1}{k!} \big((-\zeta)^k \widehat{\psi}\big) * \widehat{\varphi}^{*k}, \qquad \widehat{\mu} = \sum_{k \geq 1} h_k \widehat{\chi}^{*k}.$$

These can be viewed as formally convergent series of elements of $\mathbb{C}[[\zeta]]$, in which each term belongs to $\widehat{\mathscr{R}}_\Omega$ (by virtue of Theorem 6.18). They define holomorphic germs in $\mathbb{D}_{\rho(\Omega)}$ because they can also be seen as normally convergent series of holomorphic functions in any compact disc contained in $\mathbb{D}_{\rho(\Omega)}$ (by virtue of inequalities (5.79) and (5.81)).

To conclude, it is sufficient to check that, given a path $\gamma\colon [0,1] \to \mathbb{C} \setminus \Omega$ starting in $\mathbb{D}_{\rho(\Omega)}$, for every $t \in [0,1]$ and $R_t > 0$ such that $\overline{D(\gamma(t), R_t)} \subset \mathbb{C} \setminus \Omega$ the series of holomorphic functions

$$\sum \frac{1}{k!} \operatorname{cont}_{\gamma|[0,t]} \Big(\big((-\zeta)^k \widehat{\psi}\big) * \widehat{\varphi}^{*k} \Big) \quad \text{and} \quad \sum h_k \operatorname{cont}_{\gamma|[0,t]} \big(\widehat{\chi}^{*k}\big)$$

are normally convergent on $\overline{D(\gamma(t), R_t)}$ (indeed, this will provide families of analytic elements which analytically continue $\widehat{\lambda}$ and $\widehat{\mu}$). This follows from Lemma 6.31. □

Example 6.33. In view of Lemma 6.5, since $2\pi\mathrm{i}\mathbb{Z}$ is stable under addition, this implies that the exponential of the Stirling series $\widetilde{\lambda} = \mathrm{e}^{\widetilde{\mu}}$ is $2\pi\mathrm{i}\mathbb{Z}$-resurgent.

Recall that $\widetilde{\mathscr{G}} = \mathrm{id} + \mathbb{C}[[z^{-1}]]$ is the topological group of formal tangent-to-identity diffeomorphisms at ∞ studied in Section 5.15.

Definition 6.34. We call *Ω-resurgent tangent-to-identity diffeomorphism* any $\widetilde{f} = \mathrm{id} + \widetilde{\varphi} \in \widetilde{\mathscr{G}}$ where $\widetilde{\varphi}$ is an Ω-resurgent formal series. We use the notations

$$\widetilde{\mathscr{G}}^{\mathrm{RES}}(\Omega) := \{ \widetilde{f} = \mathrm{id} + \widetilde{\varphi} \mid \widetilde{\varphi} \in \widetilde{\mathscr{R}}_\Omega \}, \qquad \widetilde{\mathscr{G}}^{\mathrm{RES}}_\sigma(\Omega) := \widetilde{\mathscr{G}}^{\mathrm{RES}}(\Omega) \cap \widetilde{\mathscr{G}}_\sigma \ \text{ for } \sigma \in \mathbb{C}.$$

Observe that $\widetilde{\mathscr{G}}^{\mathrm{RES}}(\Omega)$ is not a closed subset of $\widetilde{\mathscr{G}}$ for the topology which was introduced in Section 5.15; in fact it is dense, since it contains the subset $\mathscr{G}$ of holomorphic tangent-to-identity germs of diffeomorphisms at ∞, which itself is dense in $\widetilde{\mathscr{G}}$.

Theorem 6.35. *The set $\widetilde{\mathscr{G}}^{\mathrm{RES}}(\Omega)$ is a subgroup of $\widetilde{\mathscr{G}}$, the set $\widetilde{\mathscr{G}}^{\mathrm{RES}}_0(\Omega)$ is a subgroup of $\widetilde{\mathscr{G}}_0$.*

Proof. The stability under group composition stems from Theorem 6.32, since $(\mathrm{id}+\widetilde{\psi})\circ(\mathrm{id}+\widetilde{\varphi}) = \mathrm{id}+\widetilde{\varphi}+\widetilde{\psi}\circ(\mathrm{id}+\widetilde{\varphi})$.

For the stability under group inversion, we only need to prove

$$\widetilde{h} = \mathrm{id}+\widetilde{\chi} \in \widetilde{\mathscr{G}}^{\mathrm{RES}}(\Omega) \quad \Longrightarrow \quad \widetilde{h}^{\circ(-1)} \in \widetilde{\mathscr{G}}^{\mathrm{RES}}(\Omega).$$

It is sufficient to prove this when $\widetilde{\chi}$ has no constant term, i.e. when $\widetilde{h} \in \widetilde{\mathscr{G}}_0^{\mathrm{RES}}(\Omega)$, since we can always write $\widetilde{h} = (\mathrm{id}+\widetilde{\chi}_1)\circ(\mathrm{id}+a)$ with a formal series $\widetilde{\chi}_1 = T_{-a}(-a+\widetilde{\chi}) \in \widetilde{\mathscr{R}}_\Omega$ which has no constant term (taking $a =$ constant term of $\widetilde{\chi}$ and using Lemma 6.14) and then $\widetilde{h}^{\circ(-1)} = (\mathrm{id}+\widetilde{\chi}_1)^{\circ(-1)} - a$.

We thus assume that $\widetilde{\chi} = \widetilde{\chi}_1 \in \widetilde{\mathscr{R}}_\Omega$ has no constant term and apply the Lagrange reversion formula (5.90) to $\widetilde{h} = \mathrm{id}+\widetilde{\chi}$. We get $\widetilde{h}^{\circ(-1)} = \mathrm{id}-\widetilde{\varphi}$ with the Borel transform of $\widetilde{\varphi}$ given by

$$\widehat{\varphi} = \sum_{k\geq 1} \frac{\zeta^{k-1}}{k!}\widehat{\chi}^{*k},$$

formally convergent series in $\mathbb{C}[[\zeta]]$, in which each term belongs to $\widehat{\mathscr{R}}_\Omega$. The holomorphy of $\widehat{\varphi}$ in $\mathbb{D}_{\rho(\Omega)}$ and its analytic continuation along the paths of $\mathbb{C}\setminus\Omega$ are obtained by invoking inequalities (5.94) and Lemma 6.31, similarly to what we did at the end of the proof of Theorem 6.32. $\square$

6.6 Singular points

When the analytic continuation of a holomorphic germ $\overset{\wedge}{\varphi}(\zeta)$ has singularities (i.e. $\overset{\wedge}{\varphi}$ does not extend to an entire function), its inverse formal Borel transform $\widetilde{\varphi} = \mathscr{B}^{-1}\overset{\wedge}{\varphi}$ is a divergent formal series, and the location and the nature of the singularities in the ζ-plane influence the growth of the coefficients of $\widetilde{\varphi}$. By analysing carefully the singularities of $\overset{\wedge}{\varphi}$, one may hope to be able to deduce subtler information on $\widetilde{\varphi}$ and, if Borel-Laplace summation is possible, on its Borel sums.

Therefore, we shall now develop a theory which allows one to study and manipulate singularities (in the case of isolated singular points).

First, recall the definition of a *singular point* in complex analysis: given f holomorphic in an open subset U of $\mathbb{C}$, a boundary point ω of U is said to be a singular point of f if one cannot find an open neighbourhood V of ω, a function g holomorphic in V, and an open subset U' of U such that $\omega \in \partial U'$ and $f_{|U'\cap V} = g_{|U'\cap V}$.

Thus this notion refers to the imposssibility of extending locally the function: even when restricting to a smaller domain U' to which ω is adherent, we cannot find an analytic continuation in a full neighbourhood of ω. Think of the example of the principal branch of logarithm: it can be defined as the holomorphic function

$$\mathrm{Log}\,\zeta := \int_1^\zeta \frac{\mathrm{d}\xi}{\xi} \quad \text{for } \zeta \in U = \mathbb{C}\setminus\mathbb{R}^-. \tag{6.25}$$

Then, for $\omega < 0$, one cannot find a holomorphic extension of $f = \operatorname{Log}$ from U to any larger open set containing ω (not even a continuous extension!), however such a point ω is not singular: if we first restrict, say, to the upper half-plane $U' := \{\Im m\,\zeta > 0\}$, then we can easily find an analytic continuation of $\operatorname{Log}_{|U'}$ to $U' \cup V$, where V is the disc $D(\omega, |\omega|)$: define g by

$$g(\zeta) = \left(\int_\gamma + \int_\omega^\zeta \right) \frac{\mathrm{d}\xi}{\xi}$$

with any path $\gamma\colon\ [0,1] \to \mathbb{C}$ such that $\gamma(0) = 1$, $\gamma\big((0,1)\big) \subset U'$ and $\gamma(1) = \omega$. In fact, for the function $f = \operatorname{Log}$, the only singular point is 0, there is no other local obstacle to analytic continuation, even though there is no holomorphic extension of this function to $\mathbb{C}^*$.

If ω is an isolated[6] singular point for a holomorphic function f, we can wonder what kind of *singularity* occurs at this point. There are certainly many ways for a point to be singular: maybe the function near ω looks like $\log(\zeta - \omega)$ (for an appropriate branch of the logarithm), or like a pole $\frac{C}{(\zeta-\omega)^m}$, and the reader can imagine many other singular behaviours (square-root branching $(\zeta - \omega)^{1/2}$, powers of logarithm $\big(\log(\zeta - \omega)\big)^m$, iterated logarithms $\log\big(\log(\zeta - \omega)\big)$, etc.). The singularity of f at ω will be defined as an equivalence class modulo regular functions in Section 6.8. Of course, by translating the variable, we can always assume $\omega = 0$. Observe that, in this text, we make a distinction between singular points and singularities (the former being the locations of the latter).

As a preliminary, we need to introduce a few notations in relation with the Riemann surface of the logarithm.

6.7 The Riemann surface of the logarithm

The Riemann surface of the logarithm $\widetilde{\mathbb{C}}$ can be defined topologically (without any reference to the logarithm!) as the universal cover of $\mathbb{C}^*$ with base point at 1. This means that we consider the set $\mathscr{P}$ of all paths[7] $\gamma\colon\ [0,1] \to \mathbb{C}^*$ with $\gamma(0) = 1$, we put on $\mathscr{P}$ the equivalence relation $\sim$ of "homotopy with fixed endpoints", i.e.

$$\gamma \sim \gamma_0 \quad \Longleftrightarrow \quad \left| \begin{array}{l} \exists H\colon\ [0,1] \times [0,1] \to \mathbb{C}^* \text{ continuous, such that} \\ H(0,t) = \gamma_0(t) \text{ and } H(1,t) = \gamma(t) \text{ for each } t \in [0,1], \\ H(s,0) = \gamma_0(0) \text{ and } H(s,1) = \gamma_0(1) \text{ for each } s \in [0,1], \end{array} \right.$$

6 As a rule, all the singular points that we shall encounter in resurgence theory will be isolated even when the same holomorphic function f is considered in various domains U (i.e. no "natural boundary" will show up). This does not mean that our functions will extend in punctured dics centred on the singular points, because there may be "monodromy": leaving the original domain of definition U' on one side of ω or the other may lead to different analytic continuations.

7 In this section, "path" means any continuous $\mathbb{C}$-valued map defined on $[0,1]$.

and we define $\widetilde{\mathbb{C}}$ as the set of all equivalence classes,

$$\widetilde{\mathbb{C}} := \mathscr{P}/\sim.$$

Observe that, if $\gamma \sim \gamma_0$, then $\gamma(1) = \gamma_0(1)$: the endpoint $\gamma(1)$ does not depend on γ but only on its equivalence class $[\gamma]$. We thus get a map

$$\pi\colon \widetilde{\mathbb{C}} \to \mathbb{C}^*, \qquad \pi(\zeta) = \gamma(1) \text{ for any } \gamma \in \mathscr{P} \text{ such that } [\gamma] = \zeta$$

(recall that the other endpoint is the same for all paths $\gamma \in \mathscr{P}$: $\gamma(0) = 1$).

Among all the representatives of an equivalence class $\zeta \in \widetilde{\mathbb{C}}$, there is a canonical one: there exists a unique pair $(r,\theta) \in (0,+\infty) \times \mathbb{R}$ such that ζ is represented by the concatenation[8] of the paths $t \in [0,1] \mapsto \mathrm{e}^{\mathrm{i}t\theta}$ and $t \in [0,1] \mapsto \big(1+t(r-1)\big)\mathrm{e}^{\mathrm{i}\theta}$. In that situation, we use the notations

$$\zeta = r\underline{\mathrm{e}}^{\mathrm{i}\theta}, \qquad r = |\zeta|, \qquad \theta = \arg\zeta, \tag{6.26}$$

so that we can write $\pi(r\underline{\mathrm{e}}^{\mathrm{i}\theta}) = r\mathrm{e}^{\mathrm{i}\theta}$. Heuristically, one may think of $\theta \mapsto \underline{\mathrm{e}}^{\mathrm{i}\theta}$ as of a non-periodic exponential: it keeps track of the number of turns around the origin, not only of the angle θ modulo 2π.

There is a simple way of defining a Riemann surface structure on $\widetilde{\mathbb{C}}$. One first defines a Hausdorff topology on $\widetilde{\mathbb{C}}$ by taking as a basis $\big\{\widetilde{D}(\zeta,R) \mid \zeta \in \widetilde{\mathbb{C}},\ 0 < R < |\pi(\zeta)|\big\}$, where $\widetilde{D}(\zeta,R)$ is the set of the equivalence classes of all paths γ obtained as concatenation of a representative of ζ and a line-segment[9] starting from $\pi(\zeta)$ and contained in $D(\pi(\zeta),R)$ (notation (6.1)). (Exercise: check that this is legitimate, i.e. that $\big\{\widetilde{D}(\zeta,R)\big\}$ is a collection of subsets of $\widetilde{\mathbb{C}}$ which meets the necessary conditions for being the basis of a topology, and check that the resulting topology satisfies the Hausdorff separation axiom.) It is easy to check that, for each basis element, the projection π induces a homeomorphism $\pi_{\zeta,R}\colon \widetilde{D}(\zeta,R) \to D(\pi(\zeta),R)$ and that, for each pair of basis elements with non-empty intersection, the transition map $\pi_{\zeta',R'} \circ \pi_{\zeta,R}^{-1}$ is the identity map on $D(\pi(\zeta),R) \cap D(\pi(\zeta'),R') \subset \mathbb{C}$, hence we get an atlas $\{\pi_{\zeta,R}\}$ which defines a Riemann surface structure on $\widetilde{\mathbb{C}}$, i.e. a 1-dimensional complex manifold structure (because the identity map is holomorphic!).

Now, why do we call $\widetilde{\mathbb{C}}$ the Riemann surface of the logarithm? This is not so apparent in the presentation that was adopted here, but in fact the above construction is related to a more general one, in which one starts with an arbitrary open connected subset U of $\mathbb{C}$ and a holomorphic function f on U, and one constructs (by quotienting a certain set of paths) a Riemann surface in which U is embedded and on which f has a holomorphic extension. We shall not give the details, but content ourselves with checking the last property for

$$U = \mathbb{C} \setminus \mathbb{R}^-$$

---

[8] Here we use the variant of concatenation in which the parametrisation interval is always $[0,1]$, as described in Remark 6.24.

[9] Using the same convention as in footnote 4 for the parametrisation of line-segments.

and $f = \text{Log}$ defined by (6.25): we shall define a holomorphic function $\overset{\vee}{\mathscr{L}} : \widetilde{\mathbb{C}} \to \mathbb{C}$ and explain why it deserves to be considered as a holomorphic extension of the logarithm.

We first observe that $\widetilde{U} := \pi^{-1}(U)$ is an open subset of $\widetilde{\mathbb{C}}$ with infinitely many connected components,

$$\widetilde{U}_m := \{ r\underline{e}^{i\theta} \in \widetilde{\mathbb{C}} \mid r > 0,\ 2\pi m - \pi < \theta < 2\pi m + \pi \}, \qquad m \in \mathbb{Z}.$$

By restriction, the projection π induces a biholomorphism

$$\pi_0 \colon \widetilde{U}_0 \xrightarrow{\sim} U$$

(it does so for any $m \in \mathbb{Z}$ but, quite arbitrarily, we choose $m = 0$ here). The *principal sheet of the Riemann surface of the logarithm* is defined to be the set $\widetilde{U}_0 \subset \widetilde{\mathbb{C}}$, which is identified to the cut plane $U \subset \mathbb{C}$ by means of π_0.

On the other hand, since the function $\xi \mapsto 1/\xi$ is holomorphic on $\mathbb{C}^*$, the Cauchy theorem guarantees that, for any $\gamma \in \mathscr{P}$, the integral $\int_\gamma \frac{d\xi}{\xi}$ depends only on the equivalence class $[\gamma]$, we thus get a function

$$\overset{\vee}{\mathscr{L}} \colon \widetilde{\mathbb{C}} \to \mathbb{C}, \qquad \overset{\vee}{\mathscr{L}}([\gamma]) := \int_\gamma \frac{d\xi}{\xi}.$$

This function is holomorphic on the whole of $\widetilde{\mathbb{C}}$, because its expression in any chart domain $\widetilde{D}(\zeta_0, R)$ is

$$\overset{\vee}{\mathscr{L}}(\zeta) = \overset{\vee}{\mathscr{L}}(\zeta_0) + \int_{\pi(\zeta_0)}^{\pi(\zeta)} \frac{d\xi}{\xi},$$

which is a holomorphic function of $\pi(\zeta)$.

Now, since any $\zeta \in \widetilde{U}_0$ can be represented by a line-segment starting from 1, we have

$$\overset{\vee}{\mathscr{L}}_{|\widetilde{U}_0} = \text{Log} \circ \pi_0.$$

In other words, if we identify U and $\widetilde{U}_0$ by means of π_0, we can view $\overset{\vee}{\mathscr{L}}$ as a holomorphic extension of Log to the whole of $\widetilde{\mathbb{C}}$.

The function $\overset{\vee}{\mathscr{L}}$ is usually denoted by log and this is the notation we will adopt from now on:

$$\log([\gamma]) := \int_\gamma \frac{d\xi}{\xi}, \qquad \gamma \in \mathscr{P}.$$

Notice that $\log(r\underline{e}^{i\theta}) = \ln r + i\theta$ for all $r > 0$ and $\theta \in \mathbb{R}$, and that $\log \colon \widetilde{\mathbb{C}} \to \mathbb{C}$ is a biholomorphism (with our notations, the inverse map is $x + iy \mapsto e^x \underline{e}^{iy}$). Notice also that there is a natural multiplication $(r_1 \underline{e}^{i\theta_1}, r_2 \underline{e}^{i\theta_2}) \mapsto r_1 r_2 \underline{e}^{i(\theta_1+\theta_2)}$ in $\widetilde{\mathbb{C}}$, for which log appears as a group isomorphism.

Observe that, if $\mathscr{L}$ is a branch of the logarithm holomorphic on an open connected set $V \subset \mathbb{C}$ in the sense of footnote 2, then $\pi^{-1}(V) \subset \widetilde{\mathbb{C}}$ has infinitely many connected components and $\mathscr{L} \circ \pi$ coincides with log on one of them.

6.8 The formalism of singularities

We are interested in holomorphic functions f for which the origin is locally the only singular point in the following sense:

Definition 6.36. We say that a function f *has spiral continuation around* 0 if it is holomorphic in an open disc D to which 0 is adherent and, for every $L>0$, there exists $\rho>0$ such that f can be analytically continued along any path of length $\leq L$ starting from $D\cap\mathbb{D}^*_\rho$ and staying in $\mathbb{D}^*_\rho$ (recall the notation (6.2)). See Figure 6.6.

In the following we shall need to single out one of the connected components of $\pi^{-1}(D)$ in $\widetilde{\mathbb{C}}$, but there is no canonical choice in general. (If one of the connected components is contained in the principal sheet of $\widetilde{\mathbb{C}}$, we may be tempted to choose this one, but this does not happen when the centre of D has negative real part and we do not want to eliminate a priori this case.) We thus choose $\zeta_0\in\widetilde{\mathbb{C}}$ such that $\pi(\zeta_0)$ is the centre of D, then the connected component of $\pi^{-1}(D)$ which contains ζ_0 is a domain $\widetilde{D}$ of the form $\widetilde{D}(\zeta_0,R_0)$ (notation of the previous section) and this will be the connected component that we single out.

Since π induces a biholomorphism $\widetilde{D}\xrightarrow{\sim}D$, we can identify f with $\overset{\vee}{f}:=f\circ\pi$ viewed as a holomorphic function on $\widetilde{D}$. Now, the spiral continuation property implies that $\overset{\vee}{f}$ extends analytically to a domain of the form

$$\mathscr{V}(h):=\{\,\zeta=r\underline{\mathrm{e}}^{\mathrm{i}\theta}\mid 0<r<h(\theta),\ \theta\in\mathbb{R}\,\}\subset\widetilde{\mathbb{C}},$$

with a continuous function $h\colon\ \mathbb{R}\to(0,+\infty)$, but in fact the precise function h is of no interest to us.[10] We are thus led to

Definition 6.37. We define the space ANA of all *singular germs* as follows: on the set of all pairs $(\overset{\vee}{f},h)$, where $h\colon\ \mathbb{R}\to(0,+\infty)$ is continuous and $\overset{\vee}{f}\colon\ \mathscr{V}(h)\to\mathbb{C}$ is holomorphic, we put the equivalence relation

$$(\overset{\vee}{f}_1,h_1)\sim(\overset{\vee}{f}_2,h_2)\quad\overset{\text{def}}{\Longleftrightarrow}\quad\overset{\vee}{f}_1\equiv\overset{\vee}{f}_2\text{ on }\mathscr{V}(h_1)\cap\mathscr{V}(h_2),$$

and we define ANA as the quotient set.

Heuristically, one may think of a singular germ as of a "germ of holomorphic function at the origin of $\widetilde{\mathbb{C}}$" (except that $\widetilde{\mathbb{C}}$ has no origin!). We shall usually make no notational difference between an element of ANA and any of its representatives. As explained above, the formula $f=\overset{\vee}{f}\circ\pi$ allows one to identify a singular germ $\overset{\vee}{f}$ with a function f which has spiral continuation around 0; however, one must be aware

[10] Observe that there is a countable infinity of choices for ζ_0 (all the possible "lifts" of the centre of D in $\widetilde{\mathbb{C}}$) thus, a priori, infinitely many different functions $\overset{\vee}{f}$ associated with the same function f; they are all of the form $\overset{\vee}{f}(\zeta\,\underline{\mathrm{e}}^{2\pi\mathrm{i}m})$, $m\in\mathbb{Z}$, where $\overset{\vee}{f}(\zeta)$ is one of them, so that if $\overset{\vee}{f}$ is holomorphic in a domain of the form $\mathscr{V}(h)$ then each of them is holomorphic in a domain of this form.

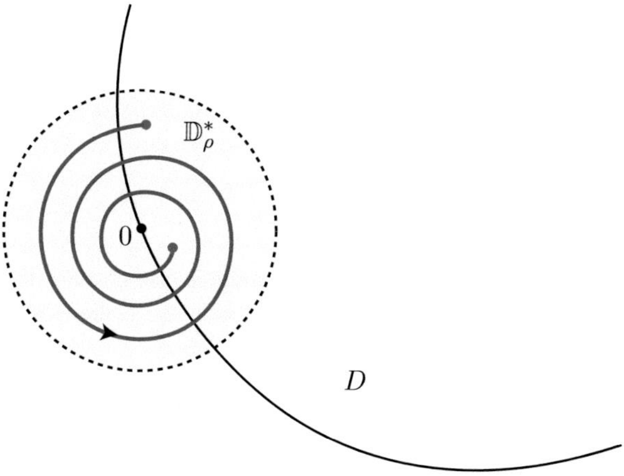

Fig. 6.6: The function f is holomorphic in D and has spiral continuation around 0.

that this presupposes an identification, by means of π, between a simply connected domain D of $\mathbb{C}^*$ (e.g. an open disc) and a subset $\widetilde{D}$ of a domain of the form $\mathscr{V}(h)$ (and, given D, there are countably many choices for $\widetilde{D}$).

Example 6.38. Suppose that f is holomorphic in the punctured disc $\mathbb{D}^*_\rho$, for some $\rho > 0$; in particular, it is holomorphic in $D = D(\frac{\rho}{2}, \frac{\rho}{2})$ and we can apply the above construction. Then, for whatever choice of a connected component of $\pi^{-1}(D)$ in $\widetilde{\mathbb{C}}$, we obtain the same $\overset{\vee}{f} := f \circ \pi$ holomorphic in $\mathscr{V}(h)$ with a constant function $h(\theta) \equiv \rho$. The corresponding element of ANA identifies itself with the Laurent series of f at 0, which is of the form

$$\sum_{n\in\mathbb{Z}} a_n \zeta^n = S(1/\zeta) + R(\zeta), \tag{6.27}$$

with $R(\zeta) := \sum_{n\geq 0} a_n \zeta^n$ of radius of convergence $\geq \rho$ and $S(\xi) := \sum_{n>0} a_{-n} \xi^n$ of infinite radius of convergence. Heuristically, the "singularity of f" is encoded by the sole term $S(1/\zeta)$; Definition 6.43 will formalize the idea of discarding the regular term $R(\zeta)$.

Example 6.39. Suppose that f is of the form $f(\zeta) = \overset{\wedge}{\varphi}(\zeta) \operatorname{Log} \zeta$, where $\overset{\wedge}{\varphi}$ is holomorphic in the disc $\mathbb{D}_\rho$, for some $\rho > 0$, and we are using the principal branch of the logarithm. Then we may define $\overset{\vee}{f}(\zeta) := \overset{\wedge}{\varphi}\big(\pi(\zeta)\big) \log \zeta$ for $\zeta \in \mathscr{V}(h)$ with a constant function $h(\theta) \equiv \rho$; this corresponds to the situation described above with $D = D(\frac{\rho}{2}, \frac{\rho}{2})$ and $\widetilde{D} =$ the connected component of $\pi^{-1}(D)$ which is contained in the principal sheet of $\widetilde{\mathbb{C}}$ (choosing some other connected component for $\widetilde{D}$ would have resulted in adding to the above $\overset{\vee}{f}$ an integer multiple of $2\pi\mathrm{i}\, \overset{\wedge}{\varphi} \circ \pi$). The corresponding element of ANA identifies itself with

$$\Big(\sum_{n\geq 0} a_n \zeta^n\Big) \log\zeta,$$

where $\sum_{n\geq 0} a_n \zeta^n$ is the Taylor series of $\overset{\wedge}{\varphi}$ at 0 (which has radius of convergence $\geq \rho$).

Example 6.40. For $\alpha \in \mathbb{C}^*$ we define "the principal branch of ζ^α" as $\mathrm{e}^{\alpha \operatorname{Log} \zeta}$ for $\zeta \in \mathbb{C} \setminus \mathbb{R}^-$. If we choose D and $\widetilde{D}$ as in Example 6.39, then the corresponding singular germ is

$$\zeta^\alpha := \mathrm{e}^{\alpha \log \zeta},$$

which extends holomorphically to the whole of $\widetilde{\mathbb{C}}$. One can easily check that 0 is a singular point for ζ^α if and only if $\alpha \notin \mathbb{N}$.

Exercise 6.41. Consider a power series $\sum_{n\geq 0} a_n \xi^n$ with *finite* radius of convergence $R > 0$ and denote by $\Phi(\xi)$ its sum for $\xi \in \mathbb{D}_R$. Prove that there exists $\rho > 0$ such that

$$f(\zeta) := \Phi(\zeta \operatorname{Log} \zeta)$$

is holomorphic in the half-disc $\mathbb{D}_\rho \cap \{\Re e\, \zeta > 0\}$ and that 0 is a singular point. Prove that f has spiral continuation around 0. Consider any function $\overset{\vee}{f}$ associated with f as above; prove that one cannot find a *constant* function h such that $\overset{\vee}{f}$ is holomorphic in $\mathscr{V}(h)$.

Exercise 6.42. Let $\alpha \in \mathbb{C}^*$ and

$$f_\alpha(\zeta) := \frac{1}{\zeta^\alpha - \zeta^{-\alpha}}$$

(notation of Example 6.40). Prove that f_α has spiral continuation around 0 if and only if $\alpha \notin \mathrm{i}\mathbb{R}$. Suppose that α is not real nor pure imaginary and consider any function $\overset{\vee}{f}_\alpha$ associated with f_α as above; prove that one cannot find a constant function h such that $\overset{\vee}{f}_\alpha$ is holomorphic in $\mathscr{V}(h)$.

The set ANA is clearly a linear space which contains $\mathbb{C}\{\zeta\}$, in the sense that there is a natural injective linear map $\mathbb{C}\{\zeta\} \hookrightarrow$ ANA (particular case of Example 6.38 with f holomorphic in a disc $\mathbb{D}_\rho$). We can thus form the quotient space:

Definition 6.43. We call *singularities* the elements of the space SING := ANA $/\mathbb{C}\{\zeta\}$. The canonical projection is denoted by sing_0 and we use the notation

$$\operatorname{sing}_0 : \begin{cases} \text{ANA} \to \text{SING} \\ \overset{\vee}{f} \;\mapsto \overset{\triangledown}{f} = \operatorname{sing}_0\big(\overset{\vee}{f}(\zeta)\big). \end{cases}$$

Any representative $\overset{\vee}{f}$ of a singularity $\overset{\triangledown}{f}$ is called a *major* of $\overset{\triangledown}{f}$.

The idea is that singular germs like $\log\zeta$ and $\log\zeta + \frac{1}{1-\zeta}$ have the same singular behaviour near 0: they are different majors for the same singularity (at the origin). Similarly, in Example 6.38, the singularity $\operatorname{sing}_0\big(\overset{\vee}{f}(\zeta)\big)$ coincides with $\operatorname{sing}_0\big(S(1/\zeta)\big)$. The simplest case is that of a simple pole or a pole of higher order, for which we introduce the notation

$$\delta := \operatorname{sing}_0\Big(\frac{1}{2\pi\mathrm{i}\zeta}\Big), \qquad \delta^{(k)} := \operatorname{sing}_0\Big(\frac{(-1)^k k!}{2\pi\mathrm{i}\zeta^{k+1}}\Big) \quad \text{for } k \geq 0. \tag{6.28}$$

The singularity of Example 6.38 can thus be written $2\pi\mathrm{i}\sum_{k=0}^{\infty}\frac{(-1)^k}{k!}a_{-k-1}\delta^{(k)}$.

Remark 6.44. In Example 6.39, a singular germ $\overset{\vee}{f}$ was defined from a holomorphic function of the form $f(\zeta) = \hat{\varphi}(\zeta)\operatorname{Log}\zeta$, with $\hat{\varphi}(\zeta) \in \mathbb{C}\{\zeta\}$, by identifying the cut plane $U = \mathbb{C}\setminus\mathbb{R}^-$ with the principal sheet $\widetilde{U}_0$ of $\widetilde{\mathbb{C}}$, and we can now regard $\overset{\vee}{f}$ as a major. Choosing some other branch of the logarithm or identiying U with some other sheet $\widetilde{U}_m$ would yield another major *for the same singularity*, because this modifies the major by an integer multiple of $2\pi\mathrm{i}\,\hat{\varphi}(\zeta)$ which is regular at 0. The notation

$$^{\flat}\hat{\varphi} := \operatorname{sing}_0\Big(\hat{\varphi}(\zeta)\frac{\log\zeta}{2\pi\mathrm{i}}\Big) \tag{6.29}$$

is sometimes used in this situation. Things are different if we replace $\hat{\varphi}$ by the Laurent series of a function which is holomorphic in a punctured disc $\mathbb{D}^*_\rho$ and not regular at 0; for instance, if we denote by $\mathscr{L}$ a branch of the logarithm in U, i.e. $\mathscr{L} = \operatorname{Log} + 2\pi\mathrm{i}k$ with some $k \in \mathbb{Z}$, then, for any choice of a connected component $\widetilde{U}_m$ of $\pi^{-1}(U)$ in $\widetilde{\mathbb{C}}$, the function $\dfrac{1}{2\pi\mathrm{i}\zeta}\mathscr{L}(\zeta)$ defines a singular germ, hence a singularity, but we change the singularity by an integer multiple of $2\pi\mathrm{i}\,\delta$ if we change k or m.

Example 6.45. Let us define

$$\overset{\triangledown}{I}_\sigma := \operatorname{sing}_0(\overset{\vee}{I}_\sigma), \qquad \overset{\vee}{I}_\sigma(\zeta) := \frac{\zeta^{\sigma-1}}{(1-\mathrm{e}^{-2\pi\mathrm{i}\sigma})\Gamma(\sigma)} \quad \text{for } \sigma \in \mathbb{C}\setminus\mathbb{Z} \tag{6.30}$$

(notation of Example 6.40). For $k \in \mathbb{N}$, in view of the poles of Euler's gamma function (cf. (5.49)), we have $(1-\mathrm{e}^{-2\pi\mathrm{i}\sigma})\Gamma(\sigma) \xrightarrow[\sigma\to -k]{} 2\pi\mathrm{i}(-1)^k/k!$, which suggests to extend the definition by setting

$$\overset{\vee}{I}_{-k}(\zeta) := \frac{(-1)^k k!}{2\pi\mathrm{i}\zeta^{k+1}}, \qquad \overset{\triangledown}{I}_{-k} := \delta^{(k)}$$

(we could have noticed as well that the reflection formula (5.62) yields $\overset{\vee}{I}_\sigma(\zeta) = \frac{1}{2\pi\mathrm{i}}\mathrm{e}^{\pi\mathrm{i}\sigma}\Gamma(1-\sigma)\zeta^{\sigma-1}$, which yields the same $\overset{\vee}{I}_{-k}$ when $\sigma = -k$). If $n \in \mathbb{N}^*$, there is no limit for $\overset{\vee}{I}_\sigma$ as $\sigma \to n$, however $\overset{\triangledown}{I}_\sigma$ can also be represented by the equivalent major $\dfrac{\zeta^{\sigma-1}-\zeta^{n-1}}{(1-\mathrm{e}^{-2\pi\mathrm{i}\sigma})\Gamma(\sigma)}$ which tends to the limit

$$\overset{\vee}{I}_n(\zeta) := \frac{\zeta^{n-1}}{(n-1)!}\frac{\log\zeta}{2\pi\mathrm{i}},$$

therefore we set $\overset{\triangledown}{I}_n := \operatorname{sing}_0\left(\dfrac{\zeta^{n-1}}{(n-1)!}\dfrac{\log\zeta}{2\pi\mathrm{i}}\right)$. We thus get a family of singularities $\left(\overset{\triangledown}{I}_\sigma\right)_{\sigma\in\mathbb{C}}$. Observe that

$$\operatorname{sing}_0(\zeta^{\sigma-1}) = (1-\mathrm{e}^{-2\pi\mathrm{i}\sigma})\Gamma(\sigma)\overset{\triangledown}{I}_\sigma, \qquad \sigma\in\mathbb{C}, \tag{6.31}$$

with the convention $(1-\mathrm{e}^{-2\pi\mathrm{i}\sigma})\Gamma(\sigma) = 2\pi\mathrm{i}(-1)^k/k!$ if $\sigma = -k \in -\mathbb{N}$ (and this singularity is 0 if and only if $\sigma = n \in \mathbb{N}^*$).

We shall not investigate deeply the structure of the space SING in this volume, but let us mention that there is a natural algebra structure on it: one can define a commutative associative product $\overset{\triangledown}{*}$ on SING, for which δ is a unit, which is compatible with the convolution law of $\mathbb{C}\{\zeta\}$ defined by Lemma 5.14 in the sense that

$$^\flat\hat{\varphi}\overset{\triangledown}{*}{}^\flat\hat{\psi} = {}^\flat(\hat{\varphi}*\hat{\psi}) \tag{6.32}$$

for any $\hat{\varphi},\hat{\psi}\in\mathbb{C}\{\zeta\}$ (with the notation (6.29)), and for which $\overset{\triangledown}{I}_\sigma\overset{\triangledown}{*}\overset{\triangledown}{I}_\tau = \overset{\triangledown}{I}_{\sigma+\tau}$ for any $\sigma,\tau\in\mathbb{C}$. See Chapter 7 of the third volume of this book [Del16] (or [Éca81], or [Sau12, §3.1–3.2]) for the details.[11] The differentiation operator $\frac{\mathrm{d}}{\mathrm{d}\zeta}$ passes to the quotient and the notation (6.28) is motivated by the relation $\delta^{(k)} = \left(\frac{\mathrm{d}}{\mathrm{d}\zeta}\right)^k\delta$. Let us also mention that $\delta^{(k)}$ can be considered as the Borel transform of z^k for $k\in\mathbb{N}$, and more generally $\overset{\triangledown}{I}_\sigma$ as the Borel transform of $z^{-\sigma}$ for any $\sigma\in\mathbb{C}$: there is in fact a version of the formal Borel transform operator with values in SING, which is defined on a class of formal objects much broader than formal expansions involving only integer powers of z (see Section 6.14.5 for an aperçu of this).

There is a well-defined monodromy[12] operator $\overset{\vee}{f}(\zeta)\in\mathrm{ANA}\mapsto\overset{\vee}{f}(\zeta\,\underline{\mathrm{e}}^{-2\pi\mathrm{i}})\in\mathrm{ANA}$ (recall that multiplication is well-defined in $\widetilde{\mathbb{C}}$), and the "variation map" $\overset{\vee}{f}(\zeta)\mapsto\overset{\vee}{f}(\zeta)-\overset{\vee}{f}(\zeta\,\underline{\mathrm{e}}^{-2\pi\mathrm{i}})$ annihilates the subspace $\mathbb{C}\{\zeta\}$, thus it passes to the quotient:

Definition 6.46. The linear map induced by the variation map $\overset{\vee}{f}(\zeta)\mapsto\overset{\vee}{f}(\zeta)-\overset{\vee}{f}(\zeta\,\underline{\mathrm{e}}^{-2\pi\mathrm{i}})$ is denoted by

[11] There, the convolution of singularitites that we have denoted by $\overset{\triangledown}{*}$ is simply denoted by $*$.

[12] The operator $\overset{\vee}{f}(\zeta)\in\mathrm{ANA}\mapsto\overset{\vee}{f}(\zeta\,\underline{\mathrm{e}}^{-2\pi\mathrm{i}})\in\mathrm{ANA}$ reflects analytic continuation along a clockwise loop around the origin for any function f holomorphic in a disc $D\subset\mathbb{C}^*$ and such that $\widetilde{f} = f\circ\pi$ on one of the connected components of $\pi^{-1}(D)$.

$$\mathrm{var}\colon \begin{cases} \mathrm{SING} \to \mathrm{ANA} \\ \overset{\triangledown}{f} = \mathrm{sing}_0(\overset{\vee}{f}) \mapsto \hat{f}(\zeta) = \overset{\vee}{f}(\zeta) - \overset{\vee}{f}(\zeta\,\underline{\mathrm{e}}^{-2\pi\mathrm{i}}). \end{cases}$$

The germ $\hat{f} = \mathrm{var}\,\overset{\triangledown}{f}$ is called the *minor* of the singularity $\overset{\triangledown}{f}$.

A simple but important example is

$$\mathrm{var}\left(\mathrm{sing}_0\left(\overset{\wedge}{\varphi}(\zeta)\frac{\log\zeta}{2\pi\mathrm{i}}\right)\right) = \overset{\wedge}{\varphi}(\zeta), \tag{6.33}$$

for any $\overset{\wedge}{\varphi}$ holomorphic in a punctured disc $\mathbb{D}_\rho^*$. Another example is provided by the singular germ of ζ^α (notation of Example 6.40): we get $\mathrm{var}\big(\mathrm{sing}_0(\zeta^\alpha)\big) = (1 - \mathrm{e}^{-2\pi\mathrm{i}\alpha})\,\mathrm{sing}_0(\zeta^\alpha)$, hence

$$\mathrm{var}\,\overset{\triangledown}{I}_\sigma = \frac{\zeta^{\sigma-1}}{\Gamma(\sigma)} \qquad \text{for all } \sigma \in \mathbb{C}\setminus(-\mathbb{N}),$$

$$\mathrm{var}\,\overset{\triangledown}{I}_{-k} = \mathrm{var}\,\delta^{(k)} = 0 \qquad \text{for all } k \in \mathbb{N}.$$

Clearly, the kernel of the linear map var consists of the singularities defined by the convergent Laurent series $\sum_{n\in\mathbb{Z}} a_n\zeta^n$ of Example 6.38.

6.9 Simple singularities at the origin

6.9.1 We retain from the previous section that, starting with a function f that admits spiral continuation around 0, by identifying a part of the domain of f with a subset of $\widetilde{\mathbb{C}}$, we get a function $\overset{\vee}{f}$ holomorphic in a domain of $\widetilde{\mathbb{C}}$ of the form $\mathscr{V}(h)$ and then a singular germ, still denoted by $\overset{\vee}{f}$ (by forgetting about the precise function h); we then capture the singularity of f at 0 by modding out by the regular germs.

The space SING of all singularities is huge. In this volume, we shall almost exclusively deal with singularities of a special kind:[13]

Definition 6.47. We call *simple singularity* any singularity of the form

$$\overset{\triangledown}{\varphi} = a\,\delta + \mathrm{sing}_0\left(\overset{\wedge}{\varphi}(\zeta)\frac{\log\zeta}{2\pi\mathrm{i}}\right)$$

with $a \in \mathbb{C}$ and $\overset{\wedge}{\varphi}(\zeta) \in \mathbb{C}\{\zeta\}$. The subspace of all simple singularities is denoted by $\mathrm{SING}^{\mathrm{simp}}$. We say that a function f *has a simple singularity at* 0 if it has spiral continuation around 0 and, for any choice of a domain $\widetilde{D} \subset \widetilde{\mathbb{C}}$ which projects injec-

[13] An exception is Section 6.14.5.

tively onto a part of the domain of f, the formula $\overset{\vee}{f} := f \circ \pi_{|\widetilde{D}}$ defines the major of a simple singularity.

In other words, $\mathrm{SING}^{\mathrm{simp}}$ is the range of the $\mathbb{C}$-linear map

$$a\,\delta + \hat{\varphi}(\zeta) \in \mathbb{C}\,\delta \oplus \mathbb{C}\{\zeta\} \mapsto a\,\delta + \mathrm{sing}_0\left(\hat{\varphi}(\zeta)\frac{\log\zeta}{2\pi\mathrm{i}}\right) \in \mathrm{SING}, \tag{6.34}$$

and a function f defined in an open disc D to which 0 is adherent has a simple singularity at 0 if and only if it can be written in the form

$$f(\zeta) = \frac{a}{2\pi\mathrm{i}\zeta} + \hat{\varphi}(\zeta)\frac{\mathscr{L}_D(\zeta)}{2\pi\mathrm{i}} + R(\zeta), \qquad \zeta \in D, \tag{6.35}$$

where $a \in \mathbb{C}$, $\hat{\varphi}(\zeta) \in \mathbb{C}\{\zeta\}$, $\mathscr{L}_D(\zeta)$ is any branch of the logarithm in D, and $R(\zeta) \in \mathbb{C}\{\zeta\}$. Notice that we need not worry about the choice of the connected component $\widetilde{D}$ of $\pi^{-1}(D)$ in this case: the various singular germs defined from f differ from one another by an integer multiple of $\hat{\varphi}$ and thus define the same singularity (as in Remark 6.44).

The map (6.34) is injective (exercise[14]); it thus induces a $\mathbb{C}$-linear isomorphism

$$\mathbb{C}\,\delta \oplus \mathbb{C}\{\zeta\} \xrightarrow{\sim} \mathrm{SING}^{\mathrm{simp}}, \tag{6.36}$$

which is also an algebra isomorphism if one takes into account the algebra structure on the space of singularities which was alluded to earlier (in view of (6.32)). This is why we shall identify $\mathrm{sing}_0\left(\hat{\varphi}(\zeta)\frac{\log\zeta}{2\pi\mathrm{i}}\right)$ with $\hat{\varphi}$ and use the notation

$$\mathrm{sing}_0\big(f(\zeta)\big) = \overset{\triangledown}{\varphi} = a\,\delta + \hat{\varphi}(\zeta) \in \mathbb{C}\,\delta \oplus \mathbb{C}\{\zeta\} \simeq \mathrm{SING}^{\mathrm{simp}} \tag{6.37}$$

in the situation described by (6.35), instead of the notation $a\,\delta + {}^{\flat}\hat{\varphi}$ which is sometimes used in other texts. (Observe that there is an abuse of notation in the left-hand side of (6.37): we should have specified a major $\overset{\vee}{f}$ holomorphic in a subset of $\widetilde{\mathbb{C}}$ and written $\mathrm{sing}_0\big(\overset{\vee}{f}(\zeta)\big)$, but there is no ambiguity here, as explained above.) The germ $\hat{\varphi}$ is the *minor* of the singularity ($\hat{\varphi} = \mathrm{var}\,\overset{\triangledown}{\varphi}$) and the complex number a is called the *constant term* of $\overset{\triangledown}{\varphi}$.

6.9.2 The convolution algebra $\mathbb{C}\,\delta \oplus \mathbb{C}\{\zeta\}$ was studied in Section 5.5 as the Borel image of the algebra $\mathbb{C}[[z^{-1}]]_1$ of 1-Gevrey formal series. Then, in Section 5.9, we defined its subalgebras $\mathbb{C}\,\delta \oplus \mathscr{N}(\mathrm{e}^{\mathrm{i}\theta}\mathbb{R}^+)$ and $\mathbb{C}\,\delta \oplus \mathscr{N}(I)$, Borel images of the subalgebras consisting of formal series 1-summable in a direction θ or in the directions of an open interval I, and studied the corresponding Laplace operators.

It is interesting to notice that the Laplace transform of a simple singularity $\overset{\triangledown}{\varphi} = a\,\delta + \hat{\varphi}(\zeta) \in \mathbb{C}\,\delta \oplus \mathscr{N}(\mathrm{e}^{\mathrm{i}\theta}\mathbb{R}^+)$ can be defined in terms of a major of $\overset{\triangledown}{\varphi}$: we choose

[14] Use (6.33).

$\overset{\vee}{\varphi}(\zeta) =$ the right-hand side of (6.35) with $R(\zeta) = 0$, or any major $\overset{\vee}{\varphi}$ of $\overset{\triangledown}{\varphi}$ for which there exist $\delta, \gamma > 0$ such that this major extends analytically to

$$\{\,\zeta \in \widetilde{\mathbb{C}} \mid \theta - \tfrac{5\pi}{2} < \arg\zeta < \theta + \tfrac{\pi}{2} \text{ and } |\zeta| < \delta\,\} \cup \widetilde{S}_\delta \cup \widetilde{S}'_\delta,$$

where $\widetilde{S}_\delta$ and $\widetilde{S}'_\delta$ are the connected components of $\pi^{-1}(S^\theta_\delta \setminus \mathbb{D}_\delta) \subset \widetilde{\mathbb{C}}$ which contain $\underline{e}^{i\theta}$ and $\underline{e}^{i(\theta-2\pi)}$ (see Figure 6.7), and satisfies

$$|\overset{\vee}{\varphi}(\zeta)| \le A\,e^{\gamma|\zeta|}, \qquad \zeta \in \widetilde{S}_\delta \cup \widetilde{S}'_\delta$$

for some positive constant A; then, for $0 < \varepsilon < \delta$ and

$$z \in \underline{e}^{-i\theta}\{\,z_0 \in \widetilde{\mathbb{C}} \mid \Re e\, z_0 > \gamma \text{ and } \arg z_0 \in (-\tfrac{\pi}{2}, \tfrac{\pi}{2})\,\},$$

we have

$$(\mathscr{S}^\theta \mathscr{B}^{-1} \overset{\triangledown}{\varphi})(z) = a + (\mathscr{L}^\theta \overset{\wedge}{\varphi})(z) = \int_{\Gamma_{\theta,\varepsilon}} e^{-z\zeta}\, \overset{\vee}{\varphi}(\zeta)\, d\zeta, \tag{6.38}$$

with an integration contour $\Gamma_{\theta,\varepsilon}$ which comes from infinity along $\underline{e}^{i(\theta-2\pi)}[\varepsilon, +\infty)$, encircles the origin by following counterclokwise the circle of radius ε, and go back to infinity along $\underline{e}^{i\theta}[\varepsilon, +\infty)$ (a kind of "Hankel contour"—see Figure 6.7). The proof is left as an exercise.[15]

The right-hand side of (6.38) is the "Laplace transform of majors". It shows why the notation $\overset{\triangledown}{\varphi} = c\,\delta + \overset{\wedge}{\varphi}$ is consistent with the notations used in the context of 1-summability and suggests far-reaching extensions of 1-summability theory, which however we shall not pursue in this volume. The reader is referred to Chapter 7 of the third volume [Del16], or [Éca81] or [Sau12, §3.2]).

6.10 Simple Ω-resurgent functions and alien operators

We now leave aside the summability issues and come back to resurgent functions. Let Ω be a non-empty closed discrete subset of $\mathbb{C}$ (for the moment we do not require it to be stable under addition). From now on, we shall regard Ω-resurgent functions as simple singularities (taking advantage of (6.36) and (6.37)):

$$\mathbb{C}\,\delta \oplus \widehat{\mathscr{R}}_\Omega \subset \mathbb{C}\,\delta \oplus \mathbb{C}\{\zeta\} \simeq \mathrm{SING}^{\mathrm{simp}},$$

where the germs of $\widehat{\mathscr{R}}_\Omega$ are characterized by Ω-continuability.

More generally, at least when $0 \in \Omega$, we define the space SING_Ω of *Ω-resurgent singularities* as the space of all $\overset{\triangledown}{\varphi} \in \mathrm{SING}$ whose minors $\overset{\wedge}{\varphi} = \operatorname{var} \overset{\triangledown}{\varphi} \in \mathrm{ANA}$ are Ω-continuable in the following sense: denoting by $\mathscr{V}(h) \subset \widetilde{\mathbb{C}}$ a domain where $\overset{\wedge}{\varphi}$ defines a holomorphic function, $\overset{\wedge}{\varphi}$ admits analytic continuation along any path $\widetilde{\gamma}$ of $\widetilde{\mathbb{C}}$

[15] Use (6.33) for $\zeta \in \underline{e}^{i\theta}[\varepsilon, +\infty)$ and then the dominated convergence theorem for $\varepsilon \to 0$.

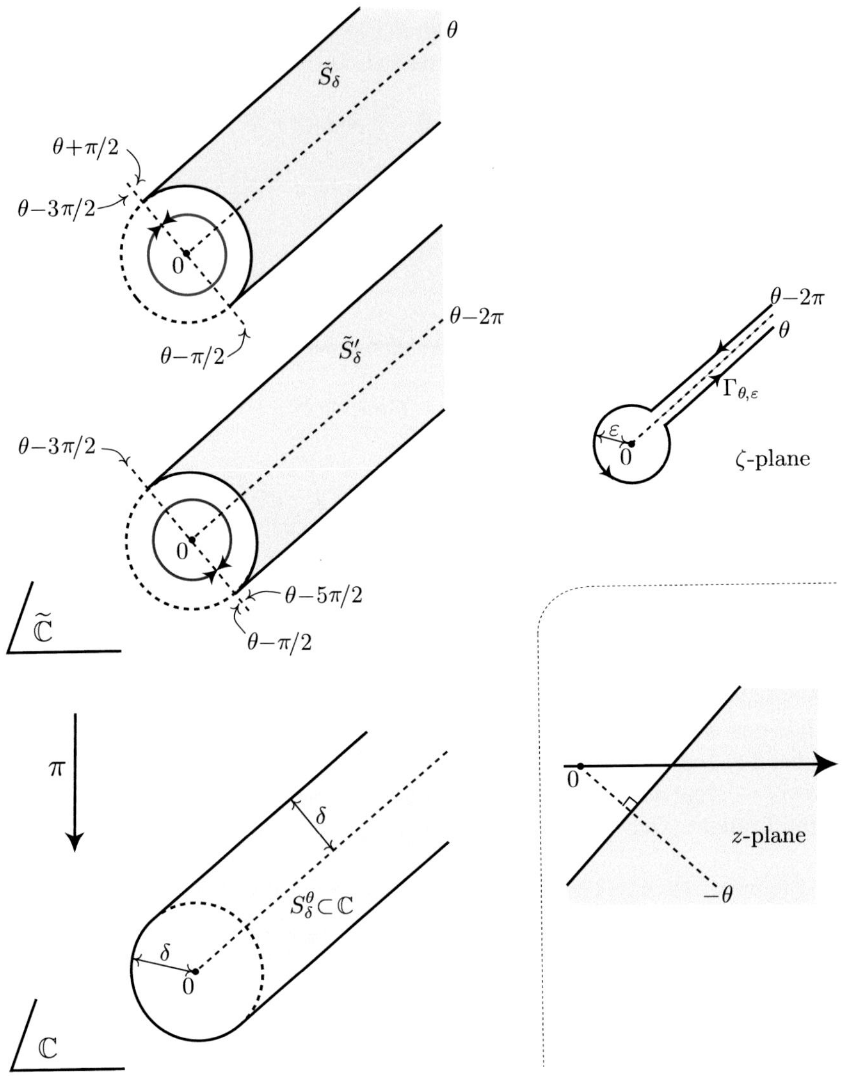

Fig. 6.7: *Laplace transform of a major.* Left: the domain of $\widetilde{\mathbb{C}}$ where $\overset{\vee}{\varphi}$ must be holomorphic and its projection S_δ^θ in $\mathbb{C}$. Right: the contour $\Gamma_{\theta,\varepsilon}$ for ζ (above) and the domain where z belongs (below).

starting in $\mathscr{V}(h)$ such that $\pi\circ\widetilde{\gamma}$ is contained in $\mathbb{C}\setminus\Omega$. We then have the following diagram:

$$\begin{array}{ccccccc} & & \mathbb{C}\,\delta\oplus\widehat{\mathscr{R}}_\Omega=\mathrm{SING}^{\mathrm{simp}}\cap\mathrm{SING}_\Omega & \hookrightarrow & \mathbb{C}\,\delta\oplus\mathbb{C}\{\zeta\}=\mathrm{SING}^{\mathrm{simp}} & \hookrightarrow & \mathrm{SING}\\ & & \wr\uparrow\mathscr{B} & & \wr\uparrow\mathscr{B} & & \\ \mathbb{C}\{z^{-1}\} & \hookrightarrow & \widetilde{\mathscr{R}}_\Omega & \hookrightarrow & \mathbb{C}[[z^{-1}]]_1 & & \end{array}$$

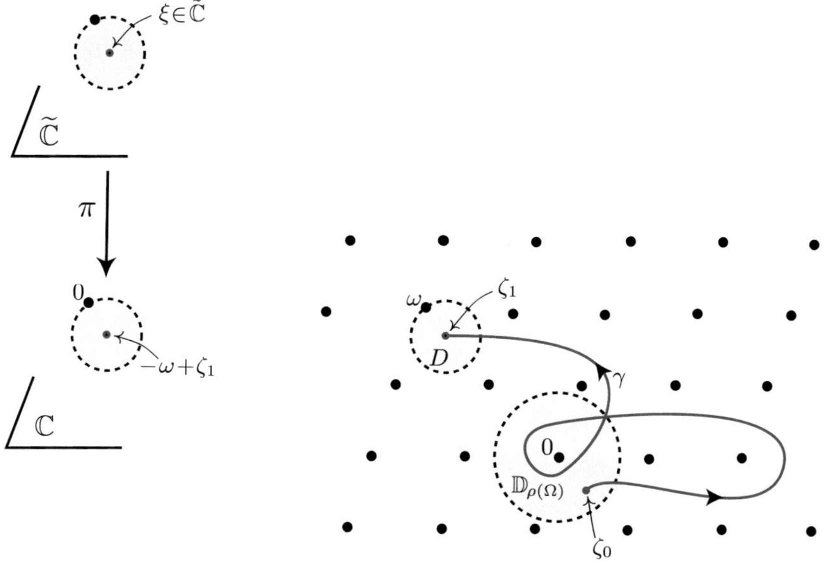

Fig. 6.8: The alien operator $\mathscr{A}^{\gamma,\xi}_\omega$ measures the singularity at ω for the analytic continuation along γ of an Ω-continuable germ.

Definition 6.48. Suppose that $\omega\in\Omega$, γ is a path of $\mathbb{C}\setminus\Omega$ starting at a point $\zeta_0\in\mathbb{D}^*_{\rho(\Omega)}$ and ending at a point ζ_1 such that there exists an open disc $D\subset\mathbb{C}\setminus\Omega$ centred at ζ_1 to which ω is adherent, and $\xi\in\widetilde{\mathbb{C}}$ satisfies $\pi(\xi)=-\omega+\zeta_1$. We then define a linear map, called the *alien operator associated with* (ω,γ,ξ),

$$\mathscr{A}^{\gamma,\xi}_\omega\colon\ \mathbb{C}\,\delta\oplus\widehat{\mathscr{R}}_\Omega\to\mathrm{SING}$$

by the formula

$$\mathscr{A}^{\gamma,\xi}_\omega(a\,\delta+\hat{\varphi}):=\mathrm{sing}_0\big(\check{f}(\zeta)\big),\qquad \check{f}(\zeta)=\mathrm{cont}_\gamma\hat{\varphi}\big(\omega+\pi(\zeta)\big)\ \text{ for }\zeta\in\widetilde{D}(\xi), \tag{6.39}$$

where $\widetilde{D}(\xi) \subset \widetilde{\mathbb{C}}$ is the connected component of $\pi^{-1}(-\omega + D)$ which contains ξ. See Figure 6.8.

This means that we follow the analytic continuation of $\overset{\wedge}{\varphi}$ along γ and get a function $\mathrm{cont}_\gamma \overset{\wedge}{\varphi}$ which is holomorphic in the disc D centred at ζ_1, of which $\omega \in \partial D$ is possibly a singular point; we then translate this picture and get a function

$$\zeta \mapsto f(\zeta) := \mathrm{cont}_\gamma \overset{\wedge}{\varphi}(\omega + \zeta)$$

which is holomorphic in the disc $-\omega + D$ centred at $-\omega + \zeta_1 = \pi(\xi)$, of which $0 \in \partial(-\omega + D)$ is possibly a singular point; the function f has spiral continuation around 0 because $\overset{\wedge}{\varphi}$ is Ω-continuable: choosing $\varepsilon > 0$ small enough so that $D(\omega, \varepsilon) \cap \Omega = \{\omega\}$ (which is possible since Ω is discrete), we see that $\mathrm{cont}_\gamma \overset{\wedge}{\varphi}$ can be continued analytically along any path starting from ζ_1 and staying in $D(\omega, \varepsilon) \cup D$, hence $\overset{\vee}{f}$ is holomorphic in $\mathscr{V}(h) \cup \widetilde{D}$ with $h(\theta) \equiv \varepsilon$ and formula (6.39) makes sense.

Remark 6.49. It is clear that the operator $\mathscr{A}_\omega^{\gamma,\xi}$ does not change if γ is replaced with a path which is homotopic (in $\mathbb{C} \setminus \Omega$, with fixed endpoints) to γ, nor if the endpoints of γ are modified in a continuous way (keeping satisfied the assumptions of Definition 6.48) provided that ξ is modified accordingly. On the other hand, modifying ξ while keeping γ unchanged results in an elementary modification of the result, in line with footnote 10 on p. 198.

In a nutshell, the idea is to measure the singularity at ω for the analytic continuation along γ of the minor $\overset{\wedge}{\varphi}$. Of course, if ω is not a singular point for $\mathrm{cont}_\gamma \overset{\wedge}{\varphi}$, then $\mathscr{A}_\omega^{\gamma,\xi} \overset{\triangledown}{\varphi} = 0$. In fact, the intersection of the kernels of all the operators $\mathscr{A}_\omega^{\gamma,\xi}$ is $\mathbb{C}\delta \oplus \mathscr{O}(\mathbb{C})$, where $\mathscr{O}(\mathbb{C})$ is the set of all entire functions. In particular,

$$\mathscr{B}^{-1} \overset{\triangledown}{\varphi} \in \mathbb{C}\{z^{-1}\} \quad \Longrightarrow \quad \mathscr{A}_\omega^{\gamma,\xi} \overset{\triangledown}{\varphi} = 0.$$

Example 6.50. We had $\overset{\wedge}{\varphi}_\alpha(\zeta) := (1+\zeta)^{\alpha-1}$ with $\alpha \in \mathbb{C}$ in Example 6.13, in connection with the incomplete gamma function. Here we can take any Ω containing -1 and we have $\mathscr{A}_\omega^{\gamma,\xi} \overset{\wedge}{\varphi}_\alpha = 0$ whenever $\omega \neq -1$, since -1 is the only possible singular point of a branch of the analytic continuation of $\overset{\wedge}{\varphi}$. For $\omega = -1$, the value of $\mathscr{A}_{-1}^{\gamma,\xi} \overset{\wedge}{\varphi}_\alpha$ depends on γ and ξ. If γ is contained in the interval $(-1, 0)$, then we find $f(\zeta) = \overset{\wedge}{\varphi}_\alpha(-1+\zeta) =$ the principal branch of $\zeta^{\alpha-1}$ and, if we choose ξ in the principal sheet of $\widetilde{\mathbb{C}}$, then

$$\mathscr{A}_{-1}^{\gamma,\xi} \overset{\wedge}{\varphi}_\alpha = (1 - \mathrm{e}^{-2\pi \mathrm{i}\alpha}) \Gamma(\alpha) \overset{\triangledown}{I}_\alpha,$$

which is 0 if and only if $\alpha \in \mathbb{N}^*$ (cf. (6.31)). If γ turns N times around -1, keeping the same endpoints for γ and the same ξ, then this result is multiplied by $\mathrm{e}^{2\pi \mathrm{i} N \alpha}$; if we multiply ξ by $\underline{\mathrm{e}}^{2\pi \mathrm{i} m}$, then the result is multiplied by $\mathrm{e}^{-2\pi \mathrm{i} m \alpha}$. (In both cases the result is unchanged if $\alpha \in \mathbb{Z}$.)

Example 6.51. Let $\overset{\wedge}{\varphi}(\zeta) = \frac{1}{\zeta}\mathrm{Log}\,(1+\zeta)$ (variant of Example 6.10) and $\Omega = \{-1,0\}$; we shall describe the logarithmic singularity which arises at -1 and the simple pole at 0 for every branch of the analytic continuation of $\overset{\wedge}{\varphi}$. Consider first a path γ contained in the interval $(-1,0)$ and ending at $\zeta_1 = -\frac{1}{2}$. For any ξ projecting onto $0+\zeta_1 = -\frac{1}{2}$, we find $\mathscr{A}_0^{\gamma,\xi}\,\overset{\wedge}{\varphi} = 0$ (no singularity at the origin for the principal branch), while for ξ projecting onto $1+\zeta_1 = \frac{1}{2}$,

$$\xi = \tfrac{1}{2}\,\underline{\mathrm{e}}^{2\pi i m} \quad\Longrightarrow\quad \mathscr{A}_{-1}^{\gamma,\xi}\,\overset{\wedge}{\varphi} = \mathrm{sing}_0\left(\frac{1}{-1+\zeta}(-2\pi i m + \log\zeta)\right) = -\frac{2\pi\mathrm{i}}{1-\zeta}$$

(using the notation (6.37)). If γ turns N times around -1, then the analytic continuation of $\overset{\wedge}{\varphi}$ is augmented by $\frac{2\pi\mathrm{i}N}{\zeta}$, which is regular at -1 but singular at 0, hence $\mathscr{A}_{-1}^{\gamma,\xi}\,\overset{\wedge}{\varphi}$ still coincides with $-\frac{2\pi\mathrm{i}}{1-\zeta}$ (the logarithmic singularity at -1 is the same for every branch) but

$$\xi = -\tfrac{1}{2}\,\underline{\mathrm{e}}^{2\pi i m} \quad\Longrightarrow\quad \mathscr{A}_{0}^{\gamma,\xi}\,\overset{\wedge}{\varphi} = \mathrm{sing}_0\left(\frac{2\pi\mathrm{i}N}{\zeta}\right) = (2\pi\mathrm{i})^2 N\delta.$$

Exercise 6.52. Consider $\overset{\wedge}{\varphi}(\zeta) = -\frac{1}{\zeta}\mathrm{Log}\,(1-\zeta)$ as in Example 6.10, with $\Omega = \{0,1\}$, and a path γ contained in $(0,1)$ and ending at $\zeta_1 = \frac{1}{2}$. Prove that $\mathscr{A}_1^{\gamma,\xi}\,\overset{\wedge}{\varphi} = -\frac{2\pi\mathrm{i}}{1+\zeta}$ for any ξ projecting onto $-1+\zeta_1 = -\frac{1}{2}$. Compute $\mathscr{A}_0^{\gamma,\xi}\,\overset{\wedge}{\varphi}$ for γ turning N times around 1 and ξ projecting onto $0+\zeta_1 = \frac{1}{2}$.

Examples 6.10 and 6.51 (but not Example 6.13 if $\alpha \notin \mathbb{N}$) are particular cases of

Definition 6.53. We call *simple Ω-resurgent function* any Ω-resurgent function $\overset{\triangledown}{\varphi}$ such that, for all (ω,γ,ξ) as in Definition 6.48, $\mathscr{A}_\omega^{\gamma,\xi}\,\overset{\triangledown}{\varphi}$ is a simple singularity. The set of all simple Ω-resurgent functions is denoted by

$$\mathbb{C}\,\delta \oplus \widehat{\mathscr{R}}_\Omega^{\mathrm{simp}},$$

where $\widehat{\mathscr{R}}_\Omega^{\mathrm{simp}}$ is the set of all simple Ω-resurgent functions without constant term. We call *simple Ω-resurgent series* any element of

$$\widetilde{\mathscr{R}}_\Omega^{\mathrm{simp}} := \mathscr{B}^{-1}\big(\mathbb{C}\,\delta \oplus \widehat{\mathscr{R}}_\Omega^{\mathrm{simp}}\big) \subset \widetilde{\mathscr{R}}_\Omega.$$

Lemma 6.54. *Let ω,γ,ξ be as in Definition 6.48. Then*

$$\overset{\triangledown}{\varphi} \in \mathbb{C}\,\delta \oplus \widehat{\mathscr{R}}_\Omega \quad\Longrightarrow\quad \mathscr{A}_\omega^{\gamma,\xi}\,\overset{\triangledown}{\varphi} \in \mathrm{SING}_{-\omega+\Omega}$$

$$\overset{\triangledown}{\varphi} \in \mathbb{C}\,\delta \oplus \widehat{\mathscr{R}}_\Omega^{\mathrm{simp}} \quad\Longrightarrow\quad \mathscr{A}_\omega^{\gamma,\xi}\,\overset{\triangledown}{\varphi} \in \mathbb{C}\,\delta \oplus \widehat{\mathscr{R}}_{-\omega+\Omega}^{\mathrm{simp}}.$$

Moreover, in the last case, $\mathscr{A}_\omega^{\gamma,\xi}\,\overset{\triangledown}{\varphi}$ does not depend on the choice of ξ in $\pi^{-1}(-\omega+\zeta_1)$; denoting it by $\mathscr{A}_\omega^{\gamma}\,\overset{\triangledown}{\varphi}$, we thus define an operator

$$\mathscr{A}^{\gamma}_{\omega}\colon \mathbb{C}\delta \oplus \widehat{\mathscr{R}}^{\text{simp}}_{\Omega} \to \mathbb{C}\delta \oplus \widehat{\mathscr{R}}^{\text{simp}}_{-\omega+\Omega}.$$

Proof of Lemma 6.54. Let $\overset{\triangledown}{\varphi} \in \mathbb{C}\delta \oplus \widehat{\mathscr{R}}_{\Omega}$ and $\overset{\triangledown}{\psi} := \mathscr{A}^{\gamma,\xi}_{\omega} \overset{\triangledown}{\varphi} \in \text{SING}$, $\hat{\psi} := \text{var}\,\overset{\triangledown}{\psi} \in$ ANA. With the notations of Definition 6.48 and ε as in the paragraph which follows it, we consider the path γ' obtained by concatenating γ and a loop of $D(\omega,\varepsilon)\cup D$ that starts and ends at ζ_1 and encircles ω clockwise (recall that path concatenation is defined in Definition 6.23). We then have $\hat{\psi} = \check{f} - \check{g}$, with

$$\check{f}(\zeta) := \text{cont}_{\gamma}\, \hat{\varphi}\big(\omega + \pi(\zeta)\big) \quad \text{and} \quad \check{g}(\zeta) := \text{cont}_{\gamma'}\, \hat{\varphi}\big(\omega + \pi(\zeta)\big) \quad \text{for } \zeta \in \widetilde{D},$$

where $\widetilde{D}$ is the connected component of $\pi^{-1}(D)$ which contains ξ.

For any path $\widetilde{\lambda}$ of $\widetilde{\mathbb{C}}$ which starts at ξ and whose projection $\lambda := \pi \circ \widetilde{\lambda}$ is contained in $\mathbb{C}\setminus(-\omega+\Omega)$, the analytic continuation of $\check{f}$ and $\check{g}$ along $\widetilde{\lambda}$ exists and is given by

$$\text{cont}_{\widetilde{\lambda}}\, \check{f}(\zeta) = \text{cont}_{\Gamma}\, \hat{\varphi}\big(\omega+\pi(\zeta)\big), \qquad \text{cont}_{\widetilde{\lambda}}\, \check{g}(\zeta) = \text{cont}_{\Gamma'}\, \hat{\varphi}\big(\omega+\pi(\zeta)\big),$$

where Γ is obtained by concatenating γ and $\omega+\lambda$, and Γ' by concatenating γ' and $\omega+\lambda$. Hence the analytic continuation of $\hat{\psi}$ along any such path $\widetilde{\lambda}$ exists, and this is sufficient to ensure that $\overset{\triangledown}{\psi} \in \text{SING}_{\Omega}$, which was the first statement to be proved.

If we suppose $\overset{\triangledown}{\varphi} \in \mathbb{C}\delta \oplus \widehat{\mathscr{R}}^{\text{simp}}_{\Omega}$, then $\overset{\triangledown}{\psi} \in \text{SING}^{\text{simp}}$ and $\overset{\triangledown}{\psi}$ itself is a simple Ω-resurgent function, and the second statement follows from Example 6.38 and Remark 6.44: changing ξ amounts to adding to $\check{f}$ an integer multiple of $\hat{\psi}$ which is now assumed to be regular at the origin, and hence does not modify $\text{sing}_0\big(\check{f}(\zeta)\big)$. Putting these facts together, we obtain $\mathscr{A}^{\gamma,\xi}_{\omega}\overset{\triangledown}{\varphi} \in \mathbb{C}\delta \oplus \widehat{\mathscr{R}}^{\text{simp}}_{-\omega+\Omega} \subset \text{SING}_{-\omega+\Omega} \cap \text{SING}^{\text{simp}}$ independent of ξ. □

In other words, an Ω-resurgent function $\overset{\triangledown}{\varphi}$ is simple if and only if all the branches of the analytic continuation of the minor $\hat{\varphi} = \text{var}\,\overset{\triangledown}{\varphi}$ have only simple singularities; the relation $\mathscr{A}^{\gamma}_{\omega}\overset{\triangledown}{\varphi} = a\delta + \hat{\psi}(\zeta)$ then means

$$\text{cont}_{\gamma}\hat{\varphi}(\omega+\zeta) = \frac{a}{2\pi \mathrm{i}\zeta} + \hat{\psi}(\zeta)\frac{\mathscr{L}(\zeta)}{2\pi\mathrm{i}} + R(\zeta) \tag{6.40}$$

for ζ close enough to 0, where $\mathscr{L}$ is any branch of the logarithm and $R(\zeta) \in \mathbb{C}\{\zeta\}$.

Notation 6.55 *We just defined an operator* $\mathscr{A}^{\gamma}_{\omega}\colon \mathbb{C}\delta \oplus \widehat{\mathscr{R}}^{\text{simp}}_{\Omega} \to \mathbb{C}\delta \oplus \widehat{\mathscr{R}}^{\text{simp}}_{-\omega+\Omega}$. *We shall denote by the same symbol the counterpart of this operator in spaces of formal series:*

$$\begin{array}{ccc} \mathbb{C}\,\delta\oplus\widehat{\mathscr{R}}^{\mathrm{simp}}_{\Omega} & \xrightarrow{\ \mathscr{A}^{\gamma}_{\omega}\ } & \mathbb{C}\,\delta\oplus\widehat{\mathscr{R}}^{\mathrm{simp}}_{-\omega+\Omega} \\ \wr\big\uparrow \mathscr{B} & & \wr\big\uparrow \mathscr{B} \\ \widetilde{\mathscr{R}}^{\mathrm{simp}}_{\Omega} & \xrightarrow{\ \mathscr{A}^{\gamma}_{\omega}\ } & \widetilde{\mathscr{R}}^{\mathrm{simp}}_{-\omega+\Omega} \end{array}$$

Definition 6.56. Let $\omega\in\Omega$. We call *alien operator at* ω any linear combination of composite operators of the form

$$\mathscr{A}^{\gamma_r}_{\omega-\omega_{r-1}}\circ\cdots\circ\mathscr{A}^{\gamma_2}_{\omega_2-\omega_1}\circ\mathscr{A}^{\gamma_1}_{\omega_1}$$

(viewed as operators $\mathbb{C}\,\delta\oplus\widehat{\mathscr{R}}^{\mathrm{simp}}_{\Omega}\to\mathbb{C}\,\delta\oplus\widehat{\mathscr{R}}^{\mathrm{simp}}_{-\omega+\Omega}$ or, equivalently, $\widetilde{\mathscr{R}}^{\mathrm{simp}}_{\Omega}\to\widetilde{\mathscr{R}}^{\mathrm{simp}}_{-\omega+\Omega}$) with any $r\geq 1$, $\omega_1,\dots,\omega_{r-1}\in\Omega$, γ_j being any path of $\mathbb{C}\setminus(-\omega_{j-1}+\Omega)$ starting in $\mathbb{D}^*_{\rho(-\omega_{j-1}+\Omega)}$ and ending in a disc $D_j\subset\mathbb{D}\setminus(-\omega_{j-1}+\Omega)$ to which $\omega_j-\omega_{j-1}$ is adherent, with the conventions $\omega_0=0$ and $\omega_r=\omega$, so that the operator $\mathscr{A}^{\gamma_j}_{\omega_j-\omega_{j-1}}:\widetilde{\mathscr{R}}^{\mathrm{simp}}_{-\omega_{j-1}+\Omega}\to\widetilde{\mathscr{R}}^{\mathrm{simp}}_{-\omega_j+\Omega}$ is well defined.

Clearly $\mathbb{C}\,\delta\oplus\mathscr{O}(\mathbb{C})\subset\mathbb{C}\,\delta\oplus\widehat{\mathscr{R}}^{\mathrm{simp}}_{\Omega}$ (since an entire function has no singularity at all!), hence

$$\mathbb{C}\{z^{-1}\}\subset\widetilde{\mathscr{R}}^{\mathrm{simp}}_{\Omega},$$

and of course all alien operators act trivially on such resurgent functions. Another easy example of simple Ω-resurgent function is provided by any meromorphic function $\widehat{\varphi}$ of ζ which is regular at 0 and whose poles are all simple and located in Ω. In this case $\mathscr{A}^{\gamma}_{\omega}\widehat{\varphi}$ does not depend on γ: its value is $2\pi\mathrm{i}c_\omega\delta$, where c_ω is the residuum of $\widehat{\varphi}$ at ω.

Example 6.57. By looking at the proof of Lemma 6.5, we see that we have meromorphic Borel transforms for the formal series associated with the names of Euler, Poincaré and Stirling, hence

$$\widetilde{\varphi}^{\mathrm{E}}\in\widetilde{\mathscr{R}}^{\mathrm{simp}}_{\{-1\}},\qquad \widetilde{\varphi}^{\mathrm{P}}\in\widetilde{\mathscr{R}}^{\mathrm{simp}}_{s+2\pi\mathrm{i}\mathbb{Z}},\qquad \widetilde{\mu}\in\widetilde{\mathscr{R}}^{\mathrm{simp}}_{2\pi\mathrm{i}\mathbb{Z}^*},$$

and we can compute

$$\mathscr{A}^{\gamma}_{-1}\widehat{\varphi}^{\mathrm{E}}=2\pi\mathrm{i}\delta,\qquad \mathscr{A}^{\gamma}_{s+2\pi\mathrm{i}k}\widehat{\varphi}^{\mathrm{P}}=2\pi\mathrm{i}\delta,\qquad \mathscr{A}^{\gamma}_{2\pi\mathrm{i}m}\widehat{\mu}=\frac{1}{m}\delta,$$

for $k\in\mathbb{Z}$, $m\in\mathbb{Z}^*$ with any γ (and correspondingly $\mathscr{A}^{\gamma}_{-1}\widetilde{\varphi}^{\mathrm{E}}=\mathscr{A}^{\gamma}_{s+2\pi\mathrm{i}k}\widetilde{\varphi}^{\mathrm{P}}=2\pi\mathrm{i}$, $\mathscr{A}^{\gamma}_{2\pi\mathrm{i}m}\widetilde{\mu}=\frac{1}{m}$). A less elementary example is $\widetilde{\lambda}=\mathrm{e}^{\widetilde{\mu}}$; we saw that $\widetilde{\lambda}\in\widetilde{\mathscr{R}}_{2\pi\mathrm{i}\mathbb{Z}}$ in Example 6.33, we shall see in Section 6.93 that it belongs to $\widetilde{\mathscr{R}}^{\mathrm{simp}}_{2\pi\mathrm{i}\mathbb{Z}}$ and that any alien operator maps $\widetilde{\lambda}$ to a multiple of $\widetilde{\lambda}$.

Here is a variant of Lemma 6.14 adapted to the case of simple resurgent functions:

Lemma 6.58. *Let Ω be any non-empty closed discrete subset of $\mathbb{C}$.*

– *If $\hat{B}$ is an entire function, then multiplication by $\hat{B}$ leaves $\widehat{\mathscr{R}}^{\text{simp}}_{\Omega}$ invariant, with*

$$\mathscr{A}^{\gamma}_{\omega}\overset{\wedge}{\varphi} = a\,\delta + \overset{\wedge}{\psi}(\zeta) \implies \mathscr{A}^{\gamma}_{\omega}(\hat{B}\overset{\wedge}{\varphi}) = \hat{B}(\omega)a\,\delta + \hat{B}(\omega+\zeta)\overset{\wedge}{\psi}(\zeta). \tag{6.41}$$

– *As a consequence, for any $c\in\mathbb{C}$, the operators $\widehat{\partial}$ and $\widehat{T}_c$ (defined by (5.21) and (5.23)) leave $\mathbb{C}\,\delta\oplus\widehat{\mathscr{R}}^{\text{simp}}_{\Omega}$ invariant or, equivalently, $\widetilde{\mathscr{R}}^{\text{simp}}_{\Omega}$ is stable by $\partial = \frac{\mathrm{d}}{\mathrm{d}z}$ and T_c; one has*

$$\widetilde{\varphi}_0 \in \widetilde{\mathscr{R}}^{\text{simp}}_{\Omega} \implies \mathscr{A}^{\gamma}_{\omega}(\partial\widetilde{\varphi}_0) = (-\omega+\partial)\mathscr{A}^{\gamma}_{\omega}\widetilde{\varphi}_0 \text{ and } \mathscr{A}^{\gamma}_{\omega}(T_c\widetilde{\varphi}_0) = \mathrm{e}^{-c\omega}T_c(\mathscr{A}^{\gamma}_{\omega}\widetilde{\varphi}_0). \tag{6.42}$$

– *If $\widetilde{\psi}\in z^{-2}\mathbb{C}\{z^{-1}\}$, then the solution in $z^{-1}\mathbb{C}[[z^{-1}]]$ of the difference equation*

$$\widetilde{\varphi}(z+1) - \widetilde{\varphi}(z) = \widetilde{\psi}(z)$$

belongs to $\widetilde{\mathscr{R}}^{\text{simp}}_{2\pi\mathrm{i}\mathbb{Z}^*}$, with $\mathscr{A}^{\gamma}_{\omega}\overset{\wedge}{\varphi} = -2\pi\mathrm{i}\overset{\wedge}{\psi}(\omega)\,\delta$ for all (ω,γ) with $\omega\in 2\pi\mathrm{i}\mathbb{Z}^*$.

Proof. Suppose that $\mathscr{A}^{\gamma}_{\omega}\overset{\wedge}{\varphi} = a\,\delta + \overset{\wedge}{\psi}(\zeta)$. Since multiplication by $\hat{B}$ commutes with analytic continuation, the relation (6.40) implies

$$\operatorname{cont}_{\gamma}\big(\hat{B}\overset{\wedge}{\varphi}\big)(\omega+\zeta) = \hat{B}(\omega+\zeta)\operatorname{cont}_{\gamma}\overset{\wedge}{\varphi}(\omega+\zeta)$$
$$= \frac{\hat{B}(\omega)a}{2\pi\mathrm{i}\zeta} + \hat{B}(\omega+\zeta)\overset{\wedge}{\psi}(\zeta)\frac{\mathscr{L}(\zeta)}{2\pi\mathrm{i}} + R^*(\zeta)$$

with $R^*(\zeta) = R(\zeta) + a\dfrac{\hat{B}(\omega+\zeta)-\hat{B}(\omega)}{2\pi\mathrm{i}\zeta} \in \mathbb{C}\{\zeta\}$, hence $\mathscr{A}^{\gamma}_{\omega}(\hat{B}\overset{\wedge}{\varphi}) = \hat{B}(\omega)a\,\delta + \hat{B}(\omega+\zeta)\overset{\wedge}{\psi}(\zeta)$.

Suppose now that $\widetilde{\varphi}_0\in\widetilde{\mathscr{R}}^{\text{simp}}_{\Omega}$ has Borel transform $\overset{\triangledown}{\varphi}_0 = \alpha\,\delta + \overset{\wedge}{\varphi}$ with $\alpha\in\mathbb{C}$ and $\overset{\wedge}{\varphi}$ as above. According to (5.21) and (5.23), we have $\widehat{\partial}\overset{\triangledown}{\varphi}_0 = -\zeta\overset{\wedge}{\varphi}(\zeta)$ and $\widehat{T}_c\overset{\triangledown}{\varphi}_0 = \alpha\,\delta + \mathrm{e}^{-c\zeta}\overset{\wedge}{\varphi}(\zeta)$; we see that both of them belong to $\mathbb{C}\,\delta\oplus\widehat{\mathscr{R}}^{\text{simp}}_{\Omega}$ by applying the first statement with $\hat{B}(\zeta) = -\zeta$ or $\mathrm{e}^{-c\zeta}$, and

$$\mathscr{A}^{\gamma}_{\omega}(\widehat{\partial}\overset{\triangledown}{\varphi}_0) = -\omega a\delta + (-\omega-\zeta)\overset{\wedge}{\psi}(\zeta) = (-\omega+\widehat{\partial})\mathscr{A}^{\gamma}_{\omega}\overset{\triangledown}{\varphi}_0$$
$$\mathscr{A}^{\gamma}_{\omega}(\widehat{T}_c\overset{\triangledown}{\varphi}_0) = \mathrm{e}^{-c\omega}a\delta + \mathrm{e}^{-c(\omega+\zeta)}\overset{\wedge}{\psi}(\zeta) = \mathrm{e}^{-c\omega}\widehat{T}_c(\mathscr{A}^{\gamma}_{\omega}\overset{\triangledown}{\varphi}_0),$$

which is equivalent to (6.42).

For the last statement, we use Corollary 5.11, according to which $\overset{\wedge}{\varphi} = \hat{B}\overset{\wedge}{\psi}$ with $\hat{B}(\zeta) = \frac{1}{\mathrm{e}^{-\zeta}-1}$ and $\overset{\wedge}{\psi}(\zeta)\in\zeta\mathscr{O}(\mathbb{C})$: the function $\overset{\wedge}{\varphi}$ is meromorphic on $\mathbb{C}$ and all its poles are simple and located in $\Omega = 2\pi\mathrm{i}\mathbb{Z}^*$, therefore it is a simple Ω-resurgent function and we get the values of $\mathscr{A}^{\gamma}_{\omega}\overset{\triangledown}{\varphi}$ by computing the residues of $\overset{\wedge}{\varphi}$ (as observed in the paragraph just before Example 6.57). □

Exercise 6.59. Given $s \in \mathbb{C}$ with $\Re e\, s > 1$, the Hurwitz zeta function[16] is defined as

$$\zeta(s,z) = \sum_{k=0}^{\infty} \frac{1}{(z+k)^s}, \qquad z \in \mathbb{C} \setminus \mathbb{R}^-$$

(using the principal branch of $(z+k)^s$ for each k). Show that, for $s \in \mathbb{N}$ with $s \geq 2$,

$$\tilde{\varphi}^{\mathrm{H}}_s(z) := \frac{1}{(s-1)z^{s-1}} + \frac{1}{2z^s} + \sum_{k=1}^{\infty} \binom{s+2k-1}{s-1} \frac{B_{2k}}{(s+2k-1)z^{s+2k-1}}$$

(where the Bernoulli numbers B_{2k} are defined in Exercise 5.42) is a simple $2\pi\mathrm{i}\,\mathbb{Z}^*$-resurgent formal series which is 1-summable in the directions of $I = (-\frac{\pi}{2}, \frac{\pi}{2})$, with

$$\zeta(s,z) = (\mathscr{S}^I \tilde{\varphi}^{\mathrm{H}}_s)(z) \sim_1 \tilde{\varphi}^{\mathrm{H}}_s(z).$$

Hint: Use Lemma 6.58 and prove that $\zeta(s,z)$ coincides with the Laplace transform of

$$\overset{\wedge}{\varphi}{}^{\mathrm{H}}_s(\zeta) = \frac{\zeta^{s-1}}{\Gamma(s)(1-\mathrm{e}^{-\zeta})}. \tag{6.43}$$

Remark 6.60. If $s \in \mathbb{C} \setminus \mathbb{N}$ has $\Re e\, s > 1$, then (6.43) is not regular at the origin but still provides an example of $2\pi\mathrm{i}\,\mathbb{Z}$-continuable minor (in the sense of the definition given in the paragraph just before Definition 6.48). In fact, there is an extension of 1-summability theory in which the Laplace transform of $\overset{\wedge}{\varphi}{}^{\mathrm{H}}_s$ in the directions of $(-\frac{\pi}{2}, \frac{\pi}{2})$ is still defined and coincides with $\zeta(s,z)$ (see Section 6.14.5).

We end this section with a look at the action of alien operators on convolution products in the "easy case" considered in Section 6.2.

Theorem 6.61. *Suppose that $\overset{\triangledown}{B}_0 \in \mathbb{C}\delta \oplus \widehat{\mathscr{R}}^{\mathrm{simp}}_{\Omega}$ with $\hat{B} := \operatorname{var} \overset{\triangledown}{B}_0$ entire. Then, for any $\omega \in \Omega$, all the alien operators $\mathbb{C}\delta \oplus \widehat{\mathscr{R}}^{\mathrm{simp}}_{\Omega} \to \mathbb{C}\delta \oplus \widehat{\mathscr{R}}^{\mathrm{simp}}_{-\omega+\Omega}$ commute with the operator of convolution with $\overset{\triangledown}{B}_0$.*

Proof. It suffices to show that, for any $\gamma \subset \mathbb{C} \setminus \Omega$ starting at a point $\zeta_0 \in \mathbb{D}^*_{\rho(\Omega)}$ and ending at the centre ζ_1 of a disc $D \subset \mathbb{C} \setminus \Omega$ to which ω is adherent, and for any $\overset{\triangledown}{\varphi}_0 \in \mathbb{C}\delta \oplus \widehat{\mathscr{R}}^{\mathrm{simp}}_{\Omega}$,

$$\mathscr{A}^{\gamma}_{\omega}(\overset{\triangledown}{B}_0 * \overset{\triangledown}{\varphi}_0) = \overset{\triangledown}{B}_0 * (\mathscr{A}^{\gamma}_{\omega} \overset{\triangledown}{\varphi}_0).$$

We can write

$$\overset{\triangledown}{B}_0 = b\delta + \hat{B}, \quad \overset{\triangledown}{\varphi}_0 = c\delta + \overset{\wedge}{\varphi}, \quad \mathscr{A}^{\gamma}_{\omega} \overset{\triangledown}{\varphi}_0 = a\delta + \overset{\wedge}{\psi}.$$

---

16 Notice that $\zeta(s,1)$ is the Riemann zeta value $\zeta(s)$.

Then $\overset{\triangledown}{B}_0 * \overset{\triangledown}{\varphi}_0 = b\overset{\triangledown}{\varphi}_0 + c\hat{B} + \hat{B} * \hat{\varphi}$ and $\mathscr{A}^{\gamma}_{\omega}(\overset{\triangledown}{B}_0 * \overset{\triangledown}{\varphi}_0) = b\,\mathscr{A}^{\gamma}_{\omega}\overset{\triangledown}{\varphi}_0 + \mathscr{A}^{\gamma}_{\omega}(\hat{B} * \hat{\varphi})$, hence we just need to prove that $\mathscr{A}^{\gamma}_{\omega}(\hat{B} * \hat{\varphi}) = \hat{B} * \mathscr{A}^{\gamma}_{\omega}\hat{\varphi}$, i.e. that

$$\mathscr{A}^{\gamma}_{\omega}(\hat{B} * \hat{\varphi}) = a\hat{B} + \hat{B} * \hat{\psi}.$$

According to Lemma 6.15, we have

$$\operatorname{cont}_{\gamma}(\hat{B} * \hat{\varphi})(\omega + \zeta) = \int_0^{\zeta_0} \hat{B}(\omega + \zeta - \xi)\hat{\varphi}(\xi)\,\mathrm{d}\xi + \int_{\gamma} \hat{B}(\omega + \zeta - \xi)\hat{\varphi}(\xi)\,\mathrm{d}\xi + \int_{\zeta_1}^{\omega+\zeta} \hat{B}(\omega + \zeta - \xi)\hat{\varphi}(\xi)\,\mathrm{d}\xi$$

for $\zeta \in -\omega + D$, where it is understood that $\hat{\varphi}(\xi)$ represents the value at ξ of the appropriate branch of the analytic continuation of $\hat{\varphi}$ (which is $\operatorname{cont}_{\gamma}\hat{\varphi}$ for the third integral). The standard theorem about an integral depending holomorphically on a parameter ensures that the sum $R_1(\zeta)$ of the first two integrals extends to an entire function of ζ. Let $\Delta := -\omega + D$ (a disc to which 0 is adherent). Performing the change of variable $\xi \to \omega + \xi$ in the third integral, we get

$$\operatorname{cont}_{\gamma}(\hat{B} * \hat{\varphi})(\omega + \zeta) = R_1(\zeta) + \int_{-\omega+\zeta_1}^{\zeta} \hat{B}(\zeta - \xi)\operatorname{cont}_{\gamma}\hat{\varphi}(\omega + \xi)\,\mathrm{d}\xi, \qquad \zeta \in \Delta.$$

Now, according to (6.40), we can write

$$\operatorname{cont}_{\gamma}\hat{\varphi}(\omega + \xi) = S(\xi) + R_2(\xi), \qquad \xi \in \Delta \cap \mathbb{D}^*_{\rho},$$

where $S(\xi) = \dfrac{a}{2\pi i \xi} + \hat{\psi}(\xi)\dfrac{\mathscr{L}_{\Delta}(\xi)}{2\pi i}$, $\mathscr{L}_{\Delta}$ being a branch of the logarithm holomorphic in Δ, $R_2(\xi) \in \mathbb{C}\{\xi\}$, and $\rho > 0$ is smaller than the radii of convergence of $\hat{\psi}$ and R_2. Let us pick $\sigma \in \Delta \cap \mathbb{D}^*_{\rho}$ and set

$$R(\zeta) := R_1(\zeta) + \int_{-\omega+\zeta_1}^{\sigma} \hat{B}(\zeta - \xi)\operatorname{cont}_{\gamma}\hat{\varphi}(\omega + \xi)\,\mathrm{d}\xi,$$

so that

$$\operatorname{cont}_{\gamma}(\hat{B} * \hat{\varphi})(\omega + \zeta) = R(\zeta) + \int_{\sigma}^{\zeta} \hat{B}(\zeta - \xi)\operatorname{cont}_{\gamma}\hat{\varphi}(\omega + \xi)\,\mathrm{d}\xi, \qquad \zeta \in \Delta.$$

We see that $R(\zeta)$ extends to an entire function of ζ and, for $\zeta \in \Delta \cap \mathbb{D}^*_{\rho}$, the last integral can be written

$$\int_{\sigma}^{\zeta} \hat{B}(\zeta - \xi)\operatorname{cont}_{\gamma}\hat{\varphi}(\omega + \xi)\,\mathrm{d}\xi = f(\zeta) + R_3(\zeta), \qquad f(\zeta) := \int_{\sigma}^{\zeta} \hat{B}(\zeta - \xi)S(\xi)\,\mathrm{d}\xi,$$

with $R_3(\zeta)$ defined by an integral involving $R_2(\xi)$ and thus extending holomorphically for $\zeta \in \mathbb{D}_\rho$. The only possibly singular term in $\operatorname{cont}_\gamma(\hat{B} * \hat{\varphi})(\omega+\zeta)$ is thus $f(\zeta)$, which is seen to admit analytic continuation along every path Γ starting from σ and contained in $\mathbb{D}_\rho^*$; indeed,

$$\operatorname{cont}_\Gamma f(\zeta) = \int_\Gamma \hat{B}(\zeta-\xi) S(\xi)\,\mathrm{d}\xi. \tag{6.44}$$

In particular, f has spiral continuation around 0. We now show that it defines a simple singularity, which is none other than $a\hat{B} + \hat{B} * \hat{\psi}$.

Let us first compute the difference $g := f^+ - f$, where we denote by f^+ the branch of the analytic continuation of f obtained by starting from $\Delta \cap \mathbb{D}_\rho^*$, turning anticlockwise around 0 and coming back to $\Delta \cap \mathbb{D}_\rho^*$. We have

$$g(\zeta) = \int_{C_\zeta} \hat{B}(\zeta-\xi) S(\xi)\,\mathrm{d}\xi, \qquad \zeta \in \Delta \cap \mathbb{D}_\rho^*,$$

where C_ζ is the circular path $t \in [0, 2\pi] \mapsto \zeta\,\mathrm{e}^{\mathrm{i}t}$. For any $\varepsilon \in (0,1)$, by the Cauchy theorem,

$$g(\zeta) = \int_{\varepsilon\zeta}^{\zeta} \hat{B}(\zeta-\xi)\hat{\psi}(\xi)\,\mathrm{d}\xi + \int_{C_{\varepsilon\zeta}} \hat{B}(\zeta-\xi) S(\xi)\,\mathrm{d}\xi,$$

because $S^+ - S = \hat{\psi}$. Keeping ζ fixed, we let ε tend to 0: the first integral clearly tends to $\hat{B} * \hat{\psi}(\zeta)$ and the second one can be written

$$a\int_{C_{\varepsilon\zeta}} \frac{\hat{B}(\zeta-\xi)}{2\pi\mathrm{i}\xi}\,\mathrm{d}\xi + \int_{C_{\varepsilon\zeta}} \hat{B}(\zeta-\xi)\hat{\psi}(\xi)\frac{\mathscr{L}_\Delta(\xi)}{2\pi\mathrm{i}}\,\mathrm{d}\xi =$$

$$a\hat{B}(\zeta) + \int_0^{2\pi} \hat{B}(\zeta-\varepsilon\zeta\,\mathrm{e}^{\mathrm{i}t})\hat{\psi}(\varepsilon\zeta\,\mathrm{e}^{\mathrm{i}t})\frac{\ln\varepsilon + \mathscr{L}_\Delta(\zeta) + \mathrm{i}t}{2\pi\mathrm{i}}\mathrm{i}\varepsilon\zeta\,\mathrm{e}^{\mathrm{i}t}\,\mathrm{d}t$$

(because the analytic continuation of $\mathscr{L}_\Delta$ is explicitly known), which tends to $a\hat{B}(\zeta)$ since the last integral is bounded in modulus by $C\varepsilon\big(C' + |\ln\varepsilon|\big)$ with appropriate constants C, C'. We thus obtain

$$g(\zeta) = a\hat{B}(\zeta) + \hat{B} * \hat{\psi}(\zeta).$$

Since this function is regular at the origin and holomorphic in $\mathbb{D}_\rho$, we can reformulate this result on $f^+ - f$ by saying that the function

$$\zeta \in \Delta \cap \mathbb{D}_\rho^* \mapsto h(\zeta) := f(\zeta) - g(\zeta)\frac{\mathscr{L}_\Delta(\zeta)}{2\pi\mathrm{i}}$$

extends analytically to a (single-valued) function holomorphic in $\mathbb{D}_\rho^*$, i.e. it can be represented by a Laurent series (6.27).

We conclude by showing that the above function h is in fact regular at the origin. For that, it is sufficient to check that, in $\mathbb{D}^*_{|\sigma|}$, it is bounded by $C\big(C'+\ln\frac{1}{|\zeta|}\big)$ with appropriate constants C, C' (indeed, this will imply $\zeta h(\zeta) \xrightarrow[\zeta\to 0]{} 0$, thus the origin will be a removable singularity for h). Observe that every point of $\mathbb{D}^*_{|\sigma|}$ can be written in the form $\zeta = \sigma u \mathrm{e}^{\mathrm{i}v}$ with $0 < u := |\zeta|/|\sigma| < 1$ and $0 \le v < 2\pi$, hence it can be reached by starting from σ and following the concatenation Γ_ζ of the circular path $t \in [0,v] \mapsto \sigma \mathrm{e}^{\mathrm{i}t}$ and the line-segment $t \in [0,1] \mapsto \sigma \mathrm{e}^{\mathrm{i}v} x(t)$ with $x(t) := 1 - t(1-u) > 0$, hence

$$(\operatorname{cont}_{\Gamma_\zeta} h)(\zeta) = (\operatorname{cont}_{\Gamma_\zeta} f)(\zeta) - \frac{1}{2\pi\mathrm{i}} g(\zeta)(\operatorname{cont}_{\Gamma_\zeta} \mathscr{L}_\Delta)(\zeta)$$
$$= \int_{\Gamma_\zeta} \hat{B}(\zeta-\xi) S(\xi)\,\mathrm{d}\xi - \frac{1}{2\pi\mathrm{i}} g(\zeta)(\mathscr{L}_\Delta(\sigma) + \ln u + \mathrm{i}v)$$

(using (6.44) for the first term). The result follows from the existence of a constant $M > 0$ such that $|\hat{B}| \le M$ on $\mathbb{D}_{2\rho}$, $|g| \le M$ on $\mathbb{D}_\rho$ and $|S(\xi)| \le M/|\xi|$ for $\xi \in \mathbb{D}_\rho$, because the first term in the above representation of $(\operatorname{cont}_{\Gamma_\zeta} h)(\zeta)$ has modulus $\le$

$$\left| \int_0^v \hat{B}(\zeta - \sigma \mathrm{e}^{\mathrm{i}t}) S(\sigma \mathrm{e}^{\mathrm{i}t}) \sigma \mathrm{i} \mathrm{e}^{\mathrm{i}t}\,\mathrm{d}t + \int_0^1 \hat{B}\big(\zeta - \sigma \mathrm{e}^{\mathrm{i}v} x(t)\big) S\big(\sigma \mathrm{e}^{\mathrm{i}v} x(t)\big) \sigma \mathrm{e}^{\mathrm{i}v} x'(t)\,\mathrm{d}t \right|$$

which is $\le M^2 v + M^2 \ln\frac{1}{u}$.

□

6.11 The alien operators Δ^+_ω and Δ_ω

We still denote by Ω a non-empty closed discrete subset of $\mathbb{C}$. We now define various families of alien operators acting on simple Ω-resurgent functions, among which the most important will be $(\Delta^+_\omega)_{\omega\in\Omega\setminus\{0\}}$ and $(\Delta_\omega)_{\omega\in\Omega\setminus\{0\}}$.

6.11.1 *Definition of $\mathscr{A}^\Omega_{\omega,\varepsilon}$, Δ^+_ω and Δ_ω*

Definition 6.62. Let $\omega \in \Omega \setminus \{0\}$. We denote by $\prec$ the total order on $[0,\omega]$ induced by $t \in [0,1] \mapsto t\omega \in [0,\omega]$ and write

$$[0,\omega] \cap \Omega = \{\omega_0, \omega_1, \ldots, \omega_{r-1}, \omega_r\}, \qquad 0 = \omega_0 \prec \omega_1 \prec \cdots \prec \omega_{r-1} \prec \omega_r = \omega \tag{6.45}$$

(with $r \in \mathbb{N}^*$ depending on ω and Ω). With any $\varepsilon = (\varepsilon_1, \ldots, \varepsilon_{r-1}) \in \{+,-\}^{r-1}$ we associate an alien operator at ω

$$\mathscr{A}^{\Omega}_{\omega,\varepsilon} \colon \widetilde{\mathscr{R}}^{\text{simp}}_{\Omega} \to \widetilde{\mathscr{R}}^{\text{simp}}_{-\omega+\Omega} \tag{6.46}$$

defined as $\mathscr{A}^{\Omega}_{\omega,\varepsilon} = \mathscr{A}^{\gamma}_{\omega}$ for any path γ chosen as follows: we pick $\delta > 0$ small enough so that the closed discs $D_j := \overline{D}(\omega_j, \delta)$, $j = 0, 1, \dots r$, are pairwise disjoint and satisfy $D_j \cap \Omega = \{\omega_j\}$, and we take a path γ connecting $]0,\omega[\cap D_0$ and $]0,\omega[\cap D_r$ by following the line-segment $]0,\omega[$ except that, for $1 \le j \le r-1$, the subsegment $]0,\omega[\cap D_j$ is replaced by one of the two half-circles which are the connected components of $]0,\omega[\cap \partial D_j$: the path γ must circumvent ω_j to the right if $\varepsilon_j = +$, to the left if $\varepsilon_j = -$. See Figure 6.9.

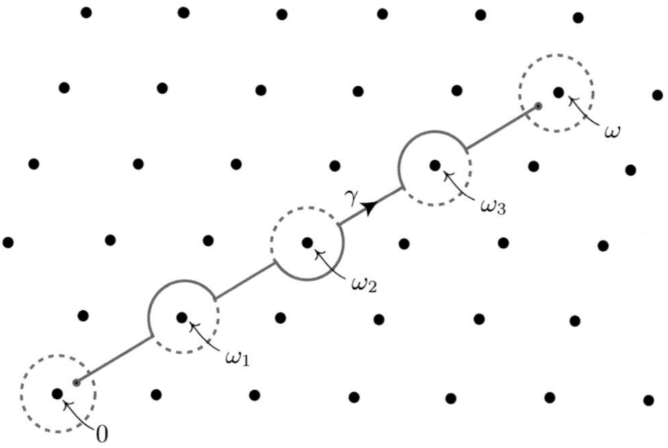

Fig. 6.9: An example of path γ used in the definition of $\mathscr{A}^{\Omega}_{\omega,\varepsilon}$, here with $\varepsilon = (-,+,-)$.

Observe that the notation (6.46) is justified by the fact that, in view of Remark 6.49, the operator $\mathscr{A}^{\Omega}_{\omega,\varepsilon}$ does not depend on δ nor on the endpoints of γ.

Definition 6.63. For any $\omega \in \Omega \setminus \{0\}$, we define two particular alien operators at ω

$$\Delta^+_{\omega},\ \Delta_{\omega} \colon \widetilde{\mathscr{R}}^{\text{simp}}_{\Omega} \to \widetilde{\mathscr{R}}^{\text{simp}}_{-\omega+\Omega}$$

by the formulas

$$\Delta^+_{\omega} := \mathscr{A}^{\Omega}_{\omega,(+,\dots,+)}, \qquad \Delta_{\omega} := \sum_{\varepsilon \in \{+,-\}^{r-1}} \frac{p(\varepsilon)!\,q(\varepsilon)!}{r!} \mathscr{A}^{\Omega}_{\omega,\varepsilon}, \tag{6.47}$$

where $r = r(\omega, \Omega)$ is defined by (6.45) and $p(\varepsilon)$ and $q(\varepsilon)$ represent the number of symbols '$+$' and '$-$' in the tuple ε (so that $p(\varepsilon) + q(\varepsilon) = r-1$).

We thus have (still with notation (6.45)), for $r = 1, 2, 3$:

$$\Delta^+_{\omega_1} = \mathscr{A}^{\Omega}_{\omega_1,()}, \qquad \Delta_{\omega_1} = \mathscr{A}^{\Omega}_{\omega_1,()},$$
$$\Delta^+_{\omega_2} = \mathscr{A}^{\Omega}_{\omega_2,(+)}, \qquad \Delta_{\omega_2} = \frac{1}{2}\mathscr{A}^{\Omega}_{\omega_2,(+)} + \frac{1}{2}\mathscr{A}^{\Omega}_{\omega_2,(-)},$$
$$\Delta^+_{\omega_3} = \mathscr{A}^{\Omega}_{\omega_3,(+,+)}, \quad \Delta_{\omega_3} = \frac{1}{3}\mathscr{A}^{\Omega}_{\omega_3,(+,+)} + \frac{1}{6}\mathscr{A}^{\Omega}_{\omega_3,(+,-)} + \frac{1}{6}\mathscr{A}^{\Omega}_{\omega_3,(-,+)} + \frac{1}{3}\mathscr{A}^{\Omega}_{\omega_3,(-,-)}.$$

Of course, the operators $\Delta^+_{\omega}, \Delta_{\omega}, \mathscr{A}^{\Omega}_{\omega,\varepsilon}$ can all be considered as operators $\mathbb{C}\,\delta \oplus \widehat{\mathscr{R}}^{\mathrm{simp}}_{\Omega} \to \mathbb{C}\,\delta \oplus \widehat{\mathscr{R}}^{\mathrm{simp}}_{-\omega+\Omega}$ as well.

Remark 6.64. Later on, in Sections 7.3–7.6, we shall assume that Ω is an additive subgroup of $\mathbb{C}$, so $-\omega + \Omega = \Omega$ and $\Delta^+_{\omega}, \Delta_{\omega}, \mathscr{A}^{\Omega}_{\omega,\varepsilon}$ are operators from $\widetilde{\mathscr{R}}^{\mathrm{simp}}_{\Omega}$ to itself; we shall see in Section 6.13.4 that, in that case, $\widetilde{\mathscr{R}}^{\mathrm{simp}}_{\Omega}$ is a subalgebra of $\widetilde{\mathscr{R}}_{\Omega}$ (which is itself a subalgebra of $\mathbb{C}[[z^{-1}]]$ by Corollary 6.19) of which each Δ_{ω} is a derivation (i.e. it satisfies the Leibniz rule). For that reason the operators Δ_{ω} are called "alien derivations".

Observe that, given $r \geq 1$, there are r possibilities for the value of $p = p(\varepsilon)$ and, for each p, there are $\binom{r-1}{p}$ tuples ε such that $p(\varepsilon) = p$; since in the definition of Δ_{ω} the coefficient in front of $\mathscr{A}^{\Omega}_{\omega,\varepsilon}$ is the inverse of $r\begin{pmatrix} r-1 \\ p(\varepsilon) \end{pmatrix}$, it follows that the sum of all these 2^{r-1} coefficients is 1. The resurgent function $\Delta_{\omega}(c\,\delta + \overset{\wedge}{\varphi})$ can thus be viewed as an average of the singularities at ω of the branches of the minor $\overset{\wedge}{\varphi}$ obtained by following the 2^{r-1} "most direct" paths from 0 to ω. The reason for this precise choice of coefficients will appear later (Theorem 6.72).

As a consequence, when the minor $\overset{\wedge}{\varphi}$ is meromorphic, both $\Delta_{\omega}(c\,\delta + \overset{\wedge}{\varphi})$ and $\Delta^+_{\omega}(c\,\delta + \overset{\wedge}{\varphi})$ coincide with $2\pi \mathrm{i} c_{\omega} \delta$, where c_{ω} is the residuum of $\overset{\wedge}{\varphi}$ at ω (cf. the paragraph just before Example 6.57). For instance, for the resurgent series associated with the names of Euler, Poincaré, Stirling and Hurwitz,

$$\Delta_{-1}\widetilde{\varphi}^{\mathrm{E}} = 2\pi\mathrm{i}, \quad \Delta_{s+2\pi \mathrm{i}k}\widetilde{\varphi}^{\mathrm{P}} = 2\pi\mathrm{i}, \quad \Delta_{2\pi \mathrm{i}m}\widetilde{\mu} = \frac{1}{m}, \quad \Delta_{2\pi \mathrm{i}m}\widetilde{\varphi}^{\mathrm{H}} = 2\pi\mathrm{i}\frac{(2\pi \mathrm{i}m)^{s-1}}{\Gamma(s)}, \tag{6.48}$$

for $k \in \mathbb{Z}$ and $\Re e\, s < 0$ in the case of Poincaré, $m \in \mathbb{Z}^*$ for Stirling, and $s \in \mathbb{N}$ with $s \geq 2$ for Hurwitz, in view of Example 6.57 and Exercise 6.59.

We note for later use an immediate consequence of formula (6.42) of Lemma 6.58:

Lemma 6.65. *Let $\widetilde{\varphi} \in \widetilde{\mathscr{R}}^{\mathrm{simp}}_{\Omega}$ and $c \in \mathbb{C}$. Then*

$$\Delta^+_{\omega}\partial\widetilde{\varphi} = (-\omega + \partial)\Delta^+_{\omega}\widetilde{\varphi}, \qquad \Delta_{\omega}\partial\widetilde{\varphi} = (-\omega + \partial)\Delta_{\omega}\widetilde{\varphi}, \tag{6.49}$$
$$\Delta^+_{\omega}T_c\widetilde{\varphi} = \mathrm{e}^{-c\omega}T_c\Delta^+_{\omega}\widetilde{\varphi}, \qquad \Delta_{\omega}T_c\widetilde{\varphi} = \mathrm{e}^{-c\omega}T_c\Delta_{\omega}\widetilde{\varphi}, \tag{6.50}$$

where $\partial = \frac{\mathrm{d}}{\mathrm{d}z}$ and T_c is defined by (5.16).

6.11.2 Dependence upon Ω

Lemma 6.66. *Suppose that we are given $\omega \in \mathbb{C}^*$ and Ω_1, Ω_2 closed discrete such that $\omega \in \Omega_1 \cap \Omega_2$. Then there are two operators "Δ^+_ω" defined by (6.47), an operator $\widetilde{\mathscr{R}}^{\mathrm{simp}}_{\Omega_1} \to \widetilde{\mathscr{R}}^{\mathrm{simp}}_{-\omega+\Omega_1}$ and an operator $\widetilde{\mathscr{R}}^{\mathrm{simp}}_{\Omega_2} \to \widetilde{\mathscr{R}}^{\mathrm{simp}}_{-\omega+\Omega_2}$, but they act the same way on $\widetilde{\mathscr{R}}^{\mathrm{simp}}_{\Omega_1} \cap \widetilde{\mathscr{R}}^{\mathrm{simp}}_{\Omega_2}$. The same is true of "$\Delta_\omega$".*

The point is that the sets $]0,\omega[\cap\Omega_1$ and $]0,\omega[\cap\Omega_2$ may differ, but their difference is constituted of points which are artificial singularities for the minor of any $\widetilde{\varphi} \in \widetilde{\mathscr{R}}^{\mathrm{simp}}_{\Omega_1} \cap \widetilde{\mathscr{R}}^{\mathrm{simp}}_{\Omega_2}$, in the sense that no branch of its analytic continuation is singular at any of these points. So Lemma 6.66 claims that, in the above situation, we get the same resurgent series for $\Delta^+_\omega\widetilde{\varphi}$ whether computing it in $\widetilde{\mathscr{R}}^{\mathrm{simp}}_{-\omega+\Omega_1}$ or in $\widetilde{\mathscr{R}}^{\mathrm{simp}}_{-\omega+\Omega_2}$.

Proof. Let $\Omega := \Omega_1$, $\Omega^* := \Omega_1 \cup \Omega_2$ and $\widetilde{\varphi} \in \widetilde{\mathscr{R}}^{\mathrm{simp}}_{\Omega_1} \cap \widetilde{\mathscr{R}}^{\mathrm{simp}}_{\Omega_2}$. As in (6.45), we write

$$]0,\omega[\cap\Omega = \{\omega_1 \prec \cdots \prec \omega_{r-1}\}, \qquad]0,\omega[\cap\Omega^* = \{\omega^*_1 \prec \cdots \prec \omega^*_{s-1}\},$$

with $1 \le r \le s$. Given $\varepsilon^* \in \{+,-\}^{s-1}$, we have $\mathscr{A}^{\Omega^*}_{\omega,\varepsilon^*}\widetilde{\varphi} = \mathscr{A}^{\Omega}_{\omega,\varepsilon}\widetilde{\varphi}$ with $\varepsilon := \varepsilon^*_{|\Omega}$, i.e. the tuple $\varepsilon \in \{+,-\}^{r-1}$ is obtained from ε^* by deleting the symbols ε^*_j corresponding to the fictitious singular points $\omega^*_j \in \Omega^* \setminus \Omega$.

In view of formula (6.47a), when $\varepsilon^* = (+,\dots,+)$ we get the same resurgent series for $\Delta^+_\omega\widetilde{\varphi}$ whether computing it in $\widetilde{\mathscr{R}}^{\mathrm{simp}}_{-\omega+\Omega^*}$ or in $\widetilde{\mathscr{R}}^{\mathrm{simp}}_{-\omega+\Omega}$, which yields the desired conclusion by exchanging the roles of Ω_1 and Ω_2.

We now compute $\Delta_\omega\widetilde{\varphi}$ in $\widetilde{\mathscr{R}}^{\mathrm{simp}}_{-\omega+\Omega^*}$ by applying formula (6.47b) and grouping together the tuples ε^* that have the same restriction ε: with the notation $c := s-r$, we get

$$\sum_{\varepsilon\in\{+,-\}^{r-1}} \sum_{\substack{\varepsilon^*\in\{+,-\}^{r+c-1}\\ \text{with } \varepsilon^*_{|\Omega}=\varepsilon}} \frac{p(\varepsilon^*)!q(\varepsilon^*)!}{(r+c)!}\mathscr{A}^{\Omega}_{\omega,\varepsilon}\widetilde{\varphi}$$

$$= \sum_{\varepsilon\in\{+,-\}^{r-1}} \sum_{c=a+b} \binom{c}{a} \frac{(p(\varepsilon)+a)!(q(\varepsilon)+b)!}{(r+c)!}\mathscr{A}^{\Omega}_{\omega,\varepsilon}\widetilde{\varphi},$$

which yields the desired result because

$$\sum_{c=a+b}\binom{c}{a}\frac{(p+a)!(q+b)!}{(r+c)!} = \frac{p!q!}{r!}$$

for any non-negative integers p,q,r with $r = p+q+1$, as is easily checked by rewriting this identity as

$$\sum_{c=a+b}\frac{(p+a)!}{a!p!}\frac{(q+b)!}{b!q!} = \frac{(r+c)!}{c!r!}$$

and observing that the generating series $\sum_{a\in\mathbb{N}} \frac{(p+a)!}{a!p!} X^a = (1-X)^{-p-1}$ satisfies $(1-X)^{-p-1}(1-X)^{-q-1} = (1-X)^{-r-1}$. □

Remark 6.67. Given $\omega \in \mathbb{C}^*$, we thus can compute $\Delta^+_\omega \widetilde{\varphi}$ or $\Delta_\omega \widetilde{\varphi}$ as soon as there exists Ω so that $\omega \in \Omega$ and $\widetilde{\varphi} \in \widetilde{\mathscr{R}}^{\text{simp}}_\Omega$, and the result does not depend on Ω. We thus have in fact a family of operators Δ^+_ω, $\Delta_\omega \colon \widetilde{\mathscr{R}}^{\text{simp}}_\Omega \to \widetilde{\mathscr{R}}^{\text{simp}}_{-\omega+\Omega}$, indexed by the closed discrete sets Ω which contain ω, and there is no need that the notation for these operators depend explicitly on Ω.

6.11.3 The operators Δ^+_ω as a system of generators

Theorem 6.68. *Let Ω be a non-empty closed discrete subset of $\mathbb{C}$ and let $\omega \in \Omega$. Any alien operator at ω can be expressed as a linear combination of composite operators of the form*

$$\Delta^+_{\eta_1,\ldots,\eta_s} := \Delta^+_{\eta_s-\eta_{s-1}} \circ \cdots \circ \Delta^+_{\eta_2-\eta_1} \circ \Delta^+_{\eta_1} \tag{6.51}$$

with $s \geq 1$, $\eta_1,\ldots,\eta_{s-1} \in \Omega$, $\eta_s = \omega$, $\eta_1 \neq 0$ and $\eta_j \neq \eta_{j+1}$ for $1 \leq j < s$, with the convention $\Delta^+_\omega := \Delta^+_\omega$ for $s = 1$ (viewing $\Delta^+_{\eta_1,\ldots,\eta_s}$ as an operator $\mathbb{C}\,\delta \oplus \widehat{\mathscr{R}}^{\text{simp}}_\Omega \to \mathbb{C}\,\delta \oplus \widehat{\mathscr{R}}^{\text{simp}}_{-\omega+\Omega}$ or, equivalently, $\widetilde{\mathscr{R}}^{\text{simp}}_\Omega \to \widetilde{\mathscr{R}}^{\text{simp}}_{-\omega+\Omega}$).

Observe that the composition (6.51) is well defined because, with the convention $\eta_0 = 0$, the operator $\Delta^+_{\eta_j-\eta_{j-1}}$ maps $\widetilde{\mathscr{R}}^{\text{simp}}_{-\eta_{j-1}+\Omega}$ into $\widetilde{\mathscr{R}}^{\text{simp}}_{-\eta_j+\Omega}$. We shall not give the proof of this theorem, but let us indicate a few examples: with the notation (6.45),

$$\mathscr{A}^\Omega_{\omega_2,(+)} = \Delta^+_{\omega_2}, \qquad \mathscr{A}^\Omega_{\omega_2,(-)} = \Delta^+_{\omega_2} - \Delta^+_{\omega_2-\omega_1} \circ \Delta^+_{\omega_1}$$

and

$$\mathscr{A}^\Omega_{\omega_3,(+,+)} = \Delta^+_{\omega_3},$$

$$\mathscr{A}^\Omega_{\omega_3,(-,+)} = \Delta^+_{\omega_3} - \Delta^+_{\omega_3-\omega_1} \circ \Delta^+_{\omega_1},$$

$$\mathscr{A}^\Omega_{\omega_3,(+,-)} = \Delta^+_{\omega_3} - \Delta^+_{\omega_3-\omega_2} \circ \Delta^+_{\omega_2},$$

$$\mathscr{A}^\Omega_{\omega_3,(-,-)} = \Delta^+_{\omega_3} - \Delta^+_{\omega_3-\omega_1} \circ \Delta^+_{\omega_1} - \Delta^+_{\omega_3-\omega_2} \circ \Delta^+_{\omega_2} + \Delta^+_{\omega_3-\omega_2} \circ \Delta^+_{\omega_2-\omega_1} \circ \Delta^+_{\omega_1}.$$

Remark 6.69. One can omit the '+' in Theorem 6.68, i.e. the family $\{\Delta_\eta\}$ is a system of generators as well. This will follow from the relation (6.53) of next section.

Exercise 6.70. Suppose that $\varepsilon, \varepsilon^* \in \{+,-\}^{r-1}$ assume the form

$$\varepsilon = a(-)b, \qquad \varepsilon^* = a(+)b, \qquad \text{with } a \in \{+,-\}^{s-1} \text{ and } b \in \{+,-\}^{r-1-s}$$

where $1 \le s \le r-1$, i.e. $(\varepsilon_1,\dots,\varepsilon_{s-1}) = (\varepsilon_1^*,\dots,\varepsilon_{s-1}^*) = a$, $\varepsilon_s = -$, $\varepsilon_s^* = +$, $(\varepsilon_{s+1},\dots,\varepsilon_{r-1}) = (\varepsilon_{s+1}^*,\dots,\varepsilon_{r-1}^*) = b$. Prove that

$$\mathscr{A}^{\Omega}_{\omega_r,\varepsilon} = \mathscr{A}^{\Omega}_{\omega_r,\varepsilon^*} - \mathscr{A}^{\Omega}_{\omega_r-\omega_s,b} \circ \mathscr{A}^{\Omega}_{\omega_s,a}$$

with the notation (6.45). Deduce the formulas given in example just above.

Remark 6.71. There is also a strong "freeness" statement for the operators Δ^+_η: consider an arbitrary finite set F of finite sequences η of elements of Ω, so that each $\eta \in F$ is of the form $(\eta_1,\dots,\eta_s)$ for some $s \in \mathbb{N}$, with $\eta_1 \neq 0$ and $\eta_j \neq \eta_{j+1}$ for $1 \le j < s$, with the convention $\eta = \emptyset$ and $\Delta^+_\emptyset = \mathrm{Id}$ for $s = 0$; then, for any non-trivial family $\big(\widetilde\psi^\eta\big)_{\eta\in F}$ of simple Ω-resurgent series,

$$\widetilde\varphi \in \mathscr{R}^{\mathrm{simp}}_\Omega \mapsto \sum_{\eta\in F} \widetilde\psi^\eta \cdot \Delta^+_\eta \widetilde\varphi$$

is a non-trivial linear map: one can construct a simple Ω-resurgent series which is not annihilated by this operator. There is a similar statement for the family $\{\Delta_\eta\}$. See [Éca85] or adapt [Sau09, §12].

6.12 The symbolic Stokes automorphism for a direction d

6.12.1 Exponential-logarithm correspondence between $\{\Delta^+_\omega\}$ and $\{\Delta_\omega\}$

For any $\omega \in \mathbb{C}^*$, we denote by $\prec$ the total order on $[0,\omega]$ induced by $t \in [0,1] \mapsto t\,\omega \in [0,\omega]$.

Theorem 6.72. *Let Ω be a non-empty closed discrete subset of $\mathbb{C}$. Then, for any $\omega \in \Omega \setminus \{0\}$,*

$$\Delta_\omega = \sum_{s\in\mathbb{N}^*} \frac{(-1)^{s-1}}{s} \sum_{(\eta_1,\dots,\eta_{s-1})\in\Sigma(s,\omega,\Omega)} \Delta^+_{\omega-\eta_{s-1}} \circ \dots \circ \Delta^+_{\eta_2-\eta_1} \circ \Delta^+_{\eta_1} \qquad (6.52)$$

$$\Delta^+_\omega = \sum_{s\in\mathbb{N}^*} \frac{1}{s!} \sum_{(\eta_1,\dots,\eta_{s-1})\in\Sigma(s,\omega,\Omega)} \Delta_{\omega-\eta_{s-1}} \circ \dots \circ \Delta_{\eta_2-\eta_1} \circ \Delta_{\eta_1} \qquad (6.53)$$

where $\Sigma(s,\omega,\Omega)$ is the set of all increasing sequences $(\eta_1,\dots,\eta_{s-1})$ of $]0,\omega[\cap\Omega$,

$$0 \prec \eta_1 \prec \dots \prec \eta_{s-1} \prec \omega,$$

with the convention that the composite operator $\Delta^+_{\omega-\eta_{s-1}} \circ \dots \circ \Delta^+_{\eta_2-\eta_1} \circ \Delta^+_{\eta_1}$ is reduced to Δ^+_ω when $s = 1$ (in which case $\Sigma(1,\omega,\Omega)$ is reduced to the empty sequence) and similarly for the composite operator appearing in (6.53).

With the notation (6.45), this means

$$\Delta_{\omega_1} = \Delta^+_{\omega_1}$$

$$\Delta_{\omega_2} = \Delta^+_{\omega_2} - \tfrac{1}{2}\Delta^+_{\omega_2-\omega_1} \circ \Delta^+_{\omega_1}$$

$$\Delta_{\omega_3} = \Delta^+_{\omega_3} - \tfrac{1}{2}\left(\Delta^+_{\omega_3-\omega_1} \circ \Delta^+_{\omega_1} + \Delta^+_{\omega_3-\omega_2} \circ \Delta^+_{\omega_2}\right) + \tfrac{1}{3}\Delta^+_{\omega_3-\omega_2} \circ \Delta^+_{\omega_2-\omega_1} \circ \Delta^+_{\omega_1}$$

$$\vdots$$

$$\Delta^+_{\omega_1} = \Delta_{\omega_1}$$

$$\Delta^+_{\omega_2} = \Delta_{\omega_2} + \tfrac{1}{2!}\Delta_{\omega_2-\omega_1} \circ \Delta_{\omega_1}$$

$$\Delta^+_{\omega_3} = \Delta_{\omega_3} + \tfrac{1}{2!}\left(\Delta_{\omega_3-\omega_1} \circ \Delta_{\omega_1} + \Delta_{\omega_3-\omega_2} \circ \Delta_{\omega_2}\right) + \tfrac{1}{3!}\Delta_{\omega_3-\omega_2} \circ \Delta_{\omega_2-\omega_1} \circ \Delta_{\omega_1}$$

$$\vdots$$

We shall obtain Theorem 6.72 in next section as a consequence of Theorem 6.73, which is in fact an equivalent formulation in term of series of homogeneous operators in a graded vector space.

6.12.2 The symbolic Stokes automorphism and the symbolic Stokes infinitesimal generator

From now on, we fix Ω and a ray $d = \{t\,\mathrm{e}^{\mathrm{i}\theta} \mid t \geq 0\}$, with some $\theta \in \mathbb{R}$, and denote by $\prec$ the total order on d induced by $t \mapsto t\,\mathrm{e}^{\mathrm{i}\theta}$. We shall be interested in the operators Δ^+_ω and Δ_ω with $\omega \in d$. Without loss of generality we can suppose that the set $\Omega \cap d$ is infinite and contains 0; indeed, if it is not the case, then we can enrich Ω and replace it say with $\Omega^* := \Omega \cup \{N\mathrm{e}^{\mathrm{i}\theta} \mid N \in \mathbb{N}\}$, and avail ourselves of Remark 6.67, observing that $\widehat{\mathscr{R}}^{\mathrm{simp}}_{\Omega} \hookrightarrow \widehat{\mathscr{R}}^{\mathrm{simp}}_{\Omega^*}$ and that any relation proved for the alien operators in the larger space induces a relation in the smaller, with $\Delta^+_{\omega^*}$ and Δ_{ω^*} annihilating the smaller space when $\omega^* \in \Omega^* \setminus \Omega$.

We can thus write $\Omega \cap d$ as an increasing sequence

$$\Omega \cap d = \{\omega_m\}_{m\in\mathbb{N}}, \qquad \omega_0 = 0 \prec \omega_1 \prec \omega_2 \prec \cdots \tag{6.54}$$

For each $\omega = \omega_m \in \Omega \cap d$, we define

- $\hat{E}_\omega(\Omega)$ as the space of all functions $\widehat{\phi}$ which are holomorphic at ω, which can be analytically continued along any path of $\mathbb{C} \setminus \Omega$ starting close enough to ω, and whose analytic continuation has at worse simple singularities;

- $\overset{\triangledown}{E}_\omega(\Omega)$ as the vector space $\mathbb{C}\,\delta_\omega \oplus \hat{E}_\omega(\Omega)$, where each δ_ω is a distinct symbol[17] analogous to the convolution unit δ;
- $\overset{\vee}{E}_\omega(\Omega,d)$ as the space of all functions $\overset{\vee}{f}$ holomorphic on the line-segment $]\omega_m,\omega_{m+1}[$ which can be analytically continued along any path of $\mathbb{C}\setminus\Omega$ starting from this line-segment and whose analytic continuation has at worse simple singularities.

We shall often use abridged notations $\overset{\triangledown}{E}_\omega$ or $\overset{\vee}{E}_\omega$. Observe that there is a linear isomorphism

$$\tau_\omega : \left| \begin{array}{lll} \mathbb{C}\,\delta \oplus \widehat{\mathscr{R}}^{\mathrm{simp}}_{-\omega+\Omega} & \xrightarrow{\sim} & \overset{\triangledown}{E}_\omega \\ a\,\delta + \widehat{\varphi} & \mapsto & a\,\delta_\omega + \widehat{\varphi}^\omega, \qquad \widehat{\varphi}^\omega(\zeta) := \widehat{\varphi}(\zeta-\omega), \end{array}\right. \tag{6.55}$$

and a linear map

$$\overset{\bullet}{\sigma} : \left| \begin{array}{lll} \overset{\vee}{E}_{\omega_m} & \to & \overset{\triangledown}{E}_{\omega_{m+1}} \\ \overset{\vee}{f} & \mapsto & \tau_{\omega_{m+1}}\overset{\triangledown}{\varphi}, \qquad \overset{\triangledown}{\varphi} := \operatorname{sing}_0\big(\overset{\vee}{f}(\omega_{m+1}+\zeta)\big). \end{array}\right.$$

The idea is that an element of $\overset{\triangledown}{E}_\omega(\Omega)$ is nothing but a simple Ω-resurgent singularity "based at ω" and that any element of $\overset{\vee}{E}_\omega(\Omega,d)$ has a well-defined simple singularity "at ω_{m+1}", i.e. we could have written $\overset{\bullet}{\sigma}\overset{\vee}{f} = \operatorname{sing}_{\omega_{m+1}}\big(\overset{\vee}{f}(\zeta)\big)$ with an obvious extension of Definition 6.43.

We also define a "minor" operator μ and two "lateral continuation" operators $\overset{\bullet}{\ell}_+$ and $\overset{\bullet}{\ell}_-$ by the formulas

$$\mu : \left| \begin{array}{lll} \overset{\triangledown}{E}_\omega & \to & \overset{\vee}{E}_\omega \\ a\,\delta_\omega + \widehat{\phi} & \mapsto & \widehat{\phi}_{|\,]\omega_m,\omega_{m+1}[} \end{array}\right. \qquad \overset{\bullet}{\ell}_\pm : \left| \begin{array}{lll} \overset{\vee}{E}_\omega & \to & \overset{\vee}{E}_{\omega_{m+1}} \\ \overset{\vee}{f} & \mapsto & \operatorname{cont}_{\gamma_\pm}\overset{\vee}{f} \end{array}\right.$$

where γ_+, resp. γ_-, is any path which connects $]\omega_m,\omega_{m+1}[$ and $]\omega_{m+1},\omega_{m+2}[$ staying in a neighbourhood of $]\omega_m,\omega_{m+2}[$ whose intersection with Ω is reduced to $\{\omega_{m+1}\}$ and circumventing ω_{m+1} to the right, resp. to the left.

Having done so for every $\omega \in \Omega \cap d$, we now "gather" the vector spaces $\overset{\triangledown}{E}_\omega$ or $\overset{\vee}{E}_\omega$ and consider the completed graded vector spaces

$$\overset{\triangledown}{E}(\Omega,d) := \widehat{\bigoplus_{\omega\in\Omega\cap d}} \overset{\triangledown}{E}_\omega(\Omega), \qquad \overset{\vee}{E}(\Omega,d) := \widehat{\bigoplus_{\omega\in\Omega\cap d}} \overset{\vee}{E}_\omega(\Omega,d)$$

[17] to be understood as a "the translate of δ from 0 to ω", or "the simple singularity at ω represented by $\frac{1}{2\pi\mathrm{i}(\zeta-\omega)}$"

(we shall often use the abridged notations $\overset{\triangledown}{E}$ or $\overset{\vee}{E}$). This means that, for instance, $\overset{\vee}{E}$ is the cartesian product of all spaces $\overset{\vee}{E}_\omega$, but with additive notation for its elements: they are infinite series

$$\overset{\vee}{\varphi} = \sum_{\omega \in \Omega \cap d} \overset{\vee}{\varphi}{}^{\omega} \in \overset{\vee}{E}, \qquad \overset{\vee}{\varphi}{}^{\omega} \in \overset{\vee}{E}_{\omega} \text{ for each } \omega \in \Omega \cap d. \tag{6.56}$$

This way $\overset{\vee}{E}_{\omega_m} \hookrightarrow \overset{\vee}{E}$ can be considered as the subspace of homogeneous elements of degree m for each m. Beware that $\overset{\vee}{\varphi} \in \overset{\vee}{E}$ may have infinitely many non-zero homogeneous components $\overset{\vee}{\varphi}{}^{\omega}$—this is the difference with the direct sum[18] $\bigoplus_{\omega \in \Omega \cap d} \overset{\vee}{E}_{\omega}$.

We get homogeneous maps

$$\mu\colon \overset{\triangledown}{E} \to \overset{\vee}{E}, \qquad \dot{\sigma}\colon \overset{\vee}{E} \to \overset{\triangledown}{E}, \qquad \dot{\ell}_{\pm}\colon \overset{\vee}{E} \to \overset{\vee}{E}$$

by setting, for instance,

$$\dot{\ell}_{+}\Big(\sum_{\omega \in \Omega \cap d} \overset{\vee}{\varphi}{}^{\omega}\Big) := \sum_{\omega \in \Omega \cap d} \dot{\ell}_{+} \overset{\vee}{\varphi}{}^{\omega}.$$

The maps $\dot{\ell}_{+}$, $\dot{\ell}_{-}$ and $\dot{\sigma}$ are 1-homogeneous, in the sense that for each m they map homogeneous elements of degree m to homogeneous elements of degree $m+1$, while μ is 0-homogeneous. Notice that

$$\mu \circ \dot{\sigma} = \dot{\ell}_{+} - \dot{\ell}_{-}. \tag{6.57}$$

For each $r \in \mathbb{N}^*$, let us define two r-homogeneous operators $\dot{\Delta}_r^+$, $\dot{\Delta}_r\colon \overset{\triangledown}{E} \to \overset{\triangledown}{E}$ by the formulas

$$\dot{\Delta}_r^+ := \dot{\sigma} \circ \dot{\ell}_{+}^{\,r-1} \circ \mu, \qquad \dot{\Delta}_r := \sum_{\varepsilon \in \{+,-\}^{r-1}} \frac{p(\varepsilon)!q(\varepsilon)!}{r!} \, \dot{\sigma} \circ \dot{\ell}_{\varepsilon_{r-1}} \circ \cdots \circ \dot{\ell}_{\varepsilon_1} \circ \mu, \tag{6.58}$$

with notations similar to those of (6.47).

Theorem 6.73. *(i) For each $m \in \mathbb{N}$ and $r \in \mathbb{N}^*$, the diagrams*

---

[18] One can define translation-invariant distances which make $\overset{\vee}{E}$ and $\overset{\triangledown}{E}$ complete metric spaces as follows: let $\operatorname{ord}\colon \overset{\vee}{E} \to \mathbb{N} \cup \{\infty\}$ be the "order function" associated with the decomposition in homogeneous components, i.e. $\operatorname{ord} \overset{\vee}{\varphi} := \min\{ m \in \mathbb{N} \mid \overset{\vee}{\varphi}{}^{\omega_m} \neq 0\}$ if $\overset{\vee}{\varphi} \neq 0$ and $\operatorname{ord} 0 = \infty$, and let $\operatorname{dist}(\overset{\vee}{\varphi}_1, \overset{\vee}{\varphi}_2) := 2^{-\operatorname{ord}(\overset{\vee}{\varphi}_2 - \overset{\vee}{\varphi}_1)}$, and similarly for $\overset{\triangledown}{E}$. This allows one to consider a series of homogeneous components as the limit of the sequence of its partial sums; we thus can say that a series like (6.56) is convergent for the topology of the formal convergence (or "formally convergent"). Compare with Section 5.3.3.

$$\begin{array}{ccc} \mathbb{C}\delta\oplus\widehat{\mathscr{R}}^{\mathrm{simp}}_{-\omega_m+\Omega} & \xrightarrow{\Delta^+_{\omega_{m+r}-\omega_m}} & \mathbb{C}\delta\oplus\widehat{\mathscr{R}}^{\mathrm{simp}}_{-\omega_{m+r}+\Omega} \\ \tau_{\omega_m}\downarrow & & \downarrow\tau_{\omega_{m+r}} \\ \overset{\triangledown}{E}_{\omega_m}(\Omega,d) & \xrightarrow{\dot{\Delta}^+_r} & \overset{\triangledown}{E}_{\omega_{m+r}}(\Omega,d) \end{array} \qquad \begin{array}{ccc} \mathbb{C}\delta\oplus\widehat{\mathscr{R}}^{\mathrm{simp}}_{-\omega_m+\Omega} & \xrightarrow{\Delta_{\omega_{m+r}-\omega_m}} & \mathbb{C}\delta\oplus\widehat{\mathscr{R}}^{\mathrm{simp}}_{-\omega_{m+r}+\Omega} \\ \tau_{\omega_m}\downarrow & & \downarrow\tau_{\omega_{m+r}} \\ \overset{\triangledown}{E}_{\omega_m}(\Omega,d) & \xrightarrow{\dot{\Delta}_r} & \overset{\triangledown}{E}_{\omega_{m+r}}(\Omega,d) \end{array}$$

commute.

(ii) The formulas $\Delta^+_d := \mathrm{Id}+\sum_{r\in\mathbb{N}^*}\dot{\Delta}^+_r$ *and* $\Delta_d := \sum_{r\in\mathbb{N}^*}\dot{\Delta}_r$ *define two operators*

$$\Delta^+_d,\Delta_d\colon \overset{\triangledown}{E}(\Omega,d)\to\overset{\triangledown}{E}(\Omega,d),$$

the first of which has a well-defined logarithm which coincides with the second, i.e.

$$\sum_{r\in\mathbb{N}^*}\dot{\Delta}_r = \sum_{s\in\mathbb{N}^*}\frac{(-1)^{s-1}}{s}\Big(\sum_{r\in\mathbb{N}^*}\dot{\Delta}^+_r\Big)^s. \tag{6.59}$$

(iii) The operator Δ_d *has a well-defined exponential which coincides with* Δ^+_d*, i.e.*

$$\mathrm{Id}+\sum_{r\in\mathbb{N}^*}\dot{\Delta}^+_r = \sum_{s\in\mathbb{N}}\frac{1}{s!}\Big(\sum_{r\in\mathbb{N}^*}\dot{\Delta}_r\Big)^s. \tag{6.60}$$

Proof of Theorem 6.73.

(i) Put together (6.47), (6.55) and (6.58).

(ii) The fact that $\Delta^+_d\colon \overset{\triangledown}{E}\to\overset{\triangledown}{E}$ and its logarithm are well-defined series of operators stems from the r-homogeneity of $\dot{\Delta}^+_r$ for every $r\geq 1$, which ensures formal convergence.
The right-hand side of (6.59) can be written

$$\sum_{\substack{r_1,\dots,r_s\geq 1\\ s\geq 1}}\frac{(-1)^{s-1}}{s}\dot{\Delta}^+_{r_1}\cdots\dot{\Delta}^+_{r_s} = \sum_{\substack{m_1,\dots,m_s\geq 0\\ s\geq 1}}\frac{(-1)^{s-1}}{s}\dot{\sigma}\dot{\ell}_+^{m_1}\mu\,\dot{\sigma}\dot{\ell}_+^{m_2}\cdots\mu\,\dot{\sigma}\dot{\ell}_+^{m_s}\mu$$
$$= \sum_{r\geq 1}\dot{\sigma}B_r\mu,$$

where we have omitted the composition symbol "$\circ$" to lighten notations, made use of (6.58), and availed ourselves of (6.57) to introduce the $(r-1)$-homogeneous operators

$$B_r := \sum_{\substack{m_1+\cdots+m_s+s=r\\ m_1,\dots,m_s\geq 0,\ s\geq 1}}\frac{(-1)^{s-1}}{s}\dot{\ell}_+^{m_1}(\dot{\ell}_+-\dot{\ell}_-)\dot{\ell}_+^{m_2}\cdots(\dot{\ell}_+-\dot{\ell}_-)\dot{\ell}_+^{m_s},$$

with the convention $B_1 = \mathrm{Id}$. It is an exercise in non-commutative algebra to check that

$$B_r = \sum_{\varepsilon\in\{+,-\}^{r-1}} \frac{p(\varepsilon)!q(\varepsilon)!}{r!} \dot{\ell}_{\varepsilon_{r-1}} \cdots \dot{\ell}_{\varepsilon_1}$$

(viewed as an identity between polynomials in two non-commutative variables $\dot{\ell}_+$ and $\dot{\ell}_-$), hence (6.58) shows that $\dot{\sigma} B_r \mu = \dot{\Delta}_r$ and we are done.

(iii) Clearly equivalent to (ii).

□

Definition 6.74. – The elements of $\overset{\triangledown}{E}(\Omega,d)$ are called *Ω-resurgent symbols with support in d.*

– The operator Δ_d^+ is called the *symbolic Stokes automorphism for the direction d.*

– The operator Δ_d is called the *symbolic Stokes infinitesimal generator for the direction d.*

The connection between Δ_d^+ and the Stokes phenomenon will be explained in next section. This operator is clearly a linear invertible map, but there is a further reason why it deserves the name "automorphism": we shall see in Section 6.13.2 that, when Ω is stable under addition, there is a natural algebra structure for which Δ_d^+ is an algebra automorphism.

Theorem 6.73 implies Theorem 6.72. Given Ω and a ray d, Theorem 6.73(i) says that

$$\dot{\Delta}_{r|\overset{\triangledown}{E}_{\omega_m}} = \tau_{\omega_{m+r}} \circ \Delta_{\omega_{m+r}-\omega_m} \circ \tau_{\omega_m}^{-1}, \qquad \dot{\Delta}^+_{r|\overset{\triangledown}{E}_{\omega_m}} = \tau_{\omega_{m+r}} \circ \Delta^+_{\omega_{m+r}-\omega_m} \circ \tau_{\omega_m}^{-1} \tag{6.61}$$

for every m and r. By restricting the identity (6.59) to $\overset{\triangledown}{E}_0$ and extracting homogeneous components we get the identity

$$\dot{\Delta}_{r|\overset{\triangledown}{E}_0} = \sum_{s\in\mathbb{N}^*} \frac{(-1)^{s-1}}{s} \sum_{r_1+\cdots+r_s=r} \dot{\Delta}^+_{r_s|\overset{\triangledown}{E}_{\omega_{r_1+\cdots+r_{s-1}}}} \circ\cdots\circ \dot{\Delta}^+_{r_2|\overset{\triangledown}{E}_{\omega_{r_1}}} \circ \dot{\Delta}^+_{r_1|\overset{\triangledown}{E}_0}$$

for each $r\in\mathbb{N}^*$, which is equivalent, by (6.61), to

$$\Delta_{\omega_r} = \sum_{s\in\mathbb{N}^*} \frac{(-1)^{s-1}}{s} \sum_{r_1+\cdots+r_s=r} \Delta^+_{\omega_r-\omega_{r_1+\cdots+r_{s-1}}} \circ\cdots\circ \Delta^+_{\omega_{r_1+r_2}-\omega_{r_1}} \circ \Delta^+_{\omega_{r_1}}. \tag{6.62}$$

Given $\omega\in\Omega\setminus\{0\}$, we can apply this with the ray $\{t\omega \mid t\geq 0\}$: the notations (6.45) and (6.54) agree for $1\leq m<r$, with $r\in\mathbb{N}^*$ defined by $\omega=\omega_r$; the identity (6.62) is then seen to be equivalent to (6.52) by the change of indices

$$\eta_1 = \omega_{r_1},\ \eta_2 = \omega_{r_1+r_2},\ \ldots\ ,\ \eta_{s-1} = \omega_{r_1+\cdots+r_{s-1}}.$$

The identity (6.53) is obtained the same way from (6.60). □

Exercise 6.75. Show that, for each $r\in\mathbb{N}^*$, the r-homogeneous component of

$$\Delta_d^- := \exp(-\Delta_d) = \left(\Delta_d^+\right)^{-1}$$

is $\dot{\Delta}_r^- := -\dot{\sigma} \circ \dot{\ell}_-^{r-1} \circ \mu$, giving rise to the family of operators $\Delta_\omega^- := -\mathscr{A}^{\Omega}_{\omega,(-,\dots,-)}$, $\omega \in \Omega \setminus \{0\}$.

6.12.3 Relation with the Laplace transform and the Stokes phenomenon

We keep the notations of the previous section, in particular $d = \{t\,\mathrm{e}^{\mathrm{i}\theta} \mid t \geq 0\}$ with $\theta \in \mathbb{R}$ fixed. With a view to use Borel-Laplace summation, we suppose that I is an open interval of length $< \pi$ which contains θ, such that the intersection of the sector $\{\,\xi\,\mathrm{e}^{\mathrm{i}\theta'} \mid \xi > 0,\ \theta' \in I\,\}$ with Ω is contained in d:

$$\Omega \cap \{\,\xi\,\mathrm{e}^{\mathrm{i}\theta'} \mid \xi \geq 0,\ \theta' \in I\,\} = \{\omega_m\}_{m \in \mathbb{N}} \subset d, \qquad \omega_0 = 0 \prec \omega_1 \prec \omega_2 \prec \cdots$$

We then set

$$I^+ := \{\,\theta^+ \in I \mid \theta^+ < \theta\,\}, \qquad I^- := \{\,\theta^- \in I \mid \theta^- > \theta\,\}$$

(mark the somewhat odd convention: the idea is that the directions of I^+ are to the right of d, and those of I^- to the left) and

$$I_\varepsilon^+ := \{\,\theta^+ \in I \mid \theta^+ < \theta - \varepsilon\,\}, \qquad I_\varepsilon^- := \{\,\theta^- \in I \mid \theta^- > \theta + \varepsilon\,\}$$

for $0 < \varepsilon < \min\left\{\frac{\pi}{2}, \operatorname{dist}(\theta, \partial I)\right\}$.

Let us give ourselves a locally bounded function $\gamma\colon I^+ \cup I^- \to \mathbb{R}$. Recall that in Section 5.9.2 we have defined the spaces $\mathscr{N}(I^\pm, \gamma)$, consisting of holomorphic germs at 0 which extend analytically to the sector $\{\,\xi\,\mathrm{e}^{\mathrm{i}\theta^\pm} \mid \xi > 0,\ \theta^\pm \in I\,\}$, with at most exponential growth along each ray $\mathbb{R}^+\mathrm{e}^{\mathrm{i}\theta^\pm}$ as prescribed by $\gamma(\theta^\pm)$, and that according to Section 5.9.3, the Laplace transform gives rise to two operators $\mathscr{L}^{I^+}$ and $\mathscr{L}^{I^-}$ defined on $\mathbb{C}\,\delta \oplus \mathscr{N}(I^+, \gamma)$ and $\mathbb{C}\,\delta \oplus \mathscr{N}(I^-, \gamma)$, producing functions holomorphic in the domains $\mathscr{D}(I^+, \gamma)$ or $\mathscr{D}(I^-, \gamma)$.

The domains $\mathscr{D}(I^+, \gamma)$ and $\mathscr{D}(I^-, \gamma)$ are sectorial neighbourhoods of ∞ which overlap: their intersection is a sectorial neighbourhood of ∞ centred on the ray $\arg z = -\theta$, with aperture π. For a formal series $\widetilde{\varphi}$ such that $\mathscr{B}\widetilde{\varphi} \in \mathbb{C}\,\delta \oplus \big(\mathscr{N}(I^+, \gamma) \cap \mathscr{N}(I^-, \gamma)\big)$, the Borel sums $\mathscr{S}^{I^+}\widetilde{\varphi} = \mathscr{L}^{I^+}\mathscr{B}\widetilde{\varphi}$ and $\mathscr{S}^{I^-}\widetilde{\varphi} = \mathscr{L}^{I^-}\mathscr{B}\widetilde{\varphi}$ may differ, but their difference is exponentially small on $\mathscr{D}(I^+, \gamma) \cap \mathscr{D}(I^-, \gamma)$. We shall investigate more precisely this difference when $\mathscr{B}\widetilde{\varphi}$ satisfies further assumptions.

To get uniform estimates, we shall restrict to a domain of the form $\mathscr{D}(I_\varepsilon^+, \gamma + \varepsilon) \cap \mathscr{D}(I_\varepsilon^-, \gamma + \varepsilon)$, which is a sectorial neighbourhood of ∞ of aperture $\pi - 2\varepsilon$ centred on the ray $\arg z = -\theta$.

Notation 6.76 *For each $m \in \mathbb{N}$, we set $\overset{\triangledown}{E}(\Omega,d,m) := \bigoplus_{j=0}^{m} \overset{\triangledown}{E}_{\omega_j}(\Omega)$ and denote by $[\cdot]_m$ the canonical projection*

$$\Phi = \sum_{\omega\in\Omega\cap d} \Phi^{\omega} \in \overset{\triangledown}{E}(\Omega,d) \mapsto [\Phi]_m := \sum_{j=0}^{m} \Phi^{\omega_j} \in \overset{\triangledown}{E}(\Omega,d,m).$$

For each $\omega \in \Omega \cap d$, we set

$$\overset{\triangledown}{E}{}^{I,\gamma}_{\omega}(\Omega) := \tau_\omega\Big(\mathbb{C}\,\delta \oplus \big(\widehat{\mathscr{R}}^{\text{simp}}_{-\omega+\Omega} \cap \mathscr{N}(I^+,\gamma) \cap \mathscr{N}(I^-,\gamma)\big)\Big) \subset \overset{\triangledown}{E}_{\omega}(\Omega).$$

We also define $\overset{\triangledown}{E}{}^{I,\gamma}(\Omega,d,m) := \bigoplus_{j=0}^{m} \overset{\triangledown}{E}{}^{I,\gamma}_{\omega_j}(\Omega) \subset \overset{\triangledown}{E}(\Omega,d,m)$, on which we define the "Laplace operators" $\mathscr{L}^+$ and $\mathscr{L}^-$ by

$$\Phi = \sum_{j=0}^{m} \Phi^{\omega_j} \mapsto \mathscr{L}^{\pm}\Phi \textit{ holomorphic in } \mathscr{D}(I\pm,\gamma),$$

$$\mathscr{L}^{\pm}\Phi(z) := \sum_{j=0}^{m} \mathrm{e}^{-\omega_j z} \mathscr{L}^{I^{\pm}}(\tau_{\omega_j}^{-1}\Phi^{\omega_j})(z).$$

Theorem 6.77. *Consider $m \in \mathbb{N}$ and real numbers ρ and ε such that $|\omega_m| < \rho < |\omega_{m+1}|$ and $0 < \varepsilon < \min\big\{\frac{\pi}{2}, \operatorname{dist}(\theta,\partial I)\big\}$. Then, for any $\Phi \in \overset{\triangledown}{E}{}^{I,\gamma}(\Omega,d,m)$ such that $[\Delta_d^+\Phi]_m \in \overset{\triangledown}{E}{}^{I,\gamma}(\Omega,d,m)$, one has*

$$\mathscr{L}^+\Phi(z) = \mathscr{L}^-[\Delta_d^+\Phi]_m(z) + O(\mathrm{e}^{-\rho\,\Re e(\mathrm{e}^{\mathrm{i}\theta}z)}) \tag{6.63}$$

uniformly for $z \in \mathscr{D}(I_\varepsilon^+,\gamma+\varepsilon) \cap \mathscr{D}(I_\varepsilon^-,\gamma+\varepsilon)$.

Proof. It is sufficient to prove it for each homogeneous component of Φ, so we can assume $\Phi = a\,\delta_{\omega_j} + \widehat{\varphi} \in \overset{\triangledown}{E}{}^{I,\gamma}_{\omega_j}(\Omega)$, with $0 \le j \le m$. Given $z \in \mathscr{D}(I_\varepsilon^+,\gamma+\varepsilon) \cap \mathscr{D}(I_\varepsilon^-,\gamma+\varepsilon)$, we choose $\theta^+ \in I_\varepsilon^+$ and $\theta^- \in I_\varepsilon^-$ so that $\zeta \mapsto \mathrm{e}^{-z\zeta}$ is exponentially decreasing on the rays $\mathbb{R}^+\mathrm{e}^{\mathrm{i}\theta^\pm}$. Then $\mathscr{L}^\pm\Phi(z)$ can be written $a\mathrm{e}^{-\omega_j z} + \int_{\omega_j}^{\mathrm{e}^{\mathrm{i}\theta^\pm}\infty} \mathrm{e}^{-z\zeta}\widehat{\varphi}(\zeta)\,\mathrm{d}\zeta$ (by the very definition of τ_{ω_j}). Decomposing the integration path as indicated on Figure 6.10, we get

$$\begin{aligned}\mathscr{L}^+\Phi(z) &= a\mathrm{e}^{-\omega_j z} + \Big(\int_{\omega_j}^{\mathrm{e}^{\mathrm{i}\theta^-}\infty} + \int_{\gamma_1} + \cdots + \int_{\gamma_{m-j}} + \int_{C_\rho}\Big)\mathrm{e}^{-z\zeta}\widehat{\varphi}(\zeta)\,\mathrm{d}\zeta \\ &= \mathscr{L}^-\Phi(z) + \sum_{r=1}^{m-j}\int_{\gamma_r} \mathrm{e}^{-z\zeta}\,\overset{\bullet}{\ell}{}_+^{r-1}\mu\Phi(\zeta)\,\mathrm{d}\zeta + \int_{C_\rho} \mathrm{e}^{-z\zeta}\,\overset{\bullet}{\ell}{}_+^{m-j-1}\mu\Phi(\zeta)\,\mathrm{d}\zeta,\end{aligned}$$

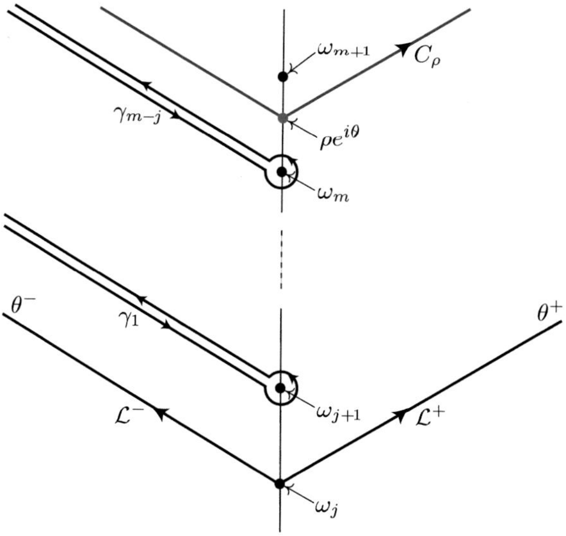

Fig. 6.10: From the Stokes phenomenon to the symbolic Stokes automorphism.

where the contour C_ρ consists of the negatively oriented half-line $[\rho\,\mathrm{e}^{\mathrm{i}\theta},\mathrm{e}^{\mathrm{i}\theta^-}\infty)$ followed by the positively oriented half-line $[\rho\,\mathrm{e}^{\mathrm{i}\theta},\mathrm{e}^{\mathrm{i}\theta^+}\infty)$. We recognize in the $m-j$ terms of the sum in the right-hand side the "Laplace transform of majors" of Section 6.9.2 applied to the homogeneous components of $[\Delta_d^+\Phi]_m$; all these integrals are convergent by virtue of our hypothesis that $[\Delta_d^+\Phi]_m \in \overset{\triangledown}{E}^{I,\gamma}(\Omega,d,m)$, and also the last term in the right-hand side is seen to be a convergent integral which yields an $O(\mathrm{e}^{-\rho\,\Re\mathrm{e}(\mathrm{e}^{\mathrm{i}\theta}z)})$ error term. □

Observe that the meaning of (6.63) for $\Phi = \widehat{\varphi} \in \overset{\triangledown}{E}{}_0^{I,\gamma}(\Omega)$, i.e. $\widehat{\varphi} \in \widehat{\mathscr{R}}_\Omega^{\mathrm{simp}} \cap \mathscr{N}(I^+,\gamma)\cap\mathscr{N}(I^-,\gamma)$, is

$$\mathscr{L}^{\theta^+}\widehat{\varphi} = \mathscr{L}^{\theta^-}\widehat{\varphi} + \mathrm{e}^{-\omega_1 z}\mathscr{L}^{\theta^-}\Delta^+_{\omega_1}\widehat{\varphi}(z) + \cdots + \mathrm{e}^{-\omega_m z}\mathscr{L}^{\theta^-}\Delta^+_{\omega_m}\widehat{\varphi}(z) + O(\mathrm{e}^{-\rho\,\Re\mathrm{e}(\mathrm{e}^{\mathrm{i}\theta}z)}).$$

The idea is that the action of the symbolic Stokes automorphism yields the exponentially small corrections needed to pass from the Borel sum $\mathscr{L}^{\theta^+}\widehat{\varphi}$ to the Borel sum $\mathscr{L}^{\theta^-}\widehat{\varphi}$. It is sometimes possible to pass to the limit $m\to\infty$ and to get rid of any error term, in which case one could be tempted to write

$$\text{“}\mathscr{L}^+ = \mathscr{L}^- \circ \Delta_d^+\text{”}. \tag{6.64}$$

Example 6.78. The simplest example of all is again provided by the Euler series, for which there is only one singular ray, $d=\mathbb{R}^-$. Taking any $\Omega\subset\mathbb{R}^-$ containing -1, we have

$$\Delta^+_{\mathbb{R}^-}\widehat{\varphi}^{\mathrm{E}} = \widehat{\varphi}^{\mathrm{E}} + 2\pi\mathrm{i}\,\delta_{-1} \tag{6.65}$$

(in view of (6.48)). If we set $I^+ := (\frac{\pi}{2},\pi)$ and $I^- := (\pi,\frac{3\pi}{2})$, then the functions $\varphi^+ = \mathscr{L}^+\widehat{\varphi}^{\mathrm{E}}$ and $\varphi^- = \mathscr{L}^-\widehat{\varphi}^{\mathrm{E}}$ coincide with those of Section 5.10. Recall that one can take $\gamma = 0$ in this case, so $\varphi^\pm$ is holomorphic in $\mathscr{D}(I^\pm,0)$ (at least) and the intersection $\mathscr{D}(I^+,0)\cap\mathscr{D}(I^-,0)$ is the half-plane $\{\Re e\, z<0\}$, on which Theorem 6.77 implies

$$\varphi^+ = \varphi^- + 2\pi\mathrm{i}\,\mathrm{e}^z,$$

which is consistent with formula (5.45).

Example 6.79. Similarly, for Poincaré's example with parameter $s\in\mathbb{C}$ of negative real part, according to Section 5.12, the singular rays are $d_k := \mathbb{R}^+\mathrm{e}^{\mathrm{i}\theta_k}$, $k\in\mathbb{Z}$, with $\omega_k = s+2\pi\mathrm{i}k$ and $\theta_k := \arg\omega_k \in (\frac{\pi}{2},\frac{3\pi}{2})$. We take any Ω contained the union of these rays and containing $s+2\pi\mathrm{i}\mathbb{Z}$. For fixed k, we can set $I^+ := J_{k-1} = (\arg\omega_{k-1},\arg\omega_k)$, $I^- := J_k = (\arg\omega_k,\arg\omega_{k+1})$, and $\gamma(\theta)\equiv\cos\theta$. Then, according to Theorem 5.50, the Borel sums $\mathscr{L}^+\widehat{\varphi}^{\mathrm{P}} = \mathscr{S}^{J_{k-1}}\widetilde{\varphi}^{\mathrm{P}}$ and $\mathscr{L}^-\widehat{\varphi}^{\mathrm{P}} = \mathscr{S}^{J_k}\widetilde{\varphi}^{\mathrm{P}}$ are well defined. In view of (6.48), we have

$$\Delta^+_{d_k}\widehat{\varphi}^{\mathrm{P}} = \widehat{\varphi}^{\mathrm{P}} + 2\pi\mathrm{i}\,\delta_{\omega_k}, \quad \text{hence} \quad \mathscr{L}^+\widehat{\varphi}^{\mathrm{P}} = \mathscr{L}^-\widehat{\varphi}^{\mathrm{P}} + 2\pi\mathrm{i}\,\mathrm{e}^{-\omega_k z}$$

by Theorem 6.77, which is consistent with (5.66).

Example 6.80. The asymptotic expansion $\widetilde{\varphi}^{\mathrm{H}}_s(z)$ of the Hurwitz zeta function was studied in Exercise 6.59. For $s\geq 2$ integer, with $I = (-\frac{\pi}{2},\frac{\pi}{2})$, we have

$$\varphi^+(z) := \sum_{k\in\mathbb{N}}(z+k)^{-s} = \mathscr{S}^I\widetilde{\varphi}^{\mathrm{H}}_s(z)$$

for $z\in\mathscr{D}(I,0) = \mathbb{C}\setminus\mathbb{R}^-$. With the help of the difference equation $\varphi(z)-\varphi(z+1) = z^{-s}$, it is an exercise to check that

$$\varphi^-(z) := -\sum_{k\in\mathbb{N}^*}(z-k)^{-s}$$

coincides with the Borel sum $\mathscr{S}^J\widetilde{\varphi}^{\mathrm{H}}_s$ defined on $\mathscr{D}(J,0) = \mathbb{C}\setminus\mathbb{R}^+$, with $J = (\frac{\pi}{2},\frac{3\pi}{2})$ or $(-\frac{3\pi}{2},-\frac{\pi}{2})$. In this case, we can take $\Omega = 2\pi\mathrm{i}\mathbb{Z}^*$ and we have two singular rays, $\mathrm{i}\mathbb{R}^+$ and $\mathrm{i}\mathbb{R}^-$, for each of which the symbolic Stokes automorphism yields infinitely many non-trivial homogeneous components: indeed, according to (6.48),

$$\Delta^+_{\mathrm{i}\mathbb{R}^+}\widehat{\varphi}^{\mathrm{H}}_s = \widehat{\varphi}^{\mathrm{H}}_s + \frac{2\pi\mathrm{i}}{\Gamma(s)}\sum_{m=1}^{\infty}(2\pi\mathrm{i}m)^{s-1}\delta_{2\pi\mathrm{i}m},$$

$$\Delta^+_{\mathrm{i}\mathbb{R}^-}\widehat{\varphi}^{\mathrm{H}}_s = \widehat{\varphi}^{\mathrm{H}}_s + \frac{2\pi\mathrm{i}}{\Gamma(s)}\sum_{m=1}^{\infty}(-2\pi\mathrm{i}m)^{s-1}\delta_{-2\pi\mathrm{i}m}.$$

Applying Theorem 6.77 with $I^+ = (0,\frac{\pi}{2})$ and $I^- = (\frac{\pi}{2},\pi)$, or with $I^+ = (-\pi,-\frac{\pi}{2})$ and $I^- = (-\frac{\pi}{2},0)$, for each $m\in\mathbb{N}$ we get

$$\Im\mathrm{m}\, z < 0 \implies \varphi^+(z) = \varphi^-(z) + \frac{2\pi\mathrm{i}}{\Gamma(s)} \sum_{j=1}^{m} (2\pi\mathrm{i} j)^{s-1} \mathrm{e}^{-2\pi\mathrm{i} jz} + O(\mathrm{e}^{-2\pi(m+\frac{1}{2})|\Im\mathrm{m}\, z|}), \tag{6.66}$$

$$\Im\mathrm{m}\, z > 0 \implies \varphi^-(z) = \varphi^+(z) + \frac{2\pi\mathrm{i}}{\Gamma(s)} \sum_{j=1}^{m} (-2\pi\mathrm{i} j)^{s-1} \mathrm{e}^{2\pi\mathrm{i} jz} + O(\mathrm{e}^{-2\pi(m+\frac{1}{2})|\Im\mathrm{m}\, z|}) \tag{6.67}$$

(and the constants implied in these estimates are uniform provided one restricts oneself to $|z| > \varepsilon$ and $|\arg z \pm \mathrm{i}\frac{\pi}{2}| < \frac{\pi}{2} - \varepsilon$). In this case we see that we can pass to the limit $m \to \infty$ because the finite sums involved in (6.66)–(6.67) are the partial sums of convergent series. In fact this could be guessed in advance: since φ^+ and φ^- satisfy the same difference equation $\varphi(z) - \varphi(z+1) = z^{-s}$, their difference yields 1-periodic functions holomorphic in the half-planes $\{\Im\mathrm{m}\, z < 0\}$ and $\{\Im\mathrm{m}\, z > 0\}$, which thus have convergent Fourier series of the form[19]

$$(\varphi^+ - \varphi^-)_{|\{\Im\mathrm{m} z<0\}} = \sum_{m\ge 0} A_m \mathrm{e}^{-2\pi\mathrm{i} mz}, \qquad (\varphi^+ - \varphi^-)_{|\{\Im\mathrm{m} z>0\}} = \sum_{m\ge 0} B_m \mathrm{e}^{2\pi\mathrm{i} mz},$$

but the finite sums in (6.66)–(6.67) are nothing but the partial sums of these series (up to sign for the second). So, in this case, the symbolic Stokes automorphism delivers the Fourier coefficients of the diffence between the two Borel sums:

$$\sum_{k\in\mathbb{Z}} (z+k)^{-s} = \begin{cases} \frac{2\pi\mathrm{i}}{\Gamma(s)} \sum_{m=1}^{\infty} (2\pi\mathrm{i} m)^{s-1} \mathrm{e}^{-2\pi\mathrm{i} mz} & \text{for } \Im\mathrm{m}\, z < 0, \\ \frac{2\pi\mathrm{i}}{\Gamma(s)} \sum_{m=1}^{\infty} (-1)^s (2\pi\mathrm{i} m)^{s-1} \mathrm{e}^{2\pi\mathrm{i} mz} & \text{for } \Im\mathrm{m}\, z > 0. \end{cases}$$

Example 6.81. The case of the Stirling series $\widetilde{\mu}$ studied in Section 5.11 is somewhat similar, with (6.48) yielding

$$\Delta^+_{\mathrm{i}\mathbb{R}^+} \widehat{\mu} = \widehat{\mu} + \sum_{m\in\mathbb{N}^*} \frac{1}{m} \delta_{2\pi\mathrm{i} m}, \qquad \Delta^+_{\mathrm{i}\mathbb{R}^-} \widehat{\mu} = \widehat{\mu} - \sum_{m\in\mathbb{N}^*} \frac{1}{m} \delta_{-2\pi\mathrm{i} m}. \tag{6.68}$$

Here we get

$$\Im\mathrm{m}\, z < 0 \implies \mu^+(z) = \mu^-(z) + \sum_{m=1}^{\infty} \tfrac{1}{m} \mathrm{e}^{-2\pi\mathrm{i} mz} = \mu^-(z) - \log(1 - \mathrm{e}^{-2\pi\mathrm{i} z}), \tag{6.69}$$

$$\Im\mathrm{m}\, z > 0 \implies \mu^-(z) = \mu^+(z) - \sum_{m=1}^{\infty} \tfrac{1}{m} \mathrm{e}^{2\pi\mathrm{i} mz} = \mu^+(z) + \log(1 - \mathrm{e}^{2\pi\mathrm{i} z}) \tag{6.70}$$

(compare with Exercise 5.46).

---

19 See Section 7.5.

6.12.4 Extension of the inverse Borel transform to Ω-resurgent symbols

In the previous section, we have defined the Laplace operators $\mathscr{L}^+$ and $\mathscr{L}^-$ on

$$\overset{\triangledown}{E}{}^{I,\gamma}(\Omega,d,m) \subset \overset{\triangledown}{E}(\Omega,d,m) \subset \overset{\triangledown}{E}(\Omega,d),$$

i.e. the Ω-resurgent symbols to which they can be applied are subjected to two constraints: finitely many non-trivial homogeneous components, with at most exponential growth at infinity for their minors. There is a natural way to define on the whole of $\overset{\triangledown}{E}(\Omega,d)$ a formal Laplace operator, which is an extension of the inverse Borel transform $\mathscr{B}^{-1}$ on $\mathbb{C}\,\delta \oplus \widehat{\mathscr{R}}^{\text{simp}}_{\Omega}$. Indeed, replacing the function $\mathrm{e}^{-\omega z} = \mathscr{L}^{\pm}\delta_{\omega}$ by a symbol $\mathrm{e}^{-\omega z}$, we define

$$\widetilde{E}_{\omega}(\Omega) := \mathrm{e}^{-\omega z}\widetilde{\mathscr{R}}^{\text{simp}}_{-\omega+\Omega} \quad \text{for } \omega \in \Omega \cap d, \qquad \widetilde{E}(\Omega,d) := \widehat{\bigoplus_{\omega \in \Omega \cap d}} \widetilde{E}_{\omega}(\Omega), \tag{6.71}$$

i.e. we take the completed graded vector space obtained as cartesian product of the spaces $\widetilde{\mathscr{R}}^{\text{simp}}_{-\omega+\Omega}$, representing its elements by formal expressions of the form $\widetilde{\Phi} = \sum_{\omega \in \Omega \cap d} \mathrm{e}^{-\omega z}\widetilde{\Phi}_{\omega}(z)$, where each $\widetilde{\Phi}_{\omega}(z)$ is a formal series and $\mathrm{e}^{-\omega z}$ is just a symbol meant to distinguish the various homogeneous components. We thus have for each $\omega \in \Omega \cap d$ a linear isomorphism

$$\widetilde{\tau}_{\omega} \colon \widetilde{\varphi}(z) \in \widetilde{\mathscr{R}}^{\text{simp}}_{-\omega+\Omega} \mapsto \mathrm{e}^{-\omega z}\widetilde{\varphi}(z) \in \widetilde{E}_{\omega}(\Omega),$$

which allow us to define

$$\mathscr{B}_{\omega} := \tau_{\omega} \circ \mathscr{B} \circ \widetilde{\tau}_{\omega}^{-1} \colon \widetilde{E}_{\omega}(\Omega) \xrightarrow{\sim} \overset{\triangledown}{E}_{\omega}(\Omega).$$

The map $\mathscr{B}_0$ can be identified with the Borel transform $\mathscr{B}$ acting on simple Ω-resurgent series; putting together the maps $\mathscr{B}_{\omega}$, $\omega \in \Omega \cap d$, we get a linear isomorphism

$$\mathscr{B} \colon \widetilde{E}(\Omega,d) \xrightarrow{\sim} \overset{\triangledown}{E}(\Omega,d),$$

which we can consider as the Borel transform acting on "Ω-resurgent transseries in the direction d", and whose inverse can be considered as the formal Laplace transform acting on Ω-resurgent symbols in the direction d.

Observe that, if $\mathrm{e}^{-\omega z}\widetilde{\varphi}(z) \in \widetilde{E}_{\omega}(\Omega)$ is such that $\widetilde{\varphi}(z)$ is 1-summable in the directions of $I^+ \cup I^-$, then $\mathscr{B}(\mathrm{e}^{-\omega z}\widetilde{\varphi}) \in \overset{\triangledown}{E}{}^{I,\gamma}_{\omega}(\Omega)$ and

$$\mathscr{L}^{\pm}\mathscr{B}(\mathrm{e}^{-\omega z}\widetilde{\varphi}) = \mathrm{e}^{-\omega z}\mathscr{S}^{I^{\pm}}\widetilde{\varphi}.$$

Beware that in the above identity, $\mathrm{e}^{-\omega z}$ is a *symbol* in the left-hand side, whereas it is a *function* in the right-hand side.

Via $\mathscr{B}$, the operators Δ_d^+ and Δ_d give rise to operators which we denote with the same symbols:

$$\Delta_d^+, \Delta_d \colon \widetilde{E}(\Omega, d) \to \widetilde{E}(\Omega, d),$$

so that we can e.g. rephrase (6.65) as

$$\Delta_{\mathbb{R}^-}^+ \widetilde{\varphi}^{\mathrm{E}} = \widetilde{\varphi}^{\mathrm{E}} + 2\pi\mathrm{i}\,\mathrm{e}^z \tag{6.72}$$

or (6.68) as

$$\Delta_{\mathrm{i}\mathbb{R}^+}^+ \widetilde{\mu} = \widetilde{\mu} + \sum_{m\in\mathbb{N}^*} \frac{1}{m} \mathrm{e}^{-2\pi\mathrm{i}mz}, \qquad \Delta_{\mathrm{i}\mathbb{R}^-}^+ \widetilde{\mu} = \widetilde{\mu} - \sum_{m\in\mathbb{N}^*} \frac{1}{m} \mathrm{e}^{2\pi\mathrm{i}mz}. \tag{6.73}$$

In Section 6.13.2, we shall see that, if Ω is stable under addition, then $\overset{\triangledown}{E}(\Omega, d)$ and thus also $\widetilde{E}(\Omega, d)$ have algebra structures, for which it is legitimate to write

$$-\log(1-\mathrm{e}^{-2\pi\mathrm{i}z}) = \sum_{m\in\mathbb{N}^*} \frac{1}{m} \mathrm{e}^{-2\pi\mathrm{i}mz}, \quad \log(1-\mathrm{e}^{2\pi\mathrm{i}z}) = -\sum_{m\in\mathbb{N}^*} \frac{1}{m} \mathrm{e}^{2\pi\mathrm{i}mz}. \tag{6.74}$$

Remark 6.82. One can always extend the definition of $\partial = \frac{\mathrm{d}}{\mathrm{d}z}$ to $\widetilde{E}(\Omega, d)$ by setting

$$\widetilde{\varphi} \in \widetilde{\mathscr{R}}^{\mathrm{simp}}_{-\omega+\Omega} \implies \partial(\mathrm{e}^{-\omega z}\widetilde{\varphi}) := \mathrm{e}^{-\omega z}(-\omega + \partial)\widetilde{\varphi}.$$

(When Ω is stable under addition ∂ will be a derivation of the algebra $\widetilde{E}(\Omega, d)$, which will thus be a differential algebra.)

On the other hand, writing as usual $\Omega \cap d = \{0 = \omega_0 \prec \omega_1 \prec \omega_2 \prec \cdots\}$, we see that the homogeneous components of Δ_d^+ and Δ_d acting on $\widetilde{E}(\Omega, d)$ (Borel counterparts of the operators $\dot{\Delta}_r^+$, $\dot{\Delta}_r \colon \overset{\triangledown}{E} \to \overset{\triangledown}{E}$ defined by (6.58)) act as follows on $\widetilde{E}_\omega(\Omega)$ for each $\omega = \omega_m \in \Omega \cap d$:

$$\widetilde{\varphi} \in \widetilde{\mathscr{R}}^{\mathrm{simp}}_{-\omega_m+\Omega} \implies \begin{cases} \dot{\Delta}_r^+(\mathrm{e}^{-\omega_m z}\widetilde{\varphi}) = \mathrm{e}^{-\omega_{m+r}z}\Delta^+_{\omega_{m+r}-\omega_m}\widetilde{\varphi}, \\ \dot{\Delta}_r(\mathrm{e}^{-\omega_m z}\widetilde{\varphi}) = \mathrm{e}^{-\omega_{m+r}z}\Delta_{\omega_{m+r}-\omega_m}\widetilde{\varphi}. \end{cases} \tag{6.75}$$

Formula (6.49) then says

$$\dot{\Delta}_r^+ \partial\phi = \partial\,\dot{\Delta}_r^+ \phi, \quad \dot{\Delta}_r \partial\phi = \partial\,\dot{\Delta}_r \phi$$

for every $\phi \in \widetilde{E}(\Omega, d)$, whence $\Delta_d^+ \circ \partial = \partial \circ \Delta_d^+$ and $\Delta_d \circ \partial = \partial \circ \Delta_d$.

6.13 The operators Δ_ω are derivations

We now investigate the way the operators Δ_ω and Δ^+_ω act on a product of two terms (convolution product or Cauchy product, according as one works with formal series or their Borel transforms).

Let Ω' and Ω'' be non-empty closed discrete subsets of $\mathbb{C}$ such that

$$\Omega := \Omega' \cup \Omega'' \cup (\Omega' + \Omega'') \tag{6.76}$$

is also closed and discrete. Recall that, according to Theorem 6.27,

$$\widetilde{\varphi} \in \widetilde{\mathscr{R}}_{\Omega'} \ \text{ and } \ \widetilde{\psi} \in \widetilde{\mathscr{R}}_{\Omega''} \quad \Longrightarrow \quad \widetilde{\varphi}\widetilde{\psi} \in \widetilde{\mathscr{R}}_{\Omega}.$$

6.13.1 Generalized Leibniz rule for the operators Δ^+_ω

We begin with the operators Δ^+_ω.

Theorem 6.83. *Let $\widetilde{\varphi} \in \widetilde{\mathscr{R}}^{\text{simp}}_{\Omega'}$ and $\widetilde{\psi} \in \widetilde{\mathscr{R}}^{\text{simp}}_{\Omega''}$. Then $\widetilde{\varphi}\widetilde{\psi} \in \widetilde{\mathscr{R}}^{\text{simp}}_{\Omega}$ and, for every $\omega \in \Omega \setminus \{0\}$,*

$$\Delta^+_\omega(\widetilde{\varphi}\widetilde{\psi}) = (\Delta^+_\omega\widetilde{\varphi})\widetilde{\psi} + \sum_{\substack{\omega=\omega'+\omega'' \\ \omega'\in\Omega'\cap]0,\omega[,\, \omega''\in\Omega''\cap]0,\omega[}} (\Delta^+_{\omega'}\widetilde{\varphi})(\Delta^+_{\omega''}\widetilde{\psi}) \;\; + \widetilde{\varphi}(\Delta^+_\omega\widetilde{\psi}). \tag{6.77}$$

Proof. **a)** The fact that $\widetilde{\varphi}\widetilde{\psi} \in \widetilde{\mathscr{R}}^{\text{simp}}_{\Omega}$ follows from the proof of formula (6.77) and Theorem 6.68, we omit the details.

b) To prove formula (6.77), we define

$$\Sigma_\omega := \{\, \eta \in]0,\omega[\mid \eta \in \Omega' \cup \Omega'' \text{ or } \omega - \eta \in \Omega' \cup \Omega'' \,\}$$

and write $\mathscr{B}\widetilde{\varphi} = a\,\delta + \widehat{\varphi}$, $\mathscr{B}\widetilde{\psi} = b\,\delta + \widehat{\psi}$, with $a, b \in \mathbb{C}$, $\widehat{\varphi} \in \widehat{\mathscr{R}}^{\text{simp}}_{\Omega'}$, $\widehat{\psi} \in \widehat{\mathscr{R}}^{\text{simp}}_{\Omega''}$,

$$\mathscr{B}\Delta^+_\eta\widetilde{\varphi} = a_\eta\,\delta + \widehat{\varphi}_\eta, \qquad a_\eta \in \mathbb{C}, \;\; \widehat{\varphi}_\eta \in \widehat{\mathscr{R}}^{\text{simp}}_{-\eta+\Omega'}, \qquad \eta \in \Sigma_\omega \cup \{\omega\} \tag{6.78}$$

$$\mathscr{B}\Delta^+_{\omega-\eta}\widetilde{\psi} = b_{\omega-\eta}\,\delta + \widehat{\psi}_{\omega-\eta}, \;\; b_{\omega-\eta} \in \mathbb{C}, \;\; \widehat{\psi}_{\omega-\eta} \in \widehat{\mathscr{R}}^{\text{simp}}_{-(\omega-\eta)+\Omega''}, \;\; \eta \in \{0\} \cup \Sigma_\omega. \tag{6.79}$$

Since $\mathscr{B}\Delta^+_\omega(\widetilde{\varphi}\widetilde{\psi}) = b\Delta^+_\omega\widehat{\varphi} + a\Delta^+_\omega\widehat{\psi} + \Delta^+_\omega(\widehat{\varphi} * \widehat{\psi})$, formula (6.77) is equivalent to

$$\Delta^+_\omega(\widehat{\varphi} * \widehat{\psi}) = \sum_{\eta\in\{0,\omega\}\cup\Sigma_\omega} (a_\eta\,\delta + \widehat{\varphi}_\eta) * (b_{\omega-\eta}\,\delta + \widehat{\psi}_{\omega-\eta}), \tag{6.80}$$

with the convention $a_0 = 0$, $\widehat{\varphi}_0 = \widehat{\varphi}$ and $b_0 = 0$, $\widehat{\psi}_0 = \widehat{\psi}$.

Consider a neighbourhood of $[0,\omega]$ of the form $U_\delta = \{\,\zeta \in \mathbb{C} \mid \operatorname{dist}\big(\zeta,[0,\omega]\big) < \delta\,\}$ with $\delta > 0$ small enough so that $U_\delta \setminus [0,\omega]$ does not meet Ω. Let $u := \omega\,\mathrm{e}^{-\mathrm{i}\alpha}$ with $0 < \alpha < \frac{\pi}{2}$, α small enough so that $u \in U_\delta$ and the line-segment $\ell := [0,u]$ can be considered as a path issuing from 0 circumventing to the right all the points of $]0,\omega[\,\cup\,\Omega$. We must show that $\operatorname{cont}_\ell(\widehat{\varphi} * \widehat{\psi})(\omega+\zeta)$ has a simple singularity at 0 and compute this singularity.

c) We shall show that, when all the numbers a_η and $b_{\omega-\eta}$ vanish,

$$f(\zeta) := \operatorname{cont}_\ell(\widehat{\varphi} * \widehat{\psi})(\omega+\zeta) = \Bigg(\sum_{\eta \in \{0,\omega\}\cup\Sigma_\omega} \widehat{\varphi}_\eta * \widehat{\psi}_{\omega-\eta}\Bigg)\frac{\mathscr{L}(\zeta)}{2\pi\mathrm{i}} + R(\zeta), \quad (6.81)$$

where $\mathscr{L}$ is a branch of the logarithm and $R(\zeta) \in \mathbb{C}\{\zeta\}$. This is sufficient to conclude, because in the general case we can write

$$\widehat{\varphi} * \widehat{\psi} = \big(\tfrac{\mathrm{d}}{\mathrm{d}\zeta}\big)^2(\widehat{\varphi}^* * \widehat{\psi}^*), \qquad \widehat{\varphi}^* := 1 * \widehat{\varphi},\ \ \widehat{\psi}^* := 1 * \widehat{\psi},$$

and, by Theorem 6.61, the anti-derivatives $\widehat{\varphi}^*$ and $\widehat{\psi}^*$ satisfy

$$\Delta^+_\eta \widehat{\varphi}^* = a_\eta + 1 * \widehat{\varphi}_\eta, \qquad \Delta^+_{\omega-\eta} \widehat{\psi}^* = b_{\omega-\eta} + 1 * \widehat{\psi}_{\omega-\eta}$$

instead of (6.78)–(6.79), whence by (6.81) $\operatorname{cont}_\ell(\widehat{\varphi}^* * \widehat{\psi}^*)(\omega+\zeta) =$

$$\Bigg(\sum_{\eta \in \{0,\omega\}\cup\Sigma_\omega} a_\eta b_{\omega-\eta}\,\zeta + a_\eta\,\zeta * \widehat{\psi}_{\omega-\eta} + b_{\omega-\eta}\,\zeta * \widehat{\varphi}_\eta + \zeta * \widehat{\varphi}_\eta * \widehat{\psi}_{\omega-\eta}\Bigg)\frac{\mathscr{L}(\zeta)}{2\pi\mathrm{i}} \mod \mathbb{C}\{\zeta\}.$$

Differentiating twice, we then get a formula whose interpretation is precisely (6.80) (because $\big(\frac{\mathrm{d}}{\mathrm{d}\zeta}(\zeta * A)\big)/\zeta$ and $(\zeta * A)/\zeta^2$ are regular at 0 for whatever regular germ A).

d) From now on, we thus suppose that all the numbers a_η and $b_{\omega-\eta}$ vanish. Our aim is to prove (6.81). We observe that $D^+ := \{\,\zeta \in D(\omega,\delta) \mid \Im\mathrm{m}(\zeta/\omega) < 0\,\}$ is a half-disc such that, for all $\zeta \in D^+$, the line-segment $[0,\zeta]$ does not meet $\Omega \setminus \{0\}$, hence $\operatorname{cont}_\ell(\widehat{\varphi} * \widehat{\psi})(\zeta) = \int_0^\zeta \widehat{\varphi}(\xi)\widehat{\psi}(\zeta-\xi)\,\mathrm{d}\xi$ for such points. We know by Section 6.4 that f has spiral continuation around 0. Following the ideas of Section 6.8, we choose a determination of $\arg\omega$ and lift the half-disc $-\omega + D^+$ to the Riemann surface of logarithm by setting $\widetilde{D}^+ := \{\,\zeta = r\underline{\mathrm{e}}^{\mathrm{i}\theta} \in \widetilde{\mathbb{C}} \mid r < \delta,\ \arg\omega - \pi < \theta < \arg\omega\,\}$. This way we can write $f = \overset{\vee}{f} \circ \pi$, where $\overset{\vee}{f}$ is a representative of a singular germ, explicitly defined on $\widetilde{D}^+$ by

$$\zeta \in \widetilde{D}^+ \implies \overset{\vee}{f}(\zeta) = \int_{\ell_{\pi(\zeta)}} \widehat{\varphi}(\xi)\widehat{\psi}(\omega+\pi(\zeta)-\xi)\,\mathrm{d}\xi \quad \text{with } \ell_{\pi(\zeta)} := [0,\omega+\pi(\zeta)]. \tag{6.82}$$

The analytic continuation of $\overset{\vee}{f}$ in

$$\widetilde{D}^- := \{\, \zeta = r\underline{e}^{i\theta} \in \widetilde{\mathbb{C}} \mid r < \delta,\ \arg\omega - 3\pi < \theta \le \arg\omega - \pi \,\}$$

is given by

$$\zeta \in \widetilde{D}^- \implies \overset{\vee}{f}(\zeta) = \int_{L_{\pi(\zeta)}} \overset{\vee}{\varphi}(\xi)\,\overset{\vee}{\psi}(\omega + \pi(\zeta) - \xi)\,\mathrm{d}\xi, \tag{6.83}$$

where the symmetrically contractible path $L_{\pi(\zeta)}$ is obtained by following the principles expounded in Section 6.4 (see particularly (6.18)); this is illustrated in Figure 6.11.

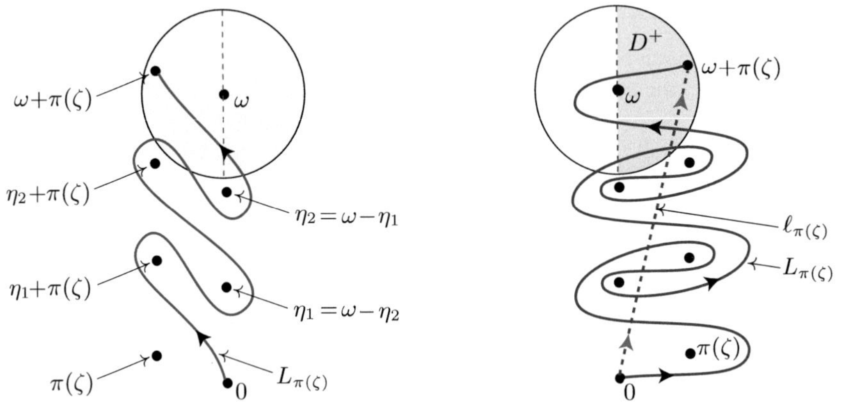

Fig. 6.11: *Integration paths for* $\widehat{\varphi} * \widehat{\psi}$. Left: $L_{\pi(\zeta)}$ for $\arg\omega - 2\pi < \arg\zeta \le \arg\omega - \pi$. Right: $\ell_{\pi(\zeta)}$ and $L_{\pi(\zeta)}$ for $\arg\omega - 3\pi < \arg\zeta \le \arg\omega - 2\pi$. (Case when Σ_ω has two elements.)

We first show that

$$\zeta \in \widetilde{D}^+ \implies \overset{\vee}{f}(\zeta) - \overset{\vee}{f}(\zeta\,\underline{e}^{-2\pi i}) = \sum_{\eta \in \{0,\omega\} \cup \Sigma_\omega} \widehat{\varphi}_\eta * \widehat{\psi}_{\omega-\eta}. \tag{6.84}$$

The point is that Σ_ω is symmetric with respect to its midpoint $\frac{\omega}{2}$, thus of the form $\{\eta_1 \prec \cdots \eta_{r-1}\}$ with $\eta_{r-j} = \omega - \eta_j$ for each j, and when ζ travels along a small circle around ω, the "moving nail" $\zeta - \eta_j$ turns around the "fixed nail" η_{r-j}, to use the language of Section 6.4.4. Thus, for $\zeta \in \widetilde{D}^+$, we can decompose the difference of paths $\ell_{\pi(\zeta)} - L_{\pi(\zeta)}$ as on Figure 6.12 and get

$$\overset{\vee}{f}(\zeta) - \overset{\vee}{f}(\zeta\,\underline{e}^{-2\pi i}) = \Big(\int_{\pi(\zeta)-\gamma} + \int_{\omega+\gamma} + \sum_{\eta\in\Sigma_\omega} \int_{\eta+\Gamma} \Big)\, \widehat{\varphi}(\xi)\,\widehat{\psi}(\omega + \pi(\zeta) - \xi)\,\mathrm{d}\xi,$$

where γ goes from $\zeta\,\underline{\mathrm{e}}^{-2\pi\mathrm{i}}$ to ζ by turning anticlockwise around 0, whereas Γ goes from $\zeta\,\underline{\mathrm{e}}^{-2\pi\mathrm{i}}$ to ζ the same way but then comes back to $\zeta\,\underline{\mathrm{e}}^{-2\pi\mathrm{i}}$ (see Figure 6.12). With an appropriate change of variable in each of these integrals, this can be rewrit-

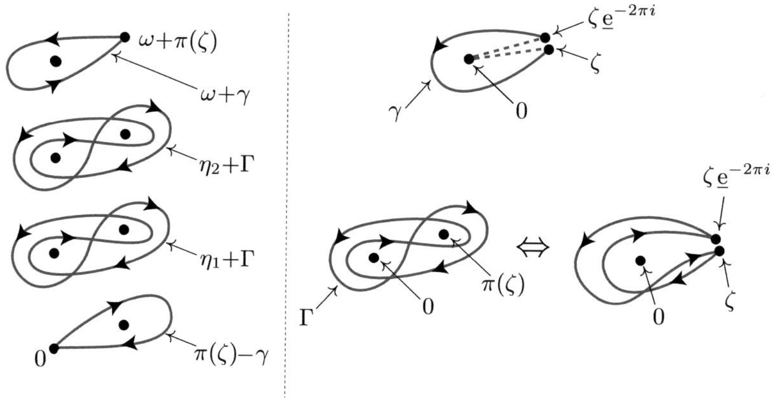

Fig. 6.12: Computation of the variation of the singularity at ω of $\widehat{\varphi} * \widehat{\psi}$.

ten as

$$\check{f}(\zeta) - \check{f}(\zeta\,\underline{\mathrm{e}}^{-2\pi\mathrm{i}}) = \int_\gamma \widehat{\varphi}(\pi(\zeta)-\xi)\widehat{\psi}(\omega+\xi)\,\mathrm{d}\xi + \int_\gamma \widehat{\varphi}(\omega+\xi)\widehat{\psi}(\pi(\zeta)-\xi)\,\mathrm{d}\xi$$
$$+ \sum_{\eta\in\Sigma_\omega} \int_\Gamma \widehat{\varphi}(\eta+\xi)\widehat{\psi}(\omega-\eta+\pi(\zeta)-\xi)\,\mathrm{d}\xi.$$

In the first two integrals, since

$$\widehat{\psi}(\omega+\xi) = \frac{1}{2\pi\mathrm{i}}\widehat{\psi}_\omega(\xi)\mathscr{L}(\xi) + R'(\xi), \qquad \widehat{\varphi}(\omega+\xi) = \frac{1}{2\pi\mathrm{i}}\widehat{\varphi}_\omega(\xi)\mathscr{L}(\xi) + R''(\xi),$$

with R' and R'' regular at 0, and we can diminish δ so that ξ and $\pi(\zeta)-\xi$ stay in a neighbourhood of 0 where $\widehat{\varphi}$, $\widehat{\psi}_\omega$, R', $\widehat{\varphi}_\omega$, R'' and $\widehat{\psi}$ are holomorphic, the Cauchy theorem cancels the contribution of R' and R'', while the contribution of the logarithms can be computed by collapsing γ onto the line-segment $[\zeta\,\underline{\mathrm{e}}^{-2\pi\mathrm{i}},0]$ followed by $[0,\zeta]$, hence the sum of the first two integrals is $\widehat{\varphi} * \widehat{\psi}_\omega + \widehat{\varphi}_\omega * \widehat{\psi}$. Similarly, by collapsing Γ as indicated on Figure 6.12,

$$\int_\Gamma \widehat{\varphi}(\eta+\xi)\widehat{\psi}(\omega-\eta+\pi(\zeta)-\xi)\,\mathrm{d}\xi =$$
$$\frac{1}{2\pi\mathrm{i}}\int_\Gamma \widehat{\varphi}(\eta+\xi)\big(\widehat{\psi}_{\omega-\eta}(\pi(\zeta)-\xi)\mathscr{L}(\xi) + R_{\omega-\eta}(\xi)\big)\,\mathrm{d}\xi$$

(with some regular germ $R_{\omega-\eta}$) is seen to coincide with $\int_\gamma \widehat{\varphi}(\eta+\xi)\widehat{\psi}_{\omega-\eta}(\pi(\zeta)-\xi)\,\mathrm{d}\xi$, which is itself seen to coincide with $\widehat{\varphi}_\eta * \widehat{\psi}_{\omega-\eta}(\zeta)$ by arguing as above. So (6.84) is proved.

e) We now observe that, since $g(\zeta) := \overset{\vee}{f}(\zeta) - \overset{\vee}{f}(\zeta\,\underline{\mathrm{e}}^{-2\pi\mathrm{i}})$ is a regular germ at 0,

$$R(\zeta) := f(\zeta) - g(\zeta)\frac{\mathscr{L}(\zeta)}{2\pi\mathrm{i}}$$

extends analytically to a (single-valued) function holomorphic in a punctured disc, i.e. it can be represented by a Laurent series (6.27). But $R(\zeta)$ can be bounded by $C\big(C' + \ln\frac{1}{|\zeta|}\big)$ with appropriate constants C, C' (using (6.82)–(6.83) to bound the analytic continuation of f), thus the origin is a removable singularity for R, which is thus regular at 0. The proof of (6.81) is now complete. □

6.13.2 Action of the symbolic Stokes automorphism on a product

Theorem 6.83 can be rephrased in terms of the symbolic Stokes automorphism Δ_d^+ of Section 6.12.2. Let us fix a ray $d = \{t\,\mathrm{e}^{\mathrm{i}\theta} \mid t \geq 0\}$, with total order $\prec$ defined as previously. Without loss of generality we can assume that both $\Omega' \cap d$ and $\Omega'' \cap d$ are infinite and contain 0. With the convention $\Delta_0^+ := \mathrm{Id}$, formula (6.77) can be rewritten

$$\Delta_\sigma^+(\widetilde{\varphi}\widetilde{\psi}) = \sum_{\substack{\sigma=\sigma'+\sigma'' \\ \sigma'\in\Omega'\cap d,\,\sigma''\in\Omega''\cap d}} (\Delta_{\sigma'}^+\widetilde{\varphi})(\Delta_{\sigma''}^+\widetilde{\psi}), \qquad \sigma \in \Omega\cap d. \tag{6.85}$$

For every $\omega' \in \Omega'\cap d$ and $\omega'' \in \Omega''\cap d$ we have commutative diagrams

$$\begin{array}{ccc} \mathbb{C}\,\delta\oplus\widehat{\mathscr{R}}^{\mathrm{simp}}_{-\omega'+\Omega'} & \hookrightarrow & \mathbb{C}\,\delta\oplus\widehat{\mathscr{R}}^{\mathrm{simp}}_{-\omega'+\Omega} \\ \tau_{\omega'}\downarrow & & \downarrow\tau_{\omega'} \\ \overset{\triangledown}{E}_{\omega'}(\Omega',d) & \hookrightarrow & \overset{\triangledown}{E}_{\omega'}(\Omega,d) \end{array} \qquad \begin{array}{ccc} \mathbb{C}\,\delta\oplus\widehat{\mathscr{R}}^{\mathrm{simp}}_{-\omega''+\Omega''} & \hookrightarrow & \mathbb{C}\,\delta\oplus\widehat{\mathscr{R}}^{\mathrm{simp}}_{-\omega''+\Omega} \\ \tau_{\omega''}\downarrow & & \downarrow\tau_{\omega''} \\ \overset{\triangledown}{E}_{\omega''}(\Omega'',d) & \hookrightarrow & \overset{\triangledown}{E}_{\omega''}(\Omega,d) \end{array}$$

hence $\overset{\triangledown}{E}(\Omega',d) = \widehat{\bigoplus\limits_{\omega'\in\Omega'\cap d}} \overset{\triangledown}{E}_{\omega'}(\Omega')$ and $\overset{\triangledown}{E}(\Omega'',d) = \widehat{\bigoplus\limits_{\omega''\in\Omega''\cap d}} \overset{\triangledown}{E}_{\omega''}(\Omega'')$ can be viewed as subspaces of $\overset{\triangledown}{E}(\Omega,d) := \widehat{\bigoplus\limits_{\omega\in\Omega\cap d}} \overset{\triangledown}{E}_\omega(\Omega)$. We shall often abbreviate the notations, writing for instance

$$\overset{\triangledown}{E}{}' \hookrightarrow \overset{\triangledown}{E}, \qquad \overset{\triangledown}{E}{}'' \hookrightarrow \overset{\triangledown}{E}.$$

The convolution law $\big(\mathbb{C}\,\delta\oplus\widehat{\mathscr{R}}^{\mathrm{simp}}_{-\omega'+\Omega'}\big)\times\big(\mathbb{C}\,\delta\oplus\widehat{\mathscr{R}}^{\mathrm{simp}}_{-\omega''+\Omega''}\big) \to \mathbb{C}\,\delta\oplus\widehat{\mathscr{R}}^{\mathrm{simp}}_{-(\omega'+\omega'')+\Omega}$ induces a bilinear map $*$ defined by

$$(\Phi,\Psi) = \Big(\sum_{\omega' \in \Omega' \cap d} \varphi^{\omega'}, \sum_{\omega'' \in \Omega'' \cap d} \psi^{\omega''} \Big) \in \overset{\triangledown}{E}' \times \overset{\triangledown}{E}'' \mapsto \sum_{\omega' \in \Omega' \cap d,\ \omega'' \in \Omega'' \cap d} \varphi^{\omega'} * \psi^{\omega''} \in \overset{\triangledown}{E}, \tag{6.86}$$

where

$$(\varphi,\psi) \in \overset{\triangledown}{E}'_{\omega'} \times \overset{\triangledown}{E}''_{\omega''} \implies \varphi * \psi := \tau_{\omega'+\omega''}\big(\tau_{\omega'}^{-1}\varphi * \tau_{\omega''}^{-1}\psi\big) \in \overset{\triangledown}{E}_{\omega'+\omega''}. \tag{6.87}$$

Theorem 6.84. *With the above notations and definitions,*

$$(\Phi,\Psi) \in \overset{\triangledown}{E}'(\Omega',d) \times \overset{\triangledown}{E}''(\Omega'',d) \implies \Delta_d^+(\Phi * \Psi) = (\Delta_d^+ \Phi) * (\Delta_d^+ \Psi). \tag{6.88}$$

Proof. It is sufficient to prove (6.88) for $(\Phi,\Psi) = (\varphi,\psi) \in \overset{\triangledown}{E}'_{\omega'} \times \overset{\triangledown}{E}''_{\omega''}$, with $(\omega',\omega'') \in \Omega' \times \Omega''$. Recall that

$$\Delta_d^+ \varphi = \sum_{\eta' \succeq \omega',\, \eta' \in \Omega' \cap d} \tau_{\eta'} \Delta^+_{\eta'-\omega'} \tau_{\omega'}^{-1} \varphi, \qquad \Delta_d^+ \psi = \sum_{\eta'' \succeq \omega'',\, \eta'' \in \Omega'' \cap d} \tau_{\eta''} \Delta^+_{\eta''-\omega''} \tau_{\omega''}^{-1} \psi. \tag{6.89}$$

Let $\omega := \omega' + \omega''$, so that $\varphi * \psi \in \overset{\triangledown}{E}_\omega$. We have

$$\Delta_d^+(\varphi * \psi) = \sum_{\eta \succeq \omega,\, \eta \in \Omega \cap d} \tau_\eta \Delta^+_{\eta-\omega} \tau_\omega^{-1}(\varphi * \psi) = \sum_{\eta \succeq \omega,\, \eta \in \Omega \cap d} \tau_\eta \Delta^+_{\eta-\omega}\big((\tau_{\omega'}^{-1}\varphi) * (\tau_{\omega''}^{-1}\psi)\big)$$

by definition of Δ_d^+ and $*$. For each η, applying (6.85) to $\sigma = \eta - \omega$, $\tau_{\omega'}^{-1}\varphi \in \mathbb{C}\delta \oplus \widehat{\mathscr{R}}^{\mathrm{simp}}_{-\omega'+\Omega'}$, $\tau_{\omega''}^{-1}\psi \in \mathbb{C}\delta \oplus \widehat{\mathscr{R}}^{\mathrm{simp}}_{-\omega''+\Omega''}$, we get

$$\Delta^+_{\eta-\omega}\big((\tau_{\omega'}^{-1}\varphi) * (\tau_{\omega''}^{-1}\psi)\big) = \sum_{\substack{\eta-\omega=\sigma'+\sigma'' \\ \sigma' \in (-\omega'+\Omega') \cap d,\, \sigma'' \in (-\omega''+\Omega'') \cap d}} \big(\Delta^+_{\sigma'} \tau_{\omega'}^{-1}\varphi\big) * \big(\Delta^+_{\sigma''} \tau_{\omega''}^{-1}\psi\big).$$

With the change of indices $(\sigma',\sigma'') \mapsto (\eta',\eta'') = (\omega'+\sigma', \omega''+\sigma'')$, this yields

$$\Delta_d^+(\varphi * \psi) = \sum_{\substack{\eta \in \Omega \cap d \\ \eta \succeq \omega}} \sum_{\substack{\eta = \eta'+\eta'' \\ \eta' \in \Omega' \cap d,\, \eta'' \in \Omega'' \cap d \\ \omega' \preceq \eta',\, \omega'' \preceq \eta''}} \tau_\eta \big((\Delta^+_{\eta'-\omega'} \tau_{\omega'}^{-1}\varphi) * (\Delta^+_{\eta''-\omega''} \tau_{\omega''}^{-1}\psi)\big).$$

By Fubini, this is

$$\Delta_d^+(\varphi * \psi) = \sum_{\substack{\eta' \in \Omega' \cap d,\, \eta'' \in \Omega'' \cap d \\ \omega' \preceq \eta',\, \omega'' \preceq \eta''}} \tau_{\eta'+\eta''} \big((\Delta^+_{\eta'-\omega'} \tau_{\omega'}^{-1}\varphi) * (\Delta^+_{\eta''-\omega''} \tau_{\omega''}^{-1}\psi)\big)$$

$$= (\Delta_d^+ \varphi) * (\Delta_d^+ \psi)$$

by definition of $*$ and (6.89). Hence (6.88) is proved. □

Remark 6.85. When Ω is stable under addition, one can take $\Omega' = \Omega'' = \Omega$. In that case, the operation $*$ makes $\overset{\triangledown}{E}(\Omega,d)$ an algebra and Theorem 6.84 implies that Δ_d^+ is an algebra automorphism. At a heuristical level, this could be guessed from (6.64), since both $\mathscr{L}^+$ and $\mathscr{L}^-$ take convolution products to pointwise products.

Remark 6.86. Via the linear isomorphism $\mathscr{B}\colon\ \widetilde{E}(\Omega,d) \xrightarrow{\sim} \overset{\triangledown}{E}(\Omega,d)$ of Section 6.12.4, the bilinear map $*$ gives rise to a bilinear map $\raisebox{0.2ex}{$\scriptscriptstyle\bullet$}\colon\ \widetilde{E}(\Omega',d)\times\widetilde{E}(\Omega'',d)\to\widetilde{E}(\Omega,d)$ which, for homogeneous components, is simply

$$\mathrm{e}^{-\omega' z}\widetilde{\varphi}(z)\raisebox{0.2ex}{$\scriptscriptstyle\bullet$}\mathrm{e}^{-\omega'' z}\widetilde{\psi}(z) = \mathrm{e}^{-(\omega'+\omega'')z}\widetilde{\varphi}(z)\widetilde{\psi}(z).$$

This justifies (6.74).

6.13.3 Leibniz rule for the symbolic Stokes infinitesimal generator and the operators Δ_ω

From Δ_d^+ we now wish to move on to its logarithm Δ_d, which will give us access to the way the operators Δ_ω act on products. We begin with a purely algebraic result, according to which, roughly speaking, "the logarithm of an automorphism is a derivation".

Lemma 6.87. *Suppose that E is a vector space over $\mathbb{Q}$, on which we have a translation-invariant distance d which makes it a complete metric space, and that $T\colon E\to E$ is a $\mathbb{Q}$-linear contraction, so that $D := \log(\mathrm{Id}+T) = \sum_{s\ge1}\frac{(-1)^{s-1}}{s}T^s$ is well defined.*

Suppose that E' and E'' are T-invariant closed subspaces and that $*\colon E'\times E''\to E$ is $\mathbb{Q}$-bilinear, with $d(\Phi*\Psi,0)\le Cd(\Phi,0)d(\Psi,0)$ for some $C>0$, and

$$(\Phi,\Psi)\in E'\times E'' \implies (\mathrm{Id}+T)(\Phi*\Psi) = \big((\mathrm{Id}+T)\Phi\big)*\big((\mathrm{Id}+T)\Psi\big). \tag{6.90}$$

Then

$$(\Phi,\Psi)\in E'\times E'' \implies D(\Phi*\Psi) = (D\Phi)*\Psi+\Phi*(D\Psi). \tag{6.91}$$

Proof. By (6.90), $T(\Phi*\Psi) = (T\Phi)*\Psi+\Phi*(T\Psi)+(T\Phi)*(T\Psi)$. Denoting by $N(s',s'',s)$ the coefficient of $X^{s'}Y^{s''}$ in the polynomial $(X+Y+XY)^s\in\mathbb{Z}[X,Y]$ for any $s',s'',s\in\mathbb{N}$, we obtain by induction

$$T^s(\Phi*\Psi) = \sum_{s',s''\in\mathbb{N}} N(s',s'',s)(T^{s'}\Phi)*(T^{s''}\Psi)$$

for every $s\in\mathbb{N}$, whence

$$D(\Phi*\Psi) = \sum_{s',s''\in\mathbb{N}}\sum_{s\in\mathbb{N}}\frac{(-1)^{s-1}}{s}N(s',s'',s)(T^{s'}\Phi)*(T^{s''}\Psi).$$

But, for every $s', s'' \in \mathbb{N}$, the number $\sum \frac{(-1)^{s-1}}{s} N(s', s'', s)$ is the coefficient of $X^{s'} Y^{s''}$ in the formal series $\sum \frac{(-1)^{s-1}}{s} (X+Y+XY)^s = \log(1+X+Y+XY) = \log(1+X) + \log(1+Y) \in \mathbb{Q}[[X,Y]]$, hence the result follows. □

The main result of this section follows easily:

Theorem 6.88. *Under the assumption* (6.76), *one has for every direction d*

$$(\Phi,\Psi) \in \overset{\triangledown}{E}'(\Omega',d) \times \overset{\triangledown}{E}''(\Omega'',d) \quad \Longrightarrow \quad \Delta_d(\Phi * \Psi) = (\Delta_d \Phi) * \Psi + \Phi * (\Delta_d \Psi) \tag{6.92}$$

and, for every $\omega \in \Omega \setminus \{0\}$,

$$(\widetilde{\varphi}, \widetilde{\psi}) \in \widetilde{\mathscr{R}}^{\mathrm{simp}}_{\Omega'} \times \widetilde{\mathscr{R}}^{\mathrm{simp}}_{\Omega''} \quad \Longrightarrow \quad \Delta_\omega(\widetilde{\varphi}\widetilde{\psi}) = (\Delta_\omega \widetilde{\varphi})\widetilde{\psi} + \widetilde{\varphi}(\Delta_\omega \widetilde{\psi}). \tag{6.93}$$

Proof. The requirements of Lemma 6.87 are satisfied by $T := \Delta_d^+ - \mathrm{Id}$ and the distance on $\overset{\triangledown}{E}$ indicated in footnote 18; since $\log \Delta_d^+ = \Delta_d$, this yields (6.92).

One gets (6.93) by evaluating (6.91) with $\Phi = \tau_0 \mathscr{B}\widetilde{\varphi} \in \overset{\triangledown}{E}'_0$ and $\Psi = \tau_0 \mathscr{B}\widetilde{\psi} \in \overset{\triangledown}{E}''_0$, and extracting the homogeneous component $\tau_\omega \Delta_\omega(\mathscr{B}\widetilde{\varphi} * \mathscr{B}\widetilde{\psi}) \in \overset{\triangledown}{E}_\omega$. □

6.13.4 The subalgebra of simple Ω-resurgent functions

We now suppose that Ω is stable under addition, so that, by Corollary 6.19, $\widetilde{\mathscr{R}}_\Omega$ is a subalgebra of $\mathbb{C}[[z^{-1}]]_1$ and $\mathbb{C}\,\delta \oplus \widehat{\mathscr{R}}_\Omega$ is a subalgebra of the convolution algebra $\mathbb{C}\,\delta \oplus \mathbb{C}\{\zeta\}$. Taking $\Omega' = \Omega'' = \Omega$ in Theorem 6.83, we get

Corollary 6.89. *If* Ω *is stable under addition, then* $\widetilde{\mathscr{R}}^{\mathrm{simp}}_\Omega$ *is a subalgebra of* $\widetilde{\mathscr{R}}_\Omega$ *and* $\mathbb{C}\,\delta \oplus \widehat{\mathscr{R}}^{\mathrm{simp}}_\Omega$ *is a subalgebra of* $\mathbb{C}\,\delta \oplus \widehat{\mathscr{R}}_\Omega$.

As anticipated in Remark 6.85, there is also for each ray d an algebra structure on $\overset{\triangledown}{E}(\Omega,d)$ given by the operation $*$ defined in (6.86), for which the symbolic Stokes automorphism Δ_d^+ is an algebra automorphism; the symbolic Stokes infinitesimal generator Δ_d now appears as a derivation, in view of formula (6.92) of Theorem 6.88 (for that reason Δ_d is sometimes called "directional alien derivation").

Remark 6.90. In particular, for each $\omega \in \Omega$ and $\Phi \in \widetilde{E}(\Omega,d)$, we have $\mathrm{e}^{-\omega z}\Phi \in \widetilde{E}(\Omega,d)$,

$$\Delta_d^+(\mathrm{e}^{-\omega z}\Phi) = \mathrm{e}^{-\omega z}\Delta_d^+\Phi, \qquad \Delta_d(\mathrm{e}^{-\omega z}\Phi) = \mathrm{e}^{-\omega z}\Delta_d\Phi \tag{6.94}$$

(because $\mathrm{e}^{-\omega z}$ is fixed by Δ_d^+ and annihilated by Δ_d).

As indicated in formula (6.93) of Theorem 6.88, the homogeneous components Δ_ω of Δ_d inherit the Leibniz rule, however it is only if $-\omega + \Omega \subset \Omega$ that $\Delta_\omega \colon \widetilde{\mathscr{R}}^{\mathrm{simp}}_\Omega \to \widetilde{\mathscr{R}}^{\mathrm{simp}}_\Omega$ is a derivation of the algebra $\widetilde{\mathscr{R}}^{\mathrm{simp}}_\Omega$, and this is the case for all

$\omega \in \Omega \setminus \{0\}$ when Ω is an additive subgroup of $\mathbb{C}$. As anticipated in Remark 6.64, the operators Δ_ω are called "alien derivations" for that reason.

Let us investigate farther the rules of "alien calculus" for non-linear operations.

Theorem 6.91. *Suppose that Ω is stable under addition. Let $\widetilde{\varphi}(z), \widetilde{\psi}(z), \widetilde{\chi}(z) \in \widetilde{\mathscr{R}}^{\,\mathrm{simp}}_{\Omega}$ and assume that $\widetilde{\chi}(z)$ has no constant term. Let $H(t) \in \mathbb{C}\{t\}$. Then*

$$\widetilde{\psi} \circ (\mathrm{id} + \widetilde{\varphi}) \in \widetilde{\mathscr{R}}^{\,\mathrm{simp}}_{\Omega}, \quad H \circ \widetilde{\chi} \in \widetilde{\mathscr{R}}^{\,\mathrm{simp}}_{\Omega}$$

and, for any $\omega \in \Omega \setminus \{0\}$, $(\Delta_\omega \widetilde{\psi}) \circ (\mathrm{id} + \widetilde{\varphi}) \in \widetilde{\mathscr{R}}^{\,\mathrm{simp}}_{-\omega+\Omega}$ and

$$\Delta_\omega\big(\widetilde{\psi} \circ (\mathrm{id} + \widetilde{\varphi})\big) = (\partial \widetilde{\psi}) \circ (\mathrm{id} + \widetilde{\varphi}) \cdot \Delta_\omega \widetilde{\varphi} + \mathrm{e}^{-\omega \widetilde{\varphi}} \cdot (\Delta_\omega \widetilde{\psi}) \circ (\mathrm{id} + \widetilde{\varphi}), \tag{6.95}$$

$$\Delta_\omega (H \circ \widetilde{\chi}) = \big(\tfrac{\mathrm{d}H}{\mathrm{d}t} \circ \widetilde{\chi}\big) \cdot \Delta_\omega \widetilde{\chi}. \tag{6.96}$$

The proof requires the following technical statement.

Lemma 6.92. *Let $U := \{ r\underline{\mathrm{e}}^{\mathrm{i}\theta} \in \widetilde{\mathbb{C}} \mid 0 < r < R,\ \theta \in I \}$, where I is an open interval of $\mathbb{R}$ of length $> 4\pi$ and $R > 0$. Suppose that, for each $k \in \mathbb{N}$, we are given a function $\overset{\vee}{\varphi}_k$ which is holomorphic in U and is the major of a simple singularity $a_k\,\delta + \overset{\wedge}{\varphi}_k$, and that the series $\sum \overset{\vee}{\varphi}_k$ converges normally on every compact subset of U.*

Then the numerical series $\sum a_k$ is absolutely convergent, the series of functions $\sum \overset{\wedge}{\varphi}_k$ converges normally on every compact subset of $\mathbb{D}_R$, and the function $\overset{\vee}{\varphi} := \sum_{k\in\mathbb{N}} \overset{\vee}{\varphi}_k$, which is holomorphic in U, is the major of the simple singularity $\big(\sum_{k\in\mathbb{N}} a_k\big)\delta + \sum_{k\in\mathbb{N}} \overset{\wedge}{\varphi}_k$.

Proof of Lemma 6.92. Pick θ_0 such that $[\theta_0, \theta_0 + 4\pi] \subset I$ and let $J := [\theta_0 + 2\pi, \theta_0 + 4\pi]$. For any $R' < R$, writing $\overset{\wedge}{\varphi}_k(\pi(\zeta)) = \overset{\vee}{\varphi}_k(\zeta) - \overset{\vee}{\varphi}_k(\zeta\,\underline{\mathrm{e}}^{-2\pi\mathrm{i}})$ for $\zeta \in U$ with $\arg\zeta \in J$ and $|\zeta| \le R'$, we get the normal convergence of $\sum \overset{\wedge}{\varphi}_k$ on $\overline{\mathbb{D}}_{R'}$.

Now, for each k, $\overset{\vee}{L}_k(\zeta) := \overset{\vee}{\varphi}_k(\zeta) - \overset{\wedge}{\varphi}_k\big(\pi(\zeta)\big)\frac{\log\zeta}{2\pi\mathrm{i}}$ is a major of $a_k\,\delta$ and is holomorphic in U; its monodromy is trivial, thus $\overset{\vee}{L}_k = L_k \circ \pi$ with L_k holomorphic in $\mathbb{D}^*_R$. For any circle C centred at 0, contained in $\mathbb{D}_R$ and positively oriented, we have $a_k = \int_C L_k(\zeta)\,\mathrm{d}\zeta$. The normal convergence of $\sum \overset{\vee}{\varphi}_k$ and $\sum \overset{\wedge}{\varphi}_k$ implies that of $\sum L_k$, hence the absolute convergence of $\sum a_k$. Moreover, for every $n \in \mathbb{N}^*$, $\int_C L_k(\zeta)\zeta^{-n}\,\mathrm{d}\zeta = 0$, hence $L := \sum_{k\in\mathbb{N}} L_k$ satisfies $\int_C L(\zeta)\zeta^{-n}\,\mathrm{d}\zeta = 0$, whence $\mathrm{sing}_0\big(L(\zeta)\big) = \big(\sum_{k\in\mathbb{N}} a_k\big)\delta$.

We conclude by observing that $\overset{\vee}{\varphi}(\zeta) = L\big(\pi(\zeta)\big) + \big(\sum_{k\in\mathbb{N}} \overset{\wedge}{\varphi}_k(\pi(\zeta))\big)\frac{\log\zeta}{2\pi\mathrm{i}}$. □

Proof of Theorem 6.91. We proceed as in the proof of Theorem 6.32, writing $\widetilde{\varphi} = a + \widetilde{\varphi}_1$, $\widetilde{\psi} = b + \widetilde{\psi}_1$, where $a, b \in \mathbb{C}$ and $\widetilde{\varphi}_1$ and $\widetilde{\psi}_1$ have no constant term, and $H(t) = \sum_{k\ge0} h_k t^k$. Thus

$$\widetilde{\psi} \circ (\mathrm{id} + \widetilde{\varphi}) = b + \widetilde{\lambda} \;\text{ with } \widetilde{\lambda} := T_a \widetilde{\psi}_1 \circ (\mathrm{id} + \widetilde{\varphi}_1), \quad H \circ \widetilde{\chi} = h_0 + \widetilde{\mu} \;\text{ with } \widetilde{\mu} := \sum_{k\ge1} h_k \widetilde{\chi}^k. \tag{6.97}$$

Both $\widetilde{\lambda}$ and $\widetilde{\mu}$ are naturally defined as formally convergent series of formal series without constant term:

$$\widetilde{\lambda} = \sum_{k\geq 0} \widetilde{\lambda}_k \quad \text{with } \widetilde{\lambda}_k := \frac{1}{k!}(\partial^k T_a \widetilde{\psi}_1)\widetilde{\varphi}_1^k, \qquad \widetilde{\mu} = \sum_{k\geq 1} \widetilde{\mu}_k \quad \text{with } \widetilde{\mu}_k := h_k \widetilde{\chi}^k.$$

By Lemma 6.65 and Theorem 6.83, each Borel transform

$$\widehat{\lambda}_k = \frac{1}{k!}\big((-\zeta)^k e^{-a\zeta}\widehat{\psi}_1\big) * \widehat{\varphi}_1^{*k}, \qquad \widehat{\mu}_k = h_k \widehat{\chi}^{*k}$$

belongs to $\widehat{\mathscr{R}}_\Omega^{\text{simp}}$, and we have checked in the proof of Theorem 6.32 that their sums $\widehat{\lambda}$ and $\widehat{\mu}$ belong to $\widehat{\mathscr{R}}_\Omega$, with their analytic continuations along the paths of $\mathbb{C}\setminus\Omega$ given by the sums of the analytic continuations of the functions $\widehat{\lambda}_k$ or $\widehat{\mu}_k$. The argument was based on Lemma 6.31; we use it again to control the behaviour of $\text{cont}_\gamma\widehat{\lambda}$ and $\text{cont}_\gamma\widehat{\mu}$ near an arbitrary $\omega\in\Omega$, for a path $\gamma\colon\ [0,1]\to\mathbb{C}\setminus\Omega$ starting close to 0 and ending close to ω. Choosing a lift ξ of $\gamma(1)-\omega$ in $\widetilde{\mathbb{C}}$, we shall then apply Lemma 6.92 to the functions $\overset{\vee}{\varphi}_k(\zeta)$ defined by $\text{cont}_\gamma\widehat{\lambda}_k\big(\omega+\pi(\zeta)\big)$ or $\text{cont}_\gamma\widehat{\mu}_k\big(\omega+\pi(\zeta)\big)$ for $\zeta\in\widetilde{\mathbb{C}}$ close to ξ.

Without loss of generality, we can suppose that $|\gamma(1)-\omega| = R/2$ with $R>0$ small enough so that $D(\omega,R)\cap\Omega=\{0\}$. Let us extend γ by a circle travelled twice, setting $\gamma(t) := \omega + \big(\gamma(1)-\omega\big)e^{2\pi i(t-1)}$ for $t\in[1,3]$. For every $t\in[1,3]$ and $R_t < R/2$, we can apply Lemma 6.31 and get the normal convergence of $\sum \text{cont}_{\gamma_{|[0,t]}}\widehat{\lambda}_k$ and $\sum \text{cont}_{\gamma_{|[0,t]}}\widehat{\mu}_k$ on $\overline{D\big(\gamma(t),R_t\big)}$. Now Lemma 6.92 shows that $\text{cont}_\gamma\widehat{\lambda}$ and $\text{cont}_\gamma\widehat{\mu}$ have simple singularities at ω. Hence $\widehat{\lambda},\widehat{\mu}\in\widehat{\mathscr{R}}_\Omega^{\text{simp}}$.

A similar argument shows that $(\Delta_\omega\widetilde{\psi})\circ(\text{id}+\widetilde{\varphi})\in\widetilde{\mathscr{R}}^{\text{simp}}_{-\omega+\Omega}$.

Lemma 6.92 also shows that $\Delta_\omega\widetilde{\lambda} = \sum_{k\geq 0}\Delta_\omega\widetilde{\lambda}_k$ and $\Delta_\omega\widetilde{\mu} = \sum_{k\geq 1}\Delta_\omega\widetilde{\mu}_k$. By means of (6.93), we compute easily $\Delta_\omega\widetilde{\mu}_k = kh_k\widetilde{\chi}^{k-1}\Delta_\omega\widetilde{\chi}$, whence

$$\Delta_\omega\widetilde{\mu} = \Big(\frac{dH}{dt}\circ\widetilde{\chi}\Big)\cdot\Delta_\omega\widetilde{\chi},$$

which yields (6.96) since (6.97) shows that $\Delta_\omega\widetilde{\mu} = \Delta_\omega(H\circ\widetilde{\chi})$.

By means of (6.49)–(6.50) and (6.93), we compute

$$\Delta_\omega\widetilde{\lambda}_k = A_k + B_k,$$

$$A_k := \frac{k}{k!}(\partial^k T_a\widetilde{\psi}_1)\widetilde{\varphi}_1^{k-1}\Delta_\omega\widetilde{\varphi}_1, \quad B_k := \frac{e^{-a\omega}}{k!}\big((-\omega+\partial)^k T_a\Delta_\omega\widetilde{\psi}_1\big)\widetilde{\varphi}_1^k,$$

hence

$$\sum_{k\geq 0} A_k = (\partial T_a \widetilde{\psi}_1)\circ(\mathrm{id}+\widetilde{\varphi}_1)\cdot\Delta_\omega\widetilde{\varphi}_1 = (\partial\widetilde{\psi}_1)\circ(\mathrm{id}+\widetilde{\varphi})\cdot\Delta_\omega\widetilde{\varphi}_1$$
$$= (\partial\widetilde{\psi})\circ(\mathrm{id}+\widetilde{\varphi})\cdot\Delta_\omega\widetilde{\varphi},$$
$$\sum_{k\geq 0} B_k = \mathrm{e}^{-a\omega}\sum_{k',k''\geq 0}\frac{(-\omega)^{k'}}{k'!k''!}(\partial^{k''}T_a\Delta_\omega\widetilde{\psi}_1)\widetilde{\varphi}_1^{k'+k''}$$
$$= \mathrm{e}^{-a\omega}\sum_{k'\geq 0}\frac{(-\omega)^{k'}}{k'!}\widetilde{\varphi}_1^{k'}\sum_{k''\geq 0}\frac{1}{k''!}(\partial^{k''}T_a\Delta_\omega\widetilde{\psi}_1)\widetilde{\varphi}_1^{k''}$$
$$= \exp(-a\omega-\omega\widetilde{\varphi}_1)\cdot(T_a\Delta_\omega\widetilde{\psi}_1)\circ(\mathrm{id}+\widetilde{\varphi}_1)$$
$$= \mathrm{e}^{-\omega\widetilde{\varphi}}\cdot(\Delta_\omega\widetilde{\psi}_1)\circ(\mathrm{id}+\widetilde{\varphi}) = \mathrm{e}^{-\omega\widetilde{\varphi}}\cdot(\Delta_\omega\widetilde{\psi})\circ(\mathrm{id}+\widetilde{\varphi}),$$

which yields (6.95) since (6.97) shows that $\Delta_\omega\big(\widetilde{\psi}\circ(\mathrm{id}+\widetilde{\varphi})\big) = \Delta_\omega\widetilde{\lambda}$. □

Example 6.93. As promised in Example 6.57, we can now study the exponential of the Stirling series $\widetilde{\mu}\in\widetilde{\mathscr{R}}_{2\pi\mathbb{Z}^*}$. Since $2\pi\mathrm{i}\mathbb{Z}^*$ is not stable under addition, we need to take at least $\Omega = 2\pi\mathrm{i}\mathbb{Z}$ to ensure $\widetilde{\lambda} = \exp\widetilde{\mu}\in\widetilde{\mathscr{R}}_\Omega^{\mathrm{simp}}$. Formulas (6.48) and (6.96) yield

$$\Delta_{2\pi\mathrm{i}m}\widetilde{\lambda} = \frac{1}{m}\widetilde{\lambda}, \qquad m\in\mathbb{Z}^*. \tag{6.98}$$

In view of Remark 6.69, this implies that *any alien operator maps* $\widetilde{\lambda}$ *to a multiple of* $\widetilde{\lambda}$. This clearly shows that the analytic continuation of the Borel transform $\mathscr{B}(\widetilde{\lambda}-1)$ is multiple-valued, since e.g. (6.98) with $m=\pm1$ says that the singularity at $\pm2\pi\mathrm{i}$ of the principal branch has a non-trivial minor. Let us show that

$$\Delta^+_{2\pi\mathrm{i}m}\widetilde{\lambda} = \begin{cases}\widetilde{\lambda} & \text{for } m=-1,+1,+2,+3,\dots\\ 0 & \text{for } m=-2,-3,\dots\end{cases} \tag{6.99}$$

(notice that the last formula implies that the analytic continuation of $\mathscr{B}(\widetilde{\lambda}-1)$ from the line-segment $(-2\pi\mathrm{i},2\pi\mathrm{i})$ to $(-2\pi\mathrm{i},-4\pi\mathrm{i})$ obtained by circumventing $-2\pi\mathrm{i}$ to the right is free of singularity in the rest of $\mathrm{i}\mathbb{R}^-$: it extends analytically to $\mathbb{C}\setminus[-2\pi\mathrm{i},+\mathrm{i}\infty)$, but that this is not the case of the analytic continuation to the left!)

Formula (6.99) could probably be obtained from the relation $\Delta^+_{2\pi\mathrm{i}m}\widetilde{\mu} = \frac{1}{m}$ by repeated use of (6.77), but it is simpler to use (6.53) and (6.98), and even better to perform the computation at the level of the symbolic Stokes automorphism and its infinitesimal generator. This time, we manipulate the multiplicative counterpart of $\Delta^+_{\mathrm{i}\mathbb{R}^\pm}$ and $\Delta_{\mathrm{i}\mathbb{R}^\pm}$ obtained through $\mathscr{B}$ as indicated in Section 6.12.4 and Remark 6.86, writing for instance

$$\Delta_{\mathrm{i}\mathbb{R}^+}\widetilde{\lambda} = \sum_{m\in\mathbb{N}^*} \frac{1}{m} \mathrm{e}^{-2\pi i m z}\widetilde{\lambda} = -\log(1-\mathrm{e}^{-2\pi \mathrm{i} z})\widetilde{\lambda},$$

$$\Delta_{\mathrm{i}\mathbb{R}^-}\widetilde{\lambda} = -\sum_{m\in\mathbb{N}^*} \frac{1}{m} \mathrm{e}^{2\pi i m z}\widetilde{\lambda} = \log(1-\mathrm{e}^{2\pi \mathrm{i} z})\widetilde{\lambda}.$$

By exponentiating in $\widetilde{E}(\Omega,\mathrm{i}\mathbb{R}^+)$ or $\widetilde{E}(\Omega,\mathrm{i}\mathbb{R}^-)$, with the help of (6.94), we get

$$\Delta^+_{\mathrm{i}\mathbb{R}^+}\widetilde{\lambda} = (1-\mathrm{e}^{-2\pi \mathrm{i} z})^{-1}\widetilde{\lambda} = \sum_{m\in\mathbb{N}} \mathrm{e}^{-2\pi i m z}\widetilde{\lambda}, \tag{6.100}$$

$$\Delta^+_{\mathrm{i}\mathbb{R}^-}\widetilde{\lambda} = (1-\mathrm{e}^{2\pi \mathrm{i} z})\widetilde{\lambda} = \widetilde{\lambda} - \mathrm{e}^{2\pi \mathrm{i} z}\widetilde{\lambda}. \tag{6.101}$$

One gets (6.99) by extracting the homogeneous components of these identities.

The Stokes phenomenon for the two Borel sums $\widetilde{\lambda}(z)$ can be described as follows: with $I := (-\frac{\pi}{2}, \frac{\pi}{2})$, we have $\lambda^+ := \lambda = \mathscr{S}^I\widetilde{\lambda}$ holomorphic in $\mathbb{C}\setminus\mathbb{R}^-$, and with $J := (\frac{\pi}{2}, \frac{3\pi}{2})$, we have $\lambda^- := \mathscr{S}^J\widetilde{\lambda}$ holomorphic in $\mathbb{C}\setminus\mathbb{R}^+$; by adapting the chain of reasoning of Example 6.81, one can deduce from (6.100)–(6.101) that

$$\Im\mathrm{m}\, z < 0 \implies \lambda^+(z) = (1-\mathrm{e}^{-2\pi \mathrm{i} z})^{-1}\lambda^-(z),$$

$$\Im\mathrm{m}\, z > 0 \implies \lambda^-(z) = (1-\mathrm{e}^{2\pi \mathrm{i} z})\lambda^+(z)$$

(one can also content oneself with exponentiating (6.69)–(6.70)), getting thus access to the exponentially small discrepancies between both Borel sums.

Observe that it follows that $\lambda^\pm$ admits a multiple-valued meromorphic continuation which gives rise to a function meromorphic in the whole of $\widetilde{\mathbb{C}}$: for instance, since $\lambda^+_{|\{\Im\mathrm{m}\, z>0\}}$ coincides with $(1-\mathrm{e}^{2\pi \mathrm{i} z})^{-1}\lambda^-$, it can be meromorphically continued to $\mathbb{C}\setminus\mathbb{R}^-$ and its anticlockwise continuation to $\{\Im\mathrm{m}\, z<0\}$ is given by $(1-\mathrm{e}^{2\pi \mathrm{i} z})^{-1}\lambda^-_{|\{\Im\mathrm{m}\, z<0\}}$, which coincides with $(1-\mathrm{e}^{2\pi \mathrm{i} z})^{-1}(1-\mathrm{e}^{-2\pi \mathrm{i} z})\lambda^+_{|\{\Im\mathrm{m}\, z<0\}}$, and can thus be anticlockwise continued to $\{\Im\mathrm{m}\, z>0\}$: we find

$$\lambda^+(\underline{\mathrm{e}}^{2\pi \mathrm{i}} z) = (1-\mathrm{e}^{2\pi \mathrm{i} z})^{-1}(1-\mathrm{e}^{-2\pi \mathrm{i} z})\lambda^+(z) = -\mathrm{e}^{-2\pi \mathrm{i} z}\lambda^+(z)$$

(compare with Remark 5.59). Since $z^{-\frac{1}{2}+z} = \mathrm{e}^{(-\frac{1}{2}+z)\log z}$ gets multiplied by $-\mathrm{e}^{2\pi \mathrm{i} z}$ after one anticlockwise turn around 0, we see that the product $\sqrt{2\pi}\,\mathrm{e}^{-z}z^{-\frac{1}{2}+z}\lambda^+(z)$ is single-valued, not a surprise in view of (5.55): this product function is none other than Euler's gamma function, which is known to be meromorphic in the whole complex plane!

6.14 Resurgent treatment of the Airy equation

We now return to the topic of analytic linear differential equations, much studied in the first part of this volume, with an example: the Airy equation. In Section 2.2, it was already studied from the differential Galois group viewpoint as Example 2.50; we will now study it from the viewpoint of Borel-Laplace summation and resurgence. In fact, the summability and resurgence properties to be established in the Airy example are a particular case of general statements to be encountered in the second volume of this book—see for instance Theorem 5.3.21 of [Lod16].

6.14.1 The Airy functions

The Airy equation reads

$$\frac{\mathrm{d}^2 y}{\mathrm{d}w^2} = wy. \tag{6.102}$$

The general theory tells us that its solutions form a 2-dimensional subspace of $\mathscr{O}(\mathbb{C})$, where $\mathscr{O}(\mathbb{C})$ is the vector space over $\mathbb{C}$ consisting of all entire functions, and that, when viewed as an analytic linear differential equation on the Riemann sphere, it has an irregular singular point at infinity.

Remark 6.94. Even though all solutions of (6.102) are entire functions of w, we shall be led to use fractional powers of w. Unless otherwise specified, expressions like w^α with $\alpha \in \mathbb{Q}$ will refer to the principal branch as in Example 6.40, defined in the Riemann surface of the logarithm $\widetilde{\mathbb{C}}$, or defined in a cut plane of the form $\mathbb{C} \setminus (\mathbb{R}^+ \mathrm{e}^{\mathrm{i}\theta})$ with $\mathrm{e}^{\mathrm{i}\theta} \notin \mathbb{R}^+$ and real positive for real positive w. A similar convention will be used for $\arg w$.

Theorem 6.95. *There exists a unique real-analytic solution y_0 to the Airy equation* (6.102) *such that, for any closed interval $I_0 \subset (-\pi, \pi)$,*

$$y_0(w) \underset{|w|\to\infty}{\sim} \frac{1}{2\sqrt{\pi}} w^{-\frac{1}{4}} \mathrm{e}^{-\frac{2}{3}w^{3/2}} \quad \textit{uniformly for } \arg w \in I_0. \tag{6.103}$$

For any $\theta \in (-\frac{\pi}{3}, \frac{\pi}{3})$, the solutions $y(w)$ such that $y(r\mathrm{e}^{\mathrm{i}\theta}) \xrightarrow[r\to+\infty]{} 0$ are exactly the multiples of y_0.

There exists a unique solution y_1 to the Airy equation (6.102) *such that, for any closed interval $I_1 \subset (-\frac{\pi}{3}, \frac{5\pi}{3})$,*

$$y_1(w) \underset{|w|\to\infty}{\sim} \frac{1}{2\mathrm{i}\sqrt{\pi}} w^{-\frac{1}{4}} \mathrm{e}^{\frac{2}{3}w^{3/2}} \quad \textit{uniformly for } \arg w \in I_1. \tag{6.104}$$

For any $\theta' \in (\frac{\pi}{3}, \pi)$, the solutions $y(w)$ such that $y(r\mathrm{e}^{\mathrm{i}\theta'}) \xrightarrow[r\to+\infty]{} 0$ are exactly the multiples of y_1.

Remark 6.96. With the notation $w = re^{i\theta}$, $r > 0$, the statement (6.103) means that the quantity

$$2\sqrt{\pi}\, y_0(re^{i\theta})\, r^{1/4} e^{i\theta/4} \exp\left(\tfrac{2}{3} r^{3/2} e^{3i\theta/2}\right)$$

tends to 1 as $r \to +\infty$ uniformly for $\theta \in I_0$, and similarly for (6.104): the quantity $2i\sqrt{\pi}\, y_1(re^{i\theta'})\, r^{1/4} e^{i\theta'/4} \exp\left(-\tfrac{2}{3} r^{3/2} e^{3i\theta'/2}\right)$ tends to 1 uniformly for $\theta' \in I_1$. Notice that, if $I_0 \subset (-\frac{\pi}{3}, \frac{\pi}{3})$ and $I_1 \subset (\frac{\pi}{3}, \pi)$, then there exists $\tau > 0$ such that

$$\theta \in I_0 \quad \Longrightarrow \quad \Re\mathrm{e}(e^{3i\theta/2}) \geq \tau,$$

$$\theta' \in I_1 \quad \Longrightarrow \quad \Re\mathrm{e}(e^{3i\theta'/2}) \leq -\tau$$

(take $\theta_0 \in (0, \frac{\pi}{3})$ such that $I_0 \subset [-\theta_0, \theta_0]$ and $I_1 \subset \left[\frac{2\pi}{3} - \theta_0, \frac{2\pi}{3} + \theta_0\right]$, and $\tau := \cos(3\theta_0/2)$); hence, in that case, $|y_0(w)|$ is uniformly bounded by a decreasing exponential for $\arg w \in I_0$ and $|y_1(w)|$ is uniformly bounded by a decreasing exponential for $\arg w \in I_1$.

The real-analytic function $y_0(w)$ is a special function, named *Airy function* (after the astronomer George Airy who used it to study the interference phenomenon known as supernumerary rainbows) and denoted by $\mathrm{Ai}(w)$. Neither $y_1(w)$ nor $iy_1(w)$ is a real-analytic function, but there is another classical real-analytic solution, called *Airy function of the second kind* and denoted by $\mathrm{Bi}(w)$, which can be defined as

$$\mathrm{Bi}(w) := i\big(2y_1(w) + y_0(w)\big) \tag{6.105}$$

(its realness will follow from exercise 6.100 later in this section). Notice that the asymptotic behaviour of $\mathrm{Bi}(re^{i\theta})$ for $r \to \infty$ and $\theta \in I_0 \subset (-\frac{\pi}{3}, \frac{\pi}{3})$ is given by the right-hand side of (6.104) multiplied by $2i$, because this expression is exponentially large while $iy_0(w)$ is exponentially small for such values of $\arg w$ (cf. Remark 6.96).

Exercise 6.97. Show that the Wronskian of (y_0, y_1) is the constant function $\frac{1}{2\pi i}$, hence the Wronskian of $(\mathrm{Ai}, \mathrm{Bi})$ is the constant function $\frac{1}{\pi}$.

Hint: By the general theory of linear differential equations, these Wronskians are known to be constant functions. Let $A_0(w)$ denote the right-hand side of (6.103) and $A_1(w)$ that of (6.104), so $y_k(w) = A_k(w)\big(1 + \varepsilon_k(w)\big)$ with $\varepsilon_k(w) \xrightarrow[|w| \to \infty]{} 0$ uniformly in $\mathscr{D}_0 := \left\{ re^{i\theta} \mid r \geq 1,\ \theta \in [-\frac{\pi}{6}, \frac{\pi}{2}] \right\}$. Use the Cauchy inequality to show that $w\varepsilon_k'(w) \xrightarrow[w \in \mathbb{R},\ w \to +\infty]{} 0$ (first choose $\kappa > 0$ such that, for any real $w \geq 2$, $\mathscr{D}_0$ contains the circle centred at w of radius $\kappa|w|$). Compute A_0' and A_1', and observe that $|\frac{A_k(w)}{wA_k'(w)}|$ is bounded on $\mathscr{D}_0$, hence $y_k'(w) = A_k'(w)\big(1 + \widetilde{\varepsilon}_k(w)\big)$ with $\widetilde{\varepsilon}_k(w) \xrightarrow[w \in \mathbb{R},\ w \to +\infty]{} 0$, and $|A_0 A_1'|$ and $|A_0' A_1|$ are bounded on $[2, +\infty)$, hence the difference between the Wronskian of (y_0, y_1) and that of (A_0, A_1) tends to 0.

6.14.2 A summability result

The proof of Theorem 6.95 will be given together with that of two more precise statements, Theorems 6.98 and 6.102.

Theorem 6.98. *(i) The formal series*

$$\widetilde{\varphi}(z) := \sum_{n\geq 0}(-1)^n c_n z^{-n}, \quad \textit{where } c_n := \frac{1}{2^{n+1}\pi n!}\Gamma\big(n+\tfrac{5}{6}\big)\Gamma\big(n+\tfrac{1}{6}\big) \ \textit{for } n\in\mathbb{N}, \tag{6.106}$$

is 1*-summable in the directions of* $J_0 := (-\pi,\pi)$*, with a formal Borel transform belonging to* $\mathbb{C}\delta\oplus\mathscr{N}(J_0,0)$ *and a Borel sum* $\mathscr{S}^{J_0}\widetilde{\varphi}$ *holomorphic in* $\widetilde{\mathscr{D}}(J_0,0) = \{z\in\widetilde{\mathbb{C}} \mid -\frac{3\pi}{2} < \arg z < \frac{3\pi}{2}\}$*. Moreover,*

$$y_0(w) = \frac{1}{2\sqrt{\pi}}w^{-\frac{1}{4}}\,\mathrm{e}^{-\frac{2}{3}w^{3/2}}\mathscr{S}^{J_0}\widetilde{\varphi}\big(Z_0(w)\big) \quad \textit{for all } w\in\mathbb{C}\setminus\mathbb{R}^-, \tag{6.107}$$

with the biholomorphism

$$Z_0\colon \mathbb{C}\setminus\mathbb{R}^- \to \widetilde{\mathscr{D}}(J_0,0), \qquad Z_0(w) := \tfrac{2}{3}w^{3/2}. \tag{6.108}$$

(ii) The formal series

$$\widetilde{\psi}(z) := \widetilde{\varphi}(-z) = \sum_{n\geq 0} c_n z^{-n} \tag{6.109}$$

is 1*-summable in the directions of* $J_1 := (-2\pi,0)$*, with a formal Borel transform belonging to* $\mathbb{C}\delta\oplus\mathscr{N}(J_1,0)$ *and a Borel sum* $\mathscr{S}^{J_1}\widetilde{\psi}$ *holomorphic in* $\widetilde{\mathscr{D}}(J_1,0) = \{z\in\widetilde{\mathbb{C}} \mid -\frac{\pi}{2} < \arg z < \frac{5\pi}{2}\}$*. Moreover,*

$$y_1(w) = \frac{1}{2\mathrm{i}\sqrt{\pi}}w^{-\frac{1}{4}}\,\mathrm{e}^{\frac{2}{3}w^{3/2}}\mathscr{S}^{J_1}\widetilde{\psi}\big(Z_1(w)\big) \quad \textit{for all } w\in\mathbb{C}\setminus\big(\mathbb{R}^+\mathrm{e}^{-\mathrm{i}\pi/3}\big), \tag{6.110}$$

with the biholomorphism

$$Z_1\colon \mathbb{C}\setminus\big(\mathbb{R}^+\mathrm{e}^{-\mathrm{i}\pi/3}\big) \to \widetilde{\mathscr{D}}(J_1,0), \qquad Z_1(w) := \tfrac{2}{3}w^{3/2}. \tag{6.111}$$

Notice that our definition of the domains $\widetilde{\mathscr{D}}(J_k,0)$ agrees with Section 5.9.4, and that the change of variable $z = \frac{2}{3}w^{3/2}$ defines two different maps Z_0 and Z_1; see Figures 6.13 and 6.14. Explicitly, any $w\in\mathbb{C}\setminus\mathbb{R}^-$ can be written $w = r\mathrm{e}^{\mathrm{i}\theta_0}$ with $r>0$ and $\theta_0\in(-\pi,\pi)$ in a unique way, then (6.107) must be interpreted as involving

$$Z_0(w) = \tfrac{2}{3}r^{3/2}\underline{\mathrm{e}}^{3\mathrm{i}\theta_0/2}, \quad w^{-1/4} = r^{-1/4}\mathrm{e}^{-\mathrm{i}\theta_0/4}, \quad \mathrm{e}^{-\frac{2}{3}w^{3/2}} = \mathrm{e}^{-\frac{2}{3}r^{3/2}\mathrm{e}^{3\mathrm{i}\theta_0/2}} \tag{6.112}$$

(with the notation (6.26) for points on the Riemann surface of the logarithm $\widetilde{\mathbb{C}}$ and because of our conventions on principal branches), whereas any $w\in\mathbb{C}\setminus\big(\mathbb{R}^+\mathrm{e}^{-\mathrm{i}\pi/3}\big)$ can be written $w = r\mathrm{e}^{\mathrm{i}\theta_1}$ with $r>0$ and $\theta_1\in(-\frac{\pi}{3},\frac{5\pi}{3})$ in a unique way, and

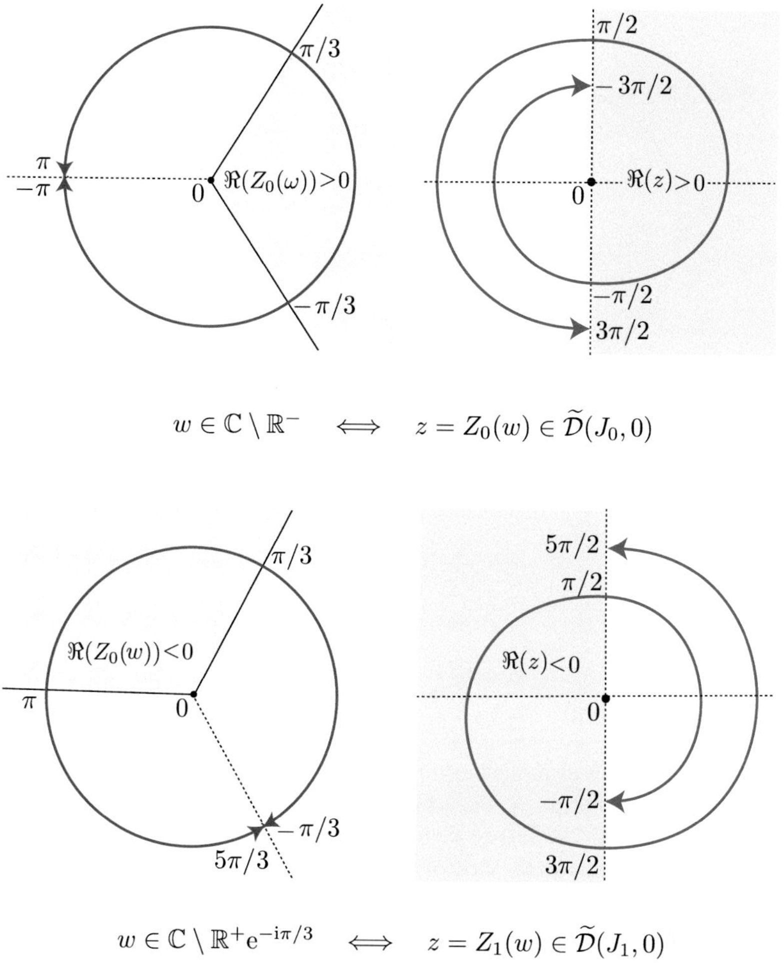

$\pi/3$
π
$-\pi$
0
$\Re(Z_0(\omega))>0$
$-\pi/3$
$\pi/2$
$-3\pi/2$
$\Re(z)>0$
0
$-\pi/2$
$3\pi/2$
$w \in \mathbb{C} \setminus \mathbb{R}^- \iff z = Z_0(w) \in \widetilde{\mathcal{D}}(J_0, 0)$
$\pi/3$
$\Re(Z_0(w))<0$
π
0
$-\pi/3$
$5\pi/3$
$5\pi/2$
$\pi/2$
$\Re(z)<0$
0
$-\pi/2$
$3\pi/2$
$w \in \mathbb{C} \setminus \mathbb{R}^+ e^{-i\pi/3} \iff z = Z_1(w) \in \widetilde{\mathcal{D}}(J_1, 0)$

Fig. 6.13: *Top: Domains for* $y_0(w)$ *and* $\mathscr{S}^{J_0}\widetilde{\varphi}(z)$*. Bottom: Domains for* $y_1(w)$ *and* $\mathscr{S}^{J_1}\widetilde{\psi}(z)$.

then (6.110) must be interpreted as involving

$$Z_1(w) = \tfrac{2}{3} r^{3/2} \underline{e}^{3i\theta_1/2}, \quad w^{-1/4} = r^{-1/4} e^{-i\theta_1/4}, \quad e^{\frac{2}{3} w^{3/2}} = e^{\frac{2}{3} r^{3/2} e^{3i\theta_1/2}}. \tag{6.113}$$

The point is that, when w belongs to the intersection of the above domains, we have $\theta_1 = \theta_0$ if $\theta_0 \in (-\frac{\pi}{3}, \pi)$ but $\theta_1 = \theta_0 + 2\pi$ if $\theta_0 \in (-\pi, -\frac{\pi}{3})$; in the latter case the value of $w^{-1/4}$ gets multiplied by $-i$ from (6.107) to (6.110), the value of $w^{3/2}$ gets multiplied by -1, and $Z_1(w) = Z_0(w)\underline{e}^{3i\pi}$.

Exercise 6.99. Use the reflection formula (5.62) to check that

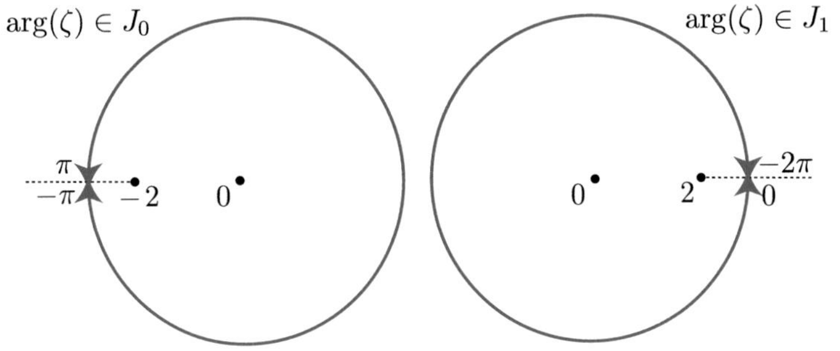

Fig. 6.14: *Summability directions in the Borel plane for $\widetilde{\varphi}(z)$ and $\widetilde{\psi}(z)$.*

$$c_0 = 1. \tag{6.114}$$

Show that, for all $n \geq 0$,

$$c_n = \frac{(2n+1)(2n+3)\cdots(6n-1)}{216^n n!} = \frac{(6n)!}{864^n (2n)!(3n)!} = \frac{\Gamma(3n+\frac{1}{2})}{54^n n! \Gamma(n+\frac{1}{2})}. \tag{6.115}$$

Theorem 6.98 gives us refined asymptotic formulas for the Airy function $\mathrm{Ai} = y_0$ and its companion solution y_1: the 1-Gevrey asymptotic expansions in the variable $z = \frac{2}{3}w^{3/2}$ give rise to "$\frac{2}{3}$-Gevrey asymptotic expansions" in the variable w. Let us explain what this means on the case of $y_0(w)$. Decomposing $\widetilde{\varphi}(z)$ into even and odd parts, we have

$$\widetilde{\varphi}(z) = \widetilde{\varphi}_0(z) - z^{-1}\widetilde{\varphi}_1(z), \qquad \widetilde{\varphi}_i(z) = \sum_{n \text{ even}} c_{n+i} z^{-n} \in \mathbb{C}[[z^{-2}]] \subset \mathbb{C}[[z^{-1}]] \tag{6.116}$$

and, correspondingly, $\widetilde{\Phi}(w) := \widetilde{\varphi}(\frac{2}{3}w^{3/2}) \in \mathbb{C}[[w^{-3/2}]]$ can be decomposed as

$$\widetilde{\Phi}(w) = \widetilde{\Phi}_0(w) - \tfrac{3}{2}w^{-3/2}\widetilde{\Phi}_1(w),$$
$$\widetilde{\Phi}_i(w) = \sum_{m \geq 0} \left(\tfrac{3}{2}\right)^{2m} c_{2m+i} w^{-3m} \in \mathbb{C}[[w^{-3}]] \subset \mathbb{C}[[w^{-1}]].$$

As elements of $\mathbb{C}[[z^{-1}]]$, $\widetilde{\varphi}_0(z)$ and $\widetilde{\varphi}_1(z)$ are 1-Gevrey, hence $\widetilde{\Phi}_0(w)$ and $\widetilde{\Phi}_1(w)$ are "$\frac{2}{3}$-Gevrey" elements of $\mathbb{C}[[w^{-1}]]$, i.e. their coefficients satisfy estimates similar to those of Definition 5.8 except that $n!$ is raised to the power $\frac{2}{3}$ (because $|c_{n+i}| \leq LM^n n!$ yields $|c_{2m+i}| \leq \widetilde{L}\widetilde{M}^{3m}(3m)!^{\frac{2}{3}}$ with the help of Stirling's formula (5.51)). Let $s := \frac{2}{3}$. We get a decomposition of the Borel sum $\mathscr{S}^{J_0}\widetilde{\varphi}\big(Z_0(w)\big) = \Phi_0(w) -$

$\frac{3}{2}w^{-3/2}\Phi_1(w)$ with

$$\Phi_i(w) := \mathscr{S}^{J_0}\widetilde{\varphi}_i\big(Z_0(w)\big) \sim_s \widetilde{\Phi}_i(w) \quad \text{uniformly for } w \in \{\, r e^{i\theta} \mid r \geq 1,\ \theta \in I_0 \,\},$$

where the symbol $\sim_s$ means a uniform asymptotic expansion in the sense of Definition 5.21 p. 136 in which one can take constants of the form $K_N = LM^N N!^s$ in the right-hand side of (5.32); this is in fact an anticipation of the notion of s-Gevrey asymptotic expansion to be found in the second volume [Lod16]. Extending the notation to half-integer power series, we write $\mathscr{S}^{J_0}\widetilde{\varphi}\big(Z_0(w)\big) \sim_s \widetilde{\Phi}_0(w) - \frac{3}{2}w^{-3/2}\widetilde{\Phi}_1(w)$, and extending it further as in footnote 5 of p. 151, we finally obtain

$$y_0(w) \sim_s \frac{1}{2\sqrt{\pi}} w^{-\frac{1}{4}} e^{-\frac{2}{3}w^{3/2}} \sum_{n\geq 0} (-1)^n \big(\tfrac{3}{2}\big)^n c_n w^{-3n/2}$$

uniformly for w in the aforementioned set. In view of (6.114), the dominant behaviour is just (6.103). Similarly,

$$y_1(w) \sim_s \frac{1}{2i\sqrt{\pi}} w^{-\frac{1}{4}} e^{\frac{2}{3}w^{3/2}} \sum_{n\geq 0} \big(\tfrac{3}{2}\big)^n c_n w^{-3n/2}$$

uniformly for $w \in \{\, r e^{i\theta'} \mid r \geq 1,\ \theta' \in I_1 \,\}$, and the dominant behaviour is just (6.104).

6.14.3 The Stokes phenomenon

We thus have a precise knowledge of the asymptotic behaviour of the Airy function $\mathrm{Ai} = y_0$ in the cut plane $\mathbb{C} \setminus \mathbb{R}^-$, with exponential decay in the sector $\arg w \in (-\frac{\pi}{3}, \frac{\pi}{3})$ and exponential growth in the sectors $\arg w \in (-\pi, -\frac{\pi}{3})$ and $\arg w \in (\frac{\pi}{3}, \pi)$ (and similarly for y_1, after a rotation by $2\pi/3$).

Since y_0 is known to be an entire function, one may wonder about its asymptotic behaviour along the negative real semi-axis. It turns out that the mere summability statement of Theorem 6.98 is sufficient to obtain this information, thanks to some symmetry properties specific to the Airy equation (6.102). This is the content of the following two exercises.

Exercise 6.100. (i) Show the following symmetry properties of equation (6.102):

$$y \text{ solution} \iff \widetilde{y} \text{ solution} \iff y^* \text{ solution},$$

where $\widetilde{y}$ and y^* are defined by $\widetilde{y}(w) := \overline{y(\overline{w})}$ and $y^*(w) := e^{-2\pi i/3} y(e^{-2\pi i/3} w)$.

(ii) Show that the formula

$$y_2(w) := -\frac{1}{2i\sqrt{\pi}} w^{-\frac{1}{4}} e^{\frac{2}{3}w^{3/2}} \mathscr{S}^{J_2}\widetilde{\psi}\big(Z_2(w)\big) \quad \text{for all } w \in \mathbb{C} \setminus \big(\mathbb{R}^+ e^{i\pi/3}\big),$$

with $J_2 := (0, 2\pi)$ and a biholomorphism $Z_2 \colon \mathbb{C} \setminus (\mathbb{R}^+ e^{i\pi/3}) \to \widetilde{\mathscr{D}}(J_2, 0) = \{ z \in \widetilde{\mathbb{C}} \mid -\frac{5\pi}{2} < \arg z < \frac{\pi}{2} \}$ defined by $Z_2(w) := \frac{2}{3} w^{3/2}$, defines a solution such that

$$\widetilde{y}_0 = y_0, \quad \widetilde{y}_1 = y_2, \quad \widetilde{y}_2 = y_1, \qquad y_0^* = y_1, \quad y_1^* = y_2, \quad y_2^* = y_0.$$

(iii) Show that

$$y_0 + y_1 + y_2 = 0. \tag{6.117}$$

Deduce that the Wronskians of (y_1, y_2) and (y_2, y_0) are equal to $\frac{1}{2\pi i}$ and that the Airy function of the second kind defined by (6.105) can also be written

$$\mathrm{Bi}(w) = i\big(y_1(w) - y_2(w)\big) \tag{6.118}$$

and is indeed real-analytic.

Hints: (ii) Show that, for $R > 0$, $\mathscr{S}^{\pi} \widetilde{\psi}(R\underline{e}^{i\pi}) = \mathscr{S}^{-\pi} \widetilde{\psi}(R\underline{e}^{-i\pi}) = \mathscr{S}^{0} \widetilde{\varphi}(R\underline{e}^{i0})$ and that $\mathscr{S}^{\frac{\pi}{2}} \widetilde{\psi}(R\underline{e}^{-i\pi/2})$ and $\mathscr{S}^{-\frac{\pi}{2}} \widetilde{\psi}(R\underline{e}^{i\pi/2})$ are complex conjugate. (iii) Compute first the Wronskian of (y_2, y_0) by means of asymptotics as in Exercise 6.97 and deduce that $y_1 + y_2$ is proportional to y_0.

Exercise 6.101. (i) Show that the formal series

$$\widetilde{\psi}_+(z) := \sum_{m \geq 0} c_{2m} z^{-2m}, \quad \widetilde{\psi}_-(z) := \sum_{m \geq 0} c_{2m+1} z^{-2m-1}$$

are 1-summable in the directions of $(-\pi, 0)$ and $(0, \pi)$. Let $\psi_\pm := \mathscr{S}^{(-\pi,0)} \widetilde{\psi}_\pm$.

(ii) Let $r > 0$, $\varepsilon \in (-\frac{2\pi}{3}, \frac{2\pi}{3})$ and $\tau := \frac{2}{3} r^{3/2} e^{3i\varepsilon/2}$. Use (6.117)–(6.118) to show that

$$\mathrm{Ai}(-re^{i\varepsilon}) = \frac{1}{\sqrt{\pi}} r^{-1/4} e^{-i\varepsilon/4} \big(\cos(\tau - \tfrac{\pi}{4}) \psi_+(i\tau) + \sin(\tau - \tfrac{\pi}{4})\, i\, \psi_-(i\tau) \big),$$

$$\mathrm{Bi}(-re^{i\varepsilon}) = \frac{1}{\sqrt{\pi}} r^{-1/4} e^{-i\varepsilon/4} \big(-\sin(\tau - \tfrac{\pi}{4}) \psi_+(i\tau) + \cos(\tau - \tfrac{\pi}{4})\, i\, \psi_-(i\tau) \big),$$

with $\psi_+(i\tau) \sim_1 \sum_{m \geq 0} (-1)^m c_{2m} \tau^{-2m}$ and $i\,\psi_-(i\tau) \sim_1 \sum_{m \geq 0} (-1)^m c_{2m+1} \tau^{-2m-1}$ in the domain $\tau \in \mathbb{C} \setminus \mathbb{R}^-$.

The second point of exercise 6.101 gives us access to the asymptotic behaviour of $\mathrm{Ai}(w) = y_0(w)$ along $\mathbb{R}^-$ or, equivalently, to the asymptotic behaviour of $\mathscr{S}^{J_0} \widetilde{\varphi}_0(z)$ along the limiting directions $\arg z = \pm \frac{3\pi}{2}$ of the sector $\widetilde{\mathscr{D}}(J_0, 0)$ of $\widetilde{\mathbb{C}}$. The key point for that was the representation of y_0 as $-y_1 - y_2$ given by (6.117). This relation can be considered as a precise description of the Stokes phenomenon for our problem—cf. Remark 5.34 on p. 144—and it allows one to compute exactly the three Stokes matrices $S_{\widetilde{\ell}_k}$ involved in the analysis performed in Section 2.2 on Example 2.50 (compare also with exercise 9 on p. 240 in the second volume [Lod16]).

Another way of deriving relation (6.117) is to prove the resurgent character of $\widetilde{\varphi}$ and $\widetilde{\psi}$ and to compute directly in the Borel plane the action of the alien derivations on $\mathscr{B}\widetilde{\varphi}$ and $\mathscr{B}\widetilde{\psi}$; this will be done in the next two sections by exploiting a very explicit formula which is available for these Borel transforms.

6.14.4 The resurgent structure

As just mentioned, it turns out that one can find a very explicit formula for $\mathscr{B}\widetilde{\varphi}$ and $\mathscr{B}\widetilde{\psi}$. It relies on a slight extension of the definition of convolution: we now use the formula $\widehat{\phi}_1 * \widehat{\phi}_2(\zeta) = \int_0^1 \widehat{\phi}_1(t\zeta)\widehat{\phi}_2\big((1-t)\zeta\big)\zeta\,\mathrm{d}t$ whenever $\zeta \in \widetilde{\mathbb{C}}$ and $\widehat{\phi}_1$ and $\widehat{\phi}_2$ are defined on a part of $\widetilde{\mathbb{C}}$ so that the functions $r \in (0,1] \mapsto \widehat{\phi}_i(r\zeta)$ are integrable,[20] e.g.

$$\frac{\zeta^{\alpha_1}}{\Gamma(\alpha_1+1)} * \frac{\zeta^{\alpha_2}}{\Gamma(\alpha_2+1)} = \frac{\zeta^{\alpha_1+\alpha_2+1}}{\Gamma(\alpha_1+\alpha_2+2)} \quad \text{if } \Re e\,\alpha_1, \Re e\,\alpha_2 > -1 \tag{6.119}$$

(using principal branches) and more generally $\widehat{\phi}_1 * \widehat{\phi}_2(\zeta) \in \zeta^{\alpha_1+\alpha_2+1}\mathbb{C}\{\zeta\}$ if $\widehat{\phi}_1(\zeta) \in \zeta^{\alpha_1}\mathbb{C}\{\zeta\}$ and $\widehat{\phi}_2(\zeta) \in \zeta^{\alpha_2}\mathbb{C}\{\zeta\}$.

Theorem 6.102. *Let*

$$\widehat{\chi}(\zeta) := C\zeta^{-5/6} * (2\zeta+\zeta^2)^{-1/6} \in \mathbb{C}\{\zeta\} \quad \textit{with } C := \frac{2^{1/6}}{\Gamma(1/6)\Gamma(5/6)}. \tag{6.120}$$

Then $\widehat{\chi}(-\zeta) = C\zeta^{-5/6} * (2\zeta-\zeta^2)^{-1/6}$ *and*

$$\mathscr{B}\widetilde{\varphi} = \delta + \widehat{\varphi}, \quad \widehat{\varphi}(\zeta) = \frac{\mathrm{d}\widehat{\chi}}{\mathrm{d}\zeta}(\zeta), \qquad \mathscr{B}\widetilde{\psi} = \delta + \widehat{\psi}, \quad \widehat{\psi}(\zeta) = -\frac{\mathrm{d}\widehat{\chi}}{\mathrm{d}\zeta}(-\zeta). \tag{6.121}$$

It follows that $\widetilde{\varphi}(z)$ is a simple $\{0,-2\}$-resurgent series, $\widetilde{\psi}(z)$ is a simple $\{0,2\}$-resurgent series, and the only non-trivial alien derivatives are

$$\Delta_{-2}\widetilde{\varphi} = -\mathrm{i}\widetilde{\psi}, \qquad \Delta_2\widetilde{\psi} = -\mathrm{i}\widetilde{\varphi}. \tag{6.122}$$

Notice that $C = \frac{1}{2^{5/6}\pi}$ (because of the reflection formula (5.62)), but the expression indicated above is more natural because, in view of (6.119), it gives directly $\widehat{\chi}(0) = 1$.

The resurgence relations (6.122) allow us to compute the Stokes phenomenon and to prove relation (6.117) in an independent and much more direct way with respect to what was done in Section 6.14.3:

Corollary 6.103. *For any $z \in \widetilde{\mathbb{C}}$,*

$$\arg z \in \left(\tfrac{\pi}{2}, \tfrac{3\pi}{2}\right) \implies \mathscr{S}^{J_0}\widetilde{\varphi}(\underline{\mathrm{e}}^{-2\pi\mathrm{i}}z) = \mathscr{S}^{J_0}\widetilde{\varphi}(z) - \mathrm{i}\,\mathrm{e}^{2z}\mathscr{S}^{J_1}\widetilde{\psi}(z) \tag{6.123}$$

$$\arg z \in \left(-\tfrac{\pi}{2}, \tfrac{\pi}{2}\right) \implies \mathscr{S}^{J_1}\widetilde{\psi}(z) = \mathscr{S}^{J_1}\widetilde{\psi}(\underline{\mathrm{e}}^{2\pi\mathrm{i}}z) - \mathrm{i}\,\mathrm{e}^{-2z}\mathscr{S}^{J_0}\widetilde{\varphi}(z). \tag{6.124}$$

[20] This is called "convolution of integrable minors"; it is part of the extension of the theory which was alluded to at the end of Section 6.8 and in Remark 6.60. See the third volume [Del16].

In the case of (6.124), rewriting the identity as

$$\mathrm{e}^{-z}\mathscr{S}^{J_0}\widetilde{\varphi}(z)+\frac{1}{\mathrm{i}}\mathrm{e}^{z}\mathscr{S}^{J_1}\widetilde{\psi}(z)-\frac{1}{\mathrm{i}}\mathrm{e}^{z}\mathscr{S}^{J_1}\widetilde{\psi}(\underline{\mathrm{e}}^{2\pi\mathrm{i}}z)=0$$

and defining $w:=(\frac{3}{2}z)^{2/3}$ and $w_*:=w\,\mathrm{e}^{4\mathrm{i}\pi/3}$ so that

$$z=Z_0(w)=Z_1(w),\qquad Z_1(w_*)=\underline{\mathrm{e}}^{2\pi\mathrm{i}}z,$$

we get from formulas (6.107) and (6.110)

$$w^{1/4}y_0(w)+w^{1/4}y_1(w)-w_*^{1/4}y_1(w_*)=0,$$

where $-w_*^{1/4}=w^{1/4}\mathrm{e}^{4\mathrm{i}\pi/3}$ (because we are required to use principal branches). Therefore,

$$y_0(w)+y_1(w)+\mathrm{e}^{4\mathrm{i}\pi/3}y_1(w\,\mathrm{e}^{4\mathrm{i}\pi/3})=0,$$

which is equivalent to (6.117).

Proof of Corollary 6.103. Let us use the symbolic Stokes infinitesimal generators Δ_d and the symbolic Stokes automorphisms Δ_d^+ introduced in Definition 6.74. The second part of Theorem 6.102, when rephrased with the notations of Section 6.12.2, yields

$$\Delta_{\mathbb{R}^-}\mathscr{B}\widetilde{\varphi}=-\mathrm{i}\tau_{-2}\mathscr{B}\widetilde{\psi},\qquad \Delta_{\mathbb{R}^+}\mathscr{B}\widetilde{\psi}=-\mathrm{i}\tau_2\mathscr{B}\widetilde{\varphi}\tag{6.125}$$

and $\Delta_{\mathrm{e}^{\mathrm{i}\theta}\mathbb{R}^+}\mathscr{B}\widetilde{\varphi}=0$ (resp. $\Delta_{\mathrm{e}^{\mathrm{i}\theta}\mathbb{R}^+}\mathscr{B}\widetilde{\psi}=0$) if $\mathrm{e}^{\mathrm{i}\theta}\neq-1$ (resp. $\mathrm{e}^{\mathrm{i}\theta}\neq1$), in view of parts (i) and (ii) of Theorem 6.73. According to part (iii) of Theorem 6.73, we pass from Δ_d to Δ_d^+ by exponentiating, which is particularly simple in this case:

$$\Delta_{\mathbb{R}^-}^+\mathscr{B}\widetilde{\varphi}=\mathscr{B}\widetilde{\varphi}-\mathrm{i}\tau_{-2}\mathscr{B}\widetilde{\psi},\qquad \Delta_{\mathbb{R}^+}^+\mathscr{B}\widetilde{\psi}=\mathscr{B}\widetilde{\psi}-\mathrm{i}\tau_{-2}\mathscr{B}\widetilde{\varphi}\tag{6.126}$$

and $\Delta_{\mathrm{e}^{\mathrm{i}\theta}\mathbb{R}^+}^+\mathscr{B}\widetilde{\varphi}=0$ (resp. $\Delta_{\mathrm{e}^{\mathrm{i}\theta}\mathbb{R}^+}^+\mathscr{B}\widetilde{\psi}=0$) if $\mathrm{e}^{\mathrm{i}\theta}\neq-1$ (resp. $\mathrm{e}^{\mathrm{i}\theta}\neq1$).

For $\arg z\in(-\frac{\pi}{2},\frac{\pi}{2})$, we apply Theorem 6.77 with $d=\mathbb{R}^+$: choosing $\theta>0$ small enough, we get $\mathscr{L}^{-\theta}\mathscr{B}\widetilde{\psi}(z)=\mathscr{L}^{\theta}\Delta_{\mathbb{R}^+}^+\mathscr{B}\widetilde{\psi}(z)$ (there is no error term in this case), hence

$$\mathscr{L}^{-\theta}\mathscr{B}\widetilde{\psi}(z)=\mathscr{L}^{\theta}\mathscr{B}\widetilde{\psi}(z)-\mathrm{i}\,\mathrm{e}^{-2z}\mathscr{L}^{\theta}\mathscr{B}\widetilde{\varphi}(z),$$

which yields (6.124) because

$$\begin{aligned}
-\theta\in J_1 &\implies \mathscr{L}^{-\theta}\mathscr{B}\widetilde{\psi}(z)=\mathscr{S}^{J_1}\widetilde{\psi}(z)\\
\theta\in J_0 &\implies \mathscr{L}^{\theta}\mathscr{B}\widetilde{\varphi}(z)=\mathscr{S}^{J_0}\widetilde{\varphi}(z)\\
-2\pi+\theta\in J_1 &\implies \mathscr{L}^{\theta}\mathscr{B}\widetilde{\psi}(z)=\mathscr{L}^{-2\pi+\theta}\mathscr{B}\widetilde{\psi}(\mathrm{e}^{2\pi\mathrm{i}}z)=\mathscr{S}^{J_1}(\mathrm{e}^{2\pi\mathrm{i}}z).
\end{aligned}$$

We leave it to the reader to treat the case $\arg z\in(\frac{\pi}{2},\frac{3\pi}{2})$ by applying Theorem 6.77 with $d=\mathbb{R}^-$. □

Remark 6.104. The relations (6.126) mean that the action of the operator $\Delta^+_{\mathbb{R}^-}$, resp. $\Delta^+_{\mathbb{R}^+}$, on the vector

$$G := \begin{pmatrix} g_1(z) \\ g_2(z) \end{pmatrix} = \begin{pmatrix} \mathrm{e}^{-z}\widetilde{\varphi}(z) \\ \mathrm{e}^{z}\widetilde{\psi}(z) \end{pmatrix}$$

coincides with the multiplication by the matrix

$$\begin{pmatrix} 1 & -\mathrm{i} \\ 0 & 1 \end{pmatrix}, \quad \text{resp.} \quad \begin{pmatrix} 1 & 0 \\ -\mathrm{i} & 1 \end{pmatrix} \tag{6.127}$$

The vector G corresponds in fact to a formal fundamental system of solutions for the linear equation deduced from (6.102) by the change of variable and unknown $z = \frac{2}{3}w^{3/2}$ and $y(w) = w^{-1/4}g(z)$. The matrices (6.127) can then be interpreted as Stokes matrices for the g-equation, associated with the directions $\mathbb{R}^-$ resp. $\mathbb{R}^+$ in the z-plane. Compare with Example 2.50 of Section 2.2 in this volume, and exercise 9 on p. 240 in the second volume [Lod16].

Exercise 6.105 (Kummer form of the confluent hypergeometric equation). Let $a, b \in \mathbb{C}$. Show that the linear differential equation

$$z\frac{\mathrm{d}^2U}{\mathrm{d}z^2} + (b-z)\frac{\mathrm{d}U}{\mathrm{d}z} - aU = 0$$

has a regular singularity at 0 and an irregular singularity at ∞. Suppose $\Re e\, a > 0$. Show that

$$\widetilde{U}_{a,b}(z) := z^{-a}\sum_{n\geq 0}(-1)^n\frac{(a)_n(1+a-b)_n}{n!}z^{-n},$$

with the notation $(a)_0 = 1$ and $(a)_n := a(a+1)\cdots(a+n-1)$ for $n \geq 1$, is a 1-summable resurgent formal solution with Borel transform

$$\widehat{U}_{a,b}(\zeta) = \frac{1}{\Gamma(a)}\zeta^{a-1}(1+\zeta)^{b-a-1}$$

and describe the resurgent structure when $\Re e\, b > \Re e\, a$ (read Section 6.14.5, or Chapter 7 of the third volume [Del16], for the slight extension of the theory needed to deal with "integrable minors" that are not regular at the origin).

6.14.5 *Proof of Theorems 6.95, 6.98 and 6.102*

The solutions of the Airy equation (6.102) are entire functions, thus they induce holomorphic functions on the whole Riemann surface of the logarithm $\widetilde{\mathbb{C}}$. We will perform the change of variable $z = \frac{2}{3}w^{3/2}$ in $\widetilde{\mathbb{C}}$ and various changes of unknown function. Since we aim at using our slightly generalised convolution to study the

Borel transforms of various formal series and their summability, it is useful to start with a similar generalisation of the formal Borel transform $\mathscr{B}$.

a) We extend the definition of the formal Borel transform by setting

$$\mathscr{B}\colon\ \tilde\phi(z)=\sum a_n z^{-n-\alpha-1}\in z^{-\alpha-1}\mathbb{C}[[z^{-1}]]\mapsto\hat\phi(\zeta)=\sum\frac{a_n\zeta^{n+\alpha}}{\Gamma(n+\alpha+1)}\in\zeta^{\alpha}\mathbb{C}[[\zeta]]$$

whenever $\Re e\,\alpha>-1$. We still have $\mathscr{B}\big(\frac{\mathrm{d}\tilde\phi}{\mathrm{d}z}\big)(\zeta)=-\zeta\mathscr{B}\tilde\phi(\zeta)$ and, if $\Re e\,\alpha_1,\Re e\,\alpha_2>-1$,

$$\mathscr{B}\tilde\phi_k=\hat\phi_k\in\zeta^{\alpha_k}\mathbb{C}\{\zeta\}\ \text{ for } k=1,2\quad\Longrightarrow\quad\mathscr{B}(\tilde\phi_1\tilde\phi_2)=\hat\phi_1*\hat\phi_2. \tag{6.128}$$

Suppose $\Re e\,\alpha>-1$, $\theta\in\mathbb{R}$ and $\gamma>0$. If $\mathscr{B}\tilde\phi=\hat\phi\in\zeta^{\alpha}\mathbb{C}\{\zeta\}$ extends analytically along the ray $\mathbb{R}^+\underline{\mathrm{e}}^{\mathrm{i}\theta}\subset\tilde{\mathbb{C}}$ with $|\hat\phi(\zeta)|=O(\mathrm{e}^{\gamma|\zeta|})$ for $|\zeta|\geq1$, then we extend the Borel-Laplace summation operator in the direction θ to that situation by setting $\mathscr{S}^\theta\tilde\phi:=\mathscr{L}^\theta\hat\phi$, which is holomorphic in $\tilde\pi_\theta\subset\tilde{\mathbb{C}}$, the lift defined in Section 5.9.4 for the half-plane Π^θ_γ (thus $\tilde\pi_\theta$ is a lifted half-plane bisected by $\underline{\mathrm{e}}^{-\mathrm{i}\theta}\mathbb{R}^+$). Then $\mathscr{S}^\theta\big(\frac{\mathrm{d}\tilde\phi}{\mathrm{d}z}\big)=\frac{\mathrm{d}}{\mathrm{d}z}\big(\mathscr{S}^\theta\tilde\phi\big)$ and, in the situation of (6.128), $\mathscr{S}^\theta(\tilde\phi_1\tilde\phi_2)=(\mathscr{S}^\theta\tilde\phi_1)(\mathscr{S}^\theta\tilde\phi_2)$. Similarly we extend the definition of the Borel-Laplace summation operator in an arc of directions J: we get a function $\mathscr{S}^J\tilde\varphi$ holomorphic in $\tilde{\mathscr{D}}(J,\gamma)$ when $|\hat\phi(r\mathrm{e}^{\mathrm{i}\theta})|=O(\mathrm{e}^{\gamma(\theta)r})$ for all $\theta\in J$ and $r\geq1$ (now γ is a positive locally bounded function).

b) A simple computation shows that $y(w)$ is a solution of (6.102) holomorphic in a domain of $\tilde{\mathbb{C}}$ if and only if

$$y(w)=wF\big(\tfrac{2}{3}w^{3/2}\big)$$

with F holomorphic in a domain of $\tilde{\mathbb{C}}$ and solution of

$$F''(z)+\tfrac{5}{3}z^{-1}F'-F=0, \tag{6.129}$$

and this holds if and only if

$$F(z)=\mathrm{e}^{-z}f(z)$$

with f holomorphic in a domain of $\tilde{\mathbb{C}}$ and solution of

$$f''(z)-2f'(z)+\tfrac{5}{3}z^{-1}(f'-f)=0. \tag{6.130}$$

One sees that the latter equation admits formal solutions in the space $z^{\nu}\mathbb{C}[[z^{-1}]]$ only if $\nu=-5/6$ (by matching the monomials proportional to $z^{\nu-1}$) and that there is a unique formal solution of the form

$$\tilde f(z)=z^{-5/6}\big(1+O(z^{-1})\big)\in z^{-5/6}\mathbb{C}[[z^{-1}]], \tag{6.131}$$

the coefficients of which can be determined by induction, but we will not need to write the induction formulas here.

c) Let $\hat{f}(\zeta) := \mathscr{B}\tilde{f}(\zeta) = \frac{\zeta^{-1/6}}{\Gamma(5/6)}\big(1+O(\zeta)\big) \in \zeta^{-1/6}\mathbb{C}[[\zeta]]$. This must be the unique formal solution of this form to the Borel transformed equation

$$(\zeta^2+2\zeta)\hat{f} = \tfrac{5}{3}1 * \big((1+\zeta)\hat{f}\big). \tag{6.132}$$

But the convergent solutions are easily found: if $\hat{f}(\zeta) \in \zeta^{-1/6}\mathbb{C}\{\zeta\}$, then both sides of (6.132) vanish as $|\zeta| \to 0$, hence the equation is equivalent to

$$\frac{\mathrm{d}}{\mathrm{d}\zeta}\big((\zeta^2+2\zeta)\hat{f}\big) = \tfrac{5}{3}(1+\zeta)\hat{f}$$

and the solutions of this first-order linear ODE are the multiples of $(\zeta^2+2\zeta)^{-1/6}$. Hence the unique formal solution of (6.132) we are interested in is nothing but the Puiseux expansion of one of them:

$$\hat{f}(\zeta) = \frac{1}{\Gamma(5/6)}\left(\zeta + \frac{\zeta^2}{2}\right)^{-1/6} = \frac{\zeta^{-1/6}}{\Gamma(5/6)}\left(1+\frac{\zeta}{2}\right)^{-1/6} \tag{6.133}$$

and we see that the radius of convergence of $\zeta^{1/6}\hat{f}(\zeta) \in \mathbb{C}[[\zeta]]$ is 2 and that $\hat{f}$ defines a function holomorphic on the "lifted cut plane"

$$\widetilde{U} := \widetilde{\mathbb{C}} \setminus \bigcup_{N\in\mathbb{Z}} 2\,\underline{\mathrm{e}}^{(2N+1)\mathrm{i}\pi}\mathbb{R}^+, \tag{6.134}$$

with $|\hat{f}(r\mathrm{e}^{\mathrm{i}\theta})| \le A_0(\theta)r^{-1/3}$ for all $\theta \in J_0 = (-\pi,\pi)$ and $r \ge 1$, for some locally bounded function A_0. It follows that $\mathscr{S}^{J_0}\tilde{f}$ is well-defined and holomorphic in $\widetilde{\mathscr{D}}(J_0,0)$ and that it is a solution to (6.130).

d) The germ $\widehat{\chi}(\zeta) \in \mathbb{C}\{\zeta\}$ defined by (6.120) can be written

$$\widehat{\chi}(\zeta) = \frac{\zeta^{-5/6}}{\Gamma(1/6)} * \hat{f}(\zeta) = \frac{\zeta^{1/6}}{\Gamma(1/6)}\int_0^1 \hat{f}(t\zeta)(1-t)^{-5/6}\,\mathrm{d}t. \tag{6.135}$$

It has a radius of convergence equal to 2 and $\widehat{\chi}(-\zeta) = C\zeta^{-5/6} * (2\zeta-\zeta^2)^{-1/6}$ because $\hat{f}(\underline{\mathrm{e}}^{\mathrm{i}\pi}\zeta) = \frac{\mathrm{e}^{-\mathrm{i}\pi/6}}{\Gamma(5/6)}\big(\zeta - \frac{\zeta^2}{2}\big)^{-1/6}$ for $|\zeta| < 2$. We observe that $\widehat{\chi}$ extends analytically to the cut plane $\mathbb{C}\setminus(-\infty,-2]$. Using $B := \max\{|\zeta|^{1/6}|\hat{f}(\zeta)|,\ |\zeta| \le 1\} < \infty$, we get $|\widehat{\chi}(\zeta)| \le A(\theta)$ with

$$A(\theta) := \frac{B}{\Gamma(1/6)}\int_0^1 t^{-1/6}(1-t)^{-5/6}\,\mathrm{d}t + \frac{A_0(\theta)}{\Gamma(1/6)}\int_0^1 t^{-1/3}(1-t)^{-5/6}\,\mathrm{d}t$$

if $\arg\zeta \in J_0$ and $|\zeta| \ge 1$, hence $\widehat{\chi} \in \mathscr{N}(J_0,0)$.

We can also bound $|\frac{\mathrm{d}\widehat{\chi}}{\mathrm{d}\zeta}|$ by means of the Cauchy inequality: for any $\theta \in J_0$, one can take $\kappa(\theta) \in (0,\frac{1}{2})$ small enough so that, for any $r \ge 2$, the circle centred at $r\mathrm{e}^{\mathrm{i}\theta}$

of radius $\kappa(\theta)r$ is contained in the set $\{|w| \geq 1, |\arg\zeta - \theta| \leq \arcsin\kappa(\theta)\}$ which is itself contained in the cut plane $\mathbb{C}\setminus(-\infty,-2]$, hence $\frac{\mathrm{d}\widehat{\chi}}{\mathrm{d}\zeta} \in \mathscr{N}(J_0,0)$.

e) Let us define $\widetilde{\chi} \in z^{-1}\mathbb{C}[[z^{-1}]]$ as the inverse formal Borel transform of $\widehat{\chi}$. Since $\widehat{\chi}(0)\delta + \frac{\mathrm{d}\widehat{\chi}}{\mathrm{d}\zeta} = \mathscr{B}(z\widetilde{\chi})$, the relations (6.121) are equivalent to

$$\widetilde{\varphi}(z) = z\widetilde{\chi}(z), \qquad \widetilde{\psi}(z) = -z\widetilde{\chi}(-z). \tag{6.136}$$

These relations do hold for $\widetilde{\varphi}(z)$ and $\widetilde{\psi}(z)$ defined by (6.106) and (6.109), as the reader may check by computing the Taylor expansion at 0 of $\widehat{\chi}(\zeta)$ (use $(1+X)^{-\beta} = \sum_{n\geq 0}(-1)^n\frac{\Gamma(n+\beta)}{\Gamma(\beta)}X^n$ to get the Taylor expansion at 0 of $(1+\frac{\zeta}{2})^{-1/6}$ and apply (6.119) then). Therefore,

$$\mathscr{B}\widetilde{\varphi} \in \mathbb{C}\delta \oplus \mathscr{N}(J_0,0), \qquad \mathscr{B}\widetilde{\psi} \in \mathbb{C}\delta \oplus \mathscr{N}(J_1,0)$$

with $J_1 = (-2\pi,0)$. But $\mathscr{S}^{J_0}\widetilde{\chi}(z) = z^{-1/6}\mathscr{S}^{J_0}\widetilde{f}(z)$, hence

$$\mathscr{S}^{J_0}\widetilde{\varphi}(z) = z\mathscr{S}^{J_0}\widetilde{\chi}(z) = z^{5/6}\mathscr{S}^{J_0}\widetilde{f}(z). \tag{6.137}$$

Moreover $\mathscr{S}^{J_1}\widetilde{\psi}(z) = \mathscr{S}^{J_0}\widetilde{\varphi}(z\underline{\mathrm{e}}^{-\mathrm{i}\pi})$ (because $\widetilde{\varphi}(z) = \widetilde{\psi}(-z)$ and $J_0 = \pi + J_1$), hence $\mathscr{S}^{J_1}\widetilde{\psi}(z)$ is proportional to $z^{5/6}\mathscr{S}^{J_0}\widetilde{f}(z\underline{\mathrm{e}}^{-\mathrm{i}\pi})$.

f) Since $\mathscr{S}^{J_0}\widetilde{f}(z)$ is a solution to (6.130), $\mathrm{e}^{-z}\mathscr{S}^{J_0}\widetilde{f}(z)$ is a solution to (6.129), and so is $\mathrm{e}^{z}\mathscr{S}^{J_0}\widetilde{f}(z\underline{\mathrm{e}}^{-\mathrm{i}\pi})$ (obvious symmetry property of equation (6.129)). We now return to the original variable by inverting our change of variable $z = \frac{2}{3}w^{3/2}$: the first solution, holomorphic for $\arg z \in (-\frac{3\pi}{2},\frac{3\pi}{2})$, yields a solution to the Airy equation (6.102) in the form $w\mathrm{e}^{-\frac{2}{3}w^{3/2}}\mathscr{S}^{J_0}\widetilde{f}(\frac{2}{3}w^{3/2})$, holomorphic for $\arg w \in (-\pi,\pi)$, and since the factor $z^{-5/6}$ is proportional to $w^{-5/4}$, (6.137) shows that this solution of the Airy equation is proportional to

$$y_0(w) := \frac{1}{2\sqrt{\pi}}w^{-1/4}\mathrm{e}^{-\frac{2}{3}w^{3/2}}\mathscr{S}^{J_0}\widetilde{\varphi}(\tfrac{2}{3}w^{3/2}).$$

We also get a solution in the form $w\mathrm{e}^{\frac{2}{3}w^{3/2}}\mathscr{S}^{J_0}\widetilde{f}(\frac{2}{3}w^{3/2}\underline{\mathrm{e}}^{-\mathrm{i}\pi})$, which is proportional to

$$y_1(w) := \frac{1}{2\mathrm{i}\sqrt{\pi}}w^{-1/4}\mathrm{e}^{\frac{2}{3}w^{3/2}}\mathscr{S}^{J_1}\widetilde{\psi}(\tfrac{2}{3}w^{3/2}).$$

Both $\mathscr{S}^{J_0}\widetilde{\varphi}(\frac{2}{3}w^{3/2})$ and $\mathscr{S}^{J_1}\widetilde{\psi}(\frac{2}{3}w^{3/2})$ tend to $c_0 = 1$ along the directions $\arg w \in (-\pi,\pi)$. In view of the behaviour of the exponential factor $\mathrm{e}^{\pm\frac{2}{3}w^{3/2}}$, this shows that

- along the directions $\arg w \in (-\pi,-\frac{\pi}{3})$, both $|y_0(w)|$ and $|y_1(w)|$ tend to ∞,
- along the directions $\arg w \in (-\frac{\pi}{3},\frac{\pi}{3})$, $|y_0(w)|$ tends to 0 and $|y_1(w)|$ tends to ∞,
- along the directions $\arg w \in (\frac{\pi}{3},\pi)$, $|y_0(w)|$ tends to ∞ and $|y_1(w)|$ tends to 0.

It follows from the last two properties that y_0 and y_1 are independent over $\mathbb{C}$ and span the linear space of all solutions of (6.102), among which only the multiples of y_0 tend to 0 along the directions $\arg w \in (-\frac{\pi}{3}, \frac{\pi}{3})$, and only the multiples of y_1 tend to 0 along the directions $\arg w \in (\frac{\pi}{3}, \pi)$.

At this stage, Theorems 6.95 and 6.98 are proved, as well as the first part of Theorem 6.102. We are only left with proving that $\widetilde{\varphi}$ and $\widetilde{\psi}$ are simple resurgent series and computing their alien derivatives.

g) Formula (6.133) shows that $\hat{f}$, which is a priori holomorphic in the lifted cut plane $\widetilde{U}$ defined by (6.134), extends analytically along any path of $\widetilde{\mathbb{C}}$ starting in $\widetilde{U}$ and avoiding the points $2\,\underline{e}^{(2N+1)i\pi}$, $N \in \mathbb{Z}$. The same is true for $\widehat{\chi}$ because of (6.135). Indeed, if $\gamma\colon [0,1] \to \widetilde{\mathbb{C}}$ is such a path, with $\gamma(0) = \zeta_0 \in \widetilde{U}$ and $\zeta := \gamma(1)$, then the analytic continuation of $\widehat{\chi}$ along γ is given by

$$\mathrm{cont}_\gamma \widehat{\chi}(\zeta) = \int_0^{\zeta_0} \hat{f}(\xi) \frac{(\zeta-\xi)^{-5/6}}{\Gamma(1/6)}\,\mathrm{d}\xi + \int_\gamma \hat{f}(\xi)\frac{(\zeta-\xi)^{-5/6}}{\Gamma(1/6)}\,\mathrm{d}\xi,$$

as can be seen by arguing as in the proof of Lemma 6.15 (easy case of resurgence of a convolution product), the only difference being the singularity of $\zeta^{-5/6}$ at the origin, which is innocuous because it is integrable.

Since $\widehat{\chi}(\zeta) \in \mathbb{C}\{\zeta\}$, this analytic continuation property exactly means that $\widehat{\chi}$ is a $\{0,-2\}$-continuable germ (cf. Definition 6.1), hence (6.121) entails that $\mathscr{B}\widetilde{\varphi}$ is a $\{0,-2\}$-resurgent function and $\mathscr{B}\widetilde{\psi}$ a $\{0,2\}$-resurgent function. We are thus left with proving that $\mathscr{B}\widetilde{\varphi}$ and $\mathscr{B}\widetilde{\psi}$ are *simple* resurgent functions and computing their alien derivatives. This follows from

Lemma 6.106.

$$\Delta_{-2}\widehat{\chi} = -i\widehat{\chi}_-, \quad \textit{with } \widehat{\chi}_-(\zeta) := \widehat{\chi}(-\zeta). \tag{6.138}$$

Indeed, Lemma 6.106 means that the principal branch of $\widehat{\chi}$ near -2 is given by

$$\widehat{\chi}(-2+\zeta) = -i\widehat{\chi}_-(\zeta)\frac{\log\zeta}{2\pi i} \mod \mathbb{C}\{\zeta\},$$

whence

$$\widehat{\varphi}(-2+\zeta) = \frac{-i}{2\pi i\zeta} - i\frac{\mathrm{d}\widehat{\chi}_-}{\mathrm{d}\zeta}(\zeta)\frac{\log\zeta}{2\pi i} \mod \mathbb{C}\{\zeta\},$$

by the first part of (6.121) (using $\widehat{\chi}_-(0) = 1$), i.e. $\Delta_{-2}\mathscr{B}\widetilde{\varphi} = -i\mathscr{B}\widetilde{\psi}$ by the second part of (6.121), which yields the first part of (6.122). The second part of (6.122) then follows from

Exercise 6.107. Suppose that $\widetilde{\varphi}(z) \in \mathbb{C}[[z^{-1}]]$ is Ω-resurgent and $\Delta_\omega\widetilde{\varphi} = \widetilde{\Phi}(z) \in \mathbb{C}[[z^{-1}]]$ for some $\omega \in \Omega$. Let $\widetilde{\psi}(z) = \widetilde{\varphi}(-z)$. Show that $\widetilde{\psi}$ is $(-\Omega)$-resurgent and

$$\Delta_{-\omega}\widetilde{\psi}(z) = \widetilde{\Phi}(-z).$$

Thus, taking Lemma 6.106 and exercise 6.107 for granted, we have that

$$\widetilde{\varphi} \text{ is } \{0,-2\}\text{-resurgent}, \qquad \Delta_{-2}\widetilde{\varphi} = -\mathrm{i}\widetilde{\psi},$$
$$\widetilde{\psi} \text{ is } \{0,2\}\text{-resurgent}, \qquad \Delta_{2}\widetilde{\psi} = -\mathrm{i}\widetilde{\varphi}.$$

Since all alien operators can be expressed as linear combinations of compositions of alien derivations (Remark 6.69), this implies that all the branches of the analytic continuation of $\widehat{\varphi}$ or $\widehat{\psi}$ have only simple singularities (and they are all linear combinations of $\mathscr{B}\widetilde{\varphi}$ and $\mathscr{B}\widetilde{\psi}$; for instance, the Borel transform of $\Delta_2 \circ \Delta_{-2}\widetilde{\varphi} = -\widetilde{\varphi}$ gives the singularity at the origin of the continuation of $\widehat{\varphi}$ along a path which turns anticlockwise around -2).

Exercise 6.107 is left to the reader. We thus conclude by proving Lemma 6.106.

h) *Proof of Lemma 6.106.* Let us write

$$\widehat{\chi}(\zeta) = \hat{I}_{1/6} * \hat{f}(\zeta), \qquad \widehat{\chi}_-(\zeta) = \hat{I}_{1/6} * \hat{f}_-(\zeta), \tag{6.139}$$
$$\hat{f}(\zeta) = c\zeta^{-1/6}(2+\zeta)^{-1/6}, \qquad \hat{f}_-(\zeta) := c\zeta^{-1/6}(2-\zeta)^{-1/6}, \tag{6.140}$$

where $\hat{I}_{1/6} := \frac{1}{\Gamma(1/6)}\zeta^{-5/6}$ and $c := \frac{2^{1/6}}{\Gamma(5/6)}$. We observe that $\hat{f}$, viewed as a singular germ in the sense of Definition 6.37, can be obtained as the minor (in the sense of Definition 6.46) of the singularity

$$\overset{\triangledown}{f} := \operatorname{sing}_0\left(c\frac{\zeta^{-1/6}}{1-\mathrm{e}^{2\mathrm{i}\pi/6}}(2+\zeta)^{-1/6}\right) \tag{6.141}$$

and, among the singularities which have the same minor, $\overset{\triangledown}{f}$ is the only one admitting a representative which is $o\big(|\zeta|^{-1}\big)$ as $|\zeta| \to 0$. The notation

$$\overset{\triangledown}{f} = {}^\flat\hat{f}$$

us used in such a situation (this is a generalisation of notation (6.29) on p. 201) and we say that $\overset{\triangledown}{f}$ is an "integrable singularity" (in accordance with Chapter 7 of the third volume of this book [Del16], or [Éca81], or [Sau12, §3.1–3.2]). Similarly, we set

$$\overset{\triangledown}{f}_- := {}^\flat\hat{f}_- = \operatorname{sing}_0\left(c\frac{\zeta^{-1/6}}{1-\mathrm{e}^{2\mathrm{i}\pi/6}}(2-\zeta)^{-1/6}\right). \tag{6.142}$$

Other examples of integrable singularity are given by ${}^\flat\widehat{\chi} := \operatorname{sing}_0\left(\widehat{\chi}(\zeta)\frac{\log\zeta}{2\pi\mathrm{i}}\right)$, or $\overset{\triangledown}{I}_\alpha$ when $\Re e\,\alpha > 0$, with the notation (6.30), so that, in that case, we can write

$$\overset{\triangledown}{I}_\alpha = {}^\flat\big(\hat{I}_\alpha\big) \quad \text{with } \hat{I}_\alpha = \operatorname{var}\overset{\triangledown}{I}_\alpha = \frac{1}{\Gamma(\alpha)}\zeta^{\alpha-1}.$$

The convolution of singularities which was alluded to in Section 6.8 can be applied to integrable singularities and it turns out that (6.32) is valid for any pair of integrable minors (using the convolution defined at the beginning of Section 6.14.4), thus we can rewrite (6.139) as

$$ {}^\flat\widehat{\chi} = \overset{\triangledown}{I}_{1/6} \overset{\triangledown}{*} \overset{\triangledown}{f}, \qquad {}^\flat\widehat{\chi}_- = \overset{\triangledown}{I}_{1/6} \overset{\triangledown}{*} \overset{\triangledown}{f}_-. \tag{6.143} $$

Now, $\overset{\triangledown}{f}$ is not a simple singularity. However, the analytic continuation property that we have shown for its minor $\hat{f}$ shows that it is a $\{0,-2\}$-resurgent singularity in the sense defined at the beginning of Section 6.10. Thus, we can make use of an extension of alien calculus for non simple resurgent functions parallel to the extension of convolution and formal Borel transform that we have already encountered in this section. We will explain how it works only on the case of $\overset{\triangledown}{f}$, this may serve as an initiation to Chapter 7 of [Del16].

The alien derivative $\Delta_{2\underline{e}^{i\pi}}\overset{\triangledown}{f}$ is defined by considering the minor $\hat{f} = \operatorname{var}\overset{\triangledown}{f}$ and following its analytic continuation along the ray $\underline{e}^{i\pi}\mathbb{R}^+ \subset \widetilde{\mathbb{C}}$: since $2\underline{e}^{i\pi}$ is the first singular point that we encounter, the definition in this case is

$$ \Delta_{2\underline{e}^{i\pi}}\overset{\triangledown}{f} = \operatorname{sing}_0\big(\overset{\vee}{g}(\zeta)\big), $$
$$ \text{where } \overset{\vee}{g}(\zeta) := \hat{f}(2\underline{e}^{i\pi}+\zeta) \text{ for } -\pi < \arg\zeta < \pi \text{ and } |\zeta| \text{ small enough.} $$

We compute

$$ \Delta_{2\underline{e}^{i\pi}}\overset{\triangledown}{f} = c\operatorname{sing}_0\big(\zeta^{-1/6}(2\underline{e}^{i\pi}+\zeta)^{-1/6}\big) = c\,e^{-i\pi/6}\operatorname{sing}_0\big(\zeta^{-1/6}(2-\zeta)^{-1/6}\big) $$
$$ = e^{-i\pi/6}(1-e^{2i\pi/6})\overset{\triangledown}{f}_- $$

in view of (6.142). Since $\sin(\pi/6) = \frac{1}{2}$, we obtain

$$ \Delta_{2\underline{e}^{i\pi}}\overset{\triangledown}{f} = -i\overset{\triangledown}{f}_-. \tag{6.144} $$

It turns out that $\Delta_{2\underline{e}^{i\pi}}$ is a derivation of the convolution algebra of resurgent singularities, and it obviously annihilates $\overset{\triangledown}{I}_{1/6}$ (whose minor is regular on the whole of $\widetilde{\mathbb{C}}$), therefore $\Delta_{2\underline{e}^{i\pi}}\big(\overset{\triangledown}{I}_{1/6} \overset{\triangledown}{*} \overset{\triangledown}{f}\big) = \overset{\triangledown}{I}_{1/6} \overset{\triangledown}{*} \big(\Delta_{2\underline{e}^{i\pi}}\overset{\triangledown}{f}\big)$. In view of (6.143)–(6.144), this yields

$$ \Delta_{2\underline{e}^{i\pi}}{}^\flat\widehat{\chi} = -i\,{}^\flat\widehat{\chi}_-, $$

which agrees with (6.138) because the integrable minor $\widehat{\chi}$ is in fact a regular minor, so one can content oneself with indexing the alien derivation by a complex number (instead of a point of $\widetilde{\mathbb{C}}$) and one then recovers the alien derivative as it was defined in Section 6.10.

This ends the proof of Lemma 6.106.

Exercise 6.108. Show directly from the formula for $\hat{f}_- = \operatorname{var}\overset{\triangledown}{f}_-$ given in (6.140) (without arguing by symmetry as in exercise 6.107) that

$$ \Delta_{2\underline{e}^{i0}}\overset{\triangledown}{f}_- = -i\overset{\triangledown}{f} $$

and hence $\Delta_2 \widehat{\chi}_- = -\mathrm{i}\,\widehat{\chi}$. (Hint: Use the property $\zeta \in \underline{\mathrm{e}}^{-\mathrm{i}\pi}(0,2) \implies \hat{f}_-(2\underline{\mathrm{e}}^{\mathrm{i}0}+\zeta) > 0$ in order to select the appropriate branch of $(-\zeta)^{-1/6}$.)

6.15 A glance at a class of non-linear differential equations

We give here a brief account of the way resurgent methods can be used to handle a specific problem, namely the so-called "saddle-node problem", object of the celebrated work by J. Martinet and J.-P. Ramis [MR82]. A resurgent approach to this problem was indicated by J. Écalle in concise manner in [Éca84], and a fully detailed exposition was given in [Sau09].

The problem can be viewed as a non-linear generalization of the Euler equation. Following [Sau09], we will illustrate alien calculus on the example of the simple $\mathbb{Z}$-resurgent series which appear in this situation. We will omit most of the proofs but try to acquaint the reader with concrete computations with alien operators; this section can be viewed as an initiation to some of the material of the third volume [Del16].

6.15.1 Let us give ourselves

$$B(z,y) = \sum_{n\in\mathbb{N}} b_n(z) y^n \in \mathbb{C}\{z^{-1},y\}$$

with $b_1(z) = 1 + O(z^{-2})$ and $b_n(z) = O(z^{-1})$ if $n \neq 1$, and consider the differential equation

$$\frac{\mathrm{d}\widetilde{\varphi}}{\mathrm{d}z} = B(z,\widetilde{\varphi}) = b_0(z) + b_1(z)\widetilde{\varphi} + b_2(z)\widetilde{\varphi}^2 + \cdots \tag{6.145}$$

(one recovers the Euler equation for $B(z,y) = -z^{-1} + y$). Observe that if $\widetilde{\varphi}(z) \in z^{-1}\mathbb{C}[[z^{-1}]]$ then $B(z,\widetilde{\varphi}(z))$ is given by a formally convergent series, so the differential equation (6.145) makes sense for formal series without constant term.

Theorem 6.109. *Equation* (6.145) *admits a unique formal solution* $\widetilde{\varphi}_0 \in z^{-1}\mathbb{C}[[z^{-1}]]$. *This formal series is* 1*-summable in the directions of* $(-\pi,\pi)$ *and*

$$\widetilde{\varphi}_0(z) \in \widetilde{\mathscr{R}}^{\mathrm{simp}}_{\mathbb{Z}^*_-}, \quad \textit{where } \mathbb{Z}^*_- := \{-1,-2,-3,\dots\}.$$

Its Borel sum $\mathscr{S}^{(-\pi,\pi)}\widetilde{\varphi}_0$ *is a particular solution of Equation* (6.145)*, defined and holomorphic in a domain of the form* $\widetilde{\mathscr{D}}\big((-\pi,\pi),\gamma\big) \subset \widetilde{\mathbb{C}}$.

We omit the proof, which can be found in [Sau09]. Let us only give a hint on why one must take $\Omega = \mathbb{Z}^*_-$. Writing $B(z,y) - y = \sum a_n(z) y^n$, we have $a_n(z) \in z^{-1}\mathbb{C}\{z^{-1}\}$ for all $n \in \mathbb{N}$, thus (6.145) can be rewritten $\frac{\mathrm{d}\widetilde{\varphi}}{\mathrm{d}z} - \widetilde{\varphi} = \sum a_n \widetilde{\varphi}^n$, which via $\mathscr{B}$ is equivalent to

$$\widehat{\varphi}_0(\zeta) = \frac{-1}{1+\zeta}\big(\widehat{a}_0 + \widehat{a}_1 * \widehat{\varphi} + \widehat{a}_2 * \widehat{\varphi}^{*2} + \cdots\big).$$

The Borel transforms $\widehat{a}_n$ are entire functions, thus it is only the division by $1+\zeta$ which is responsible for the appearance of singularities in the Borel plane: a pole at -1 in the first place, but also, because of repeated convolutions, a simple singularity at -1 rather than only a simple pole and other simple singularities at all points of the additive semigroup generated by -1.

6.15.2 The next question is: what about the Stokes phenomenon for $\widetilde{\phi}_0$ and the action of the alien operators? Let us first show how, taking for granted that $\widetilde{\phi}_0 \in \widetilde{\mathscr{R}}^{\mathrm{simp}}_{\mathbb{Z}^*_-}$, one can by elementary alien calculus see that $\Delta_\omega \widetilde{\phi}_0 = 0$ for $\omega \neq -1$ and compute $\Delta_{-1}\widetilde{\phi}_0$ up to a multiplicative factor. We just need to enrich our "alien toolbox" with two lemmas.

Notation 6.110 *Since $\partial = \frac{\mathrm{d}}{\mathrm{d}z}$ increases the standard valuation by at least one unit (recall (5.12)), the operator $\mu + \partial \colon \mathbb{C}[[z^{-1}]] \to \mathbb{C}[[z^{-1}]]$ is invertible for any $\mu \in \mathbb{C}^*$ and its inverse $(\mu+\partial)^{-1}$ is given by the formally convergent series of operators $\sum_{p\geq 0} \mu^{-p-1}(-\partial)^p$ (and its Borel counterpart is just division by $\mu - \zeta$). For $\mu = 0$, we define ∂^{-1} as the unique operator $\partial^{-1}\colon z^{-2}\mathbb{C}[[z^{-1}]] \to z^{-1}\mathbb{C}[[z^{-1}]]$ such that $\partial \circ \partial^{-1}$ on $z^{-2}\mathbb{C}[[z^{-1}]]$ (its Borel counterpart is division by $-\zeta$).*

Lemma 6.111. *Let Ω be any non-empty closed discrete subset of $\mathbb{C}$. Let $\widetilde{\varphi} \in \widetilde{\mathscr{R}}^{\mathrm{simp}}_{\Omega}$ and $\mu \in \Omega$. If $\mu = 0$ we assume $\widetilde{\varphi} \in z^{-2}\mathbb{C}[[z^{-1}]]$; if $\mu \neq 0$ we assume $\Delta_\mu \widetilde{\varphi} \in z^{-2}\mathbb{C}[[z^{-1}]]$. Then $(\mu+\partial)^{-1}\widetilde{\varphi} \in \widetilde{\mathscr{R}}^{\mathrm{simp}}_{\Omega}$ and*

$$\omega \in \Omega \setminus \{0,\mu\} \implies \Delta_\omega (\mu+\partial)^{-1}\widetilde{\varphi} = (\mu - \omega + \partial)^{-1}\Delta_\omega \widetilde{\varphi},$$

while, if $\mu \neq 0$, there exists $C \in \mathbb{C}$ such that

$$\Delta_\mu (\mu+\partial)^{-1}\widetilde{\varphi} = C + \partial^{-1}\Delta_\mu \widetilde{\varphi}.$$

Lemma 6.112. *Let $B(z,y) \in \mathbb{C}\{z^{-1},y\}$. Suppose that Ω is stable under addition and $\widetilde{\varphi}(z) \in \widetilde{\mathscr{R}}^{\mathrm{simp}}_{\Omega}$ has no constant term. Then $B\big(z,\widetilde{\varphi}(z)\big) \in \widetilde{\mathscr{R}}^{\mathrm{simp}}_{\Omega}$ and, for every $\omega \in \Omega \setminus \{0\}$,*

$$\Delta_\omega B\big(z,\widetilde{\varphi}(z)\big) = \partial_y B\big(z,\widetilde{\varphi}(z)\big) \cdot \Delta_\omega \widetilde{\varphi}.$$

The proofs of Lemmas 6.111 and 6.112 are left to the reader.

Let us come back to the solution $\widetilde{\phi}_0$ of (6.145). For $\omega \in \mathbb{Z}^*_-$, we derive a differential equation for $\widetilde{\psi} = \Delta_\omega \widetilde{\phi}_0$ by writing on the one hand $\Delta_\omega \partial_z \widetilde{\phi}_0 = \partial_z \widetilde{\psi} - \omega \widetilde{\psi}$ (by (6.49)) and, on the other hand, $\Delta_\omega\big(B(z,\widetilde{\phi}_0)\big) = \partial_y B(z,\widetilde{\phi}_0)\cdot \widetilde{\psi}$ by Lemma (6.112), thus alien differentiating Equation (6.145) yields

$$\frac{\mathrm{d}\widetilde{\psi}}{\mathrm{d}z} = \big(\omega + \partial_y B(z,\widetilde{\phi}_0)\big)\cdot \widetilde{\psi}. \tag{6.146}$$

Since $\omega + \partial_y B(z,\widetilde{\phi}_0) = \omega + 1 + O(z^{-2})$, it is immediate that the only solution of this equation in $z^{-1}\mathbb{C}[[z^{-1}]]$ is 0 when $\omega \neq -1$. This proves

$$\omega \neq -1 \implies \Delta_\omega \widetilde{\phi}_0 = 0.$$

For $\omega = -1$, Equation (6.146) reads

$$\frac{\mathrm{d}\tilde{\psi}}{\mathrm{d}z} = \tilde{\beta}_1 \tilde{\psi} \tag{6.147}$$

with $\tilde{\beta}_1(z) := -1 + \partial_y B(z, \tilde{\varphi}_0(z)) \in \tilde{\mathscr{R}}^{\mathrm{simp}}_{\mathbb{Z}^*_-}$ (still by Lemma 6.112). Since $\tilde{\beta}_1(z) = O(z^{-2})$, Lemma 6.111 implies $\tilde{\alpha} := \partial^{-1}\tilde{\beta}_1 \in \tilde{\mathscr{R}}^{\mathrm{simp}}_{\mathbb{Z}_-}$ (beware that we must replace $\mathbb{Z}^*_-$ with $\mathbb{Z}_- = \{0\} \cup \mathbb{Z}^*_-$ because a priori only the principal branch of $\hat{\alpha} := -\frac{1}{\zeta}\hat{\beta}_1(\zeta)$ is regular at 0). Then

$$\tilde{\varphi}_1 := \mathrm{e}^{\partial^{-1}\tilde{\beta}_1} = 1 + O(z^{-1}) \in \tilde{\mathscr{R}}^{\mathrm{simp}}_{\mathbb{Z}_-}$$

is a non-trivial solution of (6.147). This implies that

$$\Delta_{-1}\tilde{\varphi}_0 = C\tilde{\varphi}_1,$$

with a certain $C \in \mathbb{C}$.

6.15.3 We go on with the computation of the alien derivatives of $\tilde{\varphi}_1$. Let

$$\tilde{\beta}_2(z) := \partial_y^2 B\big(z, \tilde{\varphi}_0(z)\big) \in \tilde{\mathscr{R}}^{\mathrm{simp}}_{\mathbb{Z}^*_-},$$

so that $\Delta_{-1}\tilde{\beta}_1 = C\tilde{\beta}_2\tilde{\varphi}_1(z)$ and $\Delta_\omega\tilde{\beta}_1 = 0$ for $\omega \neq -1$ (by Lemma 6.112). Computing $\Delta_\omega(\partial^{-1}\tilde{\beta}_1)$ by Lemma 6.111 and then $\Delta_\omega\tilde{\varphi}_1$ by (6.96), we get

$$\Delta_{-1}\tilde{\varphi}_1 = 2C\tilde{\varphi}_2, \qquad \tilde{\varphi}_2 := \frac{1}{2}\tilde{\varphi}_1 \cdot (1+\partial)^{-1}(\tilde{\beta}_2\tilde{\varphi}_1) \in \tilde{\mathscr{R}}^{\mathrm{simp}}_{\mathbb{Z}_- \cup \{1\}} \tag{6.148}$$

and $\Delta_\omega\tilde{\varphi}_1 = 0$ for $\omega \neq -1$.

By the same kind of computation, we get at the next step $\Delta_\omega\tilde{\varphi}_2 = 0$ for $\omega \notin \{-1, 1\}$,

$$\Delta_{-1}\tilde{\varphi}_2 = 3C\tilde{\varphi}_3,$$

$$\tilde{\varphi}_3 := \frac{1}{3}\tilde{\varphi}_2 \cdot (1+\partial)^{-1}(\tilde{\beta}_2\tilde{\varphi}_1) + \frac{1}{6}\tilde{\varphi}_1 \cdot (2+\partial)^{-1}(\tilde{\beta}_3\tilde{\varphi}_1^2 + 2\tilde{\beta}_2\tilde{\varphi}_2) \in \tilde{\mathscr{R}}^{\mathrm{simp}}_{\mathbb{Z}_- \cup \{1,2\}}$$

with $\tilde{\beta}_3 := \partial_y^3 B\big(z, \tilde{\varphi}_0(z)\big)$. A new undetermined constant appears for $\omega = 1$: Lemma 6.111 yields a $C' \in \mathbb{C}$ such that $\Delta_1(1+\partial)^{-1}(\tilde{\beta}_2\tilde{\varphi}_1) = C' + \partial^{-1}\Delta_1(\tilde{\beta}_2\tilde{\varphi}_1) = C'$, hence (6.148) implies

$$\Delta_1\tilde{\varphi}_2 = C'\tilde{\varphi}_3.$$

We see that Equation (6.145) generates not only the formal solution $\tilde{\varphi}_0$ but also a sequence of resurgent series $(\tilde{\varphi}_n)_{n \geq 1}$, in which $\tilde{\varphi}_1$ was constructed as the unique solution of the linear homogeneous differential equation (6.147) whose constant term is 1; the other series in the sequence can be characterized by linear non-homogeneous equations: alien differentiating (6.147), we get $(1+\partial)\Delta_{-1}\widetilde{\psi} = \Delta_{-1}\partial\widetilde{\psi} = \Delta_{-1}(\widetilde{\beta}_1\widetilde{\psi}) = \widetilde{\beta}_1\Delta_{-1}\widetilde{\psi} + C\widetilde{\beta}_2\widetilde{\varphi}_1\widetilde{\psi}$, thus $\partial(\Delta_{-1}\widetilde{\varphi}_1) = (-1+\widetilde{\beta}_1)\Delta_{-1}\widetilde{\varphi}_1 + C\widetilde{\beta}_2\widetilde{\varphi}_1^2$, and it is not a surprise that $\widetilde{\varphi}_2$ is the unique formal solution of

$$\partial\widetilde{\varphi}_2 = (-1+\widetilde{\beta}_1)\widetilde{\varphi}_2 + \frac{1}{2}\widetilde{\beta}_2\widetilde{\varphi}_1^2. \tag{6.149}$$

Similarly, $\widetilde{\varphi}_3$ is the unique formal solution of

$$\partial\widetilde{\varphi}_3 = (-2+\widetilde{\beta}_1)\widetilde{\varphi}_3 + \widetilde{\beta}_2\widetilde{\varphi}_1\widetilde{\varphi}_2 + \frac{1}{6}\widetilde{\beta}_3\widetilde{\varphi}_1^3. \tag{6.150}$$

6.15.4 The previous calculations can be put into perspective with the notion of *formal integral*, i.e. a formal object which solves Equation (6.145) and is more general than a formal series like $\widetilde{\varphi}_0$. Indeed, both sides of (6.145) can be evaluated on an expression of the form

$$\widetilde{\phi}(z,u) = \sum_{n\in\mathbb{N}} u^n \mathrm{e}^{nz}\widetilde{\varphi}_n(z) = \widetilde{\varphi}_0(z) + u\mathrm{e}^z\widetilde{\varphi}_1(z) + u^2\mathrm{e}^{2z}\widetilde{\varphi}_1(z) + \dots \tag{6.151}$$

if $(\widetilde{\varphi}_n)_{n\in\mathbb{N}}$ is any sequence of formal series such that $\widetilde{\varphi}_0$ has no constant term: it is sufficient to treat $\widetilde{\phi}(z,u)$ as a formal series in u whose coefficients are transseries of a particular form and to write the left-hand side as

$$\frac{\partial\widetilde{\phi}}{\partial z}(z,u) = \sum_{n\in\mathbb{N}} u^n \mathrm{e}^{nz}(n+\partial)\widetilde{\varphi}_n$$

and the right-hand side $B\big(z,\widetilde{\phi}(z,u)\big)$ as

$$B\big(z,\widetilde{\varphi}_0(z)\big) + \sum_{r\geq 1}\frac{1}{r!}\partial_y^r B\big(z,\widetilde{\varphi}_0(z)\big) \sum_{n_1,\dots,n_r\geq 1} u^{n_1+\dots+n_r}\mathrm{e}^{(n_1+\dots+n_r)z}\widetilde{\varphi}_{n_1}\cdots\widetilde{\varphi}_{n_r}.$$

This is equivalent to setting $\widetilde{Y}(z,y) = \sum_{n\in\mathbb{N}} y^n\widetilde{\varphi}_n(z)$, so that $\widetilde{\phi}(z,u) = \widetilde{Y}(z,u\mathrm{e}^z)$, and to considering the equation

$$\partial_z\widetilde{Y} + y\partial_y\widetilde{Y} = B\big(z,\widetilde{Y}(z,y)\big). \tag{6.152}$$

for an unknown double series $\widetilde{Y} \in \mathbb{C}[[z^{-1},y]]$ without constant term.

For an expression (6.151), Equation (6.145) is thus equivalent to the sequence of equations

$$\partial\widetilde{\phi}_0 = B(z,\widetilde{\phi}_0) \qquad (E_0)$$

$$(1+\partial)\widetilde{\phi}_1 - \partial_y B(z,\widetilde{\phi}_0)\cdot\widetilde{\phi}_1 = 0 \qquad (E_1)$$

$$(n+\partial)\widetilde{\phi}_n - \partial_y B(z,\widetilde{\phi}_0)\cdot\widetilde{\phi}_n = \sum_{r\geq 2}\frac{1}{r!}\partial_y^r B(z,\widetilde{\phi}_0)\sum_{\substack{n_1,\dots,n_r\geq 1\\ n_1+\dots+n_r=n}}\widetilde{\phi}_{n_1}\cdots\widetilde{\phi}_{n_r} \quad \text{for } n\geq 2. \qquad (E_n)$$

Of course (E_0) is identical to Equation (6.145) for a formal series without constant term. The reader may check that Equation (E_1) coincides with (6.147), (E_2) with (6.149) and (E_3) with (6.150).

Theorem 6.113. *Equation* (6.145) *admits a unique solution of the form* (6.151) *for which the constant term of* $\widetilde{\phi}_0$ *is* 0 *and the constant term of* $\widetilde{\phi}_1$ *is* 1*, called "Formal Integral". The coefficients* $\widetilde{\phi}_n$ *of the formal integral are* 1*-summable in the directions of* $(-\pi,0)$ *and* $(0,\pi)$*, and*

$$\widetilde{\phi}_n(z)\in\widetilde{\mathscr{R}}^{\text{simp}}_{\mathbb{Z}^*_-\cup\{0,1,\dots,n-1\}}, \qquad n\in\mathbb{N}. \qquad (6.153)$$

The dependence on n in the exponential bounds for the Borel transforms $\widehat{\phi}_n$ *is controlled well enough to ensure the existence of locally bounded functions* γ *and* $R>0$ *on* $(-\pi,0)\cup(0,\pi)$ *such that, for* $I=(-\pi,0)$ *or* $(0,\pi)$*,*

$$Y^I(z,y) := \sum_{n\in\mathbb{N}} y^n \mathscr{S}^I\widetilde{\phi}_n(z)$$

is holomorphic in

$$\mathscr{D}(I,\gamma,R) := \{\,(z,y)\in\mathbb{C}\times\mathbb{C} \mid \exists\theta\in I \text{ such that } \Re e(z\mathrm{e}^{\mathrm{i}\theta}) > \gamma(\theta) \text{ and } |y|<R(\theta)\,\}.$$

Correspondingly, the function

$$\phi^I(z,u) := \sum_{n\in\mathbb{N}} (u\mathrm{e}^z)^n \mathscr{S}^I\widetilde{\phi}_n(z)$$

is holomorphic in $\{\,(z,u)\in\mathscr{D}(I,\gamma)\times\mathbb{C} \mid (z,u\mathrm{e}^z)\in\mathscr{D}(I,\gamma,R)\,\}$*.*

The Borel sums $\phi^{(-\pi,0)}{}_{|u=0}$ *and* $\phi^{(0,\pi)}{}_{|u=0}$ *both coincide with the particular solution of Equation* (6.145) *mentioned in Theorem 6.109. For* $I=(-\pi,0)$ *or* $(0,\pi)$ *and for each* $u\in\mathbb{C}^*$*, the function* $\phi^I(\,.\,,u)$ *is a solution of* (6.145) *holomorphic in* $\{z\in\mathscr{D}(I,\gamma) \mid \Re e\, z < \ln\frac{R}{|u|}\}$*.*

The reader is once more referred to [Sau09] for the proof.

Observe that when we see the formal integral $\widetilde{\phi}(z,u)$ as a solution of (6.145), we must think of u as of an *indeterminate*, the same way as z (or rather z^{-1}) is an indeterminate when we manipulate ordinary formal series; after Borel-Laplace summation of each $\widetilde{\phi}_n$, we get holomorphic functions of the *variable* $z\in\mathscr{D}(I,\gamma)$, coefficients of a formal expression $\sum u^n\mathrm{e}^{nz}\mathscr{S}^I\widetilde{\phi}_n(z)$; Theorem 6.113 says that, for each $z \in \mathscr{D}(I,\gamma)$, this expression is a convergent formal series, Taylor expansion of the function obtained by substituting the indeterminate u with a *variable* $u \in \mathbb{D}_{R' \mathrm{e}^{-\Re e z}}$ (with $R' > 0$ small enough depending on z).

If we think of z as of the main variable, the interpretation of the indeterminate/variable u is that of a free parameter in the solution of a first-order differential equation: $\widetilde{\phi}(z,u)$ appears as a formal 1-parameter family of formal solutions, $\phi^{(-\pi,0)}$ and $\phi^{(0,\pi)}$ as two 1-parameter families of analytic solutions.

As for the Borel sum $Y^I(z,y)$, it is an analytic solution of Equation (6.152) in its domain $\mathscr{D}(I,\gamma,R)$; this means that the vector field[21] $X_B := \frac{\partial}{\partial z} + B(z,Y)\frac{\partial}{\partial Y}$ is the direct image of $N := \frac{\partial}{\partial z} + y\frac{\partial}{\partial y}$ by the diffeormophism

$$\Theta^I \colon (z,y) \mapsto (z,Y) = \big(z, Y^I(z,y)\big).$$

We may consider N as a normal form for X_B and $\Theta^{(-\pi,0)}$ and $\Theta^{(0,\pi)}$ as two sectorial normalizations.

The results of the alien calculations of Sections 6.15.2–6.15.3 are contained in following statement (extracted from Section 10 of [Sau09]):

Theorem 6.114. *There are uniquely determined complex numbers $C_{-1}, C_1, C_2, \dots$ such that, for each $n \in \mathbb{N}$,*

$$\Delta_{-m}\widetilde{\varphi}_n = 0 \quad \textit{for } m \geq 2, \tag{6.154}$$

$$\Delta_{-1}\widetilde{\varphi}_n = (n+1)C_{-1}\widetilde{\varphi}_{n+1}, \tag{6.155}$$

$$\Delta_m\widetilde{\varphi}_n = (n-m)C_m\widetilde{\varphi}_{n-m} \quad \textit{for } 1 \leq m \leq n-1. \tag{6.156}$$

Equivalently, letting act the alien derivation Δ_ω on an expression like $\widetilde{\phi}(z,u)$ or $\widetilde{Y}(z,y)$ by declaring that it commutes with multiplication by u, e^z or y, on has

$$\Delta_m\widetilde{\phi} = C_m u^{m+1}\,\mathrm{e}^{mz}\frac{\partial\widetilde{\phi}}{\partial u} \quad \textit{or} \quad \Delta_m\widetilde{Y} = C_m y^{m+1}\frac{\partial\widetilde{Y}}{\partial y}, \qquad \textit{for } m = -1 \textit{ or } m \geq 2. \tag{6.157}$$

Equation (6.157) (either for $\widetilde{\phi}$ or for $\widetilde{Y}$) was baptized "Bridge Equation" by Écalle, in view of the bridge it establishes between ordinary differential calculus (involving ∂_u or ∂_y) and alien calculus (when dealing with the solution of an analytic equation like $\widetilde{\phi}$ or $\widetilde{Y}$).

Proof of Theorem 6.114. Differentiating (6.152) with respect to y, we get

$$(\partial_z + y\partial_y)\partial_y\widetilde{Y} = \big(-1 + \partial_y B(z,\widetilde{Y})\big)\partial_y\widetilde{Y}.$$

Alien differentiating (6.152), we get (in view of (6.49))

---

21 If we change the variable z into $x := -z^{-1}$, the vector field X_B becomes $x^2\frac{\partial}{\partial x} + B(z,Y)\frac{\partial}{\partial Y}$, which has a saddle-node singularity at $(0,0)$.

$$(\partial_z + y\partial_y)\Delta_m \widetilde{Y} = \big(m + \partial_y B(z,\widetilde{Y})\big)\Delta_m \widetilde{Y}.$$

Now $\partial_y \widetilde{Y} = 1 + O(z^{-1}, y)$ is invertible and we can consider $\widetilde{\chi} := (\partial_y \widetilde{Y})^{-1}\Delta_m \widetilde{Y} \in \mathbb{C}[[z^{-1}, y]]$, for which we get $(\partial_z + y\partial_y)\widetilde{\chi} = (m+1)\widetilde{\chi}$, and this implies the existence of a unique $C_m \in \mathbb{C}$ such that $\widetilde{\chi} = C_m y^{m+1}$. This yields the second part of (6.157), from which the first part follows, and also (6.154)–(6.156) by expanding the formula. □

6.15.5 The Stokes phenomenon for $\widetilde{\phi}(z,u)$ takes the form of two connection formulas, one for $\Re e\, z < 0$, the other for $\Re e\, z > 0$, between the two families of solutions $\phi^{(-\pi,0)}$ and $\phi^{(0,\pi)}$. For $\Re e\, z < 0$, it is obtained by analyzing the action of $\Delta^+_{\mathbb{R}^-}$, the symbolic Stokes automorphism for the direction $\mathbb{R}^-$.

Let $\Omega := \mathbb{Z}^*_-$. Since $\widetilde{\phi}_n \in \widetilde{\mathscr{R}}^{\text{simp}}_{n+\Omega}$ (by (6.153)), the formal integral $\widetilde{\phi}$ can be considered as an Ω-resurgent symbol with support in $\mathbb{R}^-$ at the price of a slight extension of the definition: we must allow our resurgent symbols to depend on the indeterminate u, so we replace (6.71) with

$$\widetilde{E}(\Omega, d) := \Big\{ \sum_{\omega \in (\Omega \cup \{0\}) \cap d} e^{-\omega z} \widetilde{\varphi}_\omega(z,u) \mid \widetilde{\varphi}_\omega(z,u) \in \widetilde{\mathscr{R}}_{-\omega+\Omega}[u] \Big\}$$

(thus restricting ourselves to a polynomial dependence on u for each homogeneous component). Then $\widetilde{\phi}(z,u) = \sum_{n\in\mathbb{N}} u^n e^{nz} \widetilde{\phi}_n(z) \in \widetilde{E}(\Omega, \mathbb{R}^-)$. According to (6.154), only one homogeneous component of $\Delta_{\mathbb{R}^-}$ needs to be taken into account, and (6.75) yields $\Delta_{\mathbb{R}^-}\widetilde{\phi}(z,u) = e^z \Delta_{-1}\widetilde{\phi}(z,u)$, whence, by (6.155),

$$\Delta_{\mathbb{R}^-}\widetilde{\phi}(z,u) = \sum_{n\geq 0}(n+1)C_{-1}u^n e^{(n+1)z}\widetilde{\phi}_{n+1}(z) = C_{-1}\frac{\partial \widetilde{\phi}}{\partial u}(z,u).$$

It follows that

$$\Delta^+_{\mathbb{R}^-}\widetilde{\phi}(z,u) = \widetilde{\phi}(z, u + C_{-1}) = \sum_{n\geq 0}(u + C_{-1})^n e^{nz}\widetilde{\phi}_n(z)$$

and one ends up with

Theorem 6.115. *For $z \in \mathscr{D}\big((-\pi,0),\gamma\big) \cap \mathscr{D}\big((0,\pi),\gamma\big)$ with $\Re e\, z < 0$,*

$$\phi^{(0,\pi)}(z,u) \equiv \phi^{(-\pi,0)}(z, u + C_{-1}), \qquad Y^{(0,\pi)}(z,y) \equiv Y^{(-\pi,0)}(z, y + C_{-1}e^z).$$

6.15.6 For $\Re e\, z > 0$, we need to inquire about the action of $\Delta^+_{\mathbb{R}^+}$, however the action of this operator is not defined on the space of resurgent symbols with support in $\mathbb{R}^-$. Luckily, we can view $\widetilde{\phi}(z,u)$ as a member of the space $\widetilde{F}(\mathbb{Z}, \mathbb{R}^-) = \widetilde{F}_0 \supset \widetilde{F}_1 \supset \widetilde{F}_2 \supset \cdots$, where

$$\widetilde{F}_p := \Big\{ \sum_{n\in\mathbb{N}} u^{n+p} e^{nz} \widetilde{\varphi}_n(z,u) \mid \widetilde{\varphi}_n(z,u) \in \widetilde{\mathscr{R}}_{\mathbb{Z}}[[u]] \text{ and}$$
$$\Delta_{m_r} \cdots \Delta_{m_1} \widetilde{\varphi}_n = 0 \text{ for } m_1,\dots,m_r \geq 1 \text{ with } m_1+\cdots+m_r > n \Big\}$$

for each $p \in \mathbb{N}$. One can check that the operator $\Delta_{\mathbb{R}^+} = \sum_{m\geq 1} e^{-mz}\Delta_m$ is well defined on $\widetilde{F}(\mathbb{Z},\mathbb{R}^+)$ and maps $\widetilde{F}_p$ in $\widetilde{F}_{p+1}$, with

$$\Delta_{\mathbb{R}^+}\Big(\sum_{n\geq 0} u^{n+p} e^{nz} \widetilde{\varphi}_n(z,u)\Big) = \sum_{n\geq 0} u^{n+p+1} e^{nz} \widetilde{\psi}_n(z,u),$$
$$\widetilde{\psi}_n(z,u) := \sum_{m\geq 1} u^{m-1} \Delta_m \widetilde{\varphi}_{m+n}(z,u),$$

therefore its exponential is well defined and coincides with $\Delta^+_{\mathbb{R}^+}$.

In the case of the formal integral $\widetilde{\phi}(z,u)$, thanks to (6.156), we find

$$\Delta_{\mathbb{R}^+}\widetilde{\phi}(z,u) = \sum_{n\geq 0,\, m\geq 1} nC_m u^{n+m} e^{nz}\widetilde{\phi}_n = \mathscr{C}\widetilde{\phi}(z,u)$$

with a new operator $\mathscr{C} := \sum_{m\geq 1} C_m u^{m+1} \frac{\partial}{\partial u}$.

One can check that $\widetilde{F}(\mathbb{Z},\mathbb{R}^-)$ is an algebra and its multiplication maps $\widetilde{F}_p \times \widetilde{F}_q$ to $\widetilde{F}_{p+q}$. Since $\mathscr{C}$ is a derivation which maps $\widetilde{F}_p$ to $\widetilde{F}_{p+1}$, its exponential $\exp\mathscr{C}$ is well defined and is an automorphism (same argument as for Lemma 6.87). Reasoning as in Exercise 5.5, one can see that there exists $\xi(u) \in u\mathbb{C}[[u]]$ such that $\exp\mathscr{C}$ coincides with the composition operator associated with $(z,u) \mapsto (z,\xi(u))$:

$$\widetilde{\varphi}(z,u) \in \widetilde{F}(\mathbb{Z},\mathbb{R}^-) \implies (\exp\mathscr{C})\widetilde{\varphi}(z,u) = \widetilde{\varphi}(z,\xi(u)).$$

In fact, there is an explicit formula

$$\xi(u) = u + \sum_{m\geq 1}\Big(\sum_{r\geq 1}\sum_{\substack{m_1,\dots,m_r\geq 1\\ m_1+\cdots+m_r=m}} \frac{1}{r!}\beta_{m_1,\dots,m_r} C_{m_1}\cdots C_{m_r}\Big) u^{m+1}$$

with the notations $\beta_{m_1} = 1$ and $\beta_{m_1,\dots,m_r} = (m_1+1)(m_1+m_2+1)\cdots(m_1+\cdots+m_{r-1}+1)$. We thus obtain

$$\Delta^+_{\mathbb{R}^+}\widetilde{\phi}(z,u) = \widetilde{\phi}(z,\xi(u)). \tag{6.158}$$

Theorem 6.116. *The series $\xi(u)$ has positive radius of convergence and, for $z \in \mathscr{D}((-\pi,0),\gamma) \cap \mathscr{D}((0,\pi),\gamma)$ with $\Re e\, z > 0$,*

$$\phi^{(-\pi,0)}(z,u) \equiv \phi^{(0,\pi)}(z,\xi(u)), \qquad Y^{(-\pi,0)}(z,y) \equiv Y^{(0,\pi)}(z,\xi(ye^{-z})e^z).$$

Sketch of proof. Let $I := [\varepsilon, \pi - \varepsilon]$, $J := [-\pi + \varepsilon, -\varepsilon]$, and consider the diffeomorphism $\theta := \left[\Theta^{(0,\pi)}\right]^{-1} \circ \Theta^{(-\pi,0)}$ in $\{z \in \mathscr{D}(I,\gamma) \cap \mathscr{D}(J,\gamma) \mid \Re e\, z > 0\} \times \{|y| < R'\}$ with $R' > 0$ small enough. It is of the form $\theta(z,y) = \big(z, \chi^+(z,y)\big)$ with $\chi^+(z,0) \equiv 0$. The direct image of $N = \frac{\partial}{\partial z} + y\frac{\partial}{\partial y}$ by θ is N, this implies that $\chi^+ = N\chi^+$, whence $\frac{1}{u\mathrm{e}^z}\chi^+(z, u\mathrm{e}^z)$ is independent of z and can be written $\frac{\xi^+(u)}{u}$ with $\xi^+(u) \in \mathbb{C}\{u\}$. Thus $\chi^+(z,y) = \xi^+(y\mathrm{e}^{-z})\mathrm{e}^z$, i.e.

$$Y^J(z,y) \equiv Y^I\big(z, \xi^+(y\mathrm{e}^{-z})\mathrm{e}^z\big).$$

To conclude, it is thus sufficient to prove that the Taylor series of $\xi^+(u)$ is $\xi(u)$. This can be done using (6.158), by arguing as in the proof of Theorem 6.77. □

Exercise 6.117 (Analytic invariants). Assume we are given two equations of the form (6.145) and, correspondingly, two vector fields $X_{B_1} = \frac{\partial}{\partial z} + B_1(z,Y)\frac{\partial}{\partial Y}$ and $X_{B_2} = \frac{\partial}{\partial z} + B_2(z,Y)\frac{\partial}{\partial Y}$ with the same assumptions as previously on $B_1, B_2 \in \mathbb{C}\{z^{-1}, y\}$. Prove that there exists a formal series $\widetilde{\chi}(z,y) \in \mathbb{C}[[z^{-1}, y]]$ such that the formula $\theta(z,y) := \big(z, \widetilde{\chi}(z,y)\big)$ defines a formal diffeomorphism which conjugates X_{B_1} and X_{B_2}. Prove that X_{B_1} and X_{B_2} are analytically conjugate, i.e. $\widetilde{\chi}(z,y) \in \mathbb{C}\{z^{-1}, y\}$, if and only both equations give rise to the same sequence $(C_{-1}, C_1, C_2, \ldots)$, or, equivalently, to the same pair $\big(C_{-1}, \xi(u)\big)$ (the latter pair is called the "Martinet-Ramis modulus").

Exercise 6.118. Study the particular case where B is of the form $B(z,y) = b_0(z) + \big(1 + b_1(z)\big)y$, with $b_0 \in z^{-1}\mathbb{C}\{z^{-1}\}$, $b_1 \in z^{-2}\mathbb{C}\{z^{-1}\}$. Prove in particular that the Borel transform of $b_0\,\mathrm{e}^{-\partial^{-1}b_1}$ is an entire function whose value at -1 is $-\frac{1}{2\pi\mathrm{i}}C_{-1}$ and that $C_m = 0$ for $m \neq -1$ in that case.

Remark 6.119. The numbers C_m, $m \in \{-1\} \cup \mathbb{N}^*$, which encode such a subtle analytic information, are usually impossible to compute in closed form. An exception is the case of the "canonical Riccati equations", for which $B(z,y) = y - \frac{1}{2\pi\mathrm{i}}(B_- + B_+y^2)z^{-1}$, with $B_-, B_+ \in \mathbb{C}$. One finds $C_m = 0$ for $m \notin \{-1, 1\}$ and

$$C_{-1} = B_-\sigma(B_-B^+), \quad C_1 = -B_+\sigma(B_-B^+)$$

with $\sigma(b) := \frac{2}{b^{1/2}} sin\frac{b^{1/2}}{2}$. See [Sau09] for the references.

References

CNP93. B. Candelpergher, J.-C. Nosmas, and F. Pham. *Approche de la résurgence*. Actualités Mathématiques. [Current Mathematical Topics]. Hermann, Paris, 1993.

Del16. E. Delabaere. *Divergent Series, summability and resurgence. Volume 3: Resurgent Methods and the First Painlevé Equation.*, volume 2155 of *Lecture Notes in Mathematics*. Springer, Heidelberg, 2016.

Éca81. J. Écalle. *Les fonctions résurgentes. Tome I*, volume 5 of *Publications Mathématiques d'Orsay 81 [Mathematical Publications of Orsay 81].* Université de Paris-Sud, Département de Mathématique, Orsay, 1981. Les algèbres de fonctions résurgentes. [The algebras of resurgent functions], With an English foreword.

Éca84. J. Écalle. *Cinq applications des fonctions résurgentes.*, volume 62 of *Publications Mathématiques d'Orsay [Mathematical Publications of Orsay].* Université de Paris-Sud, Département de Mathématiques, Orsay, 1984. [Five applications of resurgent functions.]

Éca85. J. Écalle. *Les fonctions résurgentes. Tome III*, volume 5 of *Publications Mathématiques d'Orsay [Mathematical Publications of Orsay].* Université de Paris-Sud, Département de Mathématiques, Orsay, 1985. L'équation du pont et la classification analytique des objects locaux. [The bridge equation and analytic classification of local objects].

Lod16. M. Loday-Richaud. *Divergent Series, summability and resurgence. Volume 2: Simple and multiple summability.*, volume 2154 of *Lecture Notes in Mathematics.* Springer, Heidelberg, 2016.

LR11. M. Loday-Richaud and P. Remy. Resurgence, Stokes phenomenon and alien derivatives for level-one linear differential systems. *J. Differential Equations*, 250(3):1591–1630, 2011.

MR82. J. Martinet and J.-P. Ramis. Problèmes de modules pour des équations differentielles non linéaires du premier ordre. *Inst. Hautes Etúdes Sci. Publ. Math.*, 55:63–164, 1982.

Ou10. Y. Ou. On the stability by convolution product of a resurgent algebra. *Ann. Fac. Sci. Toulouse Math. (6)*, 19(3-4):687–705, 2010.

Sau09. D. Sauzin. Mould expansions for the saddle-node and resurgence monomials. In *Renormalization and Galois theories*, volume 15 of *IRMA Lect. Math. Theor. Phys.*, pages 83–163. Eur. Math. Soc., Zürich, 2009.

Sau12. D. Sauzin. Resurgent functions and splitting problems. In *New Trends and Applications of Complex Asymptotic Analysis : around dynamical systems, summability, continued fractions*, volume 1493 of *RIMS Kokyuroku*, pages 48–117. Kyoto University, 2012.

Sau13. D. Sauzin. On the stability under convolution of resurgent functions. *Funkcial. Ekvac.*, 56(3):397–413, 2013.

Sau15. D. Sauzin. Nonlinear analysis with resurgent functions. *Ann. Sci. Éc. Norm. Supér. (4)*, 48(3):667–702, 2015.

Chapter 7
The Resurgent Viewpoint on Holomorphic Tangent-to-Identity Germs

The last chapter of this volume is concerned with germs of holomorphic tangent-to-identity diffeomorphisms. The main topics are the description of the local dynamics (describing the local structure of the orbits of the discrete dynamical system induced by a given germ) and the description of the conjugacy classes (attaching to a given germ quantities which characterize its analytic conjugacy class). We shall give a fairly complete account of the results in the simplest case, limiting ourselves to germs at ∞ of the form

$$f(z) = z + 1 + O(z^{-2}) \tag{7.1}$$

(corresponding to germs at 0 of the form $F(t) = t - t^2 + t^3 + O(t^4)$ by (5.83)–(5.84)). The reader is referred to [Éca81], [Mil06], [Lor06], [Sau12], [DS14], [DS15] for more general studies.

It turns out that formal tangent-to-identity diffeomorphisms play a prominent role, particularly those which are 1-summable and $2\pi i\mathbb{Z}$-resurgent. So the ground was prepared in Sections 5.14–5.17 and in Theorem 6.35. In fact, because of the restriction (7.1), all the resurgent functions which will appear will be simple; we thus begin with a preliminary section.

7.1 Simple Ω-resurgent tangent-to-identity diffeomorphisms

Let us give ourselves a non-empty closed discrete subset Ω of $\mathbb{C}$ which is stable under addition. Recall that, according to Section 6.5, Ω-resurgent tangent-to-identity diffeomorphisms form a group $\widetilde{\mathscr{G}}^{\mathrm{RES}}(\Omega)$ for composition (subgroup of the group $\widetilde{\mathscr{G}} = \mathrm{id} + \mathbb{C}[[z^{-1}]]$ of all formal tangent-to-identity diffeomorphisms at ∞).

Definition 7.1. We call *simple Ω-resurgent tangent-to-identity diffeomorphism* any $\tilde{f} = \mathrm{id} + \tilde{\varphi} \in \widetilde{\mathscr{G}}^{\mathrm{RES}}$ where $\tilde{\varphi}$ is a simple Ω-resurgent series. We use the notations

$$\widetilde{\mathscr{G}}^{\mathrm{simp}}(\Omega) := \{\tilde{f} = \mathrm{id} + \tilde{\varphi} \mid \tilde{\varphi} \in \widetilde{\mathscr{R}}_{\Omega}^{\mathrm{simp}}\}, \qquad \widetilde{\mathscr{G}}_{\sigma}^{\mathrm{simp}}(\Omega) := \widetilde{\mathscr{G}}^{\mathrm{simp}}(\Omega) \cap \widetilde{\mathscr{G}}_{\sigma} \text{ for } \sigma \in \mathbb{C}.$$

C. Mitschi, D. Sauzin, *Divergent Series, Summability and Resurgence I*, Lecture Notes in Mathematics 2153, DOI 10.1007/978-3-319-28736-2_7

We define $\Delta_\omega\colon \widetilde{\mathscr{G}}^{\mathrm{simp}}(\Omega) \to \widetilde{\mathscr{R}}^{\mathrm{simp}}_{-\omega+\Omega}$ for any $\omega \in \Omega$ by setting

$$\Delta_\omega(\mathrm{id}+\widetilde{\varphi}) := \Delta_\omega\widetilde{\varphi}.$$

Recall that, in Section 5.15, $\partial\widetilde{f}$ was defined as the invertible formal series $1+\partial\widetilde{\varphi}$ for any $\widetilde{f} = \mathrm{id}+\widetilde{\varphi} \in \widetilde{\mathscr{G}}$. Clearly $\widetilde{f} \in \widetilde{\mathscr{G}}^{\mathrm{simp}}(\Omega) \implies \partial\widetilde{f} \in \widetilde{\mathscr{G}}^{\mathrm{simp}}(\Omega)$.

Theorem 7.2. *The set $\widetilde{\mathscr{G}}^{\mathrm{simp}}(\Omega)$ is a subgroup of $\widetilde{\mathscr{G}}^{\mathrm{RES}}(\Omega)$, the set $\widetilde{\mathscr{G}}_0^{\mathrm{simp}}(\Omega)$ is a subgroup of $\widetilde{\mathscr{G}}_0^{\mathrm{RES}}(\Omega)$. For any $\widetilde{f},\widetilde{g} \in \widetilde{\mathscr{G}}^{\mathrm{simp}}(\Omega)$ and $\omega \in \Omega$, we have*

$$\Delta_\omega(\widetilde{g}\circ\widetilde{f}) = (\partial\widetilde{g})\circ\widetilde{f}\cdot\Delta_\omega\widetilde{f} + \mathrm{e}^{-\omega(\widetilde{f}-\mathrm{id})}\cdot(\Delta_\omega g)\circ\widetilde{f}, \tag{7.2}$$

$$\widetilde{h} = \widetilde{f}^{\circ(-1)} \implies \Delta_\omega\widetilde{h} = -\mathrm{e}^{-\omega(\widetilde{h}-\mathrm{id})}\cdot(\Delta_\omega\widetilde{f})\circ\widetilde{h}\cdot\partial\widetilde{h}. \tag{7.3}$$

Proof. The stability under group composition stems from Theorem 6.91, since $(\mathrm{id}+\widetilde{\psi})\circ(\mathrm{id}+\widetilde{\varphi}) = \mathrm{id}+\widetilde{\varphi}+\widetilde{\psi}\circ(\mathrm{id}+\widetilde{\varphi})$. The stability under group inversion is proved from Lagrange reversion formula as in the proof of Theorem 6.35, adapting the arguments of the proof of Theorem 6.91.

Formula (7.2) results from (6.95), and formula (7.3) follows by choosing $g = f^{\circ(-1)}$. □

7.2 Simple parabolic germs with vanishing resiter

We now come to the heart of the matter, giving ourselves a germ $F(t) \in \mathbb{C}\{t\}$ of holomorphic tangent-to-identity diffeomorphism at 0 and the corresponding germ $f(z) := 1/F(1/z) \in \mathscr{G}$ at ∞.

The germ F gives rise to a discrete dynamical system $F\colon U \to \mathbb{C}$, where U is an open neighbourhood of 0 on which a representative of F is holomorphic. This means that for any $t_0 \in U$ we can define a finite or infinite *forward orbit* $\{t_n = F^{\circ n}(t_0) \mid 0 \le n < N\}$, where $N \in \mathbb{N}^*\cup\{\infty\}$ is characterized by $t_1 = F(t_0) \in U, \dots,$ $t_{N-1} = F(t_{N-2}) \in U$ and $t_N = F(t_{N-1}) \notin U$ (so that apriori t_{N+1} cannot be defined), and similarly a finite or infinite *backward orbit* $\{t_{-n} = F^{\circ(-n)}(t_0) \mid 0 \le n < M\}$ with $M \in \mathbb{N}^*\cup\{\infty\}$.

We are interested in the local structure of the orbits starting close to 0, so the domain U does not matter. Moreover, the qualitative study of a such a dynamical system is insensitive to analytic changes of coordinate: we say that G is *analytically conjugate* to F if there exists an invertible $H \in t\mathbb{C}\{t\}$ such that $G = H^{\circ(-1)}\circ F\circ H$; the germ G is then itself tangent-to-identity and it should be considered as equivalent to F from the dynamical point of view (because H maps the orbits of F to those of G). The description of the analytic conjugacy classes is thus dynamically relevant.

We suppose that F is non-degenerate in the sense that $F''(0) \neq 0$. Observe that $G = H^{\circ(-1)}\circ F\circ H \implies G''(0) = H'(0)F''(0)$, thus we can rescale the variable w

so as to make the second derivative equal to -2, i.e. we assume from now on $F(t) = t - t^2 + (\rho+1)t^3 + O(t^4)$ with a certain $\rho \in \mathbb{C}$, and correspondingly

$$f(z) = z + 1 - \rho z^{-1} + O(z^{-2}) \in \mathscr{G}_1. \tag{7.4}$$

Such a germ F or f is called a *simple parabolic germ.*

Once we have done that, we should only consider tangent-to-identity changes of coordinate G, so as to maintain the condition $F''(0) = -2$. In the variable z, this means that we shall study the $\mathscr{G}$-conjugacy class $\{\, h^{\circ(-1)} \circ f \circ h \mid h \in \mathscr{G} \,\} \subset \mathscr{G}_1$.

As already alluded to, the $\widetilde{\mathscr{G}}$-conjugacy class of f in $\widetilde{\mathscr{G}}_1$ plays a role in the problem, i.e. we must also consider the formal conjugacy equivalence relation. The point is that it may happen that two *holomorphic* germs f and g are *formally* conjugate (there exists $\widetilde{h} \in \widetilde{\mathscr{G}}$ such that $f \circ \widetilde{h} = \widetilde{h} \circ g$) without being *analytically* conjugate (there exists no $h \in \mathscr{G}$ with the same property): the $\mathscr{G}$-conjugacy classes we are interested in form a finer partition of $\mathscr{G}_1$ than the $\widetilde{\mathscr{G}}$-conjugacy classes.

It turns out that the number ρ in (7.4) is invariant by formal conjugacy and that two germs with the same ρ are always formally conjugate (we omit the proof). This number is called "resiter".

We suppose further that the resiter ρ is 0, i.e. we limit ourselves to the most elementary formal conjugacy class. This implies that our f is of the form (7.1) and formally conjugate to $f_0(z) := z+1$, the most elementary simple parabolic germ with vanishing resiter, which may be considered as a formal *normal form* for all simple parabolic germs with vanishing resiter. The corresponding normal form at 0 is $F_0(t) := \frac{t}{1+t}$. The orbits of the normal form are easily computed: we have $f_0^{\circ n} = \mathrm{id} + n$ and $F_0^{\circ n}(t) = \frac{t}{1+nt}$ for all $n \in \mathbb{Z}$, thus the backward and forward orbits of a point $t_0 \neq 0$ are infinite and contained either in $\mathbb{R}$ (if $t_0 \in \mathbb{R}$) or in a circle passing through 0 centred at a point of $\mathrm{i}\mathbb{R}^*$.

In particular, all the forward orbits of F_0 converge to 0 and all its backward orbits converge in negative time to 0. If the formal conjugacy between F and F_0 happens to be convergent, then such qualitative properties of the dynamics automatically hold for the orbits of F itself (at least for those which start close enough to 0). We shall see that in general the picture is more complex…

7.3 Resurgence and summability of the iterators

Notation 7.3 *Given $\widetilde{g} \in \widetilde{\mathscr{G}}$, the operator of composition with $\widetilde{g}$ is denoted by*

$$C_{\widetilde{g}} \colon \ \widetilde{\varphi} \in \mathbb{C}[[z^{-1}]] \mapsto \widetilde{\varphi} \circ \widetilde{g} \in \mathbb{C}[[z^{-1}]].$$

The operator $C_{\mathrm{id}-1} - \mathrm{Id}$ induces an invertible map $z^{-1}\mathbb{C}[[z^{-1}]] \to z^{-2}\mathbb{C}[[z^{-1}]]$ with Borel counterpart $\widehat{\varphi}(\zeta) \in \mathbb{C}[[\zeta]] \mapsto (\mathrm{e}^{\zeta} - 1)\widehat{\varphi}(\zeta) \in \zeta\mathbb{C}[[\zeta]]$; we denote by

$$E \colon z^{-2}\mathbb{C}[[z^{-1}]] \to z^{-1}\mathbb{C}[[z^{-1}]], \qquad \widehat{E} \colon \zeta\mathbb{C}[[\zeta]] \to \mathbb{C}[[\zeta]]$$

its inverse and the Borel counterpart of its inverse, hence $(\widehat{E}\widehat{\varphi})(\zeta) = \frac{1}{e^{\zeta}-1}\widehat{\varphi}(\zeta)$ *(variant of Corollary 5.11). We also set*

$$f_0 := \mathrm{id}+1 \in \mathscr{G}_1. \tag{7.5}$$

The operator E will allow us to give a very explicit proof of the existence of a formal conjugacy between a diffeomorphism with vanishing resiter and the normal form (7.5).

Lemma 7.4. *Given a simple parabolic germ with vanishing resiter* $f \in \mathscr{G}_1$*, there is a unique* $\widetilde{v}_* \in \widetilde{\mathscr{G}}_0$ *such that*

$$\widetilde{v}_* \circ f = f_0 \circ \widetilde{v}_*. \tag{7.6}$$

It can be written as a formally convergent series

$$\widetilde{v}_* = \mathrm{id} + \sum_{k\in\mathbb{N}} \widetilde{\varphi}_k, \qquad \widetilde{\varphi}_k := (EB)^k Eb \in z^{-2k-1}\mathbb{C}[[z^{-1}]] \textit{ for each } k \in \mathbb{N}, \tag{7.7}$$

with a holomorphic germ $b := f \circ f_0^{\circ(-1)} - \mathrm{id} \in z^{-2}\mathbb{C}\{z^{-1}\}$ *and an operator* $B := C_{\mathrm{id}+b} - \mathrm{Id}$.

The solutions in $\widetilde{\mathscr{G}}$ *of the conjugacy equation* $\widetilde{v} \circ f = f_0 \circ \widetilde{v}$ *are the formal diffeomorphisms* $\widetilde{v} = \widetilde{v}_* + c$ *with arbitrary* $c \in \mathbb{C}$.

Proof. The conjugacy equation can be written $\widetilde{v} \circ f = \widetilde{v} + 1$ or, equivalently (composing with $f_0^{\circ(-1)} = \mathrm{id}-1$), $\widetilde{v} \circ (\mathrm{id}+b) = \widetilde{v} \circ (\mathrm{id}-1)+1$. Searching for a formal solution in the form $\widetilde{v} = \mathrm{id}+\widetilde{\varphi}$ with $\widetilde{\varphi} \in \mathbb{C}[[z^{-1}]]$, we get $b+\widetilde{\varphi}\circ(\mathrm{id}+b) = \widetilde{\varphi}\circ(\mathrm{id}-1)$, i.e.

$$(C_{\mathrm{id}-1} - \mathrm{Id})\widetilde{\varphi} = B\widetilde{\varphi} + b. \tag{7.8}$$

We have $\mathrm{val}\big((C_{\mathrm{id}-1} - \mathrm{Id})\widetilde{\varphi}\big) \geq \mathrm{val}(\widetilde{\varphi}) + 1$ for the standard valuation (5.10), and $\mathrm{val}(B\widetilde{\varphi}) \geq \mathrm{val}(\widetilde{\varphi}) + 3$ (because B can be written as the formally convergent series if operators $\sum_{r\geq 1} \frac{1}{r!} b^r \partial^r$ with $\mathrm{val}(b) \geq 2$ and $\mathrm{val}(\partial\widetilde{\varphi}) \geq \mathrm{val}(\widetilde{\varphi}) + 1$), thus the difference between any two formal solutions of (7.8) is a constant. If we specify $\widetilde{\varphi} \in z^{-1}\mathbb{C}[[z^{-1}]]$, then (7.8) is equivalent to

$$\widetilde{\varphi} = EB\widetilde{\varphi} + Eb,$$

where $\mathrm{val}(EB\widetilde{\varphi}) \geq \mathrm{val}(\widetilde{\varphi}) + 2$, thus the formal series $\widetilde{\varphi}_k$ of (7.7) have valuation at least $2k+1$ and yield the unique formal solution without constant term in the form $\widetilde{\varphi} = \sum_{k\in\mathbb{N}} \widetilde{\varphi}_k$. □

Definition 7.5. The unique formal diffeomorphism $\widetilde{v}_* \in \widetilde{\mathscr{G}}_0$ such that $\widetilde{v}_* \circ f = f_0 \circ \widetilde{v}_*$ is called the *iterator* of f. Its inverse $\widetilde{u}_* := \widetilde{v}_*^{\circ(-1)} \in \widetilde{\mathscr{G}}_0$ is called the *inverse iterator* of f.

We illustrate this in the following commutative diagram, including the parabolic germ at 0 defined by $F(t) := 1/f(1/t)$:

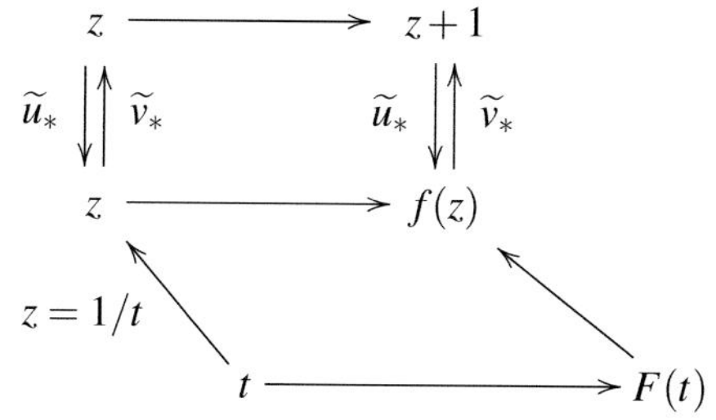

Observe that

$$f \circ \widetilde{u}_* = \widetilde{u}_* \circ f_0, \tag{7.9}$$

which can be viewed as a difference equation: $\widetilde{u}_*(z+1) = f\big(\widetilde{u}_*(z)\big)$.

Theorem 7.6. *Suppose that $f \in \mathscr{G}_1$ has vanishing resiter. Then its iterator $\widetilde{v}_*$ and its inverse interator $\widetilde{u}_*$ belong to $\widetilde{\mathscr{G}}_0^{\mathrm{simp}}(2\pi\mathrm{i}\mathbb{Z}) \cap \widetilde{\mathscr{G}}_0(I^+) \cap \widetilde{\mathscr{G}}_0(I^-)$ with $I^+ := (-\frac{\pi}{2}, \frac{\pi}{2})$ and $I^- := (\frac{\pi}{2}, \frac{3\pi}{2})$ (notations of Definitions 5.65 and 7.1).*

Moreover, the iterator can be written $\widetilde{v}_* = \mathrm{id} + \widetilde{\varphi}$ with a simple $2\pi\mathrm{i}\mathbb{Z}$-resurgent series $\widetilde{\varphi}$ whose Borel transform satifies the following: for any path γ issuing from 0 and then avoiding $2\pi\mathrm{i}\mathbb{Z}$ and ending at a point $\zeta_* \in \mathrm{i}\mathbb{R}$, or for $\gamma = \{0\}$ and $\zeta_* = 0$, there exist locally bounded functions $\alpha, \beta\colon I^+ \cup I^- \to \mathbb{R}^+$ such that

$$\Big|\mathrm{cont}_\gamma \widehat{\varphi}\big(\zeta_* + t\,\mathrm{e}^{\mathrm{i}\theta}\big)\Big| \le \alpha(\theta)\,\mathrm{e}^{\beta(\theta)t} \quad \textit{for all } t \ge 0 \textit{ and } \theta \in I^+ \cup I^- \tag{7.10}$$

(see Figure 7.1a).

Since $\widetilde{\varphi} := \widetilde{v}_* - \mathrm{id}$ is given by Lemma 7.4 in the form of the formally convergent series $\sum_{k\ge 0} \widetilde{\varphi}_k$, the statement can be proved by controlling the formal Borel transforms $\widehat{\varphi}_k$.

Lemma 7.7. *For each $k \in \mathbb{N}$ we have $\widehat{\varphi}_k := \mathscr{B}(\widetilde{\varphi}_k) \in \widehat{\mathscr{R}}^{\mathrm{simp}}_{2\pi\mathrm{i}\mathbb{Z}}$.*

Lemma 7.8. *Suppose that $0 < \varepsilon < \pi < \tau$, $0 < \kappa \le 1$ and D is a closed disc of radius ε centred at $2\pi\mathrm{i}m$ with $m \in \mathbb{Z}^*$, and let*

$$\Omega^+_{\varepsilon,\tau,D} := \{\, \zeta \in \mathbb{C} \mid \Re e\, \zeta > -\tau,\ \mathrm{dist}\big(\zeta, 2\pi\mathrm{i}\mathbb{Z}^*\big) > \varepsilon \,\} \setminus \{\, u\zeta \in \mathbb{C} \mid u \in [1,+\infty),\ \pm\zeta \in D \,\} \tag{7.11}$$

(see Figure 7.1b). Then there exist $A, M, R > 0$ such that, for any naturally parametrised path $\gamma\colon [0,\ell] \to \Omega^+_{\varepsilon,\tau,D}$ with

$$s \in [0,\varepsilon] \implies |\gamma(s)| = s, \quad s > \varepsilon \implies |\gamma(s)| > \varepsilon, \quad s \in [0,\ell] \implies |\gamma(s)| > \kappa s, \tag{7.12}$$

one has

$$\big|\mathrm{cont}_\gamma \widehat{\varphi}_k\big(\gamma(\ell)\big)\big| \le A \frac{(M\ell)^k}{k!} \mathrm{e}^{R\ell} \quad \textit{for every } k \ge 0. \tag{7.13}$$

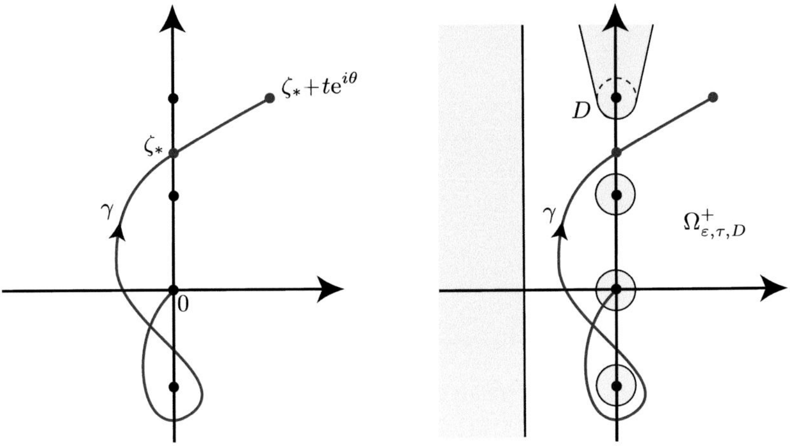

Fig. 7.1: *Resurgence of the iterator (Theorem 7.6).* Left: A path of analytic continuation for $\widehat{\varphi}$. Right: The domain $\Omega^+_{\varepsilon,\tau,D}$ of Lemma 7.8.

Lemmas 7.7 and 7.8 imply Theorem 7.6. According to notation (6.2), we denote by $\mathbb{D}_{2\pi}$ or $\mathbb{D}_1$ the open disc centred at 0 of radius 2π or 1. Lemma 7.8 implies that the series of holomorphic functions $\sum \widehat{\varphi}_k$ converges normally in any compact subset of $\mathbb{D}_{2\pi}$ (using paths γ of the form $[0,\zeta]$) and that its sum, which is $\widehat{\varphi}$, extends analytically along any naturally parametrised path γ which starts as the line-segment $[0,1]$ and then stays in $\mathbb{C}\setminus 2\pi i\mathbb{Z}$: indeed, taking ε,κ small enough and τ, m large enough, we see that Lemma 7.8 applies to γ and the neighbouring paths, so that (7.13) yields the normal convergence of $\sum_{k\geq 0}\operatorname{cont}_\gamma \widehat{\varphi}_k(\gamma(t)+\zeta) = \operatorname{cont}_\gamma \widehat{\varphi}(\gamma(t)+\zeta)$ for all t and ζ with $|\zeta|$ small enough. Therefore $\widehat{\varphi}$ is $2\pi i\mathbb{Z}$-resurgent and, combining Lemma 7.7 with the estimates (7.13), we also get $\widehat{\varphi}\in\widehat{\mathscr{R}}^{\mathrm{simp}}_{2\pi i\mathbb{Z}}$ by Lemma 6.92.

This establishes $\widetilde{v}_*\in\widetilde{\mathscr{G}}_0^{\mathrm{simp}}(2\pi i\mathbb{Z})$, whence $\widetilde{u}_*\in\widetilde{\mathscr{G}}_0^{\mathrm{simp}}(2\pi i\mathbb{Z})$ by Theorem 7.2.

For the part of (7.10) relative to I^+, we give ourselves an arbitrary $n>1$ and set $\delta_n := \frac{\pi}{2n}$, $I_n^+ := [-\frac{\pi}{2}+\delta_n, \frac{\pi}{2}-\delta_n]$. Given γ with endpoint $\zeta_*\in i\mathbb{R}$, we first replace an initial portion of γ with a line-segment of length 1 (unless γ stays in $\mathbb{D}_1$, in which case the modification of the arguments which follow is trivial) and switch to its natural parametrisation $\gamma\colon\ [0,\ell]\to\mathbb{C}$. We then choose ε_n and κ_n small enough:

$$\varepsilon_n < \min\Big\{1, \min_{[1,\ell]}|\gamma|, \operatorname{dist}\big(\gamma([0,\ell]), 2\pi i\mathbb{Z}^*\big), \operatorname{dist}\big(\zeta_*, 2\pi i\mathbb{Z}\big)\cos\delta_n\Big\},$$

$$\kappa_n < \min\Big\{\min_{[0,\ell]}\tfrac{|\gamma(s)|}{s}, \min_{t\geq 0}\tfrac{|\zeta_*+t\,e^{\pm i\delta_n}|}{\ell+t}\Big\},$$

and τ and m_n large enough:

$$\tau > -\min \Re e\, \gamma, \qquad m_n > \frac{1}{2\pi}\big(\varepsilon_n + \max|\Im m\, \gamma|\big),$$

so that Lemma 7.8 applies to the concatenation $\Gamma := \gamma \boldsymbol{\cdot} [\zeta_*, \zeta_* + t\,e^{i\theta}]$ for each $t \geq 0$ and $\theta \in I_n^+$; since Γ has length $\ell + t$, (7.13) yields

$$t \geq 0 \text{ and } \theta \in I_n^+ \implies \Big|\mathrm{cont}_\gamma\, \widehat{\varphi}\big(\zeta_* + t\,e^{i\theta}\big)\Big| = \big|\mathrm{cont}_\Gamma\, \widehat{\varphi}\big(\Gamma(\ell+t)\big)\big| \leq A_n\, e^{(M_n+R_n)(\ell+t)},$$

where A_n, M_n and R_n depend on n and γ but not on t or θ. We thus take

$$\alpha^+(\theta) := e^{\ell\beta^+(\theta)} \max\big\{ A_n \mid n \geq 1 \text{ s.t. } \theta \in I_n^+ \big\},$$

$$\beta^+(\theta) := \max\big\{ M_n + R_n \mid n \geq 1 \text{ s.t. } \theta \in I_n^+ \big\}$$

for any $\theta \in I^+$, and get

$$t \geq 0 \text{ and } \theta \in I^+ \quad \implies \quad \Big|\mathrm{cont}_\gamma\, \widehat{\varphi}\big(\zeta_* + t\,e^{i\theta}\big)\Big| \leq \alpha^+(\theta)\, e^{\beta^+(\theta)t}.$$

The part of (7.10) relative to I^- follows from the fact that $\widehat{\varphi}^-(\zeta) := \widehat{\varphi}(-\zeta)$ satisfies all the properties we just obtained for $\widehat{\varphi}(\zeta)$, since it is the formal Borel transform of $\widetilde{\varphi}^-(z) := -\widetilde{\varphi}(-z)$ which solves the equation $C_{\mathrm{id}-1}\widetilde{\varphi}^- = C_{\mathrm{id}+b^-}\widetilde{\varphi}^- + b_*^-$ associated with the simple parabolic germ $f^-(z) := -f^{-1}(-z) = z + 1 + b^-(z+1)$.

This establishes (7.10), which yields (in the particular case $\gamma = \{0\}$) $\widetilde{v}_* \in \widetilde{\mathscr{G}}_0(I^+) \cap \widetilde{\mathscr{G}}_0(I^-)$, whence $\widetilde{u}_* \in \widetilde{\mathscr{G}}_0(I^+) \cap \widetilde{\mathscr{G}}_0(I^-)$ by Theorem 5.67. □

Proof of Lemma 7.7. Since $b(z) \in z^{-2}\mathbb{C}\{\zeta\}$, its formal Borel transform is an entire function $\widehat{b}(\zeta)$ vanishing at 0, hence

$$\widehat{\varphi}_0(\zeta) = \frac{\widehat{b}(\zeta)}{e^\zeta - 1} \in \widehat{\mathscr{R}}^{\mathrm{simp}}_{2\pi i\mathbb{Z}}$$

(variant of Lemma 6.58).

We proceed by induction on k and assume $k \geq 1$ and $\widetilde{\varphi}_{k-1} \in \widetilde{\mathscr{R}}^{\mathrm{simp}}_{2\pi i\mathbb{Z}}$. By Theorem 6.91 we get $C_{\mathrm{id}+b}\widetilde{\varphi}_{k-1} \in \widetilde{\mathscr{R}}^{\mathrm{simp}}_{2\pi i\mathbb{Z}}$, thus $B\widetilde{\varphi}_{k-1} \in \widetilde{\mathscr{R}}^{\mathrm{simp}}_{2\pi i\mathbb{Z}}$, thus (since $\mathscr{B}(B\widetilde{\varphi}_{k-1})(\zeta) \in \zeta\mathbb{C}\{\zeta\}$)

$$\widehat{\varphi}_k(\zeta) = \frac{1}{e^\zeta - 1}\mathscr{B}(B\widetilde{\varphi}_{k-1})(\zeta) \in \widehat{\mathscr{R}}_{2\pi i\mathbb{Z}},$$

but is it true that all the singularities of all the branches of the analytic continuation of $\widehat{\varphi}_k$ are simple?

By repeated use of (6.95), we get

$$\Delta_{\omega_s} \cdots \Delta_{\omega_1} C_{\mathrm{id}+b}\widetilde{\varphi}_{k-1} = e^{-(\omega_1+\cdots+\omega_s)b} C_{\mathrm{id}+b}\Delta_{\omega_s} \cdots \Delta_{\omega_1}\widetilde{\varphi}_{k-1}$$

for every $s \geq 1$ and $\omega_1, \ldots, \omega_s \in 2\pi i\mathbb{Z}^*$, hence

$$\Delta_{\omega_s}\cdots\Delta_{\omega_1} B\widetilde{\varphi}_{k-1} = B_{\omega_1,\dots,\omega_s}\Delta_{\omega_s}\cdots\Delta_{\omega_1}\widetilde{\varphi}_{k-1}$$
$$\text{with } B_{\omega_1,\dots,\omega_s} := \mathrm{e}^{-(\omega_1+\cdots+\omega_s)b} C_{\mathrm{id}+b} - \mathrm{Id}.$$

Now, for any $\widetilde{\psi} \in \mathbb{C}[[z^{-1}]]$, we have

$$B_{\omega_1,\dots,\omega_s}\widetilde{\psi} = \mathrm{e}^{-(\omega_1+\cdots+\omega_s)b} B\widetilde{\psi} + (\mathrm{e}^{-(\omega_1+\cdots+\omega_s)b} - 1)\widetilde{\psi} \in z^{-2}\mathbb{C}[[z^{-1}]],$$

thus each of the simple $2\pi\mathrm{i}\mathbb{Z}$-resurgent series $\Delta_{\omega_s}\cdots\Delta_{\omega_1} B\widetilde{\varphi}_{k-1}$ has valuation ≥ 2. By Remark 6.69, the same is true of $\mathscr{A}^{\gamma}_{\omega} B\widetilde{\varphi}_{k-1}$ for every $\omega \in 2\pi\mathrm{i}\mathbb{Z}$ and every γ starting close to 0 and ending close to ω: we have

$$\mathrm{cont}_{\gamma}\mathscr{B}(B\widetilde{\varphi}_{k-1})(\omega+\zeta) = \widehat{\psi}(\zeta)\frac{\mathrm{Log}\,\zeta}{2\pi} + R(\zeta)$$

with $\widehat{\psi} \in \zeta\mathbb{C}\{\zeta\}$ and $R \in \mathbb{C}\{\zeta\}$ depending on k, ω, γ, hence $\widehat{\chi}(\zeta) := \frac{\widehat{\psi}(\zeta)}{\mathrm{e}^{\zeta}-1} \in \mathbb{C}\{\zeta\}$ and (since $\mathrm{e}^{\omega+\zeta} \equiv \mathrm{e}^{\zeta}$)

$$\mathrm{cont}_{\gamma}\widehat{\varphi}_k(\omega+\zeta) = \frac{c}{2\pi\mathrm{i}\zeta} + \widehat{\chi}(\zeta)\frac{\mathrm{Log}\,\zeta}{2\pi} + R^*(\zeta)$$

with $c := 2\pi\mathrm{i}R(0)$ and $R^*(\zeta) \in \mathbb{C}\{\zeta\}$. Therefore $\widehat{\varphi}_k$ has only simple singularities. □

Proof of Lemma 7.8. The set $\Omega^+_{\varepsilon,\tau,D}$ is such that we can find $M_0, L > 0$ so that

$$\zeta \in \Omega^+_{\varepsilon,\tau,D} \quad\Longrightarrow\quad \left|\frac{\zeta}{\mathrm{e}^{\zeta}-1}\right| \leq M_0\,\mathrm{e}^{-L|\zeta|}. \tag{7.14}$$

On the other hand, we can find $C > L$ and $R_0 > 0$ such that the entire function $\widehat{b}$ satisfies

$$|\widehat{b}(\zeta)| \leq C|\zeta|\,\mathrm{e}^{R_0|\zeta|} \quad \text{for all } \zeta \in \mathbb{C},$$

whence, by Lemma 5.53,

$$|\widehat{b}^{*k}(\zeta)| \leq C^k \frac{|\zeta|^{2k-1}}{(2k-1)!}\mathrm{e}^{R_0|\zeta|} \quad \text{for all } \zeta \in \mathbb{C} \text{ and } k \in \mathbb{N}^*.$$

Let us give ourselves a naturally parametrised path $\gamma\colon [0,\ell] \to \Omega^+_{\varepsilon,\tau,D}$ satisfying (7.12). For any $2\pi\mathrm{i}\mathbb{Z}$-resurgent series $\widetilde{\psi}$ with formal Borel transform $\widehat{\psi}$, we have $B\widetilde{\psi} \in \mathscr{R}_{2\pi\mathrm{i}\mathbb{Z}}$ by Theorem 6.32, the proof of which shows that $\widehat{B\psi} := \mathscr{B}(B\widetilde{\psi})$ can be expressed as an integral transform $\widehat{B}\widehat{\psi}(\zeta) = \int_0^{\zeta} K(\xi,\zeta)\widehat{\psi}(\xi)\,\mathrm{d}\xi$ for ζ close to 0, with kernel function

$$K(\xi,\zeta) = \sum_{k\geq 1}\frac{(-\xi)^k}{k!}\widehat{b}^{*k}(\zeta-\xi).$$

The estimates available for $\widehat{b}^{*k}$ show that K is holomorphic in $\mathbb{C}\times\mathbb{C}$, we can thus adapt the arguments of the "easy" Lemma 6.15 and get

$$\operatorname{cont}_\gamma \widehat{B}\widehat{\psi}\big(\gamma(s)\big) = \int_0^s K\big(\gamma(\sigma),\gamma(s)\big)\operatorname{cont}_\gamma \widehat{\psi}\big(\gamma(\sigma)\big)\gamma'(\sigma)\,\mathrm{d}\sigma \quad \text{for all } s\in[0,\ell].$$

The crude estimate

$$|K(\xi,\zeta)| \le C|\xi|\,\mathrm{e}^{\frac{C}{\mu}|\xi|+(R_0+\mu)|\zeta-\xi|} \quad \text{for all } (\xi,\zeta)\in\mathbb{C}\times\mathbb{C},$$

with arbitrary $\mu>1$, will allow us to bound inductively $\operatorname{cont}_\gamma\widehat{\varphi}_k = \operatorname{cont}_\gamma\widehat{E}\widehat{B}\widehat{\varphi}_{k-1}$.

Indeed, the meromorphic function $\widehat{\varphi}_0 = \frac{\widehat{b}}{\mathrm{e}^\zeta-1}$ satisfies (7.13) with $A := M_0C$ and any $R\ge R_0$. Suppose now that a $2\pi\mathrm{i}\mathbb{Z}$-resurgent function $\widehat{\psi}$ satisfies

$$\big|\operatorname{cont}_\gamma \widehat{\psi}\big(\gamma(s)\big)\big| \le \mathrm{e}^{Rs}\Psi(s) \ \text{ for all } s\in[0,\ell], \qquad \text{with } R:=R_0+\mu,\ \mu:=\frac{C}{\kappa L},$$

and a certain positive continuous function Ψ. Since $|\gamma(\sigma)|\le\sigma$ and $|\gamma(s)-\gamma(\sigma)|\le s-\sigma$, we obtain

$$\Big|\operatorname{cont}_\gamma \widehat{B}\widehat{\psi}\big(\gamma(s)\big)\Big| \le Cs\,\mathrm{e}^{(\frac{C}{\mu}+R)s}\int_0^s \Psi(\sigma)\,\mathrm{d}\sigma \ \text{ for all } s\in[0,\ell],$$

whence $\Big|\operatorname{cont}_\gamma \widehat{E}\widehat{B}\widehat{\psi}\big(\gamma(s)\big)\Big| \le M\mathrm{e}^{Rs}\int_0^s\Psi(\sigma)\,\mathrm{d}\sigma$ with $M:=\frac{CM_0}{\kappa}$ by (7.14), using $|\gamma(s)|\ge\kappa s$. We thus get $\big|\operatorname{cont}_\gamma\widehat{\varphi}_k\big(\gamma(s)\big)\big|\le A\,\mathrm{e}^{Rs}\frac{(Ms)^k}{k!}$ by induction on k. □

7.4 Fatou coordinates of a simple parabolic germ

7.4.1 For every $R>0$ and $\delta\in(0,\pi/2)$, we define

$$\Sigma^+_{R,\delta} := \{r\mathrm{e}^{\mathrm{i}\theta}\in\mathbb{C}\mid r>R,\ |\theta|<\pi-\delta\},\quad \Sigma^-_{R,\delta} := \{r\mathrm{e}^{\mathrm{i}\theta}\in\mathbb{C}\mid r>R,\ |\theta-\pi|<\pi-\delta\}.$$

Definition 7.9. A *pair of Fatou coordinates at* ∞ is a pair (v^+,v^-) of injective holomorphic maps

$$v^+ : \Sigma^+_{R,\delta}\to\mathbb{C}, \qquad v^- : \Sigma^-_{R,\delta}\to\mathbb{C},$$

with some $R>0$ and $\delta\in(0,\pi/2)$, such that

$$v^+\circ f = f_0\circ v^+, \qquad v^-\circ f = f_0\circ v^-.$$

We still assume that $f\in\mathscr{G}_1$ has vanishing resiter, with iterator $\widetilde{v}_*$ and inverse iterator $\widetilde{u}_*$. We still use the notations $I^+ = (-\frac{\pi}{2},\frac{\pi}{2})$ and $I^- = (\frac{\pi}{2},\frac{3\pi}{2})$.

Theorem 7.10. *There exists locally bounded functions* $\beta, \beta_1 \colon I^+ \cup I^- \to (0, +\infty)$ *such that* $\beta < \beta_1$ *and*

- $\tilde{v}_* \in \tilde{\mathscr{G}}_0(I^+, \beta) \cap \tilde{\mathscr{G}}_0(I^-, \beta)$ *and* $v_*^\pm := \mathscr{S}^{I^\pm} \tilde{v}_*$ *is injective on* $\mathscr{D}(I^\pm, \beta)$ *(notation of Definition 5.32);*
- $\tilde{u}_* \in \tilde{\mathscr{G}}_0(I^+, \beta_1) \cap \tilde{\mathscr{G}}_0(I^-, \beta_1)$ *and* $u_*^\pm := \mathscr{S}^{I^\pm} \tilde{u}_*$ *is injective on* $\mathscr{D}(I^\pm, \beta_1)$*, with*

$$u_*^\pm\big(\mathscr{D}(I^\pm, \beta_1)\big) \subset \mathscr{D}(I^\pm, \beta) \quad \text{and} \quad v_*^\pm \circ u_*^\pm = \mathrm{id} \ \text{ on } \mathscr{D}(I^\pm, \beta_1).$$

Moreover, the pairs of Fatou coordinates at ∞ *are the pairs* $(v_*^+ + c^+, v_*^- + c^-)$ *with arbitrary* $c^+, c^- \in \mathbb{C}$.

Remark 7.11. We may consider (v_*^+, v_*^-) as a normalized pair of Fatou coordinates. Being obtained as Borel sums of a 1-summable formal diffeomorphism, they admit a uniform 1-Gevrey asymptotic expansion in any domain $\Sigma_{R,\delta}^\pm$ with R large enough, and the same is true of the inverse Fatou coordinates u_*^+ and u_*^-. The first use of Borel-Laplace summation for obtaining Fatou coordinates is in [Éca81]. The asymptotic property without the Gevrey qualification can be found in earlier works by G. Birkhoff, G. Szekeres, T. Kimura and J. Écalle—see [Lor06] and [Lod16] for the references; see [LY14] for a recent independent proof and an application to numerical computations.

Proof of Theorem 7.10. The case $\gamma = \{0\}$ of Theorem 7.6 yields locally bounded functions $\alpha, \beta \colon I^+ \cup I^- \to \mathbb{R}^+$ such that $\tilde{v}_* \in \tilde{\mathscr{G}}_0(I^\pm, \beta, \alpha)$ (notation of Definition 5.65). In view of Theorem 5.67, we can replace β by a larger function so that $v_*^\pm$ is injective on $\mathscr{D}(I^\pm, \beta)$. We apply again Theorem 5.67: setting

$$\beta \; < \; \beta^* := \beta + 2\sqrt{\alpha} \; < \; \beta_1 := \beta + (1 + \sqrt{2})\sqrt{\alpha},$$

we get $\tilde{u}_* \in \tilde{\mathscr{G}}_0(I^\pm, \beta^*)$, hence $\tilde{u}_* \in \tilde{\mathscr{G}}_0(I^\pm, \beta_1)$, and all the desired properties follow.

By Lemma 5.35, we have $f = \mathscr{S}^{I^\pm} f$; replacing the above function β by a larger one if necessary so as to take into account the domain of definition of f, Theorem 5.66 shows that $\mathscr{S}^{I^\pm}(\tilde{v}_* \circ f) = v_*^\pm \circ f$ and $\mathscr{S}^{I^\pm}(f \circ \tilde{u}_*) = f \circ u_*^\pm$. In view of (7.6) and (7.9), this yields

$$v_*^\pm \circ f = f_0 \circ v_*^\pm, \qquad f \circ u_*^\pm = u_*^\pm \circ f_0. \tag{7.15}$$

We see that for any $\delta \in (0, \pi/2)$ there exists $R > 0$ such that $\Sigma_{R,\delta}^\pm \subset \mathscr{D}(I^\pm, \beta)$, therefore (v_*^+, v_*^-) is a pair of Fatou coordinates.

Suppose now that $v^\pm$ is holomorphic and injective on $\Sigma_{R,\delta}^\pm$. Replacing the above function β by a larger one if necessary, we may suppose $\beta \ge R$, then $\mathscr{D}(J^\pm, \beta) \subset \Sigma_{R,\delta}^\pm$ with $J^+ := (-\frac{\pi}{2} + \delta, \frac{\pi}{2} - \delta)$, $J^- := (\frac{\pi}{2} + \delta, \frac{3\pi}{2} - \delta)$. By Theorem 5.67, we have $u_*^\pm\big(\mathscr{D}(J^\pm, \beta_1)\big) \subset \mathscr{D}(J^\pm, \beta)$, thus $\Phi^\pm := v^\pm \circ u_*^\pm$ is holomorphic and injective on $\mathscr{D}(J^\pm, \beta_1)$. In view of (7.15), the equation $v^\pm \circ f = f_0 \circ v^\pm$ is equivalent to $f_0 \circ \Phi^\pm = \Phi^\pm \circ f_0$, i.e. $\Phi^\pm = \mathrm{id} + \Psi^\pm$ with $\Psi^\pm$ 1-periodic. If $\Psi^\pm$ is a constant $c^\pm$, then we find $v^\pm = v_*^\pm + c^\pm$. In general, the periodicity of $\Psi^\pm$ allows one to extend analytically $\Phi^\pm$

to the whole of $\mathbb{C}$ and we get an injective entire function; the Casorati-Weierstrass theorem shows that such a function must be of the form $az+c$, hence $\Psi^{\pm}$ is constant. □

7.4.2 Here are a few dynamical consequences of Theorem 7.10. The domain $\mathscr{D}^+ := \mathscr{D}(I^+,\beta_1)$ is invariant by the normal form $f_0 = \mathrm{id}+1$, while $\mathscr{D}^- := \mathscr{D}(I^-,\beta_1)$ is invariant by the backward dynamics $f_0^{\circ(-1)} = \mathrm{id}-1$, hence

$$\mathscr{P}^+ := u_*^+(\mathscr{D}^+) \text{ is invariant by } f, \quad \mathscr{P}^- := u_*^-(\mathscr{D}^-) \text{ is invariant by } f^{\circ(-1)}, \tag{7.16}$$

and the conjugacy relations $f = u_*^+ \circ f_0 \circ v_*^+$, $f^{\circ(-1)} = u_*^- \circ f_0^{\circ(-1)} \circ v_*^-$ yield

$$z \in \mathscr{P}^+ \Longrightarrow f^{\circ n}(z) = u_*^+\big(v_*^+(z)+n\big)$$

$$z \in \mathscr{P}^- \Longrightarrow f^{\circ(-n)}(z) = u_*^-\big(v_*^-(z)-n\big)$$

for every $n \in \mathbb{N}$. We thus see that all the forward orbits of f which start in $\mathscr{P}^+$ and all the backward orbits of f which start in $\mathscr{P}^-$ are infinite and converge to the fixed point at ∞ (we could even describe the asymptotics with respect to the discrete time n)—see Figure 7.2.

All this can be transferred to the variable $t = 1/z$ and we get for the dynamics of F a version of what is usually called the "Leau-Fatou flower theorem": we define the attracting and repelling "petals" by

$$P^+ := \{t \in \mathbb{C}^* \mid 1/t \in \mathscr{P}^+\}, \qquad P^- := \{t \in \mathbb{C}^* \mid 1/t \in \mathscr{P}^-\},$$

whose union is a punctured neighbourhood of 0, and we see that all the forward orbits of F which start in P^+ and all the backward orbits of F which start in P^- are infinite and converge to 0 (see Figure 7.2). Notice that P^+ and P^- overlap, giving rise to two families of bi-infinite orbits which are positively and negatively asymptotic to the fixed point.

We can also define Fatou coordinates and inverse Fatou coordinates at 0 as well as their formal counterparts by

$$V_*^{\pm}(t) := v_*^{\pm}(1/t), \quad U_*^{\pm}(z) := 1/u_*^{\pm}(z), \qquad \widetilde{V}_*(t) := \widetilde{v}_*(1/t), \quad \widetilde{U}_*(z) := 1/\widetilde{u}_*(z),$$

so that

$$t \in P^+ \Rightarrow V_*^+\big(F(t)\big) = V_*^+(t)+1, \qquad z \in \mathscr{D}^+ \Rightarrow F\big(U_*^+(z)\big) = U_*^+(z+1), \tag{7.17}$$

$$t \in P^- \Rightarrow V_*^-\big(F^{\circ(-1)}(t)\big) = V_*^-(t)-1, \qquad z \in \mathscr{D}^- \Rightarrow F^{\circ(-1)}\big(U_*^-(z)\big) = U_*^-(z-1). \tag{7.18}$$

Observe that, with the notation $\widetilde{v}_*(z) = z + \sum_{k\geq 1} a_k z^{-k}$, we have

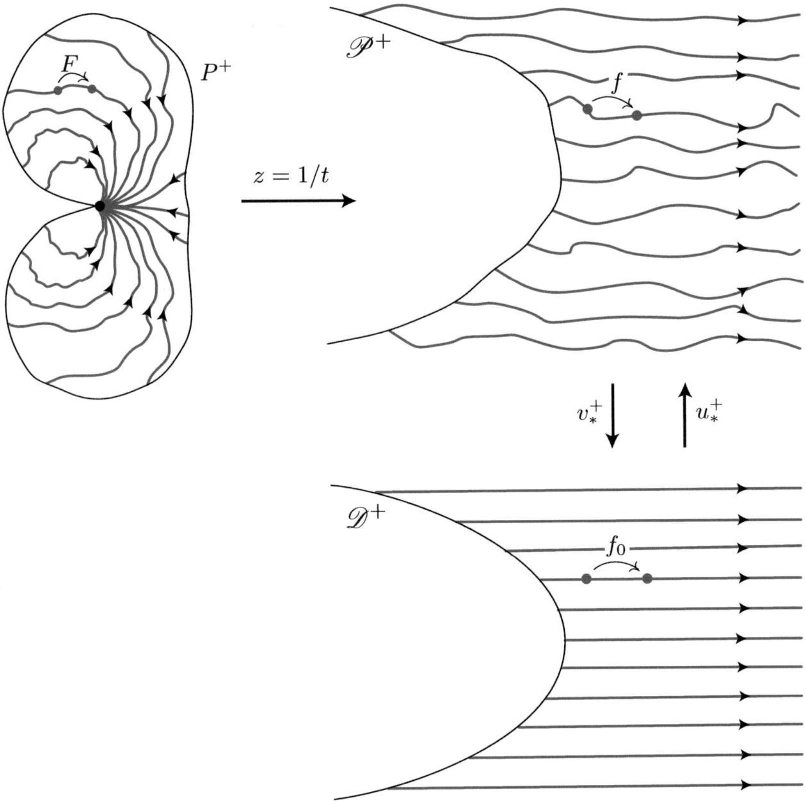

Fig. 7.2: *The dynamics in the attracting petal viewed in three coordinates.*

$$V_*^{\pm}(t) \sim \widetilde{V}_*(t) = \frac{1}{t} + \sum_{k\geq 1} a_k t^k,$$

whereas $\widetilde{u}_*(z) = z + \widetilde{\psi}(z)$ with $\widetilde{\psi}(z) = \sum_{k\geq 1} b_k z^{-k} \in z^{-1}\mathbb{C}[[z^{-1}]]$ implies

$$\widetilde{U}_*(z) = z^{-1}\big(1 + z^{-1}\widetilde{\psi}(z)\big)^{-1} \in z^{-1}\mathbb{C}[[z^{-1}]].$$

By Theorems 5.55 and 6.91, we see that $\widetilde{U}_*$ is a simple $2\pi i\mathbb{Z}$-resurgent series, which is 1-summable in the directions of I^+ and I^-, with

$$U_*^{\pm} = \mathscr{S}^{I^{\pm}} \widetilde{U}_*.$$

7.4.3 Of course it may happen that one of the formal series $\widetilde{v}_*$, $\widetilde{u}_*$, $\widetilde{V}_*$, $\widetilde{U}_*$ and thus all of them be convergent. But this is the exception rather than the rule.

There is a case in which one easily proves that all of them are divergent.

Lemma 7.12. *If $F(t)$ or $F^{\circ(-1)}(t)$ extends to an entire function, then the formal series $\widetilde{v}_*$, $\widetilde{u}_*$, $\widetilde{V}_*$, $\widetilde{U}_*$ are divergent.*

Proof. Suppose that F is entire. The function $U_*^-(z)$, intially defined and holomorphic in $\mathscr{D}^-$, which contains a left half-plane $\{\Re e\, z < -c\}$, can be analytically continued by repeated use of (7.18): for any $n \in \mathbb{N}^*$, the formula

$$U_*^-(z) = F\big(U_*^-(z-1)\big) = \cdots = F^{\circ n}\big(U_*^-(z-n)\big)$$

yields its analytic continuation in $\{\Re e(z) < -c+n\}$, hence U_*^- extends to an entire function. If $\widetilde{U}_*$ had positive radius of convergence, then we would get $U_*^- \sim_1 \widetilde{U}_*$ in a full neighbourhood of ∞ by Lemma 5.35, in particular $U_*^-(z)$ would tend to 0 as $|z| \to \infty$ and thus be uniformly bounded; then the entire function U_*^- would be constant by Liouville's theorem, which is impossible because $\widetilde{U}_*(z) = z^{-1} + O(z^{-2})$.

If it is $F^{\circ(-1)}$ that extends to an entire function, then U_*^+ extends to an entire function by virtue of (7.17) and one can argue similarly to prove that $\widetilde{U}_*$ is divergent. □

7.5 The horn maps and the analytic classification

In (7.16) we have defined $\mathscr{P}^+$ and $\mathscr{P}^-$ so that v_*^+ induces a biholomorphism $\mathscr{P}^+ \xrightarrow{\sim} \mathscr{D}^+$ and u_*^- induces a biholomorphism $\mathscr{D}^- \xrightarrow{\sim} \mathscr{P}^-$. We can thus define a holomorphic function

$$h := v_*^+ \circ u_*^- : \mathscr{D}^- \cap (u_*^-)^{-1}(\mathscr{P}^+) \to \mathscr{D}^+ \cap v_*^+(\mathscr{P}^-), \quad \text{such that } h \circ f_0 = f_0 \circ h \tag{7.19}$$

(the fact that h conjugates f_0 with itself stems from (7.15)).

Let us define, for any $R > 0$ and $\delta \in (0, \pi/2)$,

$$\mathscr{V}_{R,\delta}^{\mathrm{up}} := \{\, r\mathrm{e}^{\mathrm{i}\theta} \mid r > R,\ \delta < \theta < \pi - \delta \,\}, \quad \mathscr{V}_{R,\delta}^{\mathrm{low}} := \{\, r\mathrm{e}^{\mathrm{i}\theta} \mid r > R,\ \pi + \delta < \theta < 2\pi - \delta \,\}. \tag{7.20}$$

Since v_*^+ and u_*^- are close to identity near ∞, there exists $R > 0$ such that the domain of definition of h has a connected component which contains $\mathscr{V}_{R,\pi/4}^{\mathrm{up}}$ and a connected component which contains $\mathscr{V}_{R,\pi/4}^{\mathrm{low}}$, so that in fact formula (7.19) defines a function h^{up} and a function h^{low}.

Lemma 7.13. *There exists $\sigma > 0$ such that the function h^{up} extends analytically to the upper half-plane $\{\Im m\, z > \sigma\}$ and the function h^{low} extends analytically to the lower half-plane $\{\Im m\, z < -\sigma\}$. The functions $h^{\mathrm{up}} - \mathrm{id}$ and $h^{\mathrm{low}} - \mathrm{id}$ are 1-periodic and admit convergent Fourier expansions*

$$h_*^{\mathrm{up}}(z) - z = \sum_{m=1}^{+\infty} A_{-m} \mathrm{e}^{2\pi i m z}, \qquad h_*^{\mathrm{low}}(z) - z = \sum_{m=1}^{+\infty} A_m \mathrm{e}^{-2\pi i m z}, \tag{7.21}$$

with $A_m = O(\mathrm{e}^{\lambda|m|})$ *for every* $\lambda > 2\pi\sigma$.

Proof. The conjugacy relation $h^{\mathrm{up/low}} \circ f_0 = f_0 \circ h^{\mathrm{up/low}}$ implies that $h^{\mathrm{up/low}}$ is of the form $\mathrm{id} + P^{\mathrm{up/low}}$ with a 1-periodic holomorphic function $P^{\mathrm{up/low}}\colon \mathscr{V}_{R,\pi/4}^{\mathrm{up/low}} \to \mathbb{C}$. By 1-periodicity, $P^{\mathrm{up/low}}$ extends analytically to an upper/lower half-plane and can be written as $\chi(\mathrm{e}^{\pm 2\pi i z})$, with χ holomorphic in the punctured disc $\mathbb{D}_{2\pi\sigma}^*$. The asymptotic behaviour of v_*^+ and u_*^- at ∞ in $\mathscr{D}\big((-\frac{\pi}{4},\frac{\pi}{4}),\beta_1\big)$ shows that $h^{\mathrm{up/low}}(z) = z + o(1)$, hence $\chi(Z) \xrightarrow[Z\to 0]{} 0$. Thus χ is holomorphic in $\mathbb{D}_{2\pi\sigma}$ and vanishes at 0; its Taylor expansions yields the Fourier series of $P^{\mathrm{up/low}}$. □

Definition 7.14. We call $(h^{\mathrm{up}}, h^{\mathrm{low}})$ the pair of *lifted horn maps* of f. We call the coefficients of the sequence $(A_m)_{m\in\mathbb{Z}^*}$ the *Écalle-Voronin invariants* of f.

Theorem 7.15. *Two simple parabolic germs at* ∞ *with vanishing resiter,* f *and* g*, are analytically conjugate if and only if there exists* $c \in \mathbb{C}$ *such that their pairs of lifted horn maps* $(h_f^{\mathrm{up}}, h_f^{\mathrm{low}})$ *and* $(h_g^{\mathrm{up}}, h_g^{\mathrm{low}})$ *are related by*

$$h_g^{\mathrm{up}}(z) \equiv h_f^{\mathrm{up}}(z+c) - c, \qquad h_g^{\mathrm{low}}(z) \equiv h_f^{\mathrm{low}}(z+c) - c, \tag{7.22}$$

or, equivalently,

$$A_m(g) = \mathrm{e}^{-2\pi i m c} A_m(f) \quad \textit{for every } m \in \mathbb{Z}^*. \tag{7.23}$$

Proof. We denote by $\widetilde{v}_f$, $v_f^{\pm}$, $\widetilde{u}_f$, $u_f^{\pm}$ the iterator of f, its Borel sums and their inverses, and similarly $\widetilde{v}_g$, $v_g^{\pm}$, $\widetilde{u}_g$, $u_g^{\pm}$ for g.

Suppose that f and g are analytically conjugate, so there exists $h \in \mathscr{G}$ (convergent!) such that $g \circ h = h \circ f$. It follows that $\widetilde{v}_f \circ h^{\circ(-1)} \circ g = f_0 \circ \widetilde{v}_f \circ h^{\circ(-1)}$, hence there exists $c \in \mathbb{C}$ such that $\widetilde{v}_f \circ h^{\circ(-1)} = \widetilde{v}_g + c$ by Lemma 7.4. Let $\tau := \mathrm{id} + c$. We have $\widetilde{v}_g = \tau^{-1} \circ \widetilde{v}_f \circ h^{\circ(-1)}$ and $\widetilde{u}_g = h \circ \widetilde{u}_f \circ \tau$, whence $v_g^+ = \tau^{-1} \circ v_f^+ \circ h^{\circ(-1)}$ and $u_g^- = h \circ u_f^- \circ \tau$ by Theorem 5.66 and Lemma 5.35. This implies $v_g^+ \circ u_g^- = \tau^{-1} \circ v_f^+ \circ u_f^- \circ \tau$, i.e. $h_g^{\mathrm{up/low}} = \tau^{-1} \circ h_f^{\mathrm{up/low}} \circ \tau$, as desired.

Suppose now that there exists $c \in \mathbb{C}$ satisfying (7.22). We rewrite this relation as

$$h_g^{\mathrm{up}} = \tau^{-1} \circ h_f^{\mathrm{up}} \circ \tau, \qquad h_g^{\mathrm{low}} = \tau^{-1} \circ h_f^{\mathrm{low}} \circ \tau,$$

with $\tau = \mathrm{id} + c$. This implies

$$\tau \circ v_g^+ \circ u_g^- = v_f^+ \circ u_f^- \circ \tau \quad \text{on } \mathscr{V}_{R,\delta}^{\mathrm{up}} \cup \mathscr{V}_{R,\delta}^{\mathrm{low}}$$

with, say, $\delta = 3\pi/4$ and R large enough. Therefore

$$u_f^+ \circ \tau \circ v_g^+ = u_f^- \circ \tau \circ v_g^- \quad \text{on } \mathscr{V}_{R',\pi/4}^{\mathrm{up}} \cup \mathscr{V}_{R',\pi/4}^{\mathrm{low}}.$$

This indicates that the functions $u_f^+ \circ \tau \circ v_g^+$ and $u_f^- \circ \tau \circ v_g^-$ can be glued to form a function h holomorphic in punctured neighbourhood of ∞; the asymptotic behaviour then shows that h is holomorphic at ∞, with Taylor series $\widetilde{u}_f \circ \tau \circ \widetilde{v}_g$. The conjugacy relations $\widetilde{u}_g = g \circ \widetilde{u}_g \circ f_0^{\circ(-1)}$ and $\tau \circ \widetilde{v}_f \circ f = f_0 \circ \tau \circ \widetilde{v}_f$ imply $\widetilde{u}_g \circ \tau \circ \widetilde{v}_f \circ f = g \circ \widetilde{u}_g \circ \tau \circ \widetilde{v}_f$, hence f and g are analytically conjugate by h. □

Theorem 7.15 is just one part of Écalle-Voronin's classification result in the case of simple parabolic germs with vanishing resiter. The other part of the result (more difficult) says that any pair of Fourier series of the form

$$\Big(\sum_{m\geq 1} A_{-m} \mathrm{e}^{2\pi i m z}, \sum_{m\geq 1} A_m \mathrm{e}^{-2\pi i m z}\Big),$$

where the first (resp. second) one is holomorphic in an upper (resp. lower) half-plane, can be obtained as $(h_*^{\mathrm{up}} - \mathrm{id}, h_*^{\mathrm{low}} - \mathrm{id})$ for a simple parabolic germ f with vanishing resiter.

7.6 The Bridge Equation and the action of the symbolic Stokes automorphism

7.6.1 Let us give ourselves a simple parabolic germ at ∞ with vanishing resiter, f. So far, we have only exploited the summability statement contained in Theorem 7.10 and we have see that a deep information on the analytic conjugacy class of f is encoded by the discrepancy between the Borel sums v_*^+ and v_*^-, i.e. by the lifted horn maps. Let us now see how the analysis of this discrepancy lends itself to alien calculus, i.e. to the study of the singularities in the Borel plane.

We first use the operators Δ_ω of Sections 6.11–6.13 with $\omega \in 2\pi\mathrm{i}\mathbb{Z}^*$. They are derivations of the algebra $\widetilde{\mathscr{R}}^{\mathrm{simp}}_{2\pi\mathrm{i}\mathbb{Z}}$, and they induce operators $\Delta_\omega \colon \widetilde{\mathscr{G}}^{\mathrm{simp}}_{2\pi\mathrm{i}\mathbb{Z}} \to \widetilde{\mathscr{R}}^{\mathrm{simp}}_{2\pi\mathrm{i}\mathbb{Z}}$ defined by $\Delta_\omega(\mathrm{id}+\widetilde{\varphi}) \equiv \Delta_\omega \widetilde{\varphi}$.

Theorem 7.16. *There exists a sequence of complex numbers $(C_\omega)_{\omega\in 2\pi\mathrm{i}\mathbb{Z}^*}$ such that*

$$\Delta_\omega \widetilde{u}_* = C_\omega \partial \widetilde{u}_*, \qquad \Delta_\omega \widetilde{v}_* = -C_\omega \,\mathrm{e}^{-\omega(\widetilde{v}_* - \mathrm{id})} \tag{7.24}$$

for each $\omega \in 2\pi\mathrm{i}\mathbb{Z}^*$.

Proof. Let us apply Δ_ω to both sides of the conjugacy equation (7.9): by Theorem 7.2, since $\Delta_\omega f$ and $\Delta_\omega f_0$ vanish, we get

$$(\partial f) \circ \widetilde{u}_* \cdot \Delta_\omega \widetilde{u}_* = (\Delta_\omega \widetilde{u}_*) \circ f_0$$

(we also used the fact that $\mathrm{e}^{-\omega(f_0 - \mathrm{id})} = 1$, since $\omega \in 2\pi\mathrm{i}\mathbb{Z}^*$). By applying ∂ to (7.9), we also get

$$(\partial f) \circ \widetilde{u}_* \cdot \partial \widetilde{u}_* = (\partial \widetilde{u}_*) \circ f_0.$$

Since $\partial\widetilde{u}_* = 1 + O(z^{-2})$, this implies that the formal series $\widetilde{C} := \frac{\Delta_\omega \widetilde{u}_*}{\partial \widetilde{u}_*} \in \mathbb{C}[[z^{-1}]]$ satisfies $\widetilde{C} = \widetilde{C} \circ f_0$. Writing $\widetilde{C} \circ f_0 - \widetilde{C} = \partial\widetilde{C} + \frac{1}{2!}\partial^2\widetilde{C} + \cdots$ and reasoning on the valuation of $\partial\widetilde{C}$, we see that $\widetilde{C}$ must be constant.

We have $(\Delta_\omega \widetilde{u}_*) \circ \widetilde{v}_* \cdot \partial\widetilde{v}_* = \widetilde{C}(\partial\widetilde{u}_*) \circ \widetilde{v}_* \cdot \partial\widetilde{v}_* = \widetilde{C}\partial(\widetilde{u}_* \circ \widetilde{v}_*) = \widetilde{C}$, hence Formula (7.3) yields $\Delta_\omega \widetilde{v} = -\widetilde{C}\mathrm{e}^{-\omega(\widetilde{v}_* - \mathrm{id})}$. □

The first equation in (7.24) is called "the Bridge Equation for simple parabolic germs": like Equation (6.157), it yields a bridge between ordinary differential calculus (here involving ∂) and alien calculus (when dealing with the solution $\widetilde{u}$ of the conjugacy equation (7.9)).

7.6.2 From the operators Δ_ω we can go the operators Δ^+_ω by means of formula (6.52) of Theorem 6.72, according to which, if one sets $\Omega := 2\pi\mathrm{i}\mathbb{N}^*$ or $\Omega := -2\pi\mathrm{i}\mathbb{N}^*$, then

$$\Delta^+_\omega = \sum_{s\geq 1} \frac{1}{s!} \sum_{\substack{\omega_1,\dots,\omega_s \in \Omega \\ \omega_1+\cdots+\omega_s=\omega}} \Delta_{\omega_s} \circ \cdots \circ \Delta_{\omega_2} \circ \Delta_{\omega_1} \quad \text{for } \omega \in \Omega. \tag{7.25}$$

We also define

$$\Delta^-_\omega := \sum_{s\geq 1} \frac{(-1)^s}{s!} \sum_{\substack{\omega_1,\dots,\omega_s \in \Omega \\ \omega_1+\cdots+\omega_s=\omega}} \Delta_{\omega_s} \circ \cdots \circ \Delta_{\omega_2} \circ \Delta_{\omega_1} \quad \text{for } \omega \in \Omega. \tag{7.26}$$

The latter family of operators is related to Exercise 6.75: they correspond to the homogeneous components of $\exp(-\Delta_{\mathrm{i}\mathbb{R}^\pm})$ the same way the operators Δ^+_ω correspond to the homogeneous components of $\exp(\Delta_{\mathrm{i}\mathbb{R}^\pm})$—see formulas (6.75)–(7.29).

Corollary 7.17. *Let $\Omega := 2\pi\mathrm{i}\mathbb{N}^*$ or $\Omega := -2\pi\mathrm{i}\mathbb{N}^*$. For each $\omega \in \Omega$, define*

$$S^+_\omega := -\sum_{s\geq 1} \frac{1}{s!} \sum_{\substack{\omega_1,\dots,\omega_s \in \Omega \\ \omega_1+\cdots+\omega_s=\omega}} \Gamma_{\omega_1,\dots,\omega_s} C_{\omega_1} \cdots C_{\omega_s},$$

$$S^-_\omega := \sum_{s\geq 1} \frac{(-1)^{s-1}}{s!} \sum_{\substack{\omega_1,\dots,\omega_s \in \Omega \\ \omega_1+\cdots+\omega_s=\omega}} \Gamma_{\omega_1,\dots,\omega_s} C_{\omega_1} \cdots C_{\omega_s}$$

with $\Gamma_{\omega_1} := 1$ and $\Gamma_{\omega_1,\dots,\omega_s} := \omega_1(\omega_1+\omega_2)\cdots(\omega_1+\cdots+\omega_{s-1})$. Then

$$\Delta^+_\omega \widetilde{v}_* = S^+_\omega \mathrm{e}^{-\omega(\widetilde{v}_* - \mathrm{id})}, \qquad \Delta^-_\omega \widetilde{v}_* = S^-_\omega \mathrm{e}^{-\omega(\widetilde{v}_* - \mathrm{id})}. \tag{7.27}$$

Proof. Let $\widetilde{\varphi} := \widetilde{v}_* - \mathrm{id}$, so that the second equation in (7.24) reads $\Delta_\omega \widetilde{\varphi} = -C_\omega \mathrm{e}^{-\omega\widetilde{\varphi}}$. By repeated use of formula (6.96) of Theorem 6.91, we get

$$\Delta_{\omega_2}\Delta_{\omega_1}\widetilde{\varphi} = \omega_1 C_{\omega_1} \mathrm{e}^{-\omega_1\widetilde{\varphi}} \Delta_{\omega_2}\widetilde{\varphi} = -\omega_1 C_{\omega_1} C_{\omega_2} \mathrm{e}^{-(\omega_1+\omega_2)\widetilde{\varphi}}$$

$$\Delta_{\omega_3}\Delta_{\omega_2}\Delta_{\omega_1}\widetilde{\varphi} = \dots$$

and so on. The general formula is

$$\Delta_{\omega_s} \cdots \Delta_{\omega_1} \widetilde{\varphi} = -\Gamma_{\omega_1,\dots,\omega_s} C_{\omega_1} \cdots C_{\omega_s} \, \mathrm{e}^{-(\omega_1+\cdots+\omega_s)\widetilde{\varphi}},$$

whence the conclusion follows with the help of (7.25)–(7.26). □

In fact, in view of Remark 6.69, the above proof shows that, for every $\omega \in 2\pi\mathrm{i}\mathbb{Z}$ and for every path γ which starts close to 0 and ends close to ω, there exists $S_\omega^\gamma \in \mathbb{C}$ such that $\mathscr{A}_\omega^\gamma \widetilde{v}_* = S_\omega^\gamma \, \mathrm{e}^{-\omega(\widetilde{v}_* - \mathrm{id})}$.

7.6.3 We now wish to compute the action of the symbolic Stokes automorphism $\Delta^+_{\mathrm{i}\mathbb{R}^\pm}$ on $\widetilde{v}_*$ and to describe the Stokes phenomenon in the spirit of Section 6.12.3, so as to recover the horn maps of Section 7.5. We shall make use of the spaces

$$\widetilde{E}^\pm := \widetilde{E}(2\pi\mathrm{i}\mathbb{Z}, \mathrm{i}\mathbb{R}^\pm) = \widehat{\bigoplus_{\omega \in \pm 2\pi\mathrm{i}\mathbb{N}}} \mathrm{e}^{-\omega z} \widetilde{\mathscr{R}}^{\mathrm{simp}}_{2\pi\mathrm{i}\mathbb{Z}}$$

introduced in Section 6.12.4; since $2\pi\mathrm{i}\mathbb{Z}$ is an additive subgroup of $\mathbb{C}$, these spaces are differential algebras,

$$\widetilde{E}^+ = \widetilde{\mathscr{R}}^{\mathrm{simp}}_{2\pi\mathrm{i}\mathbb{Z}}[[\mathrm{e}^{-2\pi\mathrm{i}z}]], \qquad \widetilde{E}^- = \widetilde{\mathscr{R}}^{\mathrm{simp}}_{2\pi\mathrm{i}\mathbb{Z}}[[\mathrm{e}^{2\pi\mathrm{i}z}]], \qquad \partial = \frac{\mathrm{d}}{\mathrm{d}z},$$

on which are defined the directional alien derivation $\Delta_{\mathrm{i}\mathbb{R}^\pm}$ and the symbolic Stokes automorphism $\Delta^+_{\mathrm{i}\mathbb{R}^\pm} = \exp(\Delta_{\mathrm{i}\mathbb{R}^\pm})$. According to Remark 6.82, both operators commute with the differential ∂. So does the "inverse symbolic Stokes automorphism" $\Delta^-_{\mathrm{i}\mathbb{R}^\pm} := \exp(-\Delta_{\mathrm{i}\mathbb{R}^\pm})$.

We find it convenient to modify slightly the notation for their homogeneous components: from now on, we set

$$\omega \in 2\pi\mathrm{i}\mathbb{Z},\ m \in \mathbb{Z},\ \widetilde{\varphi} \in \widetilde{\mathscr{R}}^{\mathrm{simp}}_{2\pi\mathrm{i}\mathbb{Z}} \quad \Longrightarrow \quad \begin{cases} \dot{\Delta}_\omega(\mathrm{e}^{-2\pi\mathrm{i}mz}\widetilde{\varphi}) := \mathrm{e}^{-(2\pi\mathrm{i}m+\omega)z}\Delta_\omega\widetilde{\varphi}, \\ \dot{\Delta}^\pm_\omega(\mathrm{e}^{-2\pi\mathrm{i}mz}\widetilde{\varphi}) := \mathrm{e}^{-(2\pi\mathrm{i}m+\omega)z}\Delta^\pm_\omega\widetilde{\varphi}, \end{cases} \tag{7.28}$$

so that

$$\Delta_{\mathrm{i}\mathbb{R}^+} = \sum_{\omega \in 2\pi\mathrm{i}\mathbb{N}^*} \dot{\Delta}_\omega \quad \text{on } \widetilde{E}^+, \qquad \Delta_{\mathrm{i}\mathbb{R}^-} = \sum_{\omega \in -2\pi\mathrm{i}\mathbb{N}^*} \dot{\Delta}_\omega \quad \text{on } \widetilde{E}^-,$$

$$\Delta^\pm_{\mathrm{i}\mathbb{R}^+} = \exp(\pm\Delta_{\mathrm{i}\mathbb{R}^+}) = \mathrm{Id} + \sum_{\omega \in 2\pi\mathrm{i}\mathbb{N}^*} \dot{\Delta}^\pm_\omega, \quad \Delta^\pm_{\mathrm{i}\mathbb{R}^-} = \exp(\pm\Delta_{\mathrm{i}\mathbb{R}^-}) = \mathrm{Id} + \sum_{\omega \in -2\pi\mathrm{i}\mathbb{N}^*} \dot{\Delta}^\pm_\omega. \tag{7.29}$$

We may consider $\widetilde{v}_*$ as an element of $\mathrm{id} + \widetilde{\mathscr{R}}^{\mathrm{simp}}_{2\pi\mathrm{i}\mathbb{Z}} \subset \mathrm{id} + \widetilde{E}^\pm$. We thus set $\dot{\Delta}_\omega \,\mathrm{id} := 0$ and $\dot{\Delta}^\pm_\omega \,\mathrm{id} := 0$ so that the previous operators induce

$$\Delta_{\mathrm{i}\mathbb{R}^+}, \Delta^+_{\mathrm{i}\mathbb{R}^+}, \Delta^-_{\mathrm{i}\mathbb{R}^+} \colon \ \mathrm{id} + \widetilde{E}^+ \to \widetilde{E}^+, \qquad \Delta_{\mathrm{i}\mathbb{R}^-}, \Delta^+_{\mathrm{i}\mathbb{R}^-}, \Delta^-_{\mathrm{i}\mathbb{R}^-} \colon \ \mathrm{id} + \widetilde{E}^- \to \widetilde{E}^-.$$

This way (7.24) yields $\dot{\Delta}_\omega \widetilde{v}_* = -C_\omega\, \mathrm{e}^{-\omega\widetilde{v}_*}$ and (7.27) yields $\dot{\Delta}^{\pm}_\omega \widetilde{v}_* = S^{\pm}_\omega\, \mathrm{e}^{-\omega\widetilde{v}_*}$, and we can write

$$\Delta^{\pm}_{\mathrm{i}\mathbb{R}^+} \widetilde{v}_* = \widetilde{v}_* + \sum_{\omega\in 2\pi\mathrm{i}\mathbb{N}^*} S^{\pm}_\omega\, \mathrm{e}^{-\omega\widetilde{v}_*}, \qquad \Delta^{\pm}_{\mathrm{i}\mathbb{R}^-} \widetilde{v}_* = \widetilde{v}_* + \sum_{\omega\in -2\pi\mathrm{i}\mathbb{N}^*} S^{\pm}_\omega\, \mathrm{e}^{-\omega\widetilde{v}_*}.$$

Theorem 7.18. *We have*

$$z + \sum_{\omega\in 2\pi\mathrm{i}\mathbb{N}^*} S^{+}_\omega\, \mathrm{e}^{-\omega z} \equiv h^{\mathrm{low}}_*(z), \qquad \Delta^{+}_{\mathrm{i}\mathbb{R}^+} \widetilde{v}_* = h^{\mathrm{low}}_* \circ \widetilde{v}_*, \tag{7.30}$$

$$z + \sum_{\omega\in 2\pi\mathrm{i}\mathbb{N}^*} S^{-}_\omega\, \mathrm{e}^{-\omega z} \equiv (h^{\mathrm{low}}_*)^{\circ(-1)}(z), \qquad \Delta^{-}_{\mathrm{i}\mathbb{R}^+} \widetilde{v}_* = (h^{\mathrm{low}}_*)^{\circ(-1)} \circ \widetilde{v}_*, \tag{7.31}$$

$$z + \sum_{\omega\in -2\pi\mathrm{i}\mathbb{N}^*} S^{+}_\omega\, \mathrm{e}^{-\omega z} \equiv (h^{\mathrm{up}}_*)^{\circ(-1)}(z), \qquad \Delta^{+}_{\mathrm{i}\mathbb{R}^-} \widetilde{v}_* = (h^{\mathrm{up}}_*)^{\circ(-1)} \circ \widetilde{v}_*, \tag{7.32}$$

$$z + \sum_{\omega\in -2\pi\mathrm{i}\mathbb{N}^*} S^{-}_\omega\, \mathrm{e}^{-\omega z} \equiv h^{\mathrm{up}}_*(z), \qquad \Delta^{-}_{\mathrm{i}\mathbb{R}^-} \widetilde{v}_* = h^{\mathrm{up}}_* \circ \widetilde{v}_*. \tag{7.33}$$

In particular the Écalle-Voronin invariants $(A_m)_{m\in\mathbb{Z}^*}$ of Lemma 7.13 are given by

$$A_{-m} = S^{-}_{-2\pi\mathrm{i}m}, \quad A_m = S^{+}_{2\pi\mathrm{i}m}, \qquad m\in\mathbb{N}^*. \tag{7.34}$$

Remark 7.19. The "exponential-like" formulas which define the family of coefficients $(S^{\pm}_\omega)_{\omega\in 2\pi\mathrm{i}\mathbb{N}^*}$ from $(C_\omega)_{\omega\in 2\pi\mathrm{i}\mathbb{N}^*}$ in Corollary 7.17 are clearly invertible, and similarly $(C_\omega)_{\omega\in -2\pi\mathrm{i}\mathbb{N}^*} \mapsto (S^{\pm}_\omega)_{\omega\in 2\pi\mathrm{i}\mathbb{N}^*}$ is invertible. It follows that the coefficients C_ω of the Bridge Equation (7.24) are analytic conjugacy invariants too. However there is an important difference between the C's and the S's: Theorem 7.18 implies that there exists $\lambda > 0$ such that $S^{\pm}_{2\pi\mathrm{i}m} = O(\mathrm{e}^{\lambda|m|})$, but there are in general no estimates of the same kind for the coefficients $C_{2\pi\mathrm{i}m}$ of the Bridge Equation.

Proof. Let $I := (0,\pi)$ and $\theta := \frac{\pi}{2}$, so that $I^+ = (0,\frac{\pi}{2})$ and $I^- = (\frac{\pi}{2},\pi)$ with the notations of Section 6.12.3. Let us pick $R > 0$ large enough so that h^{low} is defined by $v^+_* \circ (v^-_*)^{\circ(-1)}$ in $\mathcal{V}^{\mathrm{low}}_{R,\pi/4}$ (recall (7.20) and the comment right after it).

For any $m\in\mathbb{N}$, we deduce from the relation $\Delta^{+}_{\mathrm{i}\mathbb{R}^+} \widetilde{v}_* = \widetilde{v}_* + \sum_{\omega\in 2\pi\mathrm{i}\mathbb{N}^*} S^{+}_\omega\, \mathrm{e}^{-\omega\widetilde{v}_*}$ that

$$[\Delta^{+}_{\mathrm{i}\mathbb{R}^+} \widetilde{v}_*]_m = \widetilde{v}_* + \sum_{j=0}^{m} S^{+}_{2\pi\mathrm{i}j}\, \mathrm{e}^{-2\pi\mathrm{i}j\widetilde{v}_*}$$

with notation 6.76. Each term $\mathrm{e}^{-2\pi\mathrm{i}j\widetilde{v}_*}$ is $2\pi\mathrm{i}\mathbb{Z}$-resurgent and 1-summable in the directions of $I^{\pm}$, with Borel sums $\mathscr{S}^{I^{\pm}}(\mathrm{e}^{-2\pi\mathrm{i}j\widetilde{v}_*}) = \mathrm{e}^{-2\pi\mathrm{i}jv^{\pm}_*}$, hence Theorem 6.77 implies that

$$z\in\mathcal{V}^{\mathrm{low}}_{R,\pi/4} \quad\Longrightarrow\quad v^+_*(z) = v^-_*(z) + \sum_{j=0}^{m} S^{+}_{2\pi\mathrm{i}j}\, \mathrm{e}^{-2\pi\mathrm{i}jv^-_*(z)} + O(\mathrm{e}^{-\rho|\Im m z|})$$

for any $\rho\in(2\pi m, 2\pi(m+1))$. It follows that

$$z \in \mathcal{V}^{\text{low}}_{R,\pi/4} \quad \Longrightarrow \quad h^{\text{low}}_*(z) = z + \sum_{j=0}^{m} S^+_{2\pi i j}\, \mathrm{e}^{-2\pi i j z} + O(\mathrm{e}^{-\rho|\Im m z|})$$

for any $\rho \in (2\pi m, 2\pi(m+1))$, whence (7.30) follows.

Formula (7.31) is obtained by the same chain of reasoning, using a variant of Theorem 6.77 relating $\mathscr{S}^-\widetilde{v}_*$ and $\mathscr{S}^+[\Delta^+_{i\mathbb{R}^+}\widetilde{v}_*]_m$.

Formulas (7.32) and (7.33) are obtained the same way, using $I^+ := (-\pi, -\frac{\pi}{2})$ and $I^- := (-\frac{\pi}{2}, 0)$, but this time $\mathscr{S}^{I^+}\widetilde{v}_* = v^-_*$ and $\mathscr{S}^{I^-}\widetilde{v}_* = v^+_*$. □

7.6.4 We conclude by computing the action of the symbolic Stokes automorphism $\Delta^+_{i\mathbb{R}^\pm}$ on $\widetilde{u}_*$.

Definition 7.20. The derivation of $\widetilde{E}^\pm$

$$D_{i\mathbb{R}^\pm} := C_{i\mathbb{R}^\pm}(z)\partial, \quad \text{where } C_{i\mathbb{R}^\pm}(z) = \sum_{\omega \in \pm 2\pi i \mathbb{N}^*} C_\omega \mathrm{e}^{-\omega z},$$

is called the *formal Stokes vector field* of f.

Such a derivation $D_{i\mathbb{R}^\pm}$ has a well-defined exponential, for the same reason by which Δ_d had one according to Theorem 6.73(iii): it increases homogeneity by at least one unit.

Lemma 7.21. *For any* $\widetilde{\phi} \in \widetilde{\mathscr{R}}^{\text{simp}}_{2\pi i \mathbb{Z}}$,

$$\exp\big(C_{i\mathbb{R}^\pm}(z)\partial\big)\widetilde{\phi} = \widetilde{\phi} \circ P_{i\mathbb{R}^\pm} \qquad \textit{with } P_{i\mathbb{R}^\pm}(z) := z + \sum_{\omega \in \pm 2\pi i \mathbb{N}^*} S^-_\omega \mathrm{e}^{-\omega z}$$

$$\exp\big(-C_{i\mathbb{R}^\pm}(z)\partial\big)\widetilde{\phi} = \widetilde{\phi} \circ Q_{i\mathbb{R}^\pm} \qquad \textit{with } Q_{i\mathbb{R}^\pm}(z) := z + \sum_{\omega \in \pm 2\pi i \mathbb{N}^*} S^+_\omega \mathrm{e}^{-\omega z}.$$

Proof. Let $\Omega = 2\pi i \mathbb{N}^*$ or $\Omega = -2\pi i \mathbb{N}^*$ and, accordingly, $C = C_{i\mathbb{R}^+}$ or $C = C_{i\mathbb{R}^-}$, $D = D_{i\mathbb{R}^+}$ or $D = D_{i\mathbb{R}^-}$. We have $C = \sum C_{\omega_1}\mathrm{e}^{-\omega_1 z}$, $DC = \sum(-\omega_1)C_{\omega_1}C_{\omega_2}\mathrm{e}^{-(\omega_1+\omega_2)z}$, $D^2C = \dots$, etc. The general formula is

$$D^{s-1}C = (-1)^{s-1} \sum_{\omega_1,\dots,\omega_s \in \Omega} \Gamma_{\omega_1,\dots,\omega_s} C_{\omega_1}\cdots C_{\omega_s}\mathrm{e}^{-(\omega_1+\dots+\omega_s)z}, \qquad s \geq 1.$$

We thus set, for every $\omega \in \Omega$,

$$S_\omega(t) := \sum_{s \geq 1} \frac{(-1)^{s-1}t^s}{s!} \sum_{\substack{\omega_1,\dots,\omega_s \in \Omega \\ \omega_1+\dots+\omega_s = \omega}} \Gamma_{\omega_1,\dots,\omega_s} C_{\omega_1}\cdots C_{\omega_s} \in \mathbb{C}[t]$$

(observe that $S_\omega(t)$ is a polynomial of degree $\leq m$ if $\omega = \pm 2\pi i m$), so that $S_\omega(1) = S^-_\omega$ and $S_\omega(-1) = S^+_\omega$, and

$$G_t(z) := \sum_{s \geq 1} \frac{t^s}{s!} D^{s-1}C = \sum_{\omega \in \Omega} S_\omega(t)\mathrm{e}^{-\omega z} \in \mathbb{C}[t][[\mathrm{e}^{\mp 2\pi i z}]].$$

We leave it to the reader to check by induction the combinatorial identity

$$D^s\widetilde{\phi} = \sum_{\substack{n\geq 1,\ s_1,\dots,s_n\geq 1\\ s_1+\cdots+s_n=s}} \frac{s!}{s_1!\cdots s_n!n!}(D^{s_1-1}C)\cdots(D^{s_n-1}C)\partial^n\widetilde{\phi}, \qquad s\geq 1$$

for any $\widetilde{\phi}\in\widetilde{\mathscr{R}}^{\mathrm{simp}}_{2\pi\mathrm{i}\mathbb{Z}}$, whence $\exp(tD)\widetilde{\phi} = \widetilde{\phi}+\sum_{n\geq 1}\frac{1}{n!}(G_t)^n\partial\widetilde{\phi} = \widetilde{\phi}\circ(\mathrm{id}+G_t)$. □

In view of Theorem 7.18, we get

Corollary 7.22.

$$\exp\big(C_{\mathrm{i}\mathbb{R}^-}(z)\partial\big)\widetilde{\phi} = \widetilde{\phi}\circ h_*^{\mathrm{up}}, \qquad \exp\big(-C_{\mathrm{i}\mathbb{R}^+}(z)\partial\big)\widetilde{\phi} = \widetilde{\phi}\circ h_*^{\mathrm{low}}$$

for every $\widetilde{\phi}\in\widetilde{\mathscr{R}}^{\mathrm{simp}}_{2\pi\mathrm{i}\mathbb{Z}}$.

Since the Bridge Equation can be rephrased as

$$\Delta_{\mathrm{i}\mathbb{R}^\pm}\widetilde{u}_* = C_{\mathrm{i}\mathbb{R}^\pm}\partial\widetilde{u}_*$$

and the operators $\Delta_{\mathrm{i}\mathbb{R}^\pm}$ and $D_{\mathrm{i}\mathbb{R}^\pm}$ commute, we obtain

Corollary 7.23.

$$\exp(t\Delta_{\mathrm{i}\mathbb{R}^\pm})\widetilde{u}_* = \exp(tC_{\mathrm{i}\mathbb{R}^\pm}\partial)\widetilde{u}_*, \qquad t\in\mathbb{C}.$$

In particular

$$\Delta^+_{\mathrm{i}\mathbb{R}^-}\widetilde{u}_* = \widetilde{u}_*\circ h_*^{\mathrm{up}}, \qquad \Delta^-_{\mathrm{i}\mathbb{R}^-}\widetilde{u}_* = \widetilde{u}_*\circ (h_*^{\mathrm{up}})^{\circ(-1)},$$

$$\Delta^+_{\mathrm{i}\mathbb{R}^+}\widetilde{u}_* = \widetilde{u}_*\circ (h_*^{\mathrm{low}})^{\circ(-1)}, \qquad \Delta^-_{\mathrm{i}\mathbb{R}^+}\widetilde{u}_* = \widetilde{u}_*\circ h_*^{\mathrm{low}}.$$

Expanding the last equation, we get

$$\omega\in 2\pi\mathrm{i}\mathbb{N}^* \quad\Longrightarrow\quad \Delta^+_\omega\widetilde{u}_* = \sum_{\substack{n\geq 1,\ \omega_1,\dots,\omega_n\geq 1\\ \omega_1+\cdots+\omega_n=\omega}} \frac{1}{n!}S^+_{\omega_1}\cdots S^+_{\omega_n}\partial^n\widetilde{u}_*.$$

We leave it to the reader to compute the formula for $\Delta^+_\omega\widetilde{u}_*$ when $\omega\in 2\pi\mathrm{i}\mathbb{N}^*$, and the formulas for $\Delta^\pm_\omega u_*$ when $\omega\in -2\pi\mathrm{i}\mathbb{N}^*$.

References

DS14. A. Dudko and D. Sauzin. The resurgent character of the Fatou coordinates of a simple parabolic germ. *C. R. Math. Acad. Sci. Paris*, 352(3):255–261, 2014.

DS15. A. Dudko and D. Sauzin. On the resurgent approach to Écalle-Voronin's invariants. *C. R. Math. Acad. Sci. Paris*, 353(3):265–271, 2015.

Éca81. J. Écalle. *Les fonctions resúrgentes. Tome II*, volume 6 of *Publications Mathématiques d'Orsay 81 [Mathematical Publications of Orsay 81]*. Université de Paris-Sud,

Département de Mathématique, Orsay, 1981. Les fonctions résurgentes appliquées à l'itération. [Resurgent functions applied to iteration].

Lor06. F. Loray. Pseudo-groupe d'une singularite de feuilletage holomorphe en dimension deux. January 2006.

Lod16. M. Loday-Richaud. *Divergent Series, summability and resurgence. Volume 2: Simple and multiple summability.*, volume 2154 of *Lecture Notes in Mathematics*. Springer, Heidelberg, 2016.

LY14. O. E. Lanford III and M. Yampolsky. *Fixed point of the parabolic renormalization operator*. Springer Briefs in Mathematics. Springer, Cham, 2014.

Mil06. J. Milnor. *Dynamics in one complex variable*, volume 160 of *Annals of Mathematics Studies*. Princeton University Press, Princeton, NJ, third edition, 2006.

Sau12. D. Sauzin. Resurgent functions and splitting problems. In *New Trends and Applications of Complex Asymptotic Analysis : around dynamical systems, summability, continued fractions*, volume 1493 of *RIMS Kokyuroku*, pages 48–117. Kyoto University, 2012.

Index

C. Mitschi, D. Sauzin, *Divergent Series, Summability and Resurgence I*,
Lecture Notes in Mathematics 2153, DOI 10.1007/978-3-319-28736-2

Printed in the United States
By Bookmasters